AF213677

NANO-OPTICS AND NANOPHOTONICS

For further volumes:
www.springer.com/series/8765

NANO-OPTICS AND NANOPHOTONICS

The Springer Series in Nano-Optics and Nanophotonics provides an expanding selection of research monographs in the area of nano-optics and nanophotonics, science- and technology-based on optical interactions of matter in the nanoscale and related topics of contemporary interest. With this broad coverage of topics, the series is of use to all research scientists, engineers and graduate students who need up-to-date reference books. The editors encourage prospective authors to correspond with them in advance of submitting a manuscript. Submission of manuscripts should be made to the editor-in-chief, one of the editors or to Springer.

Editor-in-Chief

Motoichi Ohtsu

Department of Electrical Engineering and Informations Systems, School of Engineering
The University of Tokyo
Yayoi, Bunkyo-ku 2-11-16, 113-8656 Tokyo, Japan
ohtsu@ee.t.u-tokyo.ac.jp

Editorial Board

Gunnar Björk

Department of Electronics
KTH, Electrum 229
164 40 Kista, Sweden
gbjork@kth.se

Chennupati Jagadish

Department of Electronic Materials Engineering
Research School of Physics and Engineering
Australian National University
Canberra, ACT 0200, Australia
cxj109@rsphysse.anu.edu.au

Christoph Lienau

Institut für Physik, Fakultät V
Carl von Ossietzky Universität Oldenburg
Ammerländer Heerstraße 114-118
26129 Oldenburg, Germany
christoph.lienau@uni-oldenburg.de

Lih Y. Lin

Electrical Engineering Department
University of Washington
M414 EEl Bldg., Box 352500
Seattle, WA 98195-2500, USA
lylin@uw.edu

Erich Runge

Technische Universität Ilmenau
Curiebau, Weimarer Str. 25
98693 Ilmenau, Germany
erich.runge@tu-ilmenau.de

Frank Träger

Experimentalphysik I, Universität Kassel
Heinrich-Plett-Str. 40, 34132 Kassel, Germany
traeger@physik.uni-kassel.de

Masaru Tsukada

WPI-AIMR Center, Tohoku University
2-1-1 Katahira, Aoba-ku, Sendai, 980-8577 Japan
tsukada@wpi-aimr.tohoku.ac.jp

Please view available titles in *Nano-Optics and Nanophotonics*
on series homepage http://www.springer.com/series/8765

Carsten Rockstuhl · Toralf Scharf
Editors

Amorphous Nanophotonics

With 193 Figures

Springer

Editors
Carsten Rockstuhl
Institut für Festkörpertheorie
Friedrich-Schiller-Universität
Jena, Germany

Toralf Scharf
Ecole Polytechnique
Federal de Lausanne
Neuchâtel, Switzerland

ISSN 2192-1970 ISSN 2192-1989 (electronic)
Nano-Optics and Nanophotonics
ISBN 978-3-642-32474-1 ISBN 978-3-642-32475-8 (eBook)
DOI 10.1007/978-3-642-32475-8
Springer Heidelberg New York Dordrecht London

Library of Congress Control Number: 2013931281

© Springer-Verlag Berlin Heidelberg 2013
This work is subject to copyright. All rights are reserved by the Publisher, whether the whole or part of the material is concerned, specifically the rights of translation, reprinting, reuse of illustrations, recitation, broadcasting, reproduction on microfilms or in any other physical way, and transmission or information storage and retrieval, electronic adaptation, computer software, or by similar or dissimilar methodology now known or hereafter developed. Exempted from this legal reservation are brief excerpts in connection with reviews or scholarly analysis or material supplied specifically for the purpose of being entered and executed on a computer system, for exclusive use by the purchaser of the work. Duplication of this publication or parts thereof is permitted only under the provisions of the Copyright Law of the Publisher's location, in its current version, and permission for use must always be obtained from Springer. Permissions for use may be obtained through RightsLink at the Copyright Clearance Center. Violations are liable to prosecution under the respective Copyright Law.
The use of general descriptive names, registered names, trademarks, service marks, etc. in this publication does not imply, even in the absence of a specific statement, that such names are exempt from the relevant protective laws and regulations and therefore free for general use.
While the advice and information in this book are believed to be true and accurate at the date of publication, neither the authors nor the editors nor the publisher can accept any legal responsibility for any errors or omissions that may be made. The publisher makes no warranty, express or implied, with respect to the material contained herein.

Printed on acid-free paper

Springer is part of Springer Science+Business Media (www.springer.com)

Preface

Nanophotonics is a burgeoning branch of modern optics. It promises to revolutionize many fields of physics and engineering since it is commonly understood as an enabling technology. It has impact on the evolution of science in general by supporting it with new and refined tools, e.g. novel microscopic techniques as used in biology or medicine; but allows foremost also for the implementation of applications that were unimaginable just some time ago, e.g. cloaking devices that conceal objects from external observers. Most notably, the key ability that makes nanophotonics such unique is to provide means to steer the propagation and the distribution of electromagnetic fields on length scales much smaller than the wavelength by relying on suitably tailored nanomaterials. To observe significant interactions, resonances are exploited in many cases that are evoked due to specific material properties, dimensions, symmetries, and geometrical arrangements of the involved constituents; or a combination thereof.

The technology, both in experiment and theory, to achieve and to describe nanomaterials relies in many cases on the periodic arrangement of an identical unit cell. That seems to be advantageous since it allows to eliminate scattering losses, to simulate the optical response by taking into account only a single unit cell and appropriate boundary conditions, it allows to simplify the fabrication that is accomplished in most cases by means of top-down technologies, and it permits to detect a noticeable signal from fabricated systems in the far-field by measuring quantities that are linked to an ensemble of many unit cells and not just to the response of an individual constituent. All these aspects were instrumental in the development of materials that drove the evolution of nanophotonics, but it constitutes nowadays also an obstacle that hinders its further progress. For example, it is difficult to imagine that with top-down technologies it will be possible to fabricate nanooptical devices that exploit the peculiarities of light propagation in the bulk material and not just the interaction of light with a frequency selective surface. It is difficult to imagine that a true localization of light in spatial domains significantly smaller than the wavelength using a spatially extended source can be achieved with a periodic structure that likely leads to a periodic localization in space or, as a last example, in the field of metamaterials were it is the aim to tailor effective properties as probed by an external source, it will be difficult to achieve an isotropic material response. The periodicity at which the unit cells are usually arranged in space will be similarly probed by the external source and it will leave its signature.

To overcome such limitation and to implement new functionalities, the field of *amorphous nanophotonics* is just about to emerge. The *Leitmotif* here is to exploit self-organization mechanisms that rely on bottom-up approaches for the fabrication of nanooptical systems. In result, the structures are predominantly characterized by a deterministic unit cell with tailored geometries; but the spatial arrangement of many of these unit cells forming the material is not controlled with arbitrary precision. Instead of a periodic tiling, the structures appear either amorphous or random. The goal of amorphous nanophotonics along with some explanations is illustrated at a selected example in Fig. 1. Detrimental for the immediate use of such structures in applications is the lack of a sufficiently developed language in which we can discuss on theoretical grounds the optical properties of such system, is the lack of a technology to fabricate amorphous nanophotonical systems and a lack of methods to access in an actual experiment the properties of systems alike. However, the promise of major applications with groundbreaking impact, as outlined also in depth in this book, constitutes a strong motivation to explore this field.

Since major efforts were already devoted in the past to close this gap and to establish a language in which we can access, discuss and explore amorphous nanophotonic systems, this book is intended to make the case for scientists working on or are even just interested in this field. Selected contributors to this book have been collaborating in the Nanogold project funded under the FP7 of the European Union, a project that brought together material scientists that explore bottom-up methods for the fabrication of new nanomaterials and physicists working in the field of metamaterials, to merge both streams and to advance both fields. Many more colleagues well beyond the people involved in this project contributed finally to this book to make it bold and sound.

This book is understood as a seminal reference as well as the inaugural document in which the stage is set for the field of amorphous nanophotonics. It aims at covering aspects that enable scientists to enter and work efficiently in this field and gives indications on perspective applications. The book is roughly divided into four sections; each containing a different number of chapters.

The first section contains two chapters. They provide a broader overview over selected subjects from either an experimental or a theoretical point of view, respectively. The first chapter, written by Alastair Cunningham and Thomas Bürgi, provides a detailed overview over experimental means for the bottom-up organization of metallic nanoparticles. Although not all amorphous nanostructures rely on metallic nanoparticles, their use is often preferred since the localized surface plasmon polariton they sustain guarantees a significant light matter interaction. It is moreover a prominent constituent where the isolated unit cell already sustains a strong response. However, by arranging these metallic nanoparticles by various methodologies spatially in a suitable manner, a response does emerge that is not observed in the individual element. This experimentally oriented chapter is complemented by a chapter written from a theoretical perspective by Filiberto Bilotti and Sergei

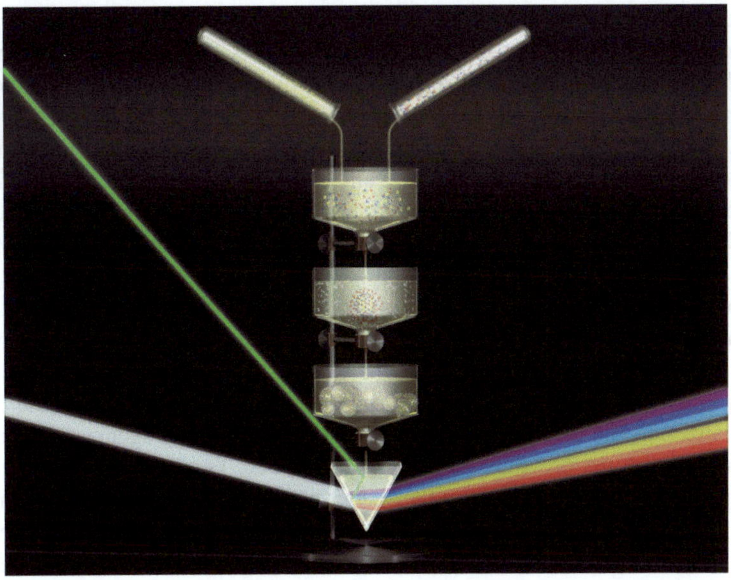

Fig. 1 Conceptual idea of how bottom-up nanofabrication techniques are used to fabricate an amorphous metamaterial; a referential example for a system from the field of amorphous nanophotonics. In this artistic view the process goes from the top to the bottom. Each stage represents a different lengths scale and should be seen as different magnification. In a first step, suitable ingredients on a molecular or nanometric scale are combined according to a chemical receipt. It leads to the formation of a unit cell that may possess a complicated geometry that can, however, be well controlled. In the present example, metallic nanoparticles decorate a dielectric core particle. The scattering response of such structure is dominated by a magnetic dipole, a response not available in nature. The geometry of this basic unit cell is the last item that can be controlled with high precision. The fabrication in solution allows for the realization of large scale quantities, i.e. an entire material is available immediately, but it comes at the expense that the arrangement of these meta-atoms in space cannot be controlled anymore. The resulting structure will be amorphous. This however, brings also a lot of advantages which are beneficially exploited. The available material, e.g., is isotropic since no directional preference exists. An isotropic material with a negative index of refraction, as achievable in principle, would be the Holy Grail that can be used in many applications. Detrimental would be only, as indicated in the figure as well, that the strongly dispersive nature would only allow for the observation of this effect at a narrow frequency range; although concepts exist to extend this spectral domain. With the present state-of-the-art we are not yet at the bottom part of this figure, but a majority of steps in this direction were already successfully gone. The documentation of this progress is one of the purposes of this book

Tretyakov. There, a broader overview on possible applications is given that base on amorphously arranged metallic nanoparticles. Emphasis in this chapter is put on metamaterials but eventually the possibilities for applications well beyond are equally documented.

The second section contains three chapters that discuss issues concerning the theoretical and numerical analysis of amorphous nanophotonical systems. If, in stark contrast to periodic structures, the constituents are no longer periodically arranged in space, novel approaches have to be put in place. Therefore, the first chapter in this section, written by Ari Sihvola and Henrik Wallén, gives an overview over homogenization techniques for amorphous nanomaterials. The insights documented in this chapter enable scientists to replace conceptually their complicated structured amorphous materials by a homogeneous one preserving its electromagnetical properties. Although, such homogenization techniques clearly have their limitations, as discussed in depth in the chapter, they allow to consider amorphous materials in the design of functional devices. The impact of such contribution cannot be estimated high enough. The second chapter in this section, written by Stefan Mühlig and Carsten Rockstuhl, outlines an approach to discuss properties of amorphous nanophotonical materials on the base of the scattering response of its constituents. The main ingredient is the expansion of the rigorously calculated scattered field into electromagnetic multipoles. Such analysis allows for a profound discussion of what properties will emerge in the bulk amorphous media. If only a few multipoles contribute, e.g. the electric and/or the magnetic dipole, homogenization methods as introduced in the previous chapter can be applied. For constituents that do not obey this requirement, i.e. where the scattering response is dominated by a multitude of multipoles, the computational strategy as discussed in depth in the third chapter of this section, written by Vassilios Yannopapas, Alexandros G. Vanakaras, and Demetri J. Photinos, can be used to explore the nanooptical system. There, it is described how the T-matrix of an individual constituent, i.e. a matrix containing information on how an arbitrary incident field is scattered into the far-field, is calculated and how it is used in a subsequent analysis of the optical properties of a media made from many such constituents. The technique they describe permits to hierarchically treat amorphous nanophotonical materials where, in analogy to Fig. 1, the view of the description is successively narrowed.

The third section contains two chapters which are devoted to the introduction of experimental means to measure in a quantitative sense the optical properties of amorphous systems. The first of these two chapters, written by Christian Helgert and Thomas Pertsch, discusses far-field characterization techniques. There, emphasis is put on the question how the amorphous character of the sample contributes to the measurable spectra. The discussion of the impact in the far-field is complemented by a chapter from Worawut Khunsin and Ralf Vogelgesang, which focuses on the near-field characterization. It is the intention to suggest that most notably the combination of both approaches provides a comprehensive picture on the optical properties of amorphous systems and how they can be used in potential applications.

This aspect, what to do actually with amorphous nanooptical systems, is discussed from a multitude of perspectives in the fourth section. It contains six further chapters. In the chapter from Keiichi Edagawa it is concisely shown under which conditions purely dielectric amorphous nanostructures do possess a photonic band gap. For frequencies inside such band gap, the propagation of electromagnetic fields is suppressed. However, it is not just the complete suppression but also the light diffusion and the ability to localize light in tiny spatial domains that are strongly affected, and actually enabled, by amorphous photonic materials. In the chapter that follows, written by Hui Cao and Heeso Noh, it is then shown how the localization of light in dielectric amorphous materials, once infiltrated with a suitable gain media, can be used to set-up novel lasers. Contrary to ordinary lasers where the light is spatially confined by mirrors forming a cavity, the formation of the cavity is promoted only by the amorphous material. In this chapter, among others, numerical and experimental results are nicely fused to provide a comprehensive picture. Applications of amorphous dielectric materials are not just found in technological devices made by mankind, but are also found in nature. The ability of amorphous nanostructures to strongly affect the visual appearance, i.e. the color an object possesses, was beneficial in the process of evolution. It led for example to their integration into the skin of various biological systems. A broad overview on this subject is given in the chapter by Stephen Luke and Peter Vukusic. A common theme that will be highlighted in this chapter is the ability of naturally occurring amorphous nanostructures to appear white, i.e. all wavelengths are scattered on a comparable strength. This ability, to strongly scatter the light of all wavelengths, is equally discussed in a chapter that follows but with a completely different point of view. This chapter, written by Franz Joseph Haug, details the scattering properties of amorphously textured surfaces with critical features down to a few tens of nanometers and their application in a thin-film solar cell to enhance the absorption of light. This is an application of paramount importance since any share of light that is additionally absorbed, promotes a higher efficiency of the solar cell. And since a solar cell cannot be optimized in its operation for just a single wavelength, the ability of these textured surface to strongly scatter the light at nearly all wavelengths is particularly important; explaining actually the success such textures do have already in commercial devices.

And to avoid the impression that only amorphous nanophotonical materials made from dielectrics are used in applications, the last two chapters in this fourth section are devoted to the use of structure containing metallic amorphous nanomaterials. In the chapter written by Roberto Caputo and co-workers from the University of Calabria, they detail how to make systems containing metallic nanoparticles integrated into self-organized soft materials active, i.e. they do change their properties upon an external stimulus. This paves the way, as outlined in this chapter, for applications such as tunable perfect absorbers and for novel sensor concepts; and their potentially use in a range of other innovative devices in diverse fields such as opto-electronics or opto-fluidics. In a final chapter Jose Dintinger and Toralf Scharf will detail the application and use of nanoparticle cluster matter to create artificial electromagnetic response with application to thin film optical devices.

It is our sincere hope that with the broad overview over all aspects related to the field of amorphous nanophotonics, we provide the necessary stimulus to significantly advance the entire field in the mid- and long-term. Our sincere thank goes at first to all the colleagues that have contributed to this book. Without their support and their passion for amorphous nanophotonics, this book and this field would not be where it is today. We would also like to thank our colleagues at Springer that supported us in all stages of the preparation of this book and its careful edition; most notably Dr. habil. Claus E. Ascheron. We would like to thank the European Union that funded some of the research in the project Nanogold that found its way into that book.

Jena, Germany
Neuchâtel, Switzerland

Carsten Rockstuhl
Toralf Scharf

Contents

Contributors

Filiberto Bilotti Department of Applied Electronics, "Roma Tre" University, Roma, Italy

Thomas Bürgi Département de Chimie Physique, Genève 4, Switzerland

Hui Cao Department of Applied Physics, Yale University, New Haven, CT, USA

Roberto Caputo LiCryL-INFM/CNR & Physics Department, University of Calabria, Rende, CS, Italy

Ugo Cataldi LiCryL-INFM/CNR & Physics Department, University of Calabria, Rende, CS, Italy

Alastair Cunningham Département de Chimie Physique, Genève 4, Switzerland

Luciano De Sio LiCryL-INFM/CNR & Physics Department, University of Calabria, Rende, CS, Italy

José Dintinger Optics & Photonics Technology Laboratory, Ecole Polytechnique Fédérale de Lausanne (EPFL), Neuchâtel, Switzerland

Keiichi Edagawa Institute of Industrial Science, The University of Tokyo, Komaba, Meguro-ku, Tokyo, Japan

Franz-Josef Haug École Polytechnique Fédérale de Lausanne (EPFL), Institute of Microengineering (IMT), Photovoltaics and Thin Film Electronics Laboratory, Neuchâtel, Switzerland

Christian Helgert Abbe Center of Photonics, Institute of Applied Physics, Friedrich-Schiller-Universität Jena, Jena, Germany

Worawut Khunsin Max Planck Institut für Festkörperforschung, Stuttgart, Germany

Stephen Luke School of Physics, University of Exeter, Exeter, UK

Stefan Mühlig Abbe Center of Photonics, Institute of Condensed Matter Theory and Solid State Optics, Friedrich-Schiller-Universität Jena, Jena, Germany

Heeso Noh Department of Nano and Electronic Physics, Kookmin University, Seoul, Korea

Thomas Pertsch Abbe Center of Photonics, Institute of Applied Physics, Friedrich-Schiller-Universität Jena, Jena, Germany

Demetri J. Photinos Department of Materials Science, University of Patras, Patras, Greece

Carsten Rockstuhl Abbe Center of Photonics, Institute of Condensed Matter Theory and Solid State Optics, Friedrich-Schiller-Universität Jena, Jena, Germany

Toralf Scharf Optics & Photonics Technology Laboratory, Ecole Polytechnique Fédérale de Lausanne (EPFL), Neuchâtel, Switzerland

Ari Sihvola Department of Radio Science and Engineering, Aalto University School of Electrical Engineering, Aalto, Finland

Sergei Tretyakov Department of Radio Science and Engineering, Alto University, Espoo, Finland

Cesare Umeton LiCryL-INFM/CNR & Physics Department, University of Calabria, Rende, CS, Italy

Alexandros G. Vanakaras Department of Materials Science, University of Patras, Patras, Greece

Ralf Vogelgesang Max Planck Institut für Festkörperforschung, Stuttgart, Germany

Peter Vukusic School of Physics, University of Exeter, Exeter, UK

Henrik Wallén Department of Radio Science and Engineering, Aalto University School of Electrical Engineering, Aalto, Finland

Vassilios Yannopapas Department of Materials Science, University of Patras, Patras, Greece

Chapter 1
Bottom-up Organisation of Metallic Nanoparticles

Alastair Cunningham and Thomas Bürgi

Abstract This chapter deals with bottom-up strategies that allow one to prepare amorphous assemblies of metal nanoparticles. Within these assemblies the nanoparticles couple to each other, affecting the effective electromagnetic properties of the materials. As a consequence, besides the properties of the individual particles, parameters such as number of individual particles within the assembly, geometry of the assembly and average distance between particles within the assembly can be used to design the optical properties of a material. It is therefore highly desirable to control these parameters with high precision, which is the art of self-assembly. Compared to top-down lithographic methods the bottom-up self-assembly approach is cheap and enables the fabrication of large area two-dimensional or three-dimensional samples, making it attractive for applications. In the following, after an introduction, different strategies that were used in the past to assemble nanoparticles into defined structures are briefly discussed. Such strategies rely on templates such as liquid crystals, DNA or surfactants. A versatile approach, which relies on charge-driven self-assembly mediated by charged surfaces and polyelectrolytes, is then discussed in more detail. This approach easily allows one to build large scale amorphous layered structures of nanoparticles with high control of parameters such as distance between particles within one layer and distance between the layers. The method is not restricted to flat surfaces and can be used to coat for example silica beads, resulting in core–shell structures. An attempt has also made to rationalise the observed optical properties in terms of coupling between particles within the different assemblies, thus paving the way to the design of materials with novel electromagnetic properties.

A. Cunningham · T. Bürgi (✉)
Département de Chimie Physique, 30 Quai Ernest-Ansermet, 1211 Genève 4, Switzerland
e-mail: Thomas.Buergi@unige.ch

C. Rockstuhl, T. Scharf (eds.), *Amorphous Nanophotonics*,
Nano-Optics and Nanophotonics, DOI 10.1007/978-3-642-32475-8_1,
© Springer-Verlag Berlin Heidelberg 2013

1.1 Introduction

1.1.1 Bottom-up vs. Top-down

The term 'nanotechnology' was first coined in 1974 [1] and since then the drive to produce structures of increasingly small dimensions and explore their unique properties has not abated. There are few aspects of modern life that do not rely on, at least in some part, the scientific breakthroughs in fabricating smaller structures that have been made in the last few decades, with nanotechnology now pervading our lives to an extent that very few, with some notable exceptions [2], could have predicted. The advances in down-scaling that have been made, have been applied to a wide range of disciplines, and have played a prominent role in the fields of information technology [3], renewable energy generation [4], medicine and health care [5]. In general, the approaches to fabricate these structures can largely be divided into two categories; top-down and bottom-up techniques [6]. The fast rate of down-scaling achieved over the last few decades, in some cases advancing at almost exponential rates, can largely be attributed to the top-down techniques, loosely defined as being the fabrication of structures from larger precursors. This can be thought of as being more of a sculpting approach, with typical examples of top-down technologies including lithographic methods, the most technologically advanced of which is arguably electron-beam lithography [7], and etching methods, which can vary from wet-chemical etches [8] to focused ion beam [9] and laser ablation [10]. While there is still an important role to be played by top-down methods, there are inherent limitations associated and these limits, if they have not been already, are on the verge of being reached. One notable drawback is the minimum feature sizes that are accessible using such techniques. In addition, such techniques tend to be cumbersome and slow, requiring prohibitively expensive equipment to produce small scale and, often uniquely, two-dimensional structures. None of these disadvantages, on the other hand, are suffered by the more novel and versatile bottom-up techniques which are increasingly being applied to produce structures of even smaller scale and more complex architecture than has previously been achievable. As the name suggests, the bottom-up approach involves the fabrication of structures from smaller units, using the properties that they possess to induce their self-assembly in the desired manner. Despite these terms first being used in connection with nanotechnology in the late 1980s this building block approach to materials synthesis, often likened to the construction of objects with Lego, is still in its nascent stage and only the surface of what is accomplishable has been scratched.

Evidence for the efficacy of these techniques can be gathered through the observation of nature, which is also seen to follow a bottom-up approach. The organisation of molecules to form progressively larger structures, from cells and DNA up to, and including, metre scale biological organisms, in a hierarchical manner show the effectiveness that such approaches can provide. Indeed the entire field of biomimetics seeks to take advantage of structures found in nature which have evolved over a long period of time and, as the name suggests, use them as models or blue-prints which can then be reproduced using innovative chemical methods. Despite being distinct from the type of bottom-up fabrication techniques which will be discussed at greater length here, the case of biomimetics also introduces another interesting aspect of structure fabrication. In general, for useful functional structures to be fabricated, a degree of organisation must be available at two discrete levels; both at the nanoscale and at the mesoscale. After the initial self-assembly step which results in the formation of a unit, it is then typically necessary to induce, chemically or otherwise, the organisation of these units into an ordered structure which can then be used in applications. It is important to note that a perfect long range order in all three dimensions of space is seldom achieved using self-assembly techniques. However, and most importantly, in the context of the materials discussed here this is not a prerequisite for obtaining functionality or attractive electromagnetic properties. The latter are more closely linked to the properties of the individual entities and their short range ordering within larger units meaning that amorphous structures hold substantial potential. A whole host of techniques exist in the ever increasing toolkit of the materials scientist that permits both the fabrication of nanoscale materials and their organisation into larger scale architectures in a controlled manner. An exhaustive list of these is, due to the inter-disciplinary and fast-moving nature of the field, extremely difficult to produce and even more difficult to discuss in great detail. However an effort will be made to provide an introduction to some of the more important bottom-up methods with particular attention being paid to the fabrication, application and organisation of metallic nanoparticles.

1.1.2 Metallic Nanoparticles and Their Applications

The wide ranging catalogue of metallic nanoparticles which can now be routinely prepared, and in many cases are now commercially available, has garnered increasing interest as their use in a broad variety of applications has become apparent. This is largely due to their particular optical properties, more specifically the fact that they support a localised surface plasmon resonance (LSPR). This effect, which has been widely studied and documented [11], describes the coherent dipolar oscillation of the surface electrons present in a nanoparticle when excited by electromagnetic radiation of a particular frequency. This manifests itself as a strong extinction of the incident radiation which, depending on the material, exists in the visible part of the electromagnetic spectrum. The effect was first modelled by Gustav Mie in 1908 who solved Maxwell's equations for light interacting with spherical metallic nanoparticles. His equation, shown below, which includes both the scattering and absorption, holds true when dipolar oscillations of the conduction electrons are considered in the limiting case where the wavelength of the incident radiation is significantly greater than the size of the metallic nanoparticles.

$$C_{\mathrm{ext}} = \frac{24\pi^2 R^3 \varepsilon_m^{3/2}}{\lambda} \frac{\varepsilon_2}{(\varepsilon_1 + 2\varepsilon_m)^2 + \varepsilon_2^2} \tag{1.1}$$

where

C_{ext} = extinction cross-section
R = particle radius
ε_m = dielectric constant of the surroundings
$\varepsilon = \varepsilon_1 + i\varepsilon_2$ = complex dielectric constant of bulk metal
λ = wavelength of incident radiation

Evidently, the maximum extinction occurs when $\varepsilon_1 = -2\varepsilon_m$ thus defining the position of the LSPR peak. The pioneering work which Mie executed is not limited to nanospheres and can also be modified to describe nanoparticles of other geometries such as rods [12].

These optical properties have made the use of metallic nanoparticles attractive to a number of fields interested in using them in applications. As shown in (1.1), the position of the LSPR peak is altered depending on the dielectric constant of the surrounding material. As such the particles can be thought of as probes that can accurately return information, observed as red or blue-shifts of the LSPR peak, on the refractive index, and therefore composition, of their surrounding medium. A host of sensing applications, where colour changes observed in the metallic nanoparticles can indicate trace changes in concentration of analytes, can be envisaged with many already both in use and development [13]. A comprehensive review of nanostructured plasmonic sensors can be found in Ref. [14].

One specific area of interest where sensing and metallic nanoparticles are concerned is in the field of surface enhanced Raman scattering (SERS). This effect was discovered in 1974 by Fleischmann [15] and relies on the extremely strong electromagnetic fields that exist in the nanogaps between metallic nanoparticles. This facilitates massive enhancements of the Raman signal, on the order of 10^{14}–10^{15} more than that observed under normal conditions [16], which is sufficient to allow the detection of single molecule analytes. This is extremely useful as the Raman signal can return detailed information on the vibrational levels of the analyte, allowing the technique to be used for both detection and structure determination [13]. SERS has now developed to the extent that spectra can be measured in vivo [17]. In principle this could facilitate faster acquisition times and was a significant breakthrough for the use of metallic nanoparticles in medical and biological applications. A more detailed discussion of the enhancement mechanism observed and specific SERS applications can be found in Refs. [18] and [19].

Biomedical studies of metallic nanoparticles began in the 1970s following the discovery of immunogold labelling [20]. Since then a wide array of biological sensing applications have been developed [5, 21]. Gold nanoparticles, used for medical diagnostics have even become commercialised where they form an integral part of certain pregnancy tests which are currently on the market. However metallic nanoparticles are not limited to sensing applications. One major reason for the widespread use of gold nanoparticles for biomedical applications is their relatively low cytotoxicity which makes them particularly suitable for in vivo applications such as drug delivery [22], bioimaging [23] and cancer therapy [24]. An innovative form of cancer therapy involves suitably modified gold nanoparticles, functionalised with antibodies that recognise antigens expressed by cancerous cells and allows them to specifically bind to these cells [24]. The excitement of the LSPR through irradiation with a suitable source results in the fast conversion of the absorbed energy to heat which, as the particles are localised at the cancerous cells, is sufficient to destroy these cells whilst leaving healthy ones undamaged. In addition to the control over the surface chemistry that is required it is also necessary to push the LSPR of the nanoparticles into the infra-red region of the electromagnetic spectrum as otherwise the incident radiation used to excite the LSPR would be absorbed by the first few millimetres of skin. Infra-red radiation, on the other hand, can penetrate significantly further and facilitates the excitement of the plasmon resonance of particles that are located at the cancerous cells. By using gold nanorods, the position of whose longitudinal plasmon resonance can be readily tuned in the growth process through altering the aspect ratio of the particle, this problem can be overcome.

As well as being able to use nanoparticles in diagnostics and in the treatment of medical conditions they have also found a niche application in measuring nanoscale distances, particularly in biological samples, and can be thought of as being 'plasmonic rulers' [25]. The system is based on the strength of the coupling between a pair of metallic nanoparticles which is itself strongly dependent on the distance between them. Stronger and weaker coupling manifests itself as red and blue shifts of the LSPR peak respectively and after adequate calibration it is possible to correlate these shifts in the optical properties to the separation of the nanoparticles. Dynamic separations of up to 70 nm can be monitored allowing information on both distance and configuration to be elucidated [25]. A number of advantages exist over other methods used to determine molecular distances such as Förster resonance energy transfer measurements. Here, low and fluctuating signal intensities along with limited observation times and upper measurement limits of approximately 10 nm mean that plasmonic solutions such as the ones described may be more appropriate in certain situations [25]. However, certain issues, including the extent to which the adsorption of nanoparticles to biomolecules affects their overall structure have not been unequivocally resolved. Despite this, the power and potential that plasmonic rulers possess with respect to structure elucidation is clear to see.

An equally important field which makes use of metallic nanoparticles is that of catalysis. Despite being historically considered as merely a chemically inert metal, supported gold nanoparticles were first found to reduce the activation energy for the oxidation of CO around two decades ago [26]. The high surface area, and therefore number of active sites, provides obvious advantages. There exists a strong dependence of catalytic activity on particle size, with a large decrease in activity for particles larger than 6–10 nm being observed [26]. Therefore one of the major challenges facing the field is finding a means to maintain particle stability under the reaction conditions required. A more recent study, and direct utilisation of the plasmonic properties of metallic nanoparticles, researches the catalytic properties that are induced upon the localised heating produced when the particles are in resonance [27].

Here, a solvent mixture of ethanol and water was passed through a microfluidic cell constructed with an array of gold nanoparticles. When excited with a low power laser the heat produced at the surface of the particles was sufficient to catalyse the breakdown of the solvents and the formation of CO_2, CO and H_2 [27].

Plasmonic structures can also be used to increase the efficiencies of photochemical and photoelectrochemical processes [28]. A number of mechanisms resulting in the increased efficacy of these reactions have been proposed. For example it has been suggested that accelerated reaction rates occur due to the enhanced electromagnetic fields that exist at the surfaces of, and particularly in the nanogaps between, metallic nanoparticles by elevating the molecules involved in the reactions to excited electronic states. The metallic nanostructures essentially act as optical antennas that harvest photons and then concentrate the electromagnetic field produced to excite target molecules in the vicinity. It has also been proposed that other effects associated with excited plasmon resonances such as localised heating and mechanical oscillation may also play a role in increasing the efficiency of these reactions [28]. The first experimental observation of this effect was reported in 1983 [29] and in more recent studies remarkable enhancements of molecular excitement have been reported [30]. Such systems have potential applications in the development of organic solar cells and photoelectrochemical biosensing devices [30].

An additional application of metallic nanoparticles structured by bottom-up techniques, and of particular pertinence to the work outlined in this book, is in the field of metamaterials, where they are of fundamental importance. This term is loosely used to describe materials which possess electromagnetic properties which do not exist in nature and the application of bottom-up techniques offers the opportunity to create structures with feature sizes that would be otherwise inaccessible with traditional top-down methods. The term metamaterial, with the prefix taken from Greek and meaning literally "beyond", first appeared in literature as recently as the year 2000 when Smith et al. published their seminal paper on a composite medium with simultaneously negative permeability and permittivity in the microwave range of the electromagnetic spectrum [31]. Since then the field has exploded, however the potential contribution that metallic nanoparticles have to make to this field is only beginning to be tapped. The down-scaling of structures which have been proven to exhibit exciting novel properties such as negative refractive indices [32], cloaking [33] and perfect lensing [34], in the microwave and infra-red regimes of the electromagnetic spectrum is necessary to push these properties into the visible range. With the obvious benefits that this would entail, increasing efforts, some of which will be outlined in the main body of this chapter, are being made to further develop the application of metallic nanoparticles to the field of metamaterials.

1.1.3 Coupling Properties of Metallic Nanoparticles

As shown, metallic nanoparticles undoubtedly have many applications, both current and future. The majority of these applications rely on the LSPR that the parti-

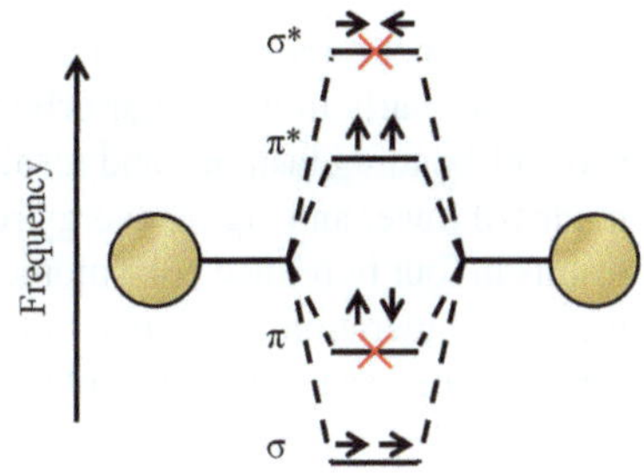

Fig. 1.1 Schematic showing hybridisation of plasmons of two approaching metallic nanoparticles. *Red crosses* indicate dark modes that only offer a weak excitation for the present configuration

cles support, allowing them to, amongst other things, sense, heat, catalyse and image. The position of the LSPR peak, as discussed above and succinctly summarised in (1.1), is dependent on a number of factors including the particle size and shape as well as the refractive index of the surrounding medium localised at the surface of the particle. However one important factor that is not considered in the equation, which analyses only a single particle, is the effect that other particles have over one another. This effect, also known as coupling, can be exploited by finding means, chemical or otherwise, to order metallic nanoparticles into specific configurations or structures. This is an additional, more flexible, method of fine-tuning the optical properties of systems and will be discussed in more detail in the following.

Coupling between metallic nanoparticles is very well summed up by plasmon hybridisation theory. This physically intuitive method, developed in 2003 by Peter Nordlander et al. [35], uses similar constructs to those used in molecular orbital theory to describe the optical changes that occur when two metallic nanoparticles are brought together to the extent where they begin to interact. The plasmonic response of a more complex structure is constructed from the interactions that exist between its constituent elements, forming hybridised modes in much the same manner that molecular orbitals are composed from the atomic wave-functions of the individual atoms that form molecules. This interaction can be considered in its simplest form as the coupling of two nearby dipoles, although depending on the size of the nanoparticles and their distance of approach higher multipole moments can also be considered. Hybridisation theory predicts two distinct classes of eigenmodes for a structure of two strongly coupled metallic spheres, i.e. a dimer. The first is associated with an in phase oscillation of the electric dipole in both spheres and is therefore termed a 'bright' eigenmode since it can radiate into the far field. The second class requires a 180° out of phase oscillation of the electric dipoles of both spheres. These eigenmodes are termed 'dark' since they cannot radiate into the far field and play an important role in the field of metamaterials as the out of phase oscillation of electric dipoles can be related in some cases to a magnetic dipole moment which is the key component in many predicted applications of metamaterials.

The method of constructing hybridised plasmonic modes through the combination of two or more resonant elements is illustrated in the plasmon hybridisation diagram shown in Fig. 1.1 and depicts the theory applied to a dimer of two metallic nanopsheres.

Shown are the individual nanoparticles with the dipoles that they support, the directions of which are dependent on the direction and polarisation of the incident electromagnetic radiation which drives the free electrons of the metal into resonance. Similarly to molecular orbital theory these dipoles are combined in bonding and anti-bonding fashions and result in the splitting of the modes to give hybridised modes of lower and higher energy respectively. The interaction of two equal spheres results in four hybridised plasmonic modes in the coupling regime. Because of symmetry considerations, only two of these modes display a net dipole moment and will show a strong scattering of light into the far-field at relevant frequencies and angles of incidence. Conversely, both the antibonding mode for incident light parallel to the main axis of the dimer (σ^*) and the bonding mode for incident light perpendicular to the main axis of the dimer (π) have no net dipole moment and are known as dark modes (noted in Fig. 1.1 with red crosses) that can only be observed by breaking the symmetry of the system and for large spheres with extremely small interparticle distances. These dark modes contribute predominately to the absorption of the system because the resulting quadrupole moments offer only a weak coupling to the far field.

Plasmon hybridisation theory can be used to model systems both quantitatively and qualitatively and has proven to be an invaluable tool both in explaining the optical changes observed, for example spectroscopically, and in designing new systems which should have specific optical properties. It is in no way restricted to simply describing nanospheres and has been applied to nanoparticles of a wide range of different forms such as asymmetric dimers, thin metallic films, systems of metallic nanoshells, nanorods and nanostars [16]. Under certain situations it has also been shown to be applicable to large-scale systems, being used to describe the coupling between cm^2 scale arrays of metallic nanoparticles [36].

1.1.4 Bottom-up Organisation of Metallic Nanoparticles

The large potential for using bottom-up methods to organise metallic nanoparticles into assemblies that could be used as functional materials with practical applications has meant that significant amounts of research effort have been focused in this direction in recent years. Some examples illustrating the principal methods used to achieve this will be given here. As will be shown, the assembly of metallic nanoparticles relies on surface chemistry and self-assembly principles using biomolecules, surfactants, mesogens or polymers and makes use of intermolecular (interparticle) interactions. It is therefore logical that the bottom-up organisation of metallic nanoparticles is a research area where the border between physics and chemistry needs to be crossed.

Self-assembly is closely linked with the liquid-crystalline state of matter and it is therefore not surprising that liquid crystals have been used in order to organise gold nanoparticles [37]. Two principal strategies can be followed. In the first one the nanoparticles are simply mixed into the liquid crystal thus forming an physical mixture. The nanoparticles need to be functionalised in order to be stable and soluble in the liquid crystal [38]. The second strategy relies on the chemical bonding of the mesogenic unit to the nanoparticle. Whereas the first strategy is easier it often

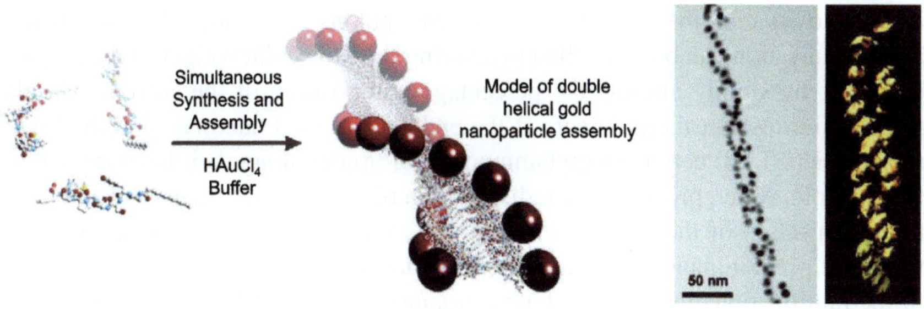

Fig. 1.2 Left-handed gold nanoparticle double helices were prepared using a method that allows simultaneous synthesis and assembly of discrete nanoparticles. *Left*: schematic principle of the method, *right*: electron microscopy images of the assembled structures. Reprinted with permission from Ref. [42]. Copyright 2008 American Chemical Society

suffers from low filling fractions. For both strategies the assembly of relatively large nanoparticles, i.e. particles that exhibit a plasmon resonance, is difficult. In some cases the metallic inclusions have been as large as 7 nm in diameter, which are large enough to support LSPRs [39].

The bottom-up assembly of metallic nanoparticles has also been further advanced through the use of biomolecules such as strands of DNA, which themselves are formed through self-assembly processes. The formation of ordered structures in such a way, using DNA as a molecular substrate for nanoparticle deposition, has facilitated the production of a number of different metallic nanoparticle architectures. One early example was the formation of dimers and trimers which was achieved through the functionalisation of gold nanocrystals with single stranded DNA oligonucleotides of defined length and sequence and assembling them on a complementary single stranded DNA template [40]. This work introduced the possibility of using oligonucleotides to self-organise metallic nanoparticles into well-defined and homogeneous structures. More recently metallic nanoparticle heterodimers comprising both gold and silver nanoparticles fabricated in a similar manner were shown to be SERS active [41]. Precise nanogap engineering was achieved through the controlled growth of silver layers on top of the original gold nanoparticles and single molecule sensitivity was shown when dyes were analysed using SERS. This advance is further evidence of such work being of fundamental importance to several fields of scientific research.

Similarly the use of a chemically modified peptide substrate to reduce a gold salt, forming gold nanoparticles along the backbone, has been shown to result in left-handed double helix arrangements of the nanoparticles (see Fig. 1.2) [42]. The introduction of chirality into structures in such a manner results in optically active arrangements of resonant elements that allow the further tuning of the optical properties of such materials.

Resting with the use of macromolecules in the ordering of metallic nanoparticles, pseudo block copolymer systems have also been employed to this end [43]. Gold nanorods selectively functionalised at their ends with polymer chains, resulting in a 'pom-pom' structure reminiscent of an ABA tri-block copolymer, can be induced to organise the nanorods either end to end or side by side. This interchangeable organisation can be controlled by altering the molecular weight of the polymer chains or by simply altering the percentage composition of the solvent. The different affinities that the polymers at the ends of the gold nanorods and the CTAB molecules (CTAB: cetyltrimethylammonium bromide) along their lengths have for, for example, water, provokes the redistribution of the polymer chains and stimulates the organisation of the nanorods in different manners as the concentration of water in the predominantly organic solvent is increased. A more in-depth discussion of the ramifications that such differing organisations of gold nanorods have on the optical properties, considered through the eyes of plasmon hybridisation theory, can be found in Ref. [44].

Additional degrees of order compared with the systems previously discussed arise from the formation of binary nanoparticle superlattices from colloidal crystallisation methods which can result in impressively large crystals containing metallic nanoparticle inclusions [45]. Here, steric repulsion and van der Waals, electrostatic, and substrate–particle interactions, amongst others, all combine to determine the crystal structure where the type of control over particle size, shape and composition outlined previously allows materials with tunable physical and chemical properties to be manufactured.

It would be easy to assume, on the other hand that less ordered systems would result from the agglomeration of nanoparticles. In general this is true and such aggregation, which can occur for a variety of reasons, is normally to be avoided. However the controlled aggregation of colloidal gold nanospheres has been used to selectively produce dimers and trimers, which were subsequently used to catalyse the formation of ZnO nanowires in solution [46]. The aggregation, occurring as a result of a reduction of the electric double layer, is achieved through the addition of HCl and the particles are simultaneously stabilised by encapsulating polymers. An optimal pH for dimer formation was found to exist and high yields could be isolated upon enrichment by density gradient centrifugation. This control over the aggregation kinetics provides a route to the fabrication of ensembles with well-defined nanoparticle inclusions.

The hierarchical organisation of plasmonic nanoparticles at different length scales can also lead to interesting optical properties [47]. Such organisation has been observed within metal nanoparticle–surfactant composites, more precisely in a mixture of gold nanoparticles with triethyleneglycolmono-11-mercaptoundecylether (EGMUDE) in water [48]. EGMUDE consists of a thiol, used to anchor to the gold nanoparticle surface, a hydrophobic alkyl chain and a hydrophilic ethyleneglycol part. After addition of EGMUDE to preformed GNPs a fast colour change from purple to dark blue and sedimentation is observed, which is usually indicative of nanoparticle agglomeration. However, after some time (several hours without shaking) the sample changed colour again. UV-visible spectra showed a strongly red-shifted plasmon resonance (see Fig. 1.3) and dynamic light scattering revealed the formation of larger amorphous assemblies of gold nanoparticles, which was then also confirmed by transmission electron microscopy (TEM, see Fig. 1.3). These entities represent hierarchical structures composed of gold nanoparticles clustered

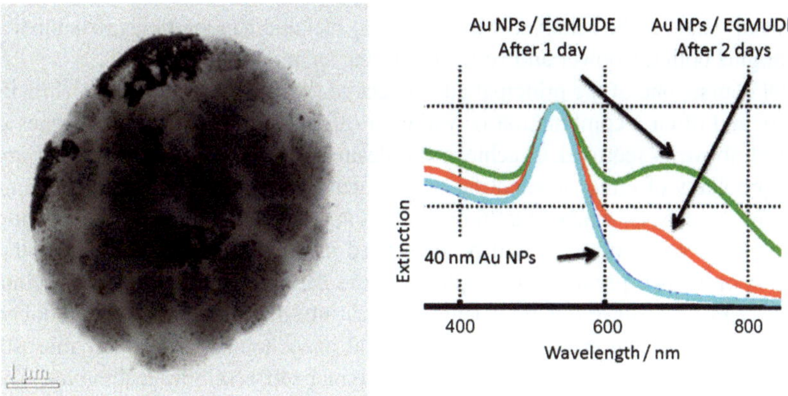

Fig. 1.3 Transmission electron microscopy (TEM) image and UV-vis spectra of gold nanoparticle (40 nm) EGMUDE assemblies [48]. Reproduced by permission of the PCCP Owner Societies

together into larger aggregates of about one micron. Several of these entities again cluster together to even larger units. The assembly of the nanoparticles is guided by the EGMUDE, which also stabilises the particles. In order to better understand the optical response calculations were performed on analogous structures by applying analytical solutions of Maxwell's equations for particles with a spherical symmetry, well-known as Mie-theory, which was extended to handle aggregates of spheres [49]. With this method one can calculate all quantities of interest for clusters consisting of arbitrarily arranged nanoparticles with a spherical shape. Such calculations very well reproduced the shifted plasmon resonance and furthermore indicated that the dominating contribution to this spectral feature arises from a magnetic dipole oscillation that is induced due to the spherical shape of the fabricated supramolecular clusters.

By combining top-down and bottom-up techniques it is also possible to exploit the advantages of both, resulting in substrates with metallic nanoparticles that self-assemble into structures that also exhibit long range order. Two-dimensional square arrays of gold nanoparticle clusters can be prepared, for example, by inducing their template directed self-assembly into holes prepared by laser interference lithography [50]. These hybrid techniques can be implemented when required and provide an additional layer of flexibility when it comes to preparing structures of a particular design.

This introduction has given a small snapshot of some of the principal bottom-up self-assembly routes towards nanoparticle organisation and is by no means comprehensive. As shown, the field of bottom-up assembly of metallic nanoparticles is considerably developed in terms of what is achievable and the wide range of methods that exist to accomplish design goals set. However, at the same time the continual introduction of new techniques and their combination with existing ones means that this is an extremely fast moving area of research that makes notable advances on an extremely quick basis. This means that only the surface has been scratched in terms of what can be achieved with respect to metallic nanoparticles structures and highlights both the power and versatility of the techniques discussed.

Of course, one of the principal advantages of the bottom-up approach is its flexibility and often a combination or variation of one or more of the techniques discussed above is required to achieve the desired results. Indeed, one could argue that problems of either a scientific or practical nature should be considered on a case by case basis when searching for bottom-up solutions. As such, the power of the bottom-up approach will be illustrated in this chapter through the detailed discussion of two particular examples in more detail which exemplify the control which can be achieved at both length scales introduced above; the nanofabrication of metallic nanoparticles using colloidal nanochemistry and their mesoscale organisation using surface chemistry and polymer self-assembly techniques. These case studies, showing the fabrication of layered arrays of metallic nanoparticles and core–shell nanoclusters, will demonstrate that both size and structure can be at least equally as important as composition in terms of tailoring the optical properties of such materials.

1.2 Manipulation of Metallic Nanoparticles at the Nanoscale

For the metallic nanoparticles discussed in the introduction to be of any practical use, for example in the construction of metamaterials with specific properties or functional optical devices, they must first be incorporated, or organised, into structures which take advantage of the particular optical properties that they possess, of which the most important is the exhibition of an LSPR [51]. The fabrication of these particles can be traced back several centuries from when they were used to give stained glass ornaments and windows their characteristic hues [52]. The beginnings of this field in true scientific terms, however, was in 1857 when Faraday, thought of by some as being the father of modern nanoscience, first prepared and characterised colloidal gold nanoparticles in a controlled manner [53]. Very similar methods [54] to that which Faraday developed are still currently used to produce nanoparticles although today a much larger gamut of preparation techniques can now be drawn upon by modern materials scientists to prepare a wide variety of nanoparticles of differing shapes, sizes and materials [55, 56]. However, merely synthesising a batch of metallic nanoparticles, whilst interesting in its own right, is not sufficient in order to fully exploit their optical properties. For this to be achieved it is first necessary for a degree of control to be assumed over a number of important material parameters, not least of which is the morphological organisation of the nanoparticles at the nanoscale. This is where the major challenge lies for nanochemists looking to make advancements in the field. The development of novel and innovative methods combined with the exploitation of pre-existing ones is vital if the manipulation of metallic nanoparticles at this scale is to be successfully realised. Yet even the large degree of control over geometrical arrangement, which can now be routinely realised with impressive results using techniques such as electron beam lithography [57] and direct laser writing [58], is not entirely sufficient where the fabrication of functional optical devices is concerned. For this to be accomplished it is also necessary to be able to expand such structures to utilisable scales in each dimension. In the modern world the importance of the efficient fabrication of large area, homogeneous structures is self-evident with the ubiquitous financial constraints that exist. In addition, the expansion of such structures into the third dimension is equally important if metamaterials with effective medium parameters are to be realised [59].

As mentioned in the introduction, a single solution to both of these problems is to use a bottom-up approach whereby small building blocks, in this case metallic nanoparticles, are coerced, through the application of chemical techniques, into self-organising into desirable structures. Of course, as with any emerging field, there remain some difficulties with employing this approach which are unavoidable. The loss, or reduction, of order that is achievable through the use of more traditional nanofabrication techniques makes the modelling of the optical properties of such materials significantly more challenging. However this loss of order can also be considered as advantageous as strong spatial dispersion hampering the unambiguous introduction of effective material parameters is no longer an issue. Additional benefits that arise through the application of the self-assembly route include the fabrication of previously inaccessible feature sizes [60], the facile and homogeneous deposition of nanoparticles onto large scale (cm^2) substrates [36] and the ready possibility of extending structures into the third dimension [36]. These advantages largely outweigh any disadvantages that exist and allow the preparation of larger scale structures constructed from smaller scale resonant elements at significantly reduced costs. Prior to devising fabrication routes it is first necessary to have an architecture in mind that would be of interest to the community. These are generally conceived in collaboration with theoreticians and this design process, which is outwith the scope of this chapter but discussed in detail in other chapters of this book, generally results in a trade-off between what is structurally desirable and what is presently technologically achievable.

As previously mentioned, the primary reason that metallic nanoparticles are of such interest is that they can support an LSPR at optical frequencies. The optical properties of an ensemble of two or more of these particles, approaching to within certain well-defined distances, can be significantly affected [16]. This effect, known as coupling, allows the LSPR to be manipulated and tuned with a high degree of control and gives rise to a new range of materials classed as plasmonic molecules. One such design that takes advantage of this effect is layered arrays of amorphously arranged metallic nanoparticles separated by well-defined and controllable distances. Structures such as this can be fabricated using true bottom-up approaches [36, 61] or, with less precision, through the use of other deposition techniques such as spin-coating [62].

1.2.1 Gold Nanoparticle Preparation

One method that can be used to deposit an array of, for example, gold nanoparticles on a large scale planar substrate such as glass or silicon is to exploit the inherent surface charge that exists on the particles when produced by the Turkevich

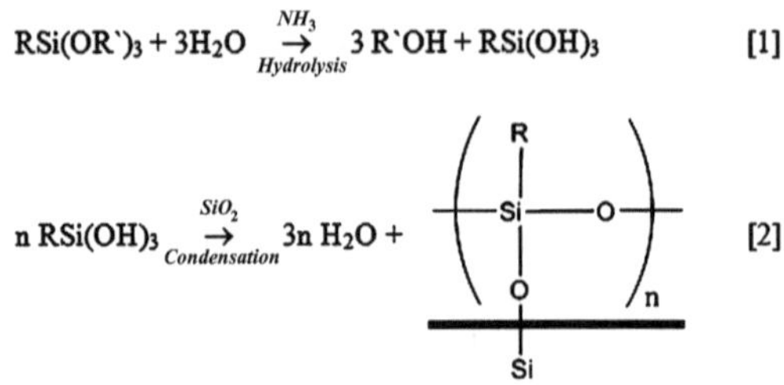

Fig. 1.4 Two step substrate functionalisation process using silica chemistry

method [63]. This method is generally accepted as being one of the most efficient means to produce spherical, monodisperse gold particles with a size range in the order of 9–120 nm in diameter [64]. Their distinctive surface charge arises from the citrate molecules which fulfil two important roles in the particle growth reaction. They firstly reduce the gold ions present in solution to produce atomic gold and secondly cap the particles that are produced, forming a protective monolayer at the surface which prevents their aggregation due to the repulsive forces generated between two like-charged particles. The negative surface charge of the particles results from deprotonated carboxylic acid groups within the molecule and is therefore clearly pH dependent. The citrate molecules, or rather their concentration relative to that of the initial gold salt, are also crucial in determining the overall size of the particles with greater relative concentrations of citrate resulting in smaller particles and *vice versa*, the greater specific surface area exhibited by smaller particles necessitating more citrate molecules to completely encapsulate all of the particles.

1.2.2 Substrate Functionalisation

In order to induce an electrostatic attraction between the gold nanoparticles and any substrate the surface of this substrate must necessarily possess an opposite charge, in this case positive. Glass and silicon, perhaps two of the most useful substrate materials, in terms of availability, cost and perhaps most importantly ease of characterisation and functionalisation, do not, however, exhibit a strong charge. As such, prior to the deposition of any particles, it is first required that the surface chemistry of the substrate be modified in such a way that a positive charge be exposed at the surface. This is readily achievable through the application of what is known as silica chemistry, a topic that has been comprehensively documented by Ralph Iler in his book: *The Chemistry of Silica: Solubility, Polymerization, Colloid and Surface Properties and Biochemistry of Silica* [65]. In brief, a suitable substrate, which has first undergone a rigorous cleaning procedure involving piranha solution, such as the ones under discussion here can be modified in a facile two-step reaction. In the first step a trifunctional organosilane for example, a trimethoxy- or trichlorosilane is hydrolysed to a silanol in what is a base-catalysed reaction. The silanol molecules then undergo a condensation reaction with the hydroxyl groups at the surface of the substrate to form a network, as illustrated in Fig. 1.4, which can be set in place using a baking step.

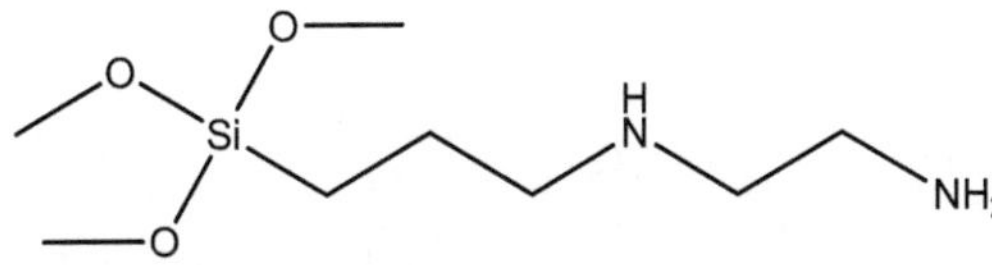

Fig. 1.5 *N*-[3-(Trimethoxysilyl)propyl]ethylenediamine—an example of a typical silane compound used to functionalise glass and silicon substrates. Following functionalisation the terminal amine group is exposed and completely alters the surface chemistry of the substrate

An enormous degree of flexibility exists here due to the wide range of silane compounds that are either commercially available or are relatively easy to prepare. In this specific case a trimethoxysilane with a terminal amine group, shown in Fig. 1.5, was chosen. At neutral pH this group is protonated, thus inducing the required electrostatic attraction between substrate and particles.

1.2.3 Fabrication of Gold Nanoparticle Arrays

When such a substrate, functionalised in such a manner, is exposed to a solution of gold nanoparticles the attractive forces between the two results in the particles adsorbing over a period of several hours until they form a complete array of well dispersed and approximately equally spaced resonant entities at the surface. A glass substrate which has such an array of gold nanoparticles adsorbed at the surface displays a characteristic pink colour which originates from the strong extinction peak that exists at around 520 nm—the LSPR wavelength. An example of this is shown in Fig. 1.6a. The colour is surprisingly strong, given that only a single layer of nanoparticles, with a diameter of around 20 nm, has been adsorbed at a relatively low density or filling fraction. The filling fraction of particles at the surface is again governed by the electrostatic repulsion that exists between them and prevents a second particle from closely approaching one that is already adsorbed at the surface. The particles, which are at least to some extent mobile at the surface, are capable of rearranging to find an organisation representing a low energy state. This is of course why, post-deposition, we observe the particles at reasonably well defined distances from one another, as can be observed in the SEM micrograph shown in Fig. 1.6b. No long range order, however, is perceived which, as can be expected, is due to the random adsorption processes. The amorphous order observed, however, is largely outweighed by other advantages. The only limitations in terms of overall substrate size that can feasibly be coated that exist are of a practical nature. The arrays are virtually defect free and as such this technique quite easily lends itself to applications on relatively large scales.

Detailed analysis of the SEM micrograph shown in Fig. 1.6b using image processing programs such as *ImageJ* [66] allow such arrays to be more fully characterised. It is possible to count thousands of particles over relatively large areas to achieve accurate average values for important material parameters such as particle

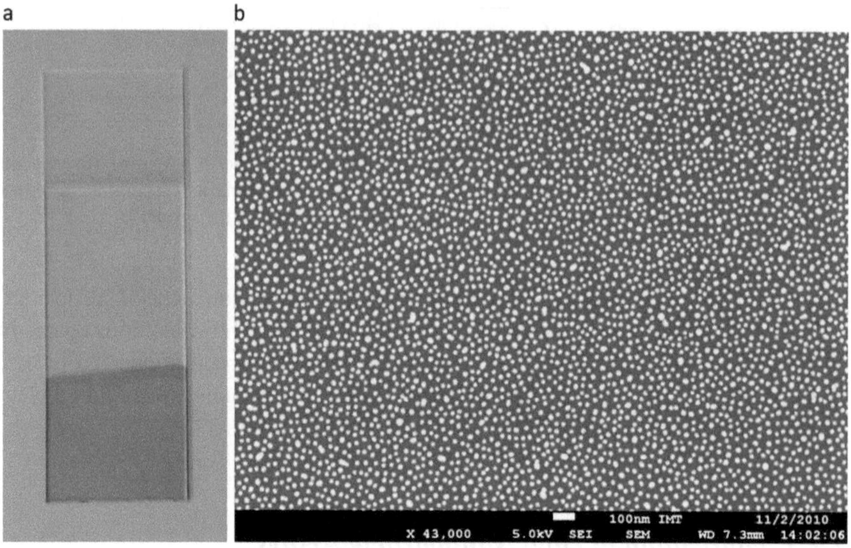

Fig. 1.6 (**a**) Photograph of single array of gold nanoparticles deposited on a functionalised glass microscope slide. (**b**) Corresponding SEM micrograph showing organisation of a single array of gold nanoparticles on a Si substrate which had been previously functionalised

density and size. Such an analysis reveals that the particles deposit at the surface with an average density of around 850 particles/μm^2 which equates to a filling fraction of 27 % [36]. While such a relatively low surface coverage can result in a strong colouration of substrates there may be instances whereby it could be desirable to have surfaces that were coated to either a greater or lesser extent. For example some simulations [67] suggest that for a sufficiently high dispersion to be achieved the filling fraction of arrays of metallic nanoparticles, such as the one under discussion here, should be higher than 30 %. For smaller filling fractions the large magnitude of the permittivity, necessary to observe Mie-resonance for sufficiently small spheres, is not achieved.

1.2.4 Controlling Nanoparticle Density

Lower density arrays are relatively straightforward to produce. As the deposition process occurs over a relatively long time-scale it is sufficient to merely reduce the number of interactions that occur between particles and substrate to create arrays with a lower filling fraction. This can be accomplished in a variety of manners, perhaps most simply by either diluting the solution from which the gold nanoparticles are deposited, or reducing the deposition time.

Creating gold nanoparticle arrays of higher density, on the other hand, is not as straightforward and requires a more considered approach. As previously mentioned, the primary barrier to a closer packing of the particles at the substrate is the negative charge that they possess, introducing an electrostatic repulsion between them. In order to induce a higher filling fraction a means of removing or mitigating, at least partially, this repulsion must be found. One option available is through the careful regulation of the electric double layer that exists in the volume surrounding each nanoparticle [68]. The electric double layer can be defined as the volume of solution containing an excess of ions which balance the surface charge. This volume can itself be considered as being composed of two distinct regimes: the *Stern layer* which denotes the molecular layer of ions of an opposite charge to that of the surface; and the *diffusive layer* which extends into the bulk of the solution. The distance that the electric double layer extends from the surface of the gold nanoparticles is governed by a number of factors including temperature, pH and, importantly, the concentration of ions in solution. Factors such as these ultimately control the maximum density of particles that can be deposited on a charged substrate. For example, the addition of salt to the gold nanoparticle solution contracts the electric double layer and arrays of higher filling fraction are obtained. On the other hand, by reducing the concentration of ions in solution using techniques such as dialysis or centrifugation, the electric double layer would expand and lower density arrays result.

Rather than reducing the mutual repulsion between nanoparticles by regulating the distance that the electric double layer extends from their surface an additional approach is the removal, at least to some extent, of the citrate molecules that cause this repulsion in the first instance. One means of realising this is through making use of the strong sulphur—gold bond and the high affinity that thiol molecules have for gold [69]. When an array of gold nanoparticles is exposed to a solution of thiol molecules a proportion of the citrate capping molecules, largely dependent on the relative concentrations of the two ligands, will be replaced in an exchange reaction. As such, the pre-existing negative charge on the particles is, at least partially, removed along with the barrier to closer approach and reorganisation at the surface [70]. This part replacement of the citrate molecules which previously formed a protective shell at the surface of the nanoparticles results in a number of changes which can be observed in terms of the organisation and morphology of the gold nanoparticle array. Disadvantageously, a number of gold nanoparticles are solubilised and desorb from the surface as their electrostatic attraction to the substrate is diminished. This, due to the high affinity that the thiol molecules have for gold, is unavoidable. However, the same removal of charge which causes this desorption of a small proportion of the nanoparticles also induces a reorganisation of the ones that remain at the surface. This has a number of consequences in terms of changing optical and morphological characteristics of the sample as well as opening up a number of possibilities with respect to a flexibility of structure fabrication and meeting design goals. Where previously, as shown in Fig. 1.6b, an array of well-separated gold nanoparticles is formed, when the barrier to reorganisation has been removed the particles increase in mobility and tend to aggregate into small chains and clusters of around 5–10 particles. This process is thought to be driven by the energetically favourable, formation of bilayers of the relatively long alkyl chains that form the second functionality of the thiol molecules employed in this work, namely

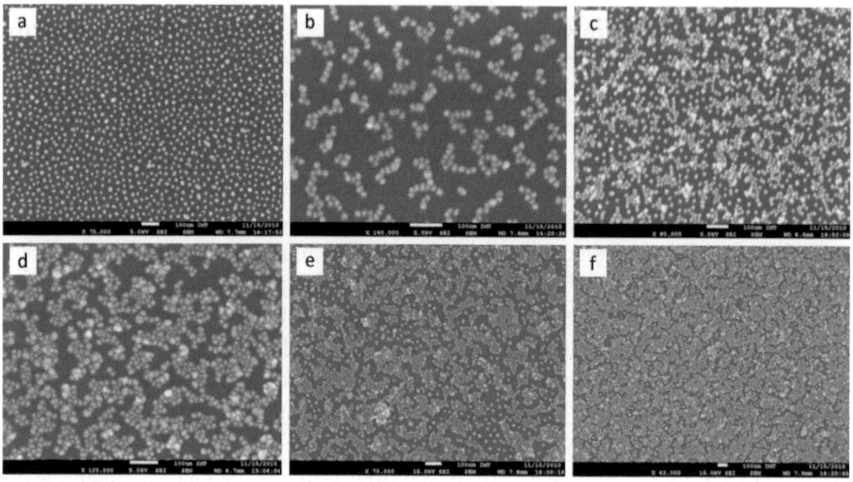

Fig. 1.7 SEM micrographs depicting various stages of the deposition and reorganisation process as *described in the main body of the text*. Shown are the morphology that could be expected after (**a**) the first deposition, (**b**) the first reorganisation, (**c**) the second deposition, (**d**) the second reorganisation, (**e**) the third deposition and (**f**) the fourth deposition. An increase in particle density as well as the introduction of a degree of order can clearly be observed

1-dodecanethiol. Crucially, any coalescence of the particles is prevented by the bilayers of the thiol molecules that are formed as they are drawn together which is essential in terms of maintaining the LSPR that they exhibit. The thiol molecules, or more importantly the length of the alkyl chain, also determine the closest distance to which the gold nanoparticles can approach one another which in this case is on the order of 1–2 nm. A by-product of forming these clusters, along with the desorption of a limited number of particles, is the appearance of large areas of void space at the surface of the substrate. Where previously the particles were well-dispersed and equally spaced indicating that a steady state had been reached and a maximum surface coverage, under these particular conditions, had been attained this is now clearly not the case. It is possible, as the void areas at the surface remain functionalised with the silane compound, to subject the substrate to a second deposition process, thus filling in these areas with arrays of particles identical in morphological form to that of the original. This array will in turn, if exposed to the same thiol solution, undergo an identical mobilisation and reorganisation process, forming larger clusters of closely packed gold nanoparticles. This process can be easily followed using scanning electron microscopy, as shown in Fig. 1.7, and if repeated several times results in a much higher density array of gold nanoparticles which, in small domains at least, are organised in a close-packed manner. This is clearly a significant improvement on the previous situation if higher density arrays of metallic nanoparticles are required to give, for example, specific optical properties.

The increase in filling fraction and order after this process is carried out is clear to the eye when SEM images are compared and contrasted. However, it is also possible to quantify these changes by analysing the micrographs using the image processing

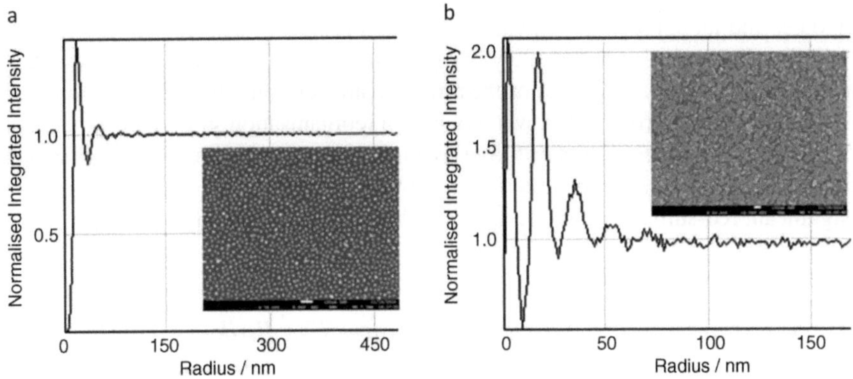

Fig. 1.8 Radial distribution functions (**a**) before and (**b**) after reorganisation and deposition processes. The corresponding SEM micrographs are shown in the *insets*

program *ImageJ* [66]. Figure 1.8a shows the radial distribution function for a low density array of gold nanoparticles before it has undergone the reorganisation and deposition process while Fig. 1.8b shows the radial distribution function after 5 cycles of reorganisation and filling have been completed. In both cases the corresponding SEM image is shown in the inset. The radial distribution function is a means of measuring the correlation between particles within a system or, more specifically, a measure of, the probability of finding a particle at a distance r away from a given reference particle. A peak indicates an increased prospect of finding particles at that given distance from a central reference particle while an intensity of 1 corresponds to the average particle density. An intensity of less than one reveals that there is a diminished probability of finding a particle at this distance. Focusing on the low-density situation shown in Fig. 1.8a and considering moving radially away from a central reference particle several important features can be noted. Firstly, the trough at short distances indicates that, as expected, there is a low probability of finding particles, which are amorphously arranged, close to others. One prominent peak, at a distance which corresponds to the average separation between any given particle and its nearest neighbours, is then observed before another trough corresponding to a second region of reduced probability is seen. Outwith this radius the distribution function tends to 1, showing that no additional order can be discerned. Two important differences can be noted, however, in Fig. 1.8b. Again moving radially away from a central reference particle it can be seen that the first peak occurs at a much smaller distance. This is a mark of closer packing and a good indicator of overall filling fraction. Secondly, several, perhaps as many as 5, peaks can now be discerned, showing that there is a considerable increase in the order observed in this sample.

The morphological changes described, at each stage of the process, result in significantly differing optical properties that can be tracked by UV-vis spectroscopy. The individual particles in the case of the low density array are suitably well-separated and have no significant influence over one another. The strength of coupling decreases exponentially as two resonant particles are moved away from one another and an appreciable interaction is only observed when they are brought to within a distance of approximately one particle diameter. As such, in the low density case the optical properties of the array are not substantially different from that of the individual particle. However, upon each reorganisation step, in the process of fabricating structures of higher filling fraction, nanoparticles are brought to within this coupling limit, therefore altering the optical properties. This is manifested as a significant red-shift of the LSPR of up to 100 nm over 5 reorganisation steps.

1.2.5 Extending into Third Dimension—The Bulk-Assembly of Polyelectrolyte Layers and Multiple Gold Nanoparticle Arrays

As has been shown, a significant amount of control is afforded over the organisation of gold nanoparticles within a single planar array. However, for a variety of applications it is also extremely desirable to extend these gold nanoparticle arrays into the third dimension. A suitable technique for achieving this is through the build-up of layers of charged polymers, or polyelectrolytes, which exploit the same electrostatic forces that were originally used to deposit the nanoparticles on the substrate [71]. A wide range of polyelectrolytes, each with their own specific properties, exist [72]. However, two typical polymers chosen for their versatility, and the ones used in this work are the anionic poly(sodium 4-styrenesulfonate) and the cationic poly(allylamine hydrochloride). The process of polyelectrolyte deposition necessitates a charged surface of some description. In the case under discussion here this charge comes from either the silane functionalised substrate or the gold nanoparticles themselves. When the surface is immersed in a solution of the polymer the polymer becomes bound through multiple electrostatic interactions. A monolayer is formed and excess charge on the exterior allows, after a washing step to remove any excess, the process to be repeated through immersion in a solution of oppositely charged polymer. This deposition and wash cycle can be repeated as many times as required with the possibility of building up truly three dimensional systems realised. Arrays of gold nanoparticles can be included at any stage of the process, the only prerequisite being, of course, that they be sandwiched by polymers of an opposite charge. The minimum array separation is therefore imposed by the thickness of one polymer layer which is thought to be on the order of 1–2 nm [36, 61, 73]. A graphical representation of a similar system to that described above is shown in Fig. 1.9.

Additionally, the finite thickness of each polymer layer, which is itself dependent on a number of factors such as polymer type, ionic strength of the solution and molecular weight, allows thin films of extremely precise thicknesses to be constructed. This has obvious implications when used in combination with gold nanoparticles of finite separation as the coupling strength, and therefore the frequency of the LSPR are largely dependent on the distance between them. Although the situation is slightly more complicated, this holds true when large scale arrays,

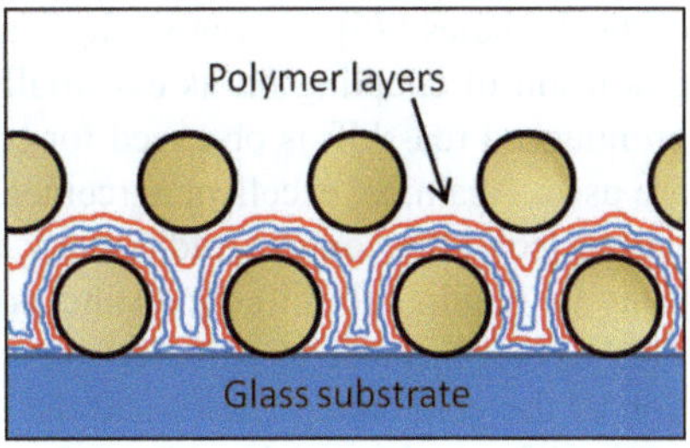

Fig. 1.9 Graphical representation of two gold nanoparticle arrays deposited on a glass substrate and separated by a well-defined number of polyelectrolytes. Adapted with permission from [36]. Copyright (2011) American Chemical Society

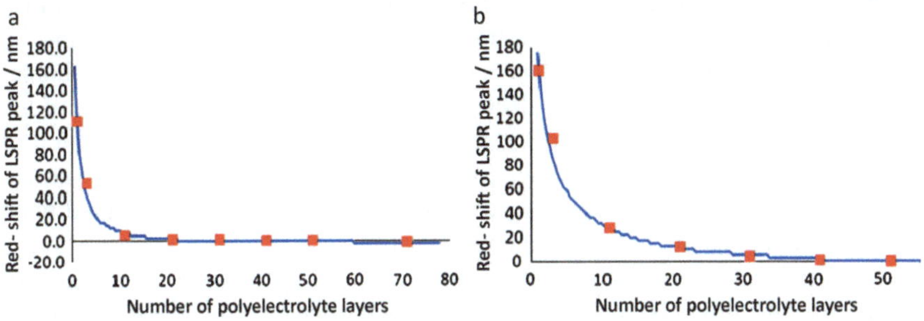

Fig. 1.10 Red-shift upon addition of a second layer of GNPs as a function of layer separation. Both experimental (*squares*) and simulated (*blue solid line*) data are shown for particles of (**a**) 10 nm radius and (**b**) 20 nm radius. Adapted with permission from [36]. Copyright (2011) American Chemical Society

rather than individual nanoparticles, are considered and offers a means of tuning the optical properties of such systems. Further advantages offered by this method include that unlike the Langmuir-Blodgett technique [74], for example, the deposition of polyelectrolytes can be applied to surfaces of virtually any topology and that the deposition occurs extremely quickly, over a period of seconds. In addition, similarly to the size of the initial substrate functionalised, only practical limitations exist both in terms of how many gold nanoparticle arrays are deposited and their relative separations. This, along with the flexibility that is offered, whereby a variety of different sample architectures can be conceived, all adds to the versatility of the method.

When two gold nanoparticle arrays are fabricated in such a manner that they are separated by well-defined distances, the magnitude of the dominant resonance is strongly dependent on the extent of that separation. This can be seen in Fig. 1.10a which shows the red-shift of the LSPR that is induced upon the deposition of a second gold nanoparticle array (both comprised of gold nanoparticles of radius 10 nm) as a function of the number of polyelectrolyte layers that separate them [36]. The red-shift is seen to decrease with increasing layer separation until the particles are moved outwith the coupling limit. An excellent agreement is observed between experimental (squares) and simulated (solid line) results.

Figure 1.10b also shows the effect that particle size has on systems such as these with larger particles, of 20 nm radius, being used to fabricate the same structure. It can be seen that the interaction between particles, as would be expected from theoretical studies [75], is even stronger and extends further for larger particles. This is a result of coupling limits essentially being a function of particle size. A more pronounced red-shift is observed for comparable separations when larger particles are used. Again, an excellent agreement between experimental (squares) and simulations (solid line) exists. Additionally, it can be seen that in this case, larger separations are required to shift the distinct arrays of particles outwith the coupling limits. This further increase of polyelectrolyte layers results in a reduction of the interaction of the gold nanoparticles and the optical response should again be dominated by that of a single particle with no observable red-shift. In addition to being affected by inter-particle coupling the LSPR that is exhibited by the gold nanoparticle arrays is also influenced by changes in the refractive index of the surrounding medium. An overall increase in dielectric constant results in the peak position shifting to higher wavelengths [76] in an effect that decreases exponentially from the surface of the particles [75]. As such the resonance frequency will shift upon the deposition of the polyelectrolyte layers themselves [36], an effect that should be taken into account when considering the resulting changes in optical properties observed upon the deposition of additional gold nanoparticle arrays.

All of the techniques, outlined in previous sections, that allow a great degree of control over the filling fraction of the particles deposited directly on a substrate, remain equally applicable to subsequent arrays that have been deposited on polyelectrolyte layers. This adds to the flexibility offered by such fabrication techniques, one of the major advantages of bottom-up technologies.

1.2.6 Observation of Plasmon Splitting

As outlined in the Introduction, the coupling of two spheres can be described with the principles of plasmon hybridisation theory, shown schematically in Fig. 1.1. The addition of a second array of gold nanoparticles results in a splitting of the single resonance into two, where the dominant one is strongly red-shifted. As the optical response within a single GNP layer was dominated by the LSPR of a single sphere, this red-shifted resonance is governed by the mutual coupling of spheres from distinct GNP layers.

When GNP array separation is extremely small, for example by depositing only a single polyelectrolyte layer between two arrays, a rather amorphous structure containing a variety of dimer orientations relative to the incident wave vector results. In such a configuration, all of the bonding and antibonding eigenmodes can be excited [77]. Importantly, only the eigenmodes with a resulting dipole moment dominate the extinction spectra. As a consequence, both such modes should be simultaneously excited and the LSPR should split into two individual and resolvable resonances. This has been both observed experimentally and predicted by simulation as can be seen in Fig. 1.11 [36]. The simulated trace is shown (dashed line, right-hand axis) in addition to both the measured optical response from a single array of GNPs (lower solid line, left-hand axis) and the splitting is observed when a second layer of GNPs is deposited with only one polyelectrolyte separating layer between (upper

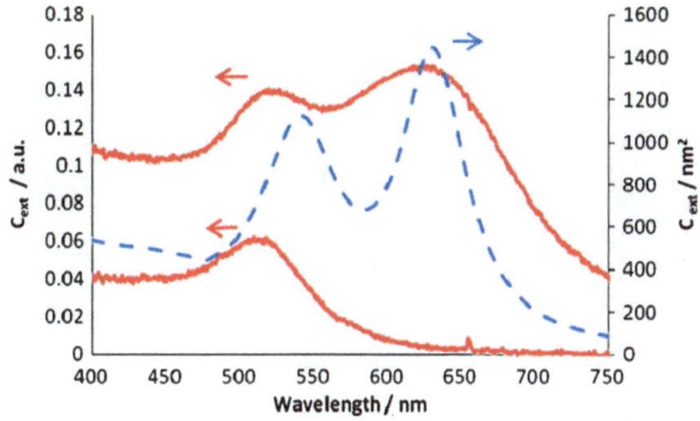

Fig. 1.11 Evolution of longitudinal plasmon peak at shorter GNP (radius 10 nm) layer separations. Shown are the optical responses of a single array of GNPs measured on a glass substrate (*lower solid line, left-hand axis*) and both the experimental (*left-hand axis, upper solid line*) and simulated (*right-hand axis, dashed line*) spectra of a double array of GNPs with one separating polyelectrolyte layer. Adapted with permission from [36]. Copyright (2011) American Chemical Society

solid line, left-hand axis). Both eigenmodes with a resulting dipole moment, π^* and σ (Fig. 1.1), can be observed with the $\pi^*(\sigma)$ coupled plasmon being observed at shorter (longer) wavelengths.

1.2.7 Asymmetric Arrays of Metallic Nanoparticles

One of the more recent applications of plasmonic nanostructures can be found in the field of metamaterials where they are increasingly being used to facilitate the down-scaling of structures that exhibit properties such as a dispersive permeability [78, 79] or permittivity that are exploited for cloaking applications [80] and negative refractive index materials. Such down-scaling is necessary if these properties are going to be routinely observed and utilised at optical frequencies. The dark eigenmodes described in the introduction to this chapter play an important role in this field as the out of phase oscillation of electric dipoles can be related in some cases to a magnetic dipole moment which is the key component in many predicted applications of metamaterials.

One way to excite dark eigenmodes is to use structures that exhibit a considerable amount of asymmetry. The majority of studies of metallic nanoparticles which show asymmetry in either size, shape, or composition has been limited to discussions, both theoretical [81–83] and experimental [84], relating to interactions between low numbers of isolated particles. However in an extension to the work with gold nanoparticles outlined in previous sections [36], it has been shown that it is possible to fabricate large-scale arrays of metallic nanoparticles which display sufficient asymmetry to allow one to excite these dark modes. A net dipole moment is now displayed in each of the four modes shown in Fig. 1.1, meaning that all modes

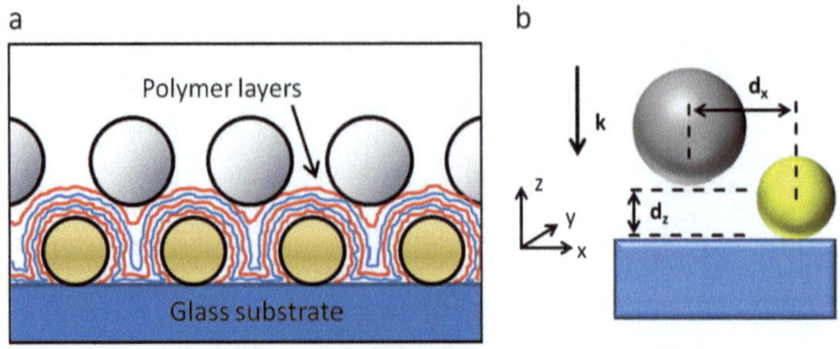

Fig. 1.12 (**a**) Schematic representation of the structure under study showing Au and Ag nanoparticle arrays separated by several polyelectrolyte layers of alternating charge. (**b**) Structure under investigation in the simulations consisting of a heterodimer of one Au and one Ag nanoparticle. The structure is illuminated by a plane wave propagating along the z direction. Geometrical parameters and polarisation of the incident field are described in the text

can be excited at normal incidence, unlike in the system constructed with homoarrays of nanoparticles. Strongly coupled gold and silver nanoparticle arrays induce the required asymmetry due to the shifted LSPR frequencies of the nanoparticles of different composition.

A representation of the principal structure under study in the experimental work is shown in Fig. 1.12a. The two strongly coupled metallic nanoparticle arrays, deposited sequentially on a substrate are separated by an odd integral number of polyelectrolyte layers. Simulations related to this work were performed on an arrangement of two nanoparticles, which could also be considered as a heterodimer, the geometrical configuration of which is shown in Fig. 1.12b.

The silver nanoparticles are prepared in a manner very similar to that of the gold nanoparticles, using a technique known as the Lee-Meisel method [85]. Essentially, a solution of silver salt is reduced by the addition of sodium citrate, which then additionally caps the particles, thus inducing a negative surface charge. Silver nanoparticles produced by the Lee-Meisel method exhibit a wider range of particle sizes and shapes, and do not form as homogeneous an array when deposited on a substrate when compared to the Turkevich method used for gold nanoparticles. This renders the spectral interpretation of the ensembles more challenging. However, the dominating resonances of the system can be fully described by considering a dimer consisting of one gold nanoparticle and one silver nanoparticle from each array, as shown in Fig. 1.12b. Therefore, the identification of peaks that correspond uniquely to the interaction between the distinct arrays of gold and silver nanoparticles and the detection of dark modes is possible for the structure under investigation.

Figure 1.13a depicts the measured extinction spectra for gold and silver nanoparticle arrays separated by a range of different numbers of polyelectrolyte layers. Two resonances with distinct spectral behaviour are observed. The resonance at longer wavelengths is shifted to the red for a decreasing number of separating polyelectrolyte layers. Furthermore, the peak amplitude of this resonance stays nearly constant. Conversely, the resonance at shorter wavelengths offers a nearly constant resonance position whereas the peak amplitude depends on the number of polyelec-

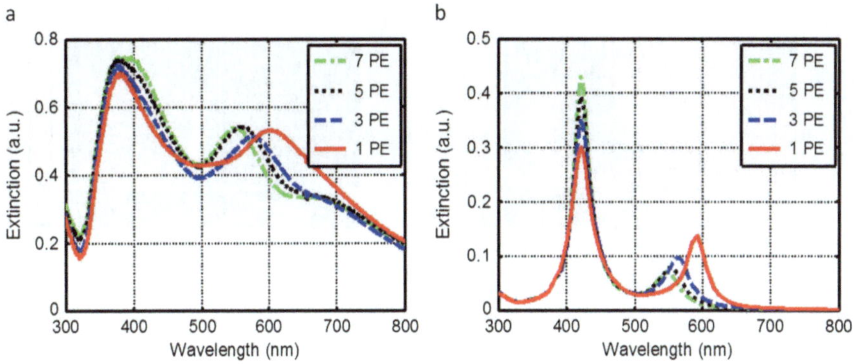

Fig. 1.13 (**a**) Measured and (**b**) simulated extinction spectra for strongly coupled gold–silver nanoparticle layers for different numbers of separating polyelectrolyte layers

trolyte layers. Increasing the number of polyelectrolyte layers, thus decreasing the coupling between particles, results in an enhancement of the peak amplitude. There is an additional resonance, which can be identified just below 700 nm, present in the experimental spectra which results from small aggregates in the silver nanoparticle array. This band, however, does not contribute to one of the four hybridised resonances, shown in Fig. 1.1, that result from the mixing and splitting of modes when two nanoparticle arrays are brought together and can therefore be neglected from subsequent discussions.

The results of the numerical simulations are shown in Fig. 1.13b. The qualitative behaviour of the two resonances as observed in the experiments is entirely reproduced in the simulations. Both simulations and experiments show an increasing peak amplitude for a greater array separation, or in other words, a reduced coupling of the spheres. One explanation of this trend could be the excitation of a dark eigenmode at this wavelength. If the particles are strongly coupled, the eigenmode cannot radiate into the far-field and therefore the amplitude of the extinction is suppressed. If the coupling is reduced the excitation into the far-field is increased and therefore the amplitude of the peak extinction should be enhanced, as is observed in both experiments and simulations.

To facilitate a clear identification of the excited eigenmodes of the structure and to reveal their properties in terms of the hybridisation scheme, shown in Fig. 1.1, the simulated results of the heterodimer can be further interpreted. In Mie theory, one has direct access to all excited multipole moments of every sphere in a system of spheres that is illuminated by an incident field. The reason is that the scattered fields of all spheres involved are decomposed into the eigenfuntions of the Helmholtz equation and this expansion is comparable to a multipole expansion in spherical coordinates. Therefore, the excited electric dipole moments, and their phase relation to each other, of the two spheres of the simulated heterodimer can be revealed directly from Mie theory. A detailed discussion can be found in Refs. [86] and [87]. Sketches of the electric dipoles at the two resonance positions for a single separating polyelectrolyte layer are shown in Fig. 1.14c and Fig. 1.14d.

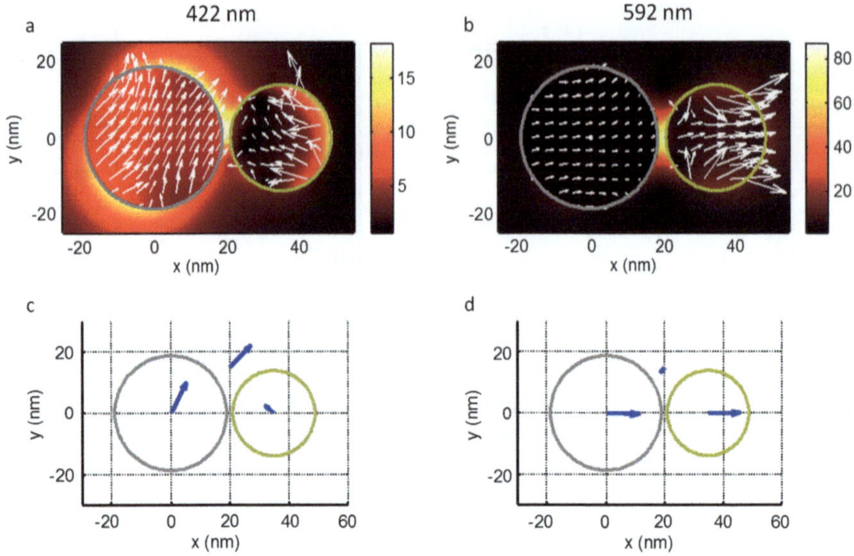

Fig. 1.14 Illustration of the asymmetric (**a**, **c**) and symmetric (**b**, **d**) eigenmodes for the dimer structure separated by one polyelectrolyte layer. (**a**) and (**b**) show the magnitude of the electric field (normalised to the incident field) for both resonances as well as the vectorial character of the internal fields. The lengths of the *white arrows* decode the magnitude of the internal field at this position and the direction shows the contribution of the x- and the y-component of the internal field to its magnitude. (**c**) and (**d**) depict the excited electric dipoles at both resonances of the gold and the silver sphere as they contribute to the scattered field. These dipoles are sketched as *blue arrows* coinciding with the origin of the respective sphere. The *third arrow* (at $x = 20$ nm) shows the polarisation of the incident field at the respective time where the snapshot is taken. All xy-cross sections shown include the centre of the gold sphere. One polyelctrolyte layer, as defined in Fig. 1.12b and in the text, has been chosen to separate the particles. Illumination is in the z-direction, i.e. normal to the plane shown

For the long wavelength resonance, shown in Fig. 1.14d, the in-phase oscillation of the electric dipoles is clearly observed which indicates the excitation of the bright σ eigenmode. However, the excited electric dipoles are not strictly oscillating along the connection line of the spheres, but are also rotating around the origin of the spheres. This is an effect of the polarisation of the incident field which is parallel to a diagonal in the xy-plane. This excited eigenmode is dominated by an in-phase oscillation of both electric dipoles and is therefore a bright mode. In addition, as is well-known for σ eigenmodes of dimers, the electric field is largely enhanced in-between both spheres as can be seen in Fig. 1.14b which shows the magnitude of the electric field and the vectorial character of the internal electric fields of both spheres. The in-phase oscillation of the σ eigenmode can also be observed. The short wavelength resonance, shown in Fig. 1.14c, offers a completely different picture of the excited dipoles in the silver and the gold sphere. Their oscillation can be understood as an interference of the σ^* and π^* eigenmode from Fig. 1.1. Here, the excited dipoles in Fig. 1.14c are oscillating 180° out of phase along the connection line of the spheres (the x direction) and in-phase parallel to the connection line (the y direction). This interference is again an effect of the chosen polarisation of the incident field which allows both types of eigenmodes to be excited at the same time. As such, in the experiments, where the dimers exhibit an amorphous orientation, both eigenmodes (σ^* and π^*) should interfere for every heterodimer at this wavelength. The overall behaviour of the eigenmode excited at shorter wavelengths offers an out-of phase oscillation of the electric dipoles of the involved spheres, and is therefore identified as a dark eigenmode. Furthermore, the asymmetric character of this eigenmode is clearly seen in Fig. 1.14a, which shows the magnitude of the electric field and the vectorial internal fields. The explanation, given previously, of the increasing peak amplitude in extinction (see Fig. 1.13) for an increasing number of polyelectrolyte layers is in full agreement with these conclusions.

1.2.8 Core–Shell Nanoclusters

It is not only planar organisations of metallic nanoparticles which are of interest to the metamaterials community. Spherical arrangements, otherwise known as core–shell nanoclusters, have also stimulated a great deal of discussion in the literature, both of a theoretical [47, 78, 80] and experimental [88–90] nature. Such materials have garnered intense interest as they have highly tunable optical properties and it has been proposed that they allow advancements towards materials with double negative properties [78]

The usefulness of such structures to the metamaterials community is largely due to the electromagnetic properties that they possess. As has been already outlined, when two coupled plasmonic nanospheres are considered using the hybridisation model described in Fig. 1.1 both symmetric and antisymmetric hybridised modes result. The antisymmetric resonances are strongly sub-radiant and will be perceived as magnetic dipoles in the far-field. This magnetic response is an important element where many further applications are concerned [91, 92]. While structures with strong magnetic dipolar responses can be fabricated by top-down techniques [93, 94], the disadvantages which this entails, not least of which are the difficulties involved in creating bulk materials and assigning effective material parameters, indicate that bottom-up techniques should, at least, be considered as a means of making advancements in the field.

While several bottom-up preparative routes resulting in such structures exist [88–90], given all of the advantages related to electrostatic self-assembly outlined in previous sections, perhaps the most facile and flexible method is an extension of that used to construct planar arrays of gold nanoparticles. In a development to techniques that have already been outlined, by replacing the large-scale planar substrates with smaller scale spherical ones it is possible to create solutions of such core–shell nanoclusters. It is also possible to apply the principles described in these syntheses to other systems, for example those including semiconductor and magnetic nanoparticles.

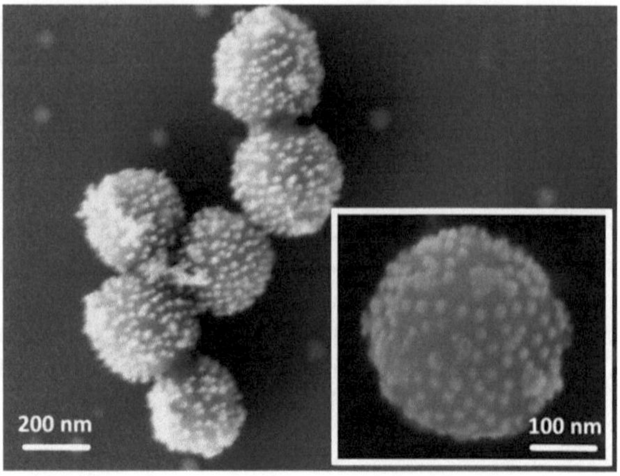

Fig. 1.15 Scanning electron micrographs of fabricated core–shell clusters. An amorphous arrangement of core–shell clusters is shown with the *inset* depicting a zoomed in view showing a single core–shell cluster. Adapted with permission from [79]. Copyright (2011) American Chemical Society

In order to induce the electrostatic attraction between substrate and particle it is first necessary, as before, to functionalise the surface of the substrate. This is regardless of the size or form of the substrate. Fortunately, it is possible to functionalise microspheres of silica using the same reaction as was described in previous sections for planar glass or silicon substrates, albeit under slightly different conditions [79]. This colloidal nanochemistry approach requires additional purification steps, such as centrifugation, however large parallels remain between the two methods. Figure 1.15 shows SEM images of core–shell nanoclusters that have been fabricated by such a method [79].

It can be seen that the dielectric spheres are coated with a large number of isolated, non-touching gold nanoparticles. The organisation of the gold nanoparticles results in large changes in the optical properties when compared to isolated units which are uncoupled to others. This is shown in Fig. 1.16 which highlights both experimental (a) and simulated (b) results [79]. All traces in Fig. 1.16 are normalised to their respective maxima in the region under study.

Immediately apparent in the experimental spectra shown in Fig. 1.16a is the large red-shift, from around 520 nm to around 670 nm, of the LSPR that is induced upon the organisation of the gold nanoparticles, shown as solid red traces, into a spherical geometry, depicted in the extinction spectra as dashed blue traces. This is in excellent agreement with the simulated spectra shown in Fig. 1.16b where the extent of the red-shift is reproduced almost exactly. Any slight deviations could be explained by the fact that in the simulations only a single core–shell nanocluster is considered, compared to the situation in the case of the experimental work where a large ensemble in solution are measured spectroscopically. Here, there will be a certain degree of polydispersity in relation to both the dielectric core and metallic shell spheres. Additionally the organisation of the metallic nanoparticles, while to a large extent comparable, is not identical on each of the individual structures. These slight differences in geometrical parameters can clearly not be taken into account in a simulation of a solitary core–shell nanocluster. Despite any discrepancy between the experimental and simulated spectra, with respect to the strongly red-shifted res-

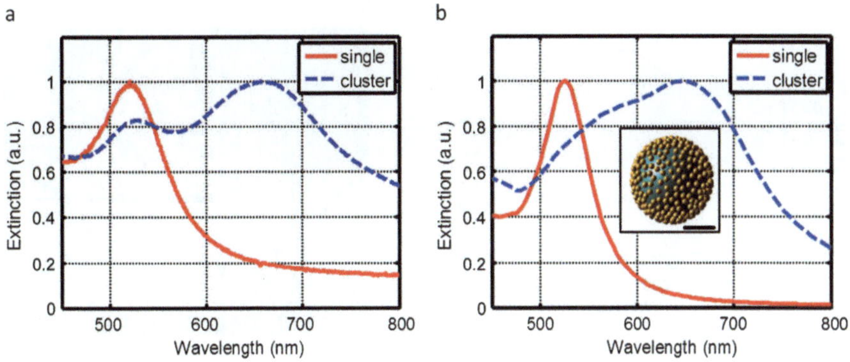

Fig. 1.16 (**a**) Measured extinction spectrum of fabricated core–shell clusters in solution (*blue dashed trace*). For comparison, the extinction spectrum of a solution of gold nanospheres is shown (*red solid trace*). (**b**) Simulated extinction spectra of core–shell clusters in solution (*blue dashed trace*) and a single gold nanosphere (*red solid trace*). The *inset* shows a sketch of the structure under investigation (*scale bar* = 100 nm). Adapted with permission from [79]. Copyright (2011) American Chemical Society

onance, the agreement is convincing. However, one major difference between the two traces can be discerned. A peak in the experimental spectra at the same wavelength as the LSPR of an isolated particle is not reproduced in the simulations. This is caused by excess gold nanoparticles in solution which result from the fabrication process. It is thought that to ensure the cores remain covered an excess of 'free' gold nanoparticles must be present. This could be as a result of an equilibrium existing between the gold nanoparticles at the surface of the dielectric cores and those in solution. These particles are not considered in the simulations and account for this difference between the experimental and simulated data.

It has been proposed that such structures exhibit a strong isotropic magnetic response [78]. This magnetism can be explained by assuming that the shell of metallic nanospheres acts, in effect, as a medium with an extremely high permittivity at wavelengths slightly above the collective plasmonic resonance. The large permittivity in turn evokes Mie resonances. For the lowest order one, the electric displacement field rotates in a plane perpendicular to the polarisation of the incident magnetic field, meaning that this mode can be associated with a magnetic dipole contribution. That the core–shell nanoclusters exhibit artificial isotropic magnetism can be confirmed through simulations, which have proven to be in excellent agreement with experimental results, by deconstructing the extinction spectra and examining the contribution of the respective multipole moments to the scattered field of the structures. Further details of this process, which is outwith the scope of this chapter, can be found in Ref. [78].

The fabrication of these structures is extremely flexible and a number of possibilities to fine tune the architecture, and therefore the optical properties, are available. For instance, a wide range of sizes of both the dielectric core and the surrounding gold nanoparticles can be accessed. Additionally, while the fabrication above relates to a coating of gold nanoparticles on a SiO_2 core sphere different elements

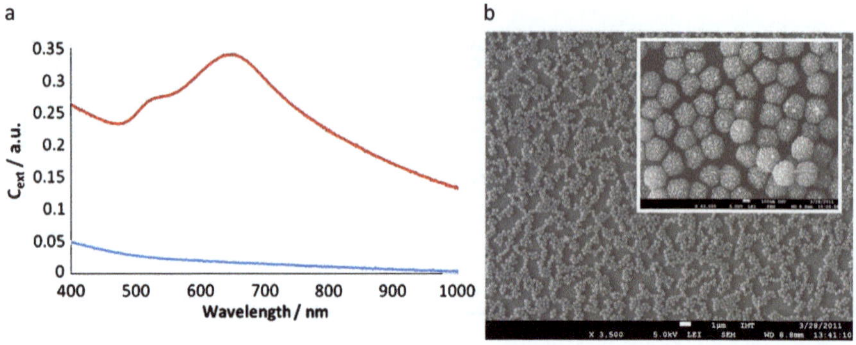

Fig. 1.17 (**a**) Extinction spectra showing an array of SiO_2 microshperes deposited on a charged glass substrate both pre- (*blue trace*) and post- (*red trace*) functionalisation with gold nanoparticles. (**b**) SEM micrographs of an array of core–shell nanoclusters deposited on a charged silicon substrate

can be brought together in a similar manner. Structures using polymer, or metallic, core spheres coated with nanoparticles of a variety of different materials and functionalities can be envisaged. The only prerequisite for their fabrication is that an attraction, of either a physical or chemical nature, exists between the two. In this work the widespread applicability of electrostatic interactions has been highlighted.

This electrostatic approach can also be used to produce large-scale arrays of core–shell nanoclusters deposited on planar substrates. Such a step is required if systems like this are to be incorporated into functional optical devices. In addition, pre-depositing functionalised SiO_2 cores on a planar substrate prior to their coating with metallic nanoparticles could prove to be a promising route to reducing or eliminating the excess particles in solution which existed in the previous fabrication approach. Shown in Fig. 1.17a are extinction spectra which again highlight the possibilities of creating systems with tunable optical properties simply through the organisation and ordering of metallic nanoparticles. The lower, blue, trace corresponds to a single layer of silica microspheres which have been adsorbed on a substrate. The deposition of the spheres, which are charged positively after a functionalisation step, self-assemble at the glass substrate surface which has itself been coated in a negatively charged polymer. The upper, red, trace corresponds to the same array of microspheres after gold nanoparticles have been deposited at the surface. The arguments, given previously, for the exhibition of a strong isotropic magnetic response and for the large red-shift of the LSPR, when compared to that of an isolated particle, remain unchanged. However, the large step that has been taken, from structures that exist purely in colloidal form to the formation of large scale single layers show the versatility and applicability of the bottom-up techniques used. The two SEM images shown in Fig. 1.17b give an idea of what is achievable. The larger image shows that arrays can be created on a suitably large scale while the image in the inset shows that, at least in small domains, a degree of order can be achieved. Additionally, it should also be possible to combine such methods with the layer by layer assembly outlined previously, allowing the construction of truly bulk optical materials.

1.3 Conclusions

The examples of bottom-up techniques used to organise metallic nanoparticles into structures with tailored optical properties described in detail in the previous section highlight the process of taking a design and discovering ways to realise it. Additionally, the large extent to which simulations can now be relied upon to give an accurate reflection of the true experimental situation means that these simulations can now be used as an effective design tool, allowing theoreticians to propose structures with desirable properties that can then be realised through chemical methods. In most cases there will exist several fabrication routes, exploiting different interactions of a chemical and physical nature, which will result in the same structure. This fabrication process, where the myriad of combinations that exist mean that, within certain practical limits, possible structures are only really restricted by the imagination. After different fabrication routes have been identified the onus is then placed on the materials scientist to determine which course provides the best scientific results while also taking into consideration other factors such as cost effectiveness and time efficiency.

An attempt has been made to give an introduction to the wide variety of bottom-up fabrication techniques that can be used to organise metallic nanoparticles into structures that could be put to practical use in functional devices or applications. Additionally, an overview of the areas to which nanoparticles such as these can be applied has also been given. The wide range of these applications, from purely decorational through to biomedical and metamaterials, shows the versatility and flexibility of such systems and that they can be put to practical use in a broad array of areas, the full scope of which is yet to be fully realised.

In particular, a wide range of applications could be envisaged for the two systems discussed in more detail in the last section, more specifically in the fields of metamaterials and plasmonics. The extent to which the optical properties of such systems can be tailored to meet particular requirements, as well as the means used to modify them has been described in detail. The flexibility of several material parameters of the structure, whereby metallic nanoparticles of different size, composition and shape can be deposited on functionalised substrates of varying form and the ability to control the distance between them to within almost nanometre precision means that such systems could have roles to play in a variety of different applications. For example, the suitability of the planar arrays of metallic nanoparticles for applications in SERS has been investigated and it has been shown that the distance between the two arrays of particles, which can be controlled by changing the number of polyelectrolyte layers deposited between them, has a large effect on the enhanced Raman signal. As the distance between the two arrays is increased the electromagnetic field focused in the nanogap, which is the principal cause of enhancements observed, decreases and the Raman signal is seen to decrease accordingly.

Alternatively, taking the structure of radio frequency Yagi-Uda antennas as inspiration it has been proposed that the close correspondence between this and the stratified arrays of metallic nanoparticles could give rise to waveguides for single photon and single plasmon sources [95]. Similar comparisons have been drawn in other publications [96, 97], taking well-developed technology and down-scaling it to achieve nano-scale equivalents with plasmonic inclusions that result in structures that could be included in nanophotonic and nanoelectronic devices. The ability to prepare many layers of metallic nanoparticles with controllable distances between them makes these applications a very real possibility. Taking motivation from other related theoretical studies, nanolenses constructed of chains of nanoparticles progressively diminishing in size and separation could also be made a reality. The focus of this nanolens or 'hottest spot' would exist in the gap separating the two smallest nanoparticles where the electromagnetic fields are further enhanced compared to what is normally seen in a simple dimer due to the multiplicative, cascade effect of its geometry [98]. Other possible applications of the layered arrays of nanoparticles include superlensing, one of the holy grails of metamaterials research, whereby the diffraction limit, which is an inherent limit present in all conventional optical devices, is broken. One of the first experimental observations of superlensing came in 2004 whereby images significantly narrower that those predicted by the diffraction limit were formed [34]. However this work was conducted in the microwave region of the electromagnetic spectrum and as has been outlined throughout this text, the control offered by novel bottom-up techniques permits these structures to be down-scaled and similar properties to be observed in the visible regime.

A massive range of potential sensing applications for such systems also exists. The sensitivity of the LSPR to localised changes in refractive index means that a huge variety of analytes can be accurately and precisely measured. When combined with the almost limitless catalogue of surface chemistry modifications that are accessible which introduces the additional dimension of selectivity. Other cutting edge sensing applications using structures such as the one under discussion could include the incorporation of, for example, thermally responsive polymers, which would result in a plasmonic system with temperature dependent optical properties [99]. Alternatively, by preparing the layered arrays of gold nanoparticles on softer substrates such as PDMS rather than glass or silicon pressure sensors, where the optical response varies with stretching or compression, resulting in a deformation of the nanoparticle organisation and shifts in the optical properties, could be envisaged.

These practical applications are in addition to the pure scientific insight that can be achieved from studying different architectures of gold nanoparticles. As was shown in the detailed study of core–shell nanoclusters [79] they exhibit artificial isotropic magnetism in the visible spectral domain—a key element of many metamaterials applications. One such application is the construction of cloaking devices [80]. It has been shown theoretically that very similar structures to the core–shell nanoclusters shown in the main body of this chapter should, to a certain extent, attenuate the scattering response of the internal dielectric core sphere. For this to be achieved it is necessary for the system to be constructed using relatively small core spheres decorated with silver nanoparticles, whose optical properties are significantly different from their gold counterparts.

This chapter has given a glimpse of what is currently achievable in the field of bottom-up self-assembly of metallic nanoparticle structures with a degree of focus given to their applicability in the preparation of metamaterials at optical frequencies. Specific methods of creating a wide variety of metallic nanoparticles with a wide range of optical properties have been introduced along with the means to organise them and control their optical properties. These means can be built upon, improved and combined with others to make other structures with exciting possibilities equally achievable. This flexibility, allowing for the inclusion of almost any charged species in the layered arrays of metallic nanoparticles for example, is a signature element of the bottom-up approach and, along with the high degree of control achievable, is one of the major advantages of this approach. Using self-assembly at all levels, from the nanoscale to the macroscale, several material parameters, such as particle composition, size and separation, can be manipulated with impressive precision. In addition, the simplicity and robustness of the approach, coupled with the large-scale of the substrates that can be produced, opens up the possibility of integrating such structures into functional devices and components, which should be the goal if the fruits of such research is to continue having a positive impact on our lives.

Acknowledgement Financial support from the University of Geneva and the European Union FP7 (project NANOGOLD) is kindly acknowledged.

References

1. C. Lok, Nanotechnology: small wonders. Nature **467**(7311), 18–21 (2010)
2. R.P. Feynman, There's plenty of room at the bottom. CALTECH Eng. Sci. **23**(5), 22–36 (1960)
3. S.E. Thompson, S. Parthasarathy, Moore's law: the future of Si microelectronics. Mater. Today **9**(6), 20–25 (2006)
4. X. Li et al., Solar cells and light sensors based on nanoparticle-grafted carbon nanotube films. ACS Nano **4**(4), 2142–2148 (2010)
5. E. Boisselier et al., How to very efficiently functionalize gold nanoparticles by "click" chemistry. Chem. Commun. **2008**(44), 5788–5790 (2008)
6. M. Shimomura, T. Sawadaishi, Bottom-up strategy of materials fabrication: a new trend in nanotechnology of soft materials. Curr. Opin. Colloid Interface Sci. **6**(1), 11–16 (2001)
7. C. Acikgoz et al., Polymers in conventional and alternative lithography for the fabrication of nanostructures. Eur. Polym. J. **47**(11), 2033–2052 (2011)
8. M. Zhu et al., Structural and optical characteristics of silicon nanowires fabricated by wet chemical etching. Chem. Phys. Lett. **511**(1–3), 106–109 (2011)
9. S. Reyntjens, R. Puers, A review of focused ion beam applications in microsystem technology. J. Micromech. Microeng. **11**(4), 287 (2001)
10. H. Zeng et al., Nanomaterials via laser Ablation/Irradiation in liquid: a review. Adv. Funct. Mater. **22**(7), 1333–1353 (2012)
11. N.J. Halas, Plasmonics: an emerging field fostered by nano letters. Nano Lett. **10**(10), 3816–3822 (2010)
12. R. Gans, Form of ultramicroscopic particles of silver. Ann. Phys. **47**, 270 (1915)
13. R. Wilson, The use of gold nanoparticles in diagnostics and detection. Chem. Soc. Rev. **37**(9), 2028–2045 (2008)
14. M.E. Stewart et al., Nanostructured plasmonic sensors. Chem. Rev. **108**(2), 494–521 (2008)
15. M. Fleischmann, P.J. Hendra, A.J. McQuillan, Raman spectra of pyridine adsorbed at a silver electrode. Chem. Phys. Lett. **26**(2), 163–166 (1974)

16. N.J. Halas et al., Plasmons in strongly coupled metallic nanostructures. Chem. Rev. **111**(6), 3913–3961 (2011)
17. G.R. Souza et al., *In vivo* detection of gold-imidazole self-assembly complexes: NIR-SERS signal reporters. Anal. Chem. **78**(17), 6232–6237 (2006)
18. J.M. Romo-Herrera, R.A. Alvarez-Puebla, L.M. Liz-Marzan, Controlled assembly of plasmonic colloidal nanoparticle clusters. Nanoscale **3**, 1304–1315 (2011)
19. R. Jin, Nanoparticle Clusters Light Up in SERS. Angewandte Chemie International Edition **49**(16), 2826–2829 (2010)
20. W. Page Faulk, G. Malcolm Taylor, Communication to the editors: an immunocolloid method for the electron microscope. Immunochemistry **8**(11), 1081–1083 (1971)
21. R.A. Sperling et al., Biological applications of gold nanoparticles. Chem. Soc. Rev. **37**(9), 1896–1908 (2008)
22. Q. Wei, J. Ji, J. Shen, Synthesis of near-infrared responsive gold nanorod/PNIPAAm core/shell nanohybrids via surface initiated ATRP for smart drug delivery. Macromol. Rapid Commun. **29**, 645–650 (2008)
23. D.A. Giljohann et al., Gold nanoparticles for biology and medicine. Angew. Chem., Int. Ed. **49**(19), 3280–3294 (2010)
24. E.C. Dreaden et al., Beating cancer in multiple ways using nanogold. Chem. Soc. Rev. **40**(7), 3391–3404 (2011)
25. C. Sonnichsen et al., A molecular ruler based on plasmon coupling of single gold and silver nanoparticles. Nat. Biotechnol. **23**(6), 741–745 (2005)
26. M. Chen, D.W. Goodman, Catalytically active gold on ordered titania supports. Chem. Soc. Rev. **37**(9), 1860–1870 (2008)
27. J.R. Adleman et al., Heterogenous catalysis mediated by plasmon heating. Nano Lett. **9**(12), 4417–4423 (2009)
28. H. Nabika et al., Toward plasmon-induced photoexcitation of molecules. J. Phys. Chem. Lett. **1**(16), 2470–2487 (2010)
29. C.J. Chen, R.M. Osgood, Direct observation of the local-field-enhanced surface photochemical reactions. Phys. Rev. Lett. **50**(21), 1705–1708 (1983)
30. T. Arakawa et al., Effects of silver nanoparticles on photoelectrochemical responses of organic dyes. J. Phys. Chem. C **113**(27), 11830–11835 (2009)
31. D.R. Smith et al., Composite medium with simultaneously negative permeability and permittivity. Phys. Rev. Lett. **84**(18), 4184–4187 (2000)
32. R.A. Shelby, D.R. Smith, S. Schultz, Experimental verification of a negative index of refraction. Science **292**(5514), 77–79 (2001)
33. D. Rainwater et al., Experimental verification of three-dimensional plasmonic cloaking in free-space. New J. Phys. **14**(1), 013054 (2012)
34. A. Grbic, G.V. Eleftheriades, Overcoming the diffraction limit with a planar left-handed transmission-line lens. Phys. Rev. Lett. **92**(11), 117403 (2004)
35. E. Prodan et al., A hybridization model for the plasmon response of complex nanostructures. Science **302**(5644), 419–422 (2003)
36. A. Cunningham et al., Coupling of plasmon resonances in tunable layered arrays of gold nanoparticles. J. Phys. Chem. C **115**(18), 8955–8960 (2011)
37. X. Zeng, 3D ordered gold strings by coating nanoparticles with mesogens. Adv. Mater. **21**, 1746 (2009)
38. H. Qi et al., Effects of hydrophilic and hydrophobic gold nanoclusters on the stability and ordering of bolaamphiphilic liquid crystals. J. Mater. Chem. **17**(20), 2139–2144 (2007)
39. R. Pratibha, W. Park, I.I. Smalyukh, Colloidal gold nanosphere dispersions in smectic liquid crystals and thin nanoparticle-decorated smectic films. J. Appl. Phys. **107**(6), 063511 (2010)
40. A.P. Alivisatos et al., Organization of 'nanocrystal molecules' using DNA. Nature **382**(6592), 609–611 (1996)
41. D.D.-K. Lim, Nanogap-engineerable Raman-active nanodumbbells for single-molecule detection. Nat. Mater. **9**(1), 60 (2010)

42. C.-L. Chen, P. Zhang, N.L. Rosi, A new peptide-based method for the design and synthesis of nanoparticle superstructures: construction of highly ordered gold nanoparticle double helices. J. Am. Chem. Soc. **130**(41), 13555–13557 (2008)
43. Nie et al., "Supramolecular" assembly of gold nanorods end-terminated with polymer "pom-poms": effect of pom-pom structure on the association modes. J. Am. Chem. Soc. **130**(11), 3683–3689 (2008)
44. P.K. Jain, S. Eustis, M.A. El-Sayed, Plasmon coupling in nanorod assemblies: optical absorption, discrete dipole approximation simulation, and exciton-coupling model. J. Phys. Chem. B **110**(37), 18243–18253 (2006)
45. E.V. Shevchenko et al., *Structural Diversity in Binary Nanoparticle Superlattices* (Nature Publishing Group, London, 2006), pp. 55–59
46. X. Wang et al., Polymer-encapsulated gold-nanoparticle dimers: facile preparation and catalytical application in guided growth of dimeric ZnO-nanowires. Nano Lett. **8**(9), 2643–2647 (2008)
47. C. Rockstuhl et al., Design of an artificial three-dimensional composite metamaterial with magnetic resonances in the visible range of the electromagnetic spectrum. Phys. Rev. Lett. **99**(1), 017401 (2007)
48. N. Shalkevich et al., Reversible formation of gold nanoparticle-surfactant composite assemblies for the preparation of concentrated colloidal solutions. Phys. Chem. Chem. Phys. **11**(43), 10175–10179 (2009)
49. S. Mühlig et al., Optical properties of a fabricated self-assembled bottom-up bulk metamaterial. Opt. Express **19**(10), 9607–9616 (2011)
50. J.H. Lee, Q. Wu, W. Park, Metal nanocluster metamaterial fabricated by the colloidal self-assembly. Opt. Lett. **34**(4), 443–445 (2009)
51. V. Myroshnychenko et al., Modelling the optical response of gold nanoparticles. Chem. Soc. Rev. **37**(9), 1792–1805 (2008)
52. I. Freestone, The lycurgus cup—a Roman nanotechnology. Gold Bull. **40**(4), 270–277 (2007)
53. M. Faraday, The bakerian lecture: experimental relations of gold (and other metals) to light. Philos. Trans. R. Soc. Lond. **147**, 145–181 (1857)
54. M. Brust et al., Synthesis of thiol-derivatised gold nanoparticles in a two-phase liquid-liquid system. J. Chem. Soc., Chem. Commun. **7**, 801–802 (1994)
55. S.J. Tan et al., Building plasmonic nanostructures with DNA. Nat. Nanotechnol. **6**(5), 268–276 (2011)
56. M. Rycenga et al., Controlling the synthesis and assembly of silver nanostructures for plasmonic applications. Chem. Rev. **111**(6), 3669–3712 (2011)
57. L. Jiang et al., Patterning of plasmonic nanoparticles into multiplexed one-dimensional arrays based on spatially modulated electrostatic potential. ACS Nano **5**(10), 8288–8294 (2011)
58. J.K. Gansel et al., Gold helix photonic metamaterial as broadband circular polarizer. Science **325**(5947), 1513–1515 (2009)
59. C.M. Soukoulis, M. Wegener, Optical metamaterials—more bulky and less lossy. Science **330**(6011), 1633–1634 (2010)
60. K.L. Kelly et al., The optical properties of metal nanoparticles: the influence of size, shape, and dielectric environment. J. Phys. Chem. B **107**(3), 668–677 (2002)
61. J. Schmitt et al., Metal nanoparticle/polymer superlattice films: fabrication and control of layer structure. Adv. Mater. **9**(1), 61–65 (1997)
62. Z. Feng, F. Yan, Preparation and tribological studies of nanocomposite films fabricated using spin-assisted layer-by-layer assembly. Surf. Coat. Technol. **202**(14), 3290–3297 (2008)
63. B.V. Enustun, J. Turkevich, Coagulation of colloidal gold. J. Am. Chem. Soc. **85**(21), 3317–3328 (1963)
64. J. Kimling et al., Turkevich method for gold nanoparticle synthesis revisited. J. Phys. Chem. B **110**(32), 15700–15707 (2006)
65. R.K. Iler, *The Chemistry of Silica: Solubility, Polymerization, Colloid and Surface Properties and the Biochemistry* (Wiley, New York, 1979)

66. M.D. Abràmoff, P.J. Magalhães, S.J. Ram, Image processing with image. J. Biophotonics Int. **7**, 7 (2004)
67. J. Dintinger et al., A bottom-up approach to fabricate optical metamaterials by self-assembled metallic nanoparticles. Opt. Mater. Express **2**(3), 269–278 (2012)
68. D.J. Shaw, *Introduction to Colloid and Surface Chemistry* (Elsevier, Amsterdam, 1989)
69. J.C. Love et al., Self-assembled monolayers of thiolates on metals as a form of nanotechnology. Chem. Rev. **105**(4), 1103–1170 (2005)
70. S.D.T. Brown, B.F.G. Johnson, Nucleation and growth of nano-gold colloidal lattices. Chem. Commun. **11**, 1007–1008 (1997)
71. G. Decher, Fuzzy nanoassemblies: toward layered polymeric multicomposites. Science **277**(5330), 1232–1237 (1997)
72. P. Lavalle et al., Dynamic aspects of films prepared by a sequential deposition of species: perspectives for smart and responsive materials. Adv. Mater. **23**(10), 1191–1221 (2011)
73. A. Tronin, Ellipsometry and x-ray reflectometry characterization of self-assembly process of polystyrensulfonate and polyallylamine. Colloid Polym. Sci. **272**, 1317–1321 (1994)
74. X. Chen et al., Langmuir–Blodgett patterning: a bottom–up way to build mesostructures over large areas. Acc. Chem. Res. **40**(6), 393–401 (2007)
75. A.M. Funston et al., Plasmon coupling of gold nanorods at short distances and in different geometries. Nano Lett. **9**(4), 1651–1658 (2009)
76. G.A. Ozin, A.C. Arsenault, *Nanochemistry—A Chemical Approach to Nanomaterials* (RSC Publishing, Cambridge, 2005)
77. A. Christ et al., Symmetry breaking in a plasmonic metamaterial at optical wavelength. Nano Lett. **8**(8), 2171–2175 (2008)
78. C.R. Simovski, S.A. Tretyakov, Model of isotropic resonant magnetism in the visible range based on core–shell clusters. Phys. Rev. B **79**(4), 045111 (2009)
79. S. Mühlig et al., Self-assembled plasmonic core–shell clusters with an isotropic magnetic dipole response in the visible range. ACS Nano **5**(8), 6586–6592 (2011)
80. S. Mühlig et al., Cloaking dielectric spherical objects by a shell of metallic nanoparticles. Phys. Rev. B **83**(19), 195116 (2011)
81. O. Peña-Rodríguez et al., Enhanced Fano resonance in asymmetrical Au: Ag heterodimers. J. Phys. Chem. C (2011)
82. G. Bachelier et al., Fano profiles induced by near-field coupling in heterogeneous dimers of gold and silver nanoparticles. Phys. Rev. Lett. **101**(19), 197401 (2008)
83. E.R. Encina, E.A. Coronado, On the far field optical properties of Ag-Au nanosphere pairs. J. Phys. Chem. C **114**(39), 16278–16284 (2010)
84. L.V. Brown et al., Heterodimers: plasmonic properties of mismatched nanoparticle pairs. ACS Nano **4**(2), 819–832 (2010)
85. P.C. Lee, D. Meisel, Adsorption and surface-enhanced Raman of dyes on silver and gold sols. J. Phys. Chem. **86**(17), 3391–3395 (1982)
86. C. Rockstuhl et al., Scattering properties of meta-atoms. Phys. Rev. B **83**(24), 245119 (2011)
87. S. Mühlig et al., Multipole analysis of meta-atoms. Metamaterials **5**(2–3), 64–73 (2011)
88. F. Caruso et al., Multilayer assemblies of silica-encapsulated gold nanoparticles on decomposable colloid templates. Adv. Mater. **13**(14), 1090–1094 (2001)
89. B. Sadtler, A. Wei, Spherical ensembles of gold nanoparticles on silica: electrostatic and size effects. Chem. Commun. **15**, 1604–1605 (2002)
90. I. Pastoriza-Santos et al., Optical properties of metal nanoparticle coated silica spheres: a simple effective medium approach. Phys. Chem. Chem. Phys. **6**(21), 5056–5060 (2004)
91. J.B. Pendry, Negative refraction makes a perfect lens. Phys. Rev. Lett. **85**(18), 3966–3969 (2000)
92. U. Leonhardt, T.G. Philbin, Transformation optics and the geometry of light, in *Progress in Optics*, ed. by E. Wolf, 1st edn. (Elsevier, Amsterdam, 2009)
93. C. Enkrich et al., Magnetic metamaterials at telecommunication and visible frequencies. Phys. Rev. Lett. **95**(20), 203901 (2005)

94. J. Valentine et al., Three-dimensional optical metamaterial with a negative refractive index. Nature **455**(7211), 376–379 (2008)
95. A.F. Koenderink, Plasmon nanoparticle array waveguides for single photon and single plasmon sources. Nano Lett. **9**(12), 4228–4233 (2009)
96. A. Manjavacas, F.J. Garcí a de Abajo, Robust plasmon waveguides in strongly interacting nanowire arrays. Nano Lett. **9**(4), 1285–1289 (2008)
97. N. Engheta, Circuits with light at nanoscales: optical nanocircuits inspired by metamaterials. Science **317**(5845), 1698–1702 (2007)
98. K. Li, M.I. Stockman, D.J. Bergman, Self-similar chain of metal nanospheres as an efficient nanolens. Phys. Rev. Lett. **91**(22), 227402 (2003)
99. M. Karg et al., Nanorod-coated PNIPAM microgels: thermoresponsive optical properties. Small **3**(7), 1222–1229 (2007)

Chapter 2
Amorphous Metamaterials and Potential Nanophotonics Applications

Filiberto Bilotti and Sergei Tretyakov

Abstract In this chapter we discuss potential applications of amorphous electromagnetic metamaterials. We have selected applications where the amorphous nature (random positioning of inclusions or random variations of inclusion dimensions and shape) does not lead to significant performance degradation or even leads to some advantageous properties. Special emphasis is on applications in cloaking and in superlensing devices. The potential use of amorphous metamaterials has the obvious practical advantage such as the possibility to manufacture the necessary nanostructures using large-scale self-assembly techniques that are described to some extent in other chapters of this book.

2.1 Introduction

In this chapter, we summarize the useful properties of applications which can be potentially realized on the base of amorphous metamaterials. We have identified a number of applications where the amorphous nature of metamaterials does not lead to performance degradation (as compared to metamaterials formed by regular, periodical lattices of inclusions) or even brings improvements of performance. For each of the foreseen applications, we present: (a) the physical mechanism behind the operation, (b) the analytical validation of the physical expectations, (c) and, whenever possible, the numerical validation of the analytical models through proper full-wave simulations.

In the last decade, a number of metamaterials showing unusual and useful electromagnetic properties have been realized and tested. These structures are usually composed of small individual resonant inclusions arranged in two- or three-dimensional periodic arrays. However, most recently, random or amorphous

F. Bilotti
Department of Applied Electronics, "Roma Tre" University, Roma, Italy
e-mail: filiberto.bilotti@uniroma3.it

S. Tretyakov (✉)
Department of Radio Science and Engineering, Alto University, Espoo, Finland
e-mail: sergei.tretyakov@aalto.fi

C. Rockstuhl, T. Scharf (eds.), *Amorphous Nanophotonics*,
Nano-Optics and Nanophotonics, DOI 10.1007/978-3-642-32475-8_2,

© Springer-Verlag Berlin Heidelberg 2013

metamaterials start attracting more attention. This is due to novel technological possibilities to manufacture amorphous structures cheaply and on a large scale with self-assembly techniques. It is generally accepted that the fundamental electromagnetic properties of both regular and random arrays are quite similar if the distances between inclusions are electrically small in comparison with the wavelength of waves propagating or decaying in the structure. The main difference in electromagnetic response is assumed to come from extra scattering loss in amorphous composites due to scattering on inhomogeneities of the lattice. This apparently results in additional loss, and for this reason regular metamaterial lattices have been the preferred choice if low-loss response is desired. However, if a random array is effectively homogeneous at the wavelength scale, there are no additional scattering losses, just like in ideally periodical grids. Moreover, as it will be clear from the following, in some situations the electromagnetic properties of random structures may bring advantages in practical applications of these materials.

The first section is a presentation of spherical and cylindrical invisibility cloaks, allowing strong reduction of the visibility of objects with small and moderate electrical (optical) sizes. Among the different known approaches to cloaking, we have focused our attention to the *scattering cancellation technique*, which can be easily implemented with amorphous metamaterials. Some applications of the cloaking devices that can be realistically obtained with the available materials are presented in turn, with particular emphasis on the reduction of the noise in near-field scanning optical microscopy systems. The second section is the presentation of planar spatial filters consisting of metamaterial slabs exhibiting near-zero values of the real part of the permittivity function. Such filters can be used to convert a broad optical beam into a directive one and are, thus, potentially useful for beaming applications. The third section presents novel sub-wavelength imaging devices. Particularly, the design and implementation of a real-time superlens capable to resolve the details of a source which are finer than the illuminating light wavelength is presented.

2.2 Cloaking Devices

2.2.1 General Concepts

In this section, we describe possible use of amorphous metamaterial structures as cloaking devices to reduce the visibility of nanosized objects. Cloaking of a particle means to put a cover around it in order to reduce its total scattering cross section (i.e., the ideal cloak reduces the total power scattered in all directions to zero, which implies also absence of absorption). According to this definition, thus, a cloaked object becomes invisible.

Several approaches to cloaking based on different physical mechanisms have been proposed in the last few years [1–15]. One of these approaches, referred to as *scattering cancellation*, consists in the compensation of the scattering produced by an illuminated object by covering it with a cover made of a suitable material. If properly designed, the scattering by the cover, in fact, compensates the one due to the object, making the object itself invisible. As was demonstrated in [5], to cloak a dielectric object, the material of the cover should be characterized by proper values of the real part of the relative permittivity function at the operating frequency (smaller than unity, usually negative or near-zero). However, whatever the material constituting the object (i.e. either a plasmonic or a dielectric one) is, it has been demonstrated that the cover can be made of a material exhibiting near-zero values (i.e. the modulus is between 0 and 1) of the real part of the relative permittivity function.

2.2.2 Cloak Design Using Metamaterials

Such *cloaking materials* can be fabricated as composites formed by metal nanoparticles. Let us consider as a first example an appropriate arrangement of gold nanospheres. Let us assume that we are allowed to describe the dispersive permittivity of gold in the visible frequency range by the Drude dispersion function with the following effective bulk properties: $\varepsilon_b = 9$, $\omega_p = 1.38 \times 10^{16}$ rad/s, $\tau = 1.2 \times 10^{-14}$ s. ε_b is the permittivity at infinite frequency. Of course, it is not the actual physical value of the relative permittivity at infinite frequency, but rather a fitting parameter used to match the Drude dispersion with the gold permittivity within the visible frequency range. From the physical point of view, this parameter models low-frequency contributions from all oscillators with resonances at higher frequencies. ω_p is the plasma frequency, depending on the electron density of bulk gold; τ is the average time between two consecutive collisions of free electrons in bulk gold. However, considering that the size of the gold spheres is in the nanometer scale, free electrons experience a different collision frequency with respect to their bulk counterparts, especially when the size of the sphere becomes comparable to the electron mean free path. Therefore, the complex relative permittivity function of gold for the nanosphere case could be expressed, in the first approximation, as follows [16]:

$$\varepsilon(\omega) = \varepsilon_b - \frac{\omega_p^2}{\omega(\omega + i\tau^{-1} + iv_F/R)} \tag{2.1}$$

where $v_F = 1.4 \times 10^6$ m/s the Fermi velocity of gold and R the radius of the nanospheres.

Now, let us assume that the sphere radius is small compared to the operation wavelength and consider an arrangement of such spheres in a non-absorbing background medium, whose refraction index is n_m. With f being the fraction of the background volume occupied by the gold nanospheres and assuming that f is sufficiently small, it is possible to homogenize the arrangement of the gold nanospheres according to the Maxwell-Garnett model as an effective medium with the following equivalent permittivity [17]:

$$\varepsilon_e(\omega) = n_m^2 \frac{\varepsilon(\omega)(1+2f) + 2n_m^2(1-f)}{\varepsilon(\omega)(1-f) + n_m^2(2+f)} \tag{2.2}$$

Here ε is the permittivity function of the single nanosphere, as introduced above.

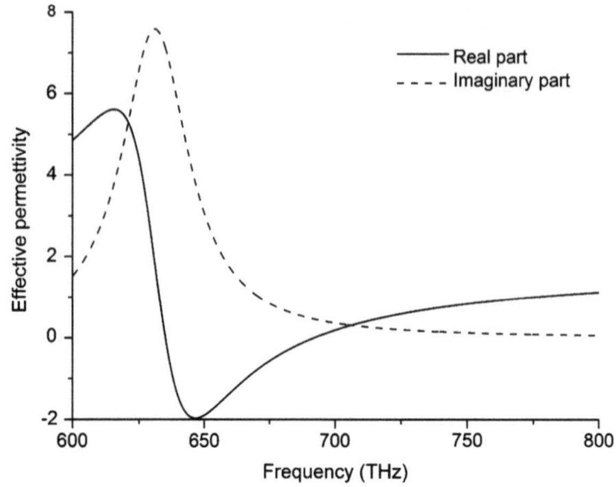

Fig. 2.1 Real part (*continuous line*) and imaginary part (*dashed line*) of the effective permittivity of a gold nanosphere arrangement with $f = 0.27$

In order to show an example, let us assume that the host medium is simply the vacuum and that we would like to obtain for the nanosphere arrangement a real part of the relative permittivity equal to $\varepsilon_e = 0.2$ at a given frequency in the visible. Fixing the frequency at 700 THz, the corresponding filling factor, needed to match the required value of the real permittivity, is $f = 0.27$. The corresponding real and imaginary parts of the effective permittivity of the gold sphere arrangement are reported in Fig. 2.1.

With this preliminary step, we have demonstrated that, by using an arrangement (not necessarily ordered) of gold nanospheres, it is possible to implement a material with near-zero values of the real relative permittivity in the optical frequency range.

In the following, we apply the material design procedure now outlined to the design of a cloak for a cylindrical nanoparticle. Let us consider a dielectric object ($\varepsilon_d = 2$) with a cylindrical shape ($a = 50$ nm, $L = 500$ nm), as reported in Fig. 2.2a. The total scattering cross section of such a cylinder, when illuminated by a plane-wave with a radial wave vector and an electric field directed along the cylinder axis, is reported in the continuous line plot of Fig. 2.3. As evident, the total scattering cross section of the cylinder increases with the frequency, since the object becomes electrically larger.

In order to minimize the total scattering cross section of the cylindrical object, it has been demonstrated that the same cloaking rules of the unbounded cylinder can be used [7]. According to these rules, assuming that the cloak is a cylindrical shell with the external radius a_c and internal radius a (see Fig. 2.2a), provided that a_c is electrically small, the relation between the geometrical dimensions a, a_c and the real relative permittivities of the object and the cloak ε_d, ε_c is given by [5]:

$$\frac{a}{a_c} = \sqrt{\frac{\varepsilon_c - 1}{\varepsilon_c - \varepsilon_d}} \tag{2.3}$$

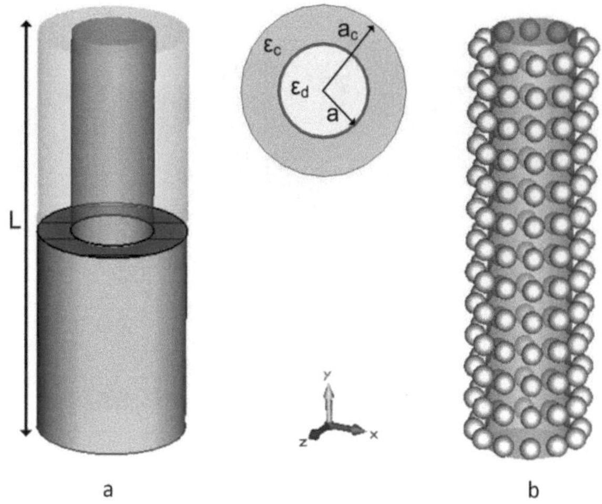

Fig. 2.2 Finite-length dielectric cylinder covered by a homogeneous (**a**) and a nanosphere based (**b**) cloak. $\varepsilon_d = 2$, $a = 50$ nm, $L = 484$ nm (around one wavelength at 700 THz)

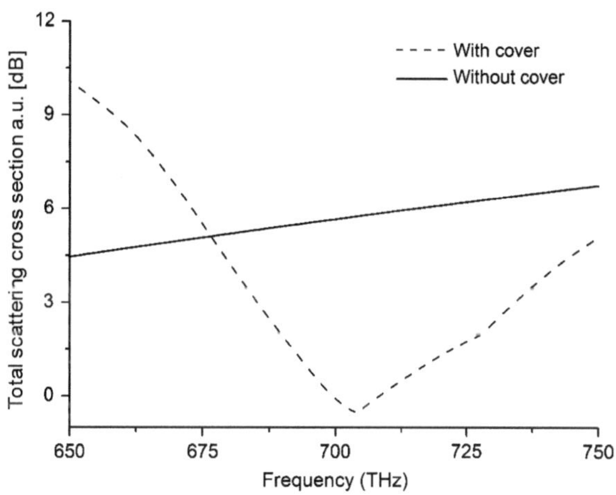

Fig. 2.3 Total scattering cross section of the bare dielectric cylinder (*continuous line*) and of the cylinder covered by a cloak consisting of 156 gold nanospheres (*dashed line*)

Assuming now that $a_c = 75$ and $a = 50$ nm, which leads to cloak thickness of 25 nm, the cloaking condition is verified if $\varepsilon_c = 0.2$. Therefore, in order to cloak the given cylinder at a given frequency, it is necessary to cover it with a material cover having thickness 25 nm and the real relative permittivity $\varepsilon_c = 0.2$.

Considering the results we have shown above for the arrangement of gold nanospheres for the filling fraction of $f = 0.27$, we are able to cloak the cylinder at around 700 THz. The cover can be designed by considering an ordered distribution of spheres as reported in Fig. 2.2b so that the spheres, having the diameter of 25 nm, occupy the fraction $f = 0.27$ of the total volume of the cover. In order to do that, it is possible to use 156 nanospheres arranged in 12 columns with 13 spheres for each column and with the angular separation of 30° between two adjacent columns. Considering the previous homogenized model of the nanosphere, the structure reported in Fig. 2.2b is equivalent to the one reported in Fig. 2.2a.

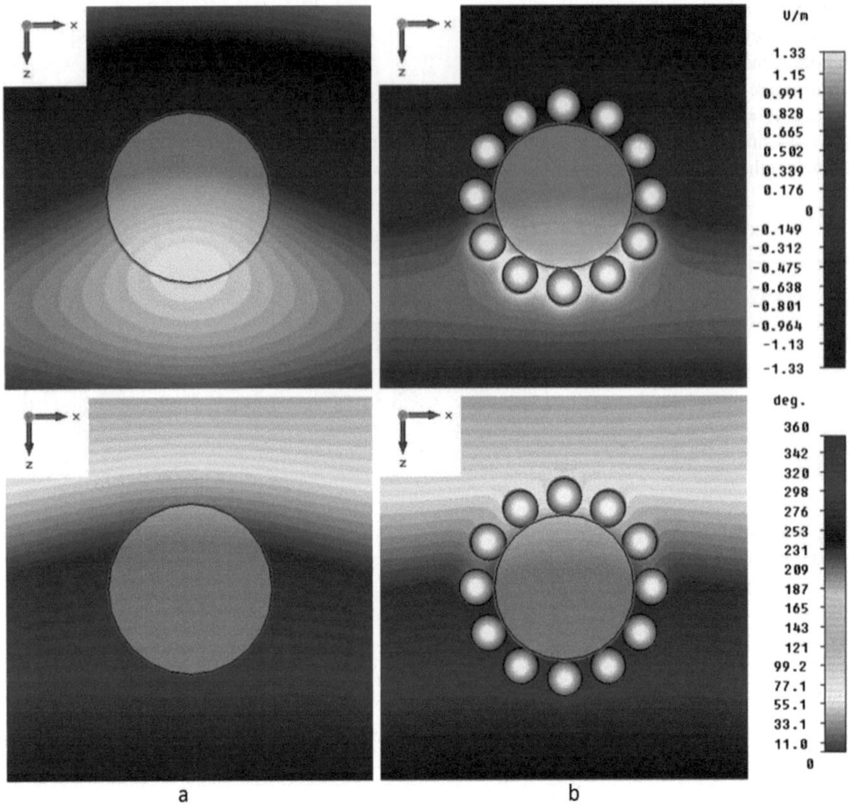

Fig. 2.4 Time-domain distribution of the y-component of the electric field at a given time instant (*up*) and electric field phase distribution (*down*) at the cloak frequency (around 700 THz) on the $y = 0$ plane for the (**a**) bare and (**b**) covered cylinder

The structure has been simulated by launching a plane-wave with the electric field directed along the cylinder axis and the obtained total scattering cross section versus frequency is reported in Fig. 2.3. As expected, at around the cloak design frequency of 700 THz the observability of the cylinder is highly reduced, due to the presence of the gold nanospheres. The amplitude and phase field maps reported in Fig. 2.4 confirm the reduced observability of the cylinder. Field maps, in fact, reveal a field distribution outside the cylinder similar to the one of the impinging plane-wave.

In addition, in order to consider the effect of possible misalignments due to fabrication errors/tolerances or due to the amorphous arrangement of gold nanospheres around the cylinder, we show in Fig. 2.5 three different geometries with increasing level of disorder in the sphere arrangement. We have divided the 156 spheres in seven independent subsets. The first subset has not been moved with respect to its original position. The second and the third subsets have been shifted along the y-axis of a quantity $\pm\Delta$, respectively. The fourth and the fifth subsets have been

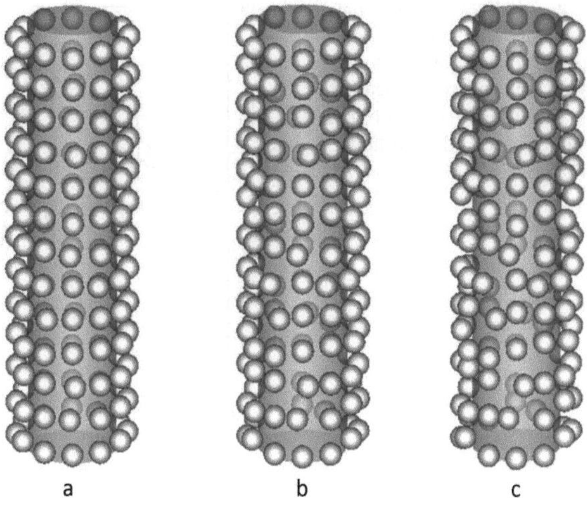

Fig. 2.5 Covered cylinder after the geometrical perturbation of the nanosphere arrangement for (**a**) $\Delta = 3$ nm, (**b**) $\Delta = 6$ nm, and (**c**) $\Delta = 9$ nm

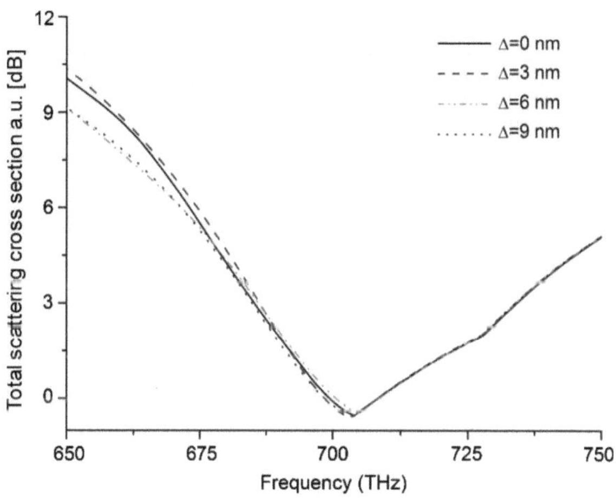

Fig. 2.6 Total scattering cross section of the covered cylinder for different values of Δ

shifted, in the same direction, by $\pm\Delta/2$, respectively. Finally, the last two subsets have been rotated with respect to their original position by an angular quantity corresponding to an arc of $\pm\Delta$, respectively. Since the homogenization conditions are preserved in all the three cases, the behavior of the cloak does not have to depend on the ordered alignment of the sphere but rather on their filling fraction. The results reported in Fig. 2.6 confirm this behavior, opening the door to the design of visible cloaks for cylindrical objects by the amorphous arrangement of gold nanospheres.

So far, we have considered gold as described by a Drude model and assumed that the homogenization rules are valid for the used filling fraction. However, measured real gold electrical properties differ from the ones of the Drude model, especially for what concerns the loss level at the frequency range around and above 700 THz (similar considerations apply also for silver). In addition, the role of the filling fraction and its validity limitations in homogenization models need further discussion.

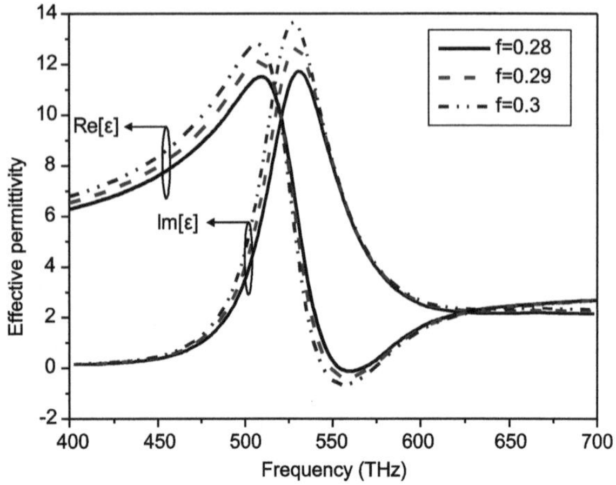

Fig. 2.7 Real and imaginary parts of the effective permittivity for an array of gold nanospheres embedded in a silica matrix for different values of the filling volume fraction

For a more appropriate analysis, in the following we replace Drude permittivity with measured dispersive permittivity for some common noble metals [18] and we repeat the previous exercise. Complex effective permittivities of arrays of gold nanospheres with different values of the filling volume fraction and embedded in a silica matrix are shown in Fig. 2.7 (please, note, that vacuum used in the previous examples has been replaced here by a more realistic dielectric support). With the indicated filling volume fractions (compatible with the limitations presented in [19]), we are able to bring the required epsilon near-zero behavior down to optical frequencies. As evident from Fig. 2.7 (see the imaginary part plot), however, losses are clearly too high to guarantee good performances of the devices realized through such an artificial material and, thus, the proposed solution is to be discarded. This unlucky exercise, however, tells us about the role of the filling volume fraction. Higher values of f are, in fact, not compatible with the limitations reported in [19] and lead to an epsilon near-zero behavior at higher frequencies. Lower values of f, instead, reduce the resonance effect of the plasmonic spheres and, consequently, the values of the real part of the effective permittivity are not in the epsilon near-zero range anymore. Therefore, the values of the filling volume fraction here considered are the ones optimizing the two opposite trends now described and keeping the required near-zero behavior at optical frequencies.

As documented in [18], a way to reduce the losses at optical frequencies is to replace gold with silver. The results presented in Fig. 2.8 confirm that losses are significantly reduced but the near-zero frequency band is now shifted towards the near-UV. Please note that, due to the different plasma frequencies of gold and silver, the values of the filling volume fraction required to have the near-zero behavior at optical frequencies are now rather low compared to the ones required in the case of gold nanospheres.

A possible solution to bring the required near-zero behavior to optical frequencies with a reasonable level of losses is reported in [20] and shown in the following. We may use an array of core–shell nanospheres with a plasmonic shell and a dielectric core. Such a structure allows lowering the near-zero frequency range, due to

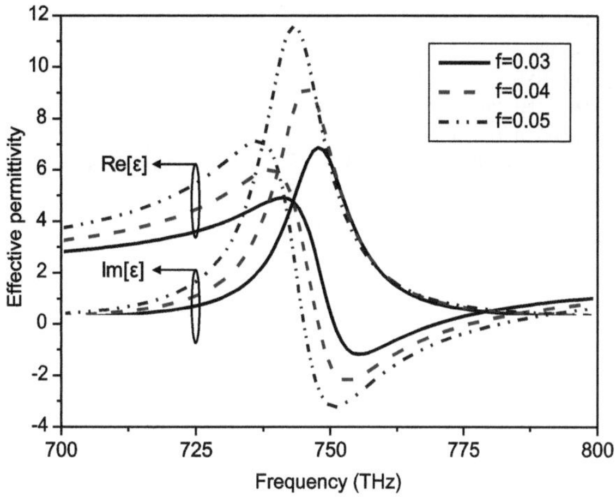

Fig. 2.8 Real and imaginary parts of the effective permittivity for an array of silver nanospheres embedded in a silica matrix for different values of the filling volume fraction

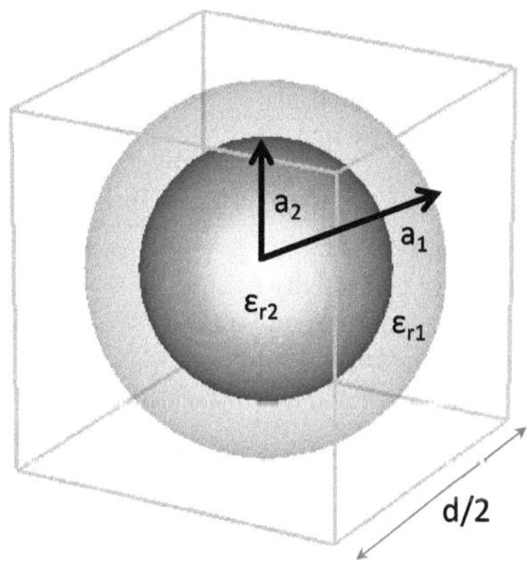

Fig. 2.9 Unit-cell of the designed artificial material slab. The thickness of the slab is $d = 0.17\lambda_0$ at the design frequency

the hybridization of the two plasmons arising at the two metal-dielectric interfaces. This phenomenon leads, in fact, to two new resonances of the particle, one of which is at a lower frequency compared to that of the plasmonic material itself [21]. The core–shell resonances can be tuned varying the shell thickness, as well as changing the material of the core (and of the shell). The complex effective permittivity of the artificial material can be analytically expressed by inserting the polarizability of a single core–shell nanoparticle into the Clausius-Mossotti homogenization formula as:

$$\begin{aligned}\varepsilon_{eff} = &-\big(\varepsilon_m\big\{2(f-1)\varepsilon_m\big[(\beta+2)\varepsilon_{r1}+\varepsilon_{r2}-\beta\varepsilon_{r2}\big]+(2f+1)\varepsilon_{r1}\big[2(\beta-1)\varepsilon_{r1}\\ &-(2\beta+1)\varepsilon_{r2}\big]\big\}\big)/\big((f+2)\varepsilon_m\big[(\beta+2)\varepsilon_{r1}+\varepsilon_{r2}-\beta\varepsilon_{r2}\big]\\ &+(f-1)\varepsilon_{r1}\big[2(\beta-1)\varepsilon_{r1}-(2\beta+1)\varepsilon_{r2}\big]\big)^{-1},\end{aligned} \tag{2.4}$$

where ε_{r2} and ε_{r1} are core and shell relative permittivities, respectively, a_2 and a_1 the core and the shell radius, respectively, and $\beta = a_2/a_1$ (see Fig. 2.9).

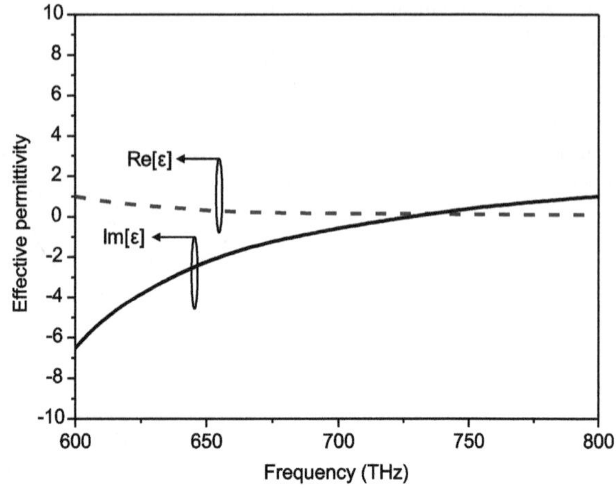

Fig. 2.10 Real and imaginary part of the effective permittivity for an array of core–shell nanospheres embedded in a silica matrix with $\beta = 0.296$ and $f = 0.3$

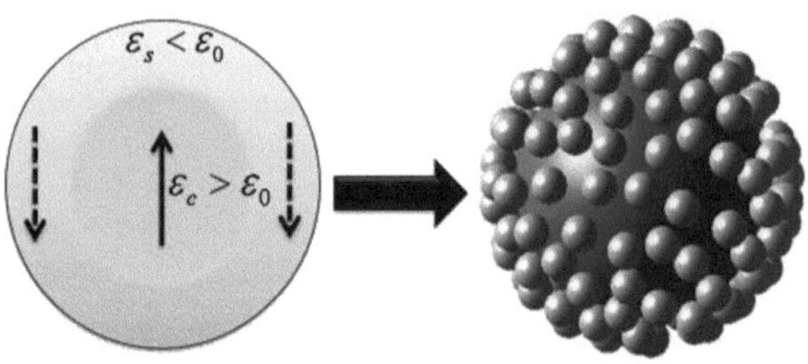

Fig. 2.11 (From [12]) Schematic of the dielectric sphere to be cloaked surrounded by metallic nanoparticles (*right*) forming an effective invisibility shell (*left*) described by its effective permittivity ε_s and effective polarization vector (*dashed arrows*)

By using this formula, we can now design an artificial material consisting of an array of nanospheres formed by a silver shell and a silica core, embedded in a silica matrix. The single nanoparticle has an overall radius of 15 nm and the β factor is assumed equal to 0.296 in order to guarantee a reasonable thickness of the silver shell. Silver is modeled by using its measured bulk permittivity [18], while a constant permittivity equal to 2.137 is assumed for the silica. The resulting complex effective permittivity is shown in Fig. 2.10. We observe that the near-zero behavior (with a reasonable amount of losses) is obtained quite far away from the resonance (around 740 THz) and, thus, the analytical model can be reasonably used.

The homogenized material here designed can be now successfully used to cloak the cylinder of Fig. 2.2. Further details about the actual cloak design by using this new kind of homogenized material are reported in [22].

The results presented so far, of course, are also valid in the case of electrically small spherical objects, where appropriate shell-cluster structures may act as a cloaking device for a finite range of frequencies [12]. Let's consider, for instance, clusters, which consist of an amorphous arrangement of metallic (silver) nanoparticles, which could be approximated in the quasistatic limit by an effective medium, having interesting properties such as a negative or very low permittivity in the optical domain with moderate losses. The effective properties of a shell made of such small spheres using the Maxwell–Garnett formula have been discussed before. In

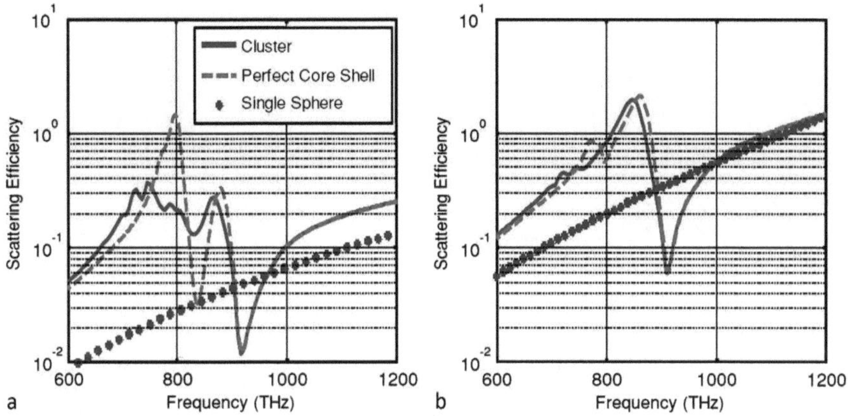

Fig. 2.12 Numerical calculation of the total scattering efficiency for two different permittivities of the dielectric core sphere as a function of frequency [(**a**) $\varepsilon_c = 2.25$, (**b**) $\varepsilon_c = 8$]. The *lines* then have following meaning: *red dotted line*, bare sphere; *blue solid line*, core–shell system rigorously calculated where the fine details of the structure are accounted for; *green dashed line*, homogeneous core–shell system where the shell is described by the Maxwell–Garnett approximation. Note the very good agreement in the frequency domain of effective cloaking around 900 THz. Reprinted with permission from [12], Copyright 2011, ACS

[12] is has been numerically shown that a dielectric core sphere can be made almost invisible around the frequency of 910 THz (where near-zero behavior and reasonable losses can be obtained) with a scattering reduction of more than 75 percent. The considered geometry is reported in Fig. 2.11, while the corresponding scattering performance in Fig. 2.12.

2.2.3 Cloaking Applications

Electromagnetic cloaking opens the door to several interesting applications, including resolution enhancement of near-field scanning optical microscope (NSOM) systems and manipulation of optical forces. In the following we will briefly describe such applications with some details.

NSOM systems allow to go beyond the diffraction limit and to obtain high-resolution images of nanoparticles, optically small samples, corrugated surfaces, etc. NSOM operation relies on a metallic tip scanning the sample/surface to get the high-resolution image. Depending on the tip classes, the image acquisition is performed by using different techniques. Basically, there exist two different probe types: *aperture* and *aperture-less* tips [23]. Here we are interested in the latter, consisting of a solid metallic tip working in the so-called *scattering mode* [23]: when illuminated by an external source, the strong scattered field due to the tip end interacts with the sample and enhances its near-field features, allowing the acquisition of a high-resolution image in the far field. However, since the illuminating beam cannot be collimated at scales comparable to the tip end dimensions, the whole probe

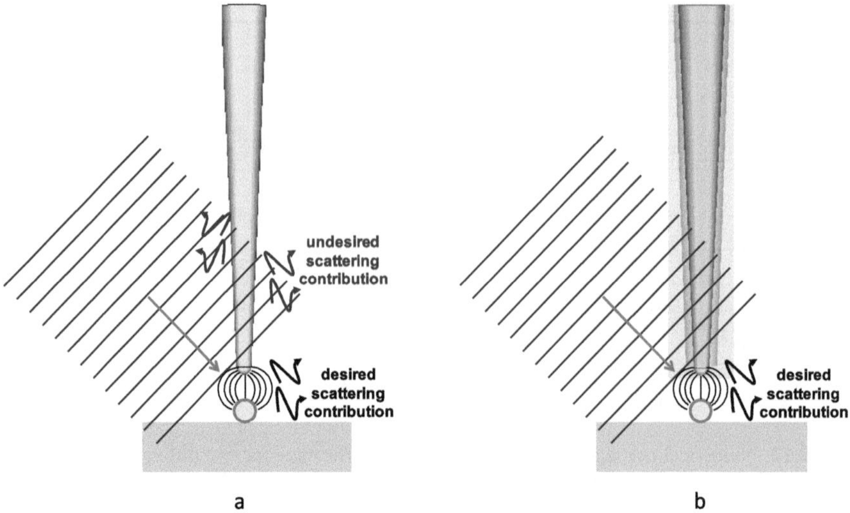

Fig. 2.13 Operation of (**a**) a regular aperture-less NSOM tip and (**b**) an NSOM tip partially covered by a material exhibiting a near-zero value of the real permittivity. Reprinted with permission from [25], Copyright 2011, OSA

body produces an undesired scattering contribution, strongly affecting the image resolution (see Fig. 2.13a). At different scales, this is true in the case of both passive and active samples, whose scattered field is either at the same or at a different frequency with respect to the illuminating source, respectively [24]. In [25] it has been shown that the scattering cancellation approach can be successfully used to enhance the resolution of NSOM systems based on aperture-less tips, due to its peculiarity to preserve the interaction between the illuminating field and the object to hide.

The operation of the NSOM tip proposed in [25] is sketched in Fig. 2.13b. The cloak covers the aperture-less NSOM probe, except its very tip, in order to preserve the needed strong tip-sample interaction. In this way, while keeping the operation mechanism of the probe, it is possible to cancel out the unwanted scattering from the whole probe, increasing, thus, the signal-to-noise ratio in the far-field. In order to prove this concept, the authors of [25] decided to work at mid-infrared frequencies and precisely at around 10.3 μm, where silicon carbide (SiC), which is a material widely used in electronics, naturally exhibits the required near-zero valued real permittivity [26]. In addition, at such frequencies, a few light sources are also available, such as the CO_2 laser, whose operation can be tuned from around 9.2 up to 11.4 μm. In Fig. 2.14, the field maps at the design frequency in the cases of covered and uncovered tips are reported. The reduction of the unwanted scattering by the tip and the preserved interaction between the tip end and a possible object to be placed in its proximity are evident.

In Fig. 2.15, the desired and the unwanted field components are plotted as functions of the wavelength in the cases of uncovered and partially covered tips. Again,

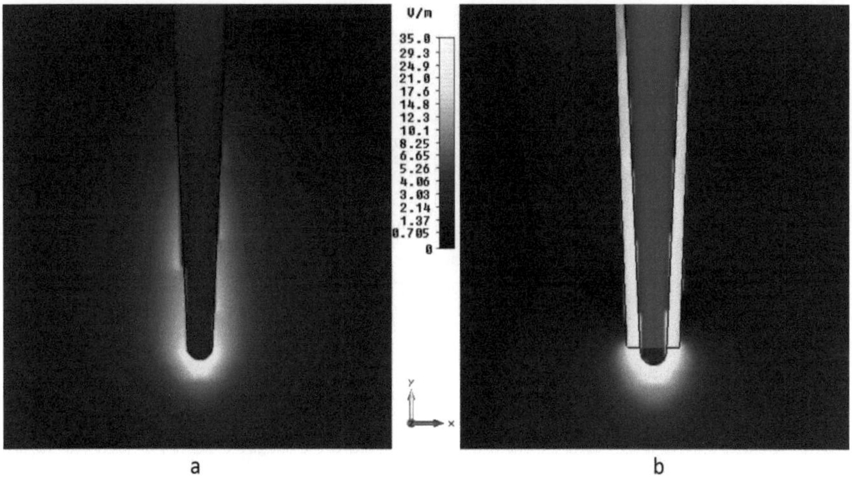

Fig. 2.14 Electric field amplitude distribution due to an unit-amplitude uniform plane-wave illuminating the regular aperture-less gold NSOM tip (**a**) and a covered gold NSOM tip (**b**). The field amplitude is plotted at the design frequency of 10.3 μm. Reprinted with permission from [25], Copyright 2011, OSA

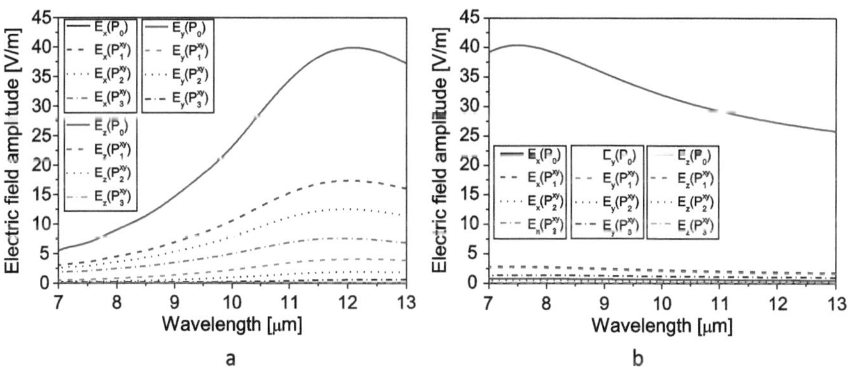

Fig. 2.15 Amplitudes of the total (incident + scattered) electric field components at some reference points along the tip body in the case of a regular aperture-less gold NSOM tip (**a**) and a covered tip (**b**) as a function of the wavelength. Reprinted with permission from [25], Copyright 2011, OSA

the amplitude of the undesired field components is dramatically reduced by using the cloak, while the desired field component is still the dominant one.

The same concept can be demonstrated at different wavelengths, such as near-infrared (NIR) and visible frequencies, but in that case, due to the lack of natural materials with a near-zero permittivity, the design requires fabrication of proper artificial materials, as shown in the previous examples of the cloaked cylinders.

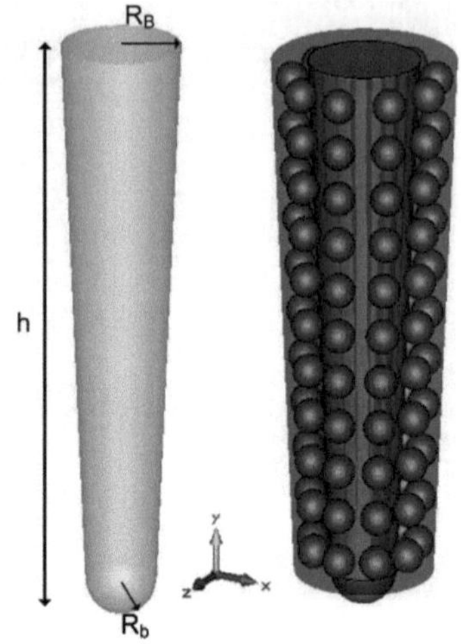

Fig. 2.16 Bare (*left*) and cloaked (*right*) cone. Geometrical parameters: $h = 484$ nm, $R_B = 50$ nm and $R_b = 30$ nm. Reprinted with permission from [20], Copyright 2012, Elsevier

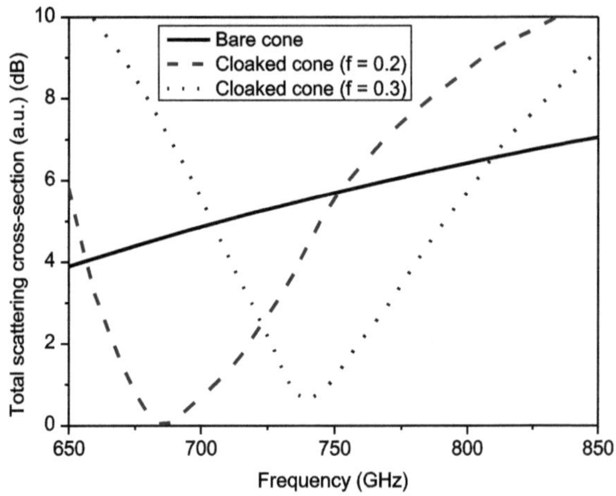

Fig. 2.17 Total scattering cross-section of the bare cone (*black solid line*) and of the cloaked one for different values of the volume filling fraction f. Reprinted with permission from [20], Copyright 2012, Elsevier

A possible design has been recently presented in [20] by using the homogenized material made of the core–shell silver nanospheres documented above. The corresponding geometry and results are shown in Figs. 2.16, 2.17, and 2.18.

Another interesting application field of metamaterial cloaking is in the reduction and manipulation of optical forces and torques. When object dimensions approach the nanoscale, optical forces, which are usually neglected in everyday experience, come into play, and their contribution cannot be neglected any more. Typically, light forces exerted on electrically small particles are explained in terms of gradient and scattering forces. The former are directly related to the interaction between the external field and the induced dipole, which is drawn by the field intensity gradients. The latter are related to the momentum transfer between the scattered field and the illuminated particle, being proportional to the Poynting vector. Since both these forces are inherently associated to the scattering prop-

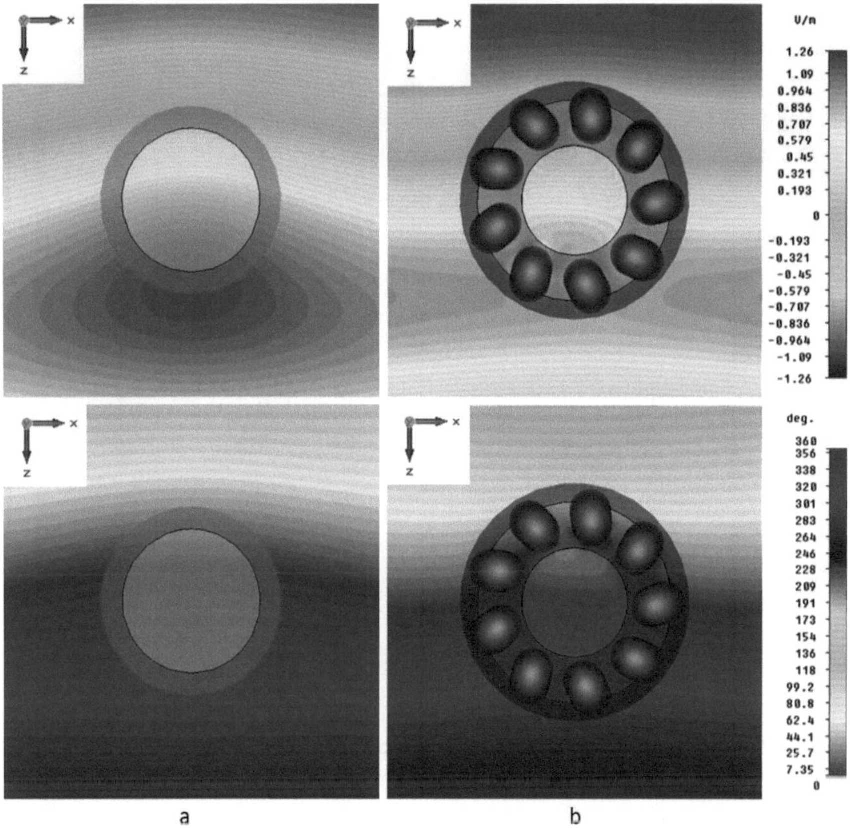

Fig. 2.18 Time-domain distribution of the y-component of the electric field at a given time instant (*up*) and electric field phase distribution (*down*) for the cases of the bare (**a**) and cloaked cone (**b**). The *plots* show the field distribution at the cloak frequency of 740 THz on a plane parallel to the xz plane placed at the half height of the cone. Reprinted with permission from [20], Copyright 2012, Elsevier

erties of the object, a cloak placed around a nanoparticle and causing the reduction, or even suppression, of the total scattered field, may open the door to unusual particle manipulation. In [27], it has been recently shown that, following the formulation based on Maxwell's stress tensor, the total optical force exerted on a nanoparticle can be written as the sum of three contributions: the radiation pressure, the gradient force and another term which is usually related to the electromagnetic field spin density. These three contributions are proportional to either the real part or the imaginary part of the object polarizability. Therefore, if we are able through a proper cloak to reduce or even suppress both the real and the imaginary parts of the polarizability, we are able in principle to reduce or suppress all the optical forces exerted on the nanoparticle. As an example, we show in Fig. 2.19 what happens to a cloaked spherical particle immersed in a field distribution generated by two orthogonal standing waves. The uncloaked

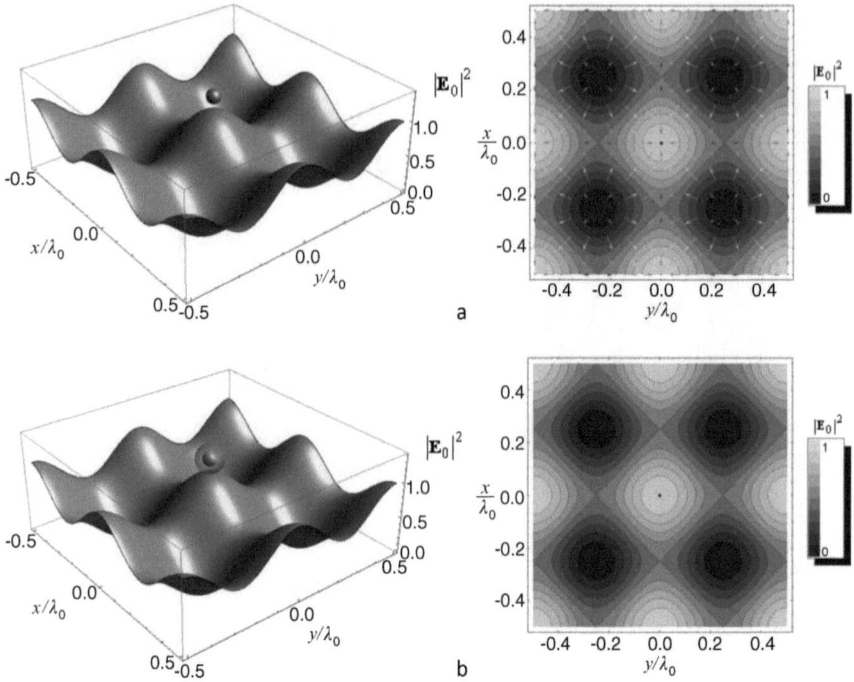

Fig. 2.19 (**a**) Spherical nanoparticle immersed in a field distribution generated by a pair of orthogonal standing waves. *Arrows* represent the strength and direction of the gradient force exerted on the sphere. (**b**) Cloaked spherical nanoparticle immersed in the same field distribution

particle experiences a gradient force which tends to move the particle towards the light spots, while the cloaked particle does not feel any force and keeps its motion state.

Metamaterial cloaks can be also used to manipulate the torque exerted on elongated nanoparticles. The idea is briefly sketched in Fig. 2.20. Elongated nanoparticles are anisotropic structures characterized by a polarizability tensor. The polarizability along the longitudinal direction is greater than the ones in the transverse plane and, consequently, when the particle is illuminated by a given electromagnetic field, it tends to be aligned with the major axis along the electric field vector. This represents the stable equilibrium position for the particle, while the orientations along the minor axes (i.e. the ones in the transverse plane) represent unstable equilibrium positions. In [27], it has been shown that metamaterial cloaks can be successfully used to obtain unusual equilibrium positions, making stable positions unstable ones, and vice versa. Particularly, it has been shown that the material properties of the cloak can be used to make the longitudinal polarizability (in both real and imaginary parts) of an elongated particle smaller than the transverse ones. In this way, the stable equilibrium position of the particle is

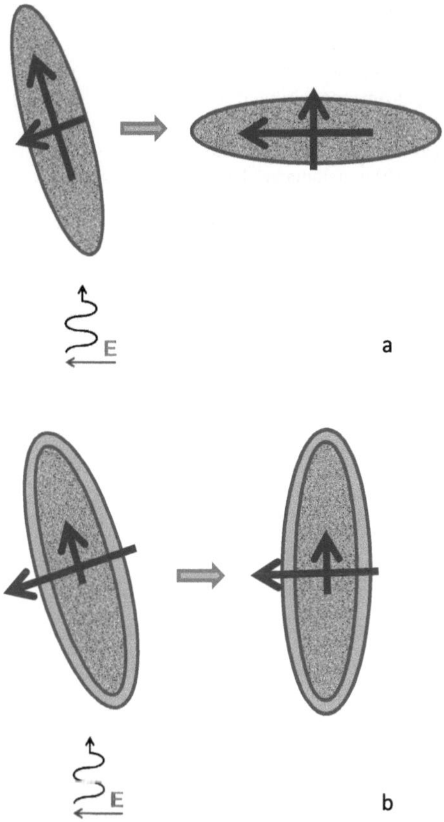

Fig. 2.20 (**a**) Rotation and stable equilibrium position of an elongated nanoparticle illuminated by an electromagnetic field. (**b**) Unusual rotation and stable equilibrium position of a cloaked elongated nanoparticle illuminated by an electromagnetic field

the one with a minor axis directed along the electric field. A simple example is sketched in Fig. 2.20, while the interested reader may find the analytical details in [27].

2.3 Beaming and Spatial-Filtering Devices

2.3.1 General Concepts

Metamaterials exhibiting near-zero values of the real permittivity have been recently proposed as good candidates to obtain innovative spatial filtering and beaming devices. The very basic intuitive concept is sketched in Fig. 2.21, where a ray approximation has been used [28]. An electrically small aperture is assumed as a light source radiating in presence of a dielectric slab. If a regular dielectric is used (see Fig. 2.21a), the individual spectral components experience reflection and transmission, according to Snell's laws, the transmitted rays are rather sparse, and the resulting radiation pattern is rather broad. In contrast, when the dielectric slab is made by an artificial material whose real part of the permittivity function is characterized

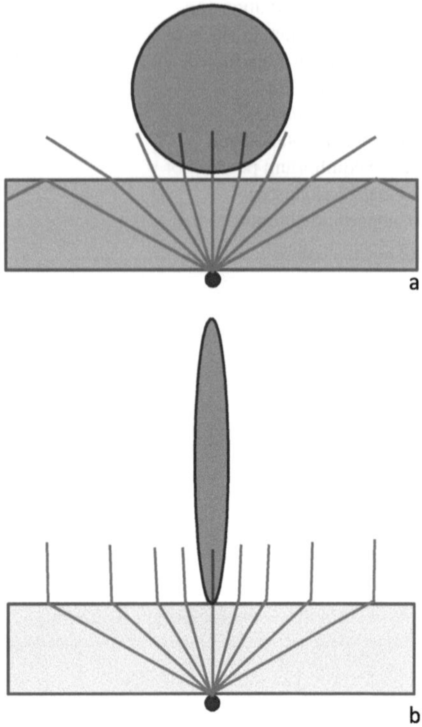

Fig. 2.21 Heuristic ray-theory interpretation of the behavior of an electromagnetic source radiating in presence of (**a**) a regular dielectric slab and (**b**) a metamaterial slab characterized by near-zero values of the effective real permittivity

by near-zero values, the refraction index is also near zero and Snell's laws return a transmission angle close to broadside, whatever the spectral component is (see Fig. 2.21b). In other words, only a few spectral components around the broadside emerge from the slab, resulting in a very directive beam.

An alternative way to look at the same phenomenon is the following: though all the rays generated by the source traverse different geometrical distances within the slab, according to the different angles, the electrical distances are, instead, nearly the same, due the near-zero value of the refraction index. Therefore, the different ray contributions emerging at the upper boundary between the slab and the vacuum are characterized by the same phase and, thus, radiate coherently towards a given direction, resulting in a very directive beam.

This concept can be easily generalized to the case reported in Fig. 2.22. In principle, with the use of near-zero permittivity materials, it is possible to manipulate a given impinging phase front and transform its phase distribution into a desired shape by properly tailoring the exit side of the metamaterial slab.

This, of course, may have important implications in imaging and communications technology, since it can be speculated that, by designing the conformal face of the exit side, it is possible to engineer the phase front shape to a given desired pattern, independent of the form of the impinging wave. Typical applications may be found in the design of beaming devices (i.e. devices capable to manipulate the radiation properties of the sources in such a way to produce directive beams point-

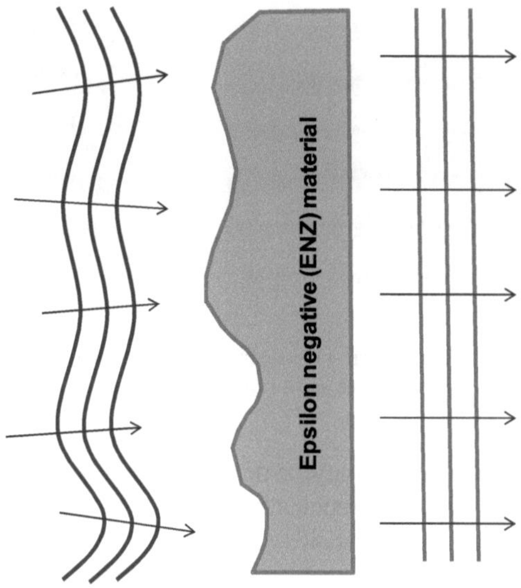

Fig. 2.22 A curved phase front impinges on an epsilon-near-zero (ENZ) material in its entrance side (*left*). At the exit side (*right*), the phase front is conformal to the exit surface, in this case a planar phase front, due to the small phase variation inside the material

ing in a desired direction), spatial filters (i.e. devices which allow transmission of some angular components of the source spectrum, reflecting back all the remaining ones), phase pattern forming devices (i.e. devices capable to tailor the phase-front of an electromagnetic source), lenses, etc.

2.3.2 Spatial Filters and Beaming Devices Obtained Using Amorphous Metamaterials

Going beyond the ray approximation, in [29] it has been analytically demonstrated that metamaterial slabs characterized by near-zero values of the real permittivity, can be effectively used to create arbitrary phase patterns for a given polarization. In particular, considering a planar metamaterial slab, the working polarization is the TM one, characterized by the magnetic field parallel to the slab interfaces. Under this assumption, in [29] it has been demonstrated that the slab allows the transmission of a few angular components around the broadside direction, while reflects back in-phase (i.e. as a perfect magnetic conductor does) all the remaining angular components. The dual phenomenon has been verified in [30], considering the case of a metamaterial slab characterized by near-zero values of the real permeability illuminated by a TE polarized wave (i.e. having the electric field parallel to slab interfaces). In this case, the slab reflects back out-of-phase (i.e. as a perfect electric conductor does) all the angular components not allowed to pass through the material.

In the extreme case of exactly zero values of either permittivity or permeability, it has been demonstrated that only the normal angular component is allowed to pass through the slab and be transmitted. By playing with the near-zero values

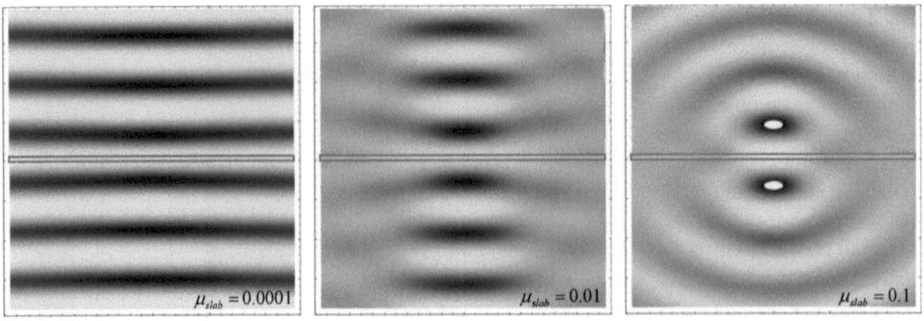

Fig. 2.23 Electric field distribution generated by an electric current line placed at the interface between free-space and a tiny slab characterized by different values of near-zero real permeability

of the permittivity or permeability it is possible to control the range of the passing angular components, as shown in Fig. 2.23. Additional degrees of freedom are given by the distance between the source and the first slab interface and by the slab thickness.

In the following, we show that by using the materials produced with self-assembly techniques it is possible to implement filtering/beaming devices. In Fig. 2.24a, the magnetic field distribution produced by a vertical electric dipole in free space at the frequency of 700 THz is reported. When placing in front of the source a finite slab made of an isotropic, homogeneous material characterized by near-zero values of the real permittivity around 700 THz, the pattern distribution is the one reported in Fig. 2.24b. The phase front is now significantly changed compared to the one of the source alone and the beam is more directive, according to the concepts presented in the previous sub-section.

As already mentioned in the case of cloaking devices, in order to implement a structure behaving as the one reported in Fig. 2.24b, it is possible to use a proper arrangement of spherical gold particles, as the one shown in Fig. 2.24c. More in details, 5184 gold nanospheres with the diameter of 25 nm are arranged in 9 layers. Each layer is made of 576 spheres, organized in a 24×24 2D array, with both periodicities equal to 6 nm. The distance between two adjacent layers is set to 2 nm. By using the Drude model approximation for gold, according to the homogenization rule reported in Sect. 2.1, this finite 3D arrangement of gold nanoparticles exhibits a real permittivity crossing the zero value at around 700 THz.

As evident from the results of Fig. 2.24c, the behavior of the finite 3D arrangement of gold nanospheres is very similar to the one of the ideal homogeneous metamaterial slab, opening the door to the possible implementation of working filtering/beaming devices operating at optical frequencies.

Also in this case, we can consider more realistic designs by involving the measured permittivity of gold. As it happened for the cloak case, when we do that, losses completely kill the function of the homogenized material. However, as documented in the previous sections, it is possible to replace gold with silver and the nanospheres with core–shell nanospheres and obtain a near-zero behavior with reasonable losses within the visible. The same artificial material proposed in [20] and used for the

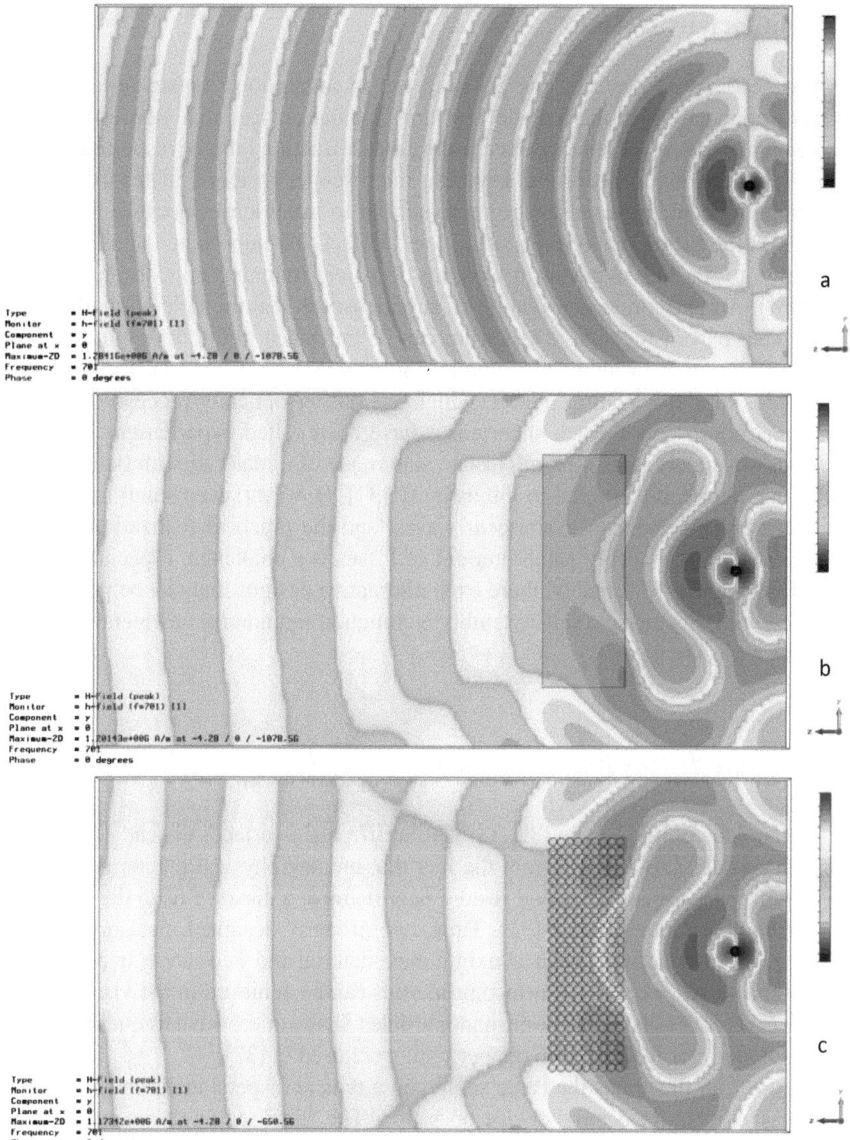

Fig. 2.24 Magnetic field distribution generated by an electric dipole in the case of the source radiating (**a**) in free space at 700 THz; (**b**) in close proximity of an ideal finite 3D metamaterial slab made of a material characterized by a near-zero real permittivity at 700 THz; (**c**) in close proximity of a real 3D arrangement of gold nanospheres behaving as a near-zero permittivity material at the design frequency of 700 THz

cloaks reported in the previous section, has been used also to design a spatial filter. The interested reader may refer to [20] for further design details and implementation examples.

2.4 Superlens Devices

In this section, we describe the possible use of amorphous metamaterials in superlens applications. Superlens devices are imaging devices that can resolve subwavelength features of objects, overcoming the diffraction limit of resolution. Functionally analogous to near-field scanning microscopes in terms of resolving subwavelength detail, superlenses are more similar to usual lenses because they form images in the image plane like a usual lens. In fact, the importance of superlenses is related to the improvement of existing near-field scanning optical microscopes (NSOM). Another important feature of superlenses is that the field strength of the evanescent part of the source spectrum can be enhanced in the device due to resonant excitation of localized eigenmodes. Some superlenses can transform evanescent fields into propagating modes, which are further optically processed by conventional microscopes. Such superlenses have been called hyperlenses. The most well-known superlens is a parallel-plate slab made of a material with both permittivity and permeability equal to minus unity [31]. However, even small losses mitigate the enhancement of evanescent waves, and the practical realization of such materials with reasonably small amount of losses is a challenge, especially at optical frequencies. Fortunately, there exist alternative designs that can be potentially realized with inexpensive self-assembly techniques and amorphous metamaterials, which we discuss next.

2.4.1 Superlenses Based on Two Grids of Plasmonic Nanoparticles

The main idea of this design [32, 33] is to utilize meta-surfaces instead of artificial bulk materials. This follows from the fact that the key physical phenomenon used to enhance evanescent fields and restore near fields at a distance from the source is a surface plasmon resonance [32]. Thus, two properly designed metasurfaces can replace two interfaces between an exotic metamaterial and free space. In paper [33] it was shown that necessary surface properties can be achieved in the visible using planar grids of resonant plasmonic nanoparticles. In the microwave frequency range, resonant particles made of thin copper wire were used in [32].

Figure 2.25 illustrates the basic setup and a typical dispersion curve of one grid which is suitable for desired device operation. The "flat" region of the curve is instrumental for the target superlens functionality: In this case eigenwaves with a wide range of large spatial frequencies can be excited by a subwavelength source at the corresponding frequency.

Numerical estimations (see Fig. 2.26) show an example of near-field distributions in the image plane when the lens is excited by two point sources separated by 162.5 nm (about 0.3λ). The lens dimensions and parameters are the same as in Ref. [33]: silver spherical nanoparticles of the radius 28 nm, the grid period is 65 nm.

The comparison of the performance of the device with ideally periodical arrays of identical particles (left) with the case of randomly positioned inclusions shows

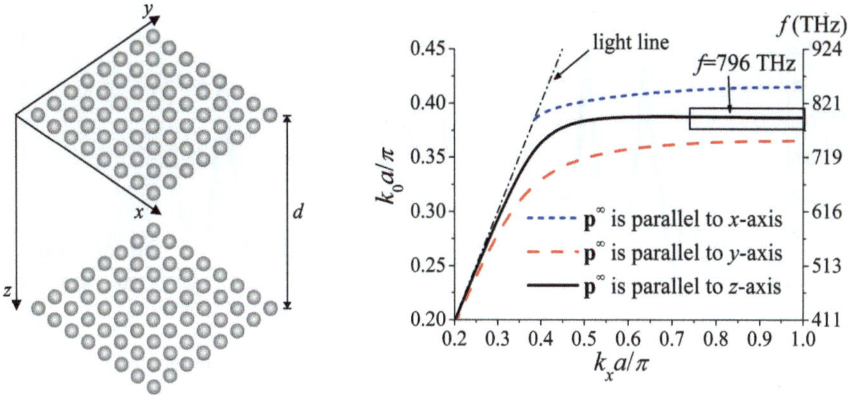

Fig. 2.25 Conceptual geometry of a superlens based on grids on plasmonic nanoparticles (reprinted with permission from [33], Copyright 2006, ACS) and typical dispersion curve of each of the two grids (reprinted with permission from [37], Copyright 2011, ACS)

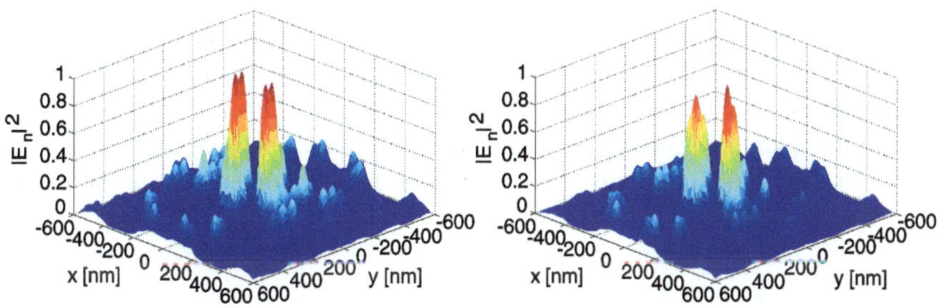

Fig. 2.26 Calculated electric field distributions (normalized field intensity) in the image plane of a superlens formed by two parallel grids of plasmonic nanoparticles. The source is a set of two point electric dipoles orthogonal to the lens plane. Fields are normalized to the maximum value. *Left*: Regular grids; *Right*: Positions of particles in the arrays have been randomized with the standard deviation equal to 1 nm

no degradation in performance. This is further illustrated by Fig. 2.27, where the image-plane field distribution is plotted in the $y = 0$ plane for regular grids (thick grey curve) and for a number of realizations of random grids (standard deviation of the position is 2 nm in this example). This picture also shows the field distribution at the same distance from the sources in case of the absence of the grids (red dashed curve). It is obvious that the lens significantly enhances the field strength in the image plane due to resonant excitation of surface modes of the arrays. We can conclude that for this application the use of amorphous grids of plasmonic nanoparticles is most appropriate. In fact we expect that with proper optimization the performance of superlenses with random grids may even be better than with ideal regular arrays. Reference [33] provides analytical formulas for calculation of dispersion curves and describes the numerical technique implemented as a MATLAB code which can be used for numerical optimization of the structure. Superimaging properties of arrays

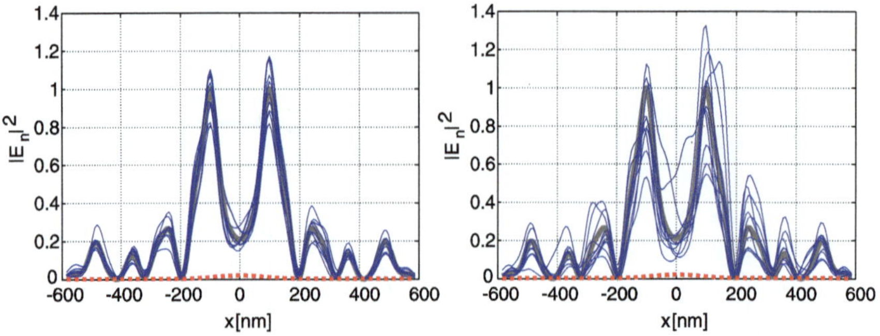

Fig. 2.27 Calculated electric field distributions (normalized intensity) in the image plane of a superlens formed by two parallel grids of plasmonic nanoparticles (the plane $y = 0$). The source is a set of two point electric dipoles orthogonal to the lens plane, same as in Fig. 2.26. *Thick grey curve*: regular grids; *Thin lines*: several realizations of random grids: *Red dashed curve*: the fields in the absence of the lens. The standard deviation of the particle position is 1 nm in the *left panel* and 2 nm in the *right panel*

of plasmonic nanoparticles were more recently confirmed by full-wave simulations of realistic structures [34, 35]. Other periodic or amorphous sub-wavelength patterned surfaces can also have flat regions in the surface mode dispersion curves, making them suitable for superimaging. For example, superimaging was demonstrated in Ref. [36] (in simulations and microwave experiments) with the use of fractal patterning.

2.4.2 Near Field Enhancement and Focusing Using Multilayer Arrays of Plasmonic Nanoparticles

Also multi-layered 3D arrays of plasmonic nanoparticles have important potential applications. They can be possibly used for two purposes: near-field enhancement of electromagnetic fields and sub-wavelength focusing and concentration of fields. Recently, paper [37] analyzed and stressed certain properties which show how to control these phenomena. Although such 3D arrays of nanoparticles do not form imaging devices in the usual sense of this term (measuring the field distribution in the "image plane" it is not possible to uniquely reconstruct the distribution of possibly volumetrically distributed sources), the fields can be effectively concentrated in sub-wavelength areas far away from the field sources. Presence of interference maxima at some lower level also make the super-resolution application still challenging. On the other hand, resonant enhancement and translation of sub-wavelength features at electrically long distances has been clearly demonstrated numerically.

Figure 2.28 shows a numerical example of a structure capable for transformation of strong near-field spikes at electrically long distances.

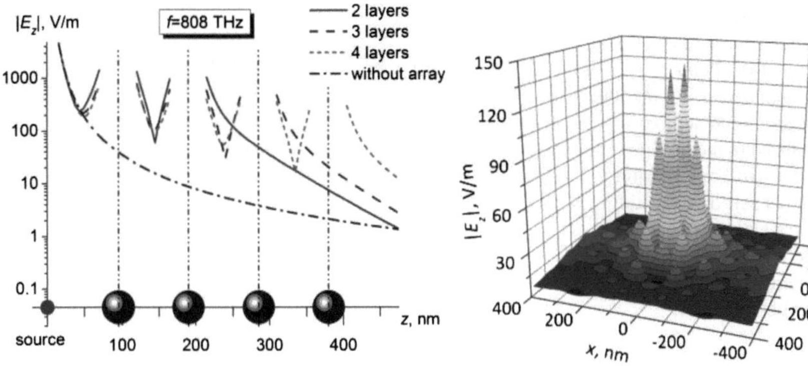

Fig. 2.28 Typical dependence of the electric field strength along the structure depth (*left*) and typical field pattern in the "image plane" for the case of two point sources in the source plane. Note strongly enhanced near fields and the presence of interference maxima. Reprinted with permission from [37], Copyright 2011, ACS

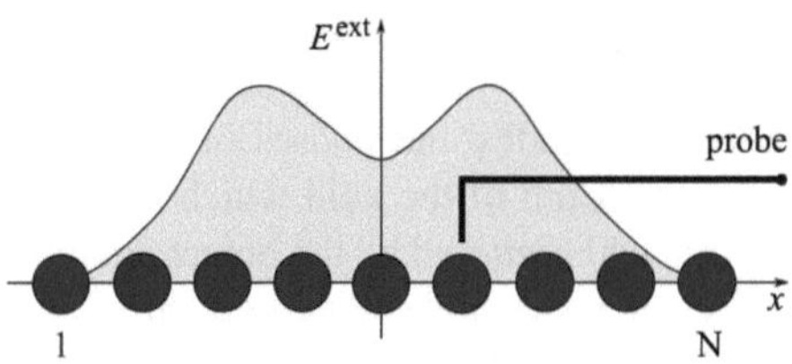

Fig. 2.29 A linear array of plasmonic nanoparticles as a subwavelength microscope sensor. The *line* shows the electric-field profile to be determined by the microscope. The probe is stationary. Reprinted with permission from [38], Copyright 2008, AIP

2.4.3 Real-Time Sub-wavelength Imaging Using a Single Grid of Plasmonic Nanoparticles

In the known sub-wavelength imaging devices the only means to actually measure the image-plane near-field distributions at the sub-wavelength scale is the scanning near-field microscope that uses a small moving probe (an exception is the hyperlenses, which transform evanescent modes into propagating ones). The main fundamental practical problem associated to this property is that it is not possible to acquire an image in real time, as it is necessary to make a mechanical scan over the whole surface, and this is a slow process.

Here we describe one recently proposed technique which is capable to provide *real-time* superimaging and can be potentially realized with both regular and random grids of nanoparticles.

The idea of the device [38] is to use the known (pre-calculated or pre-calibrated) eigenmode spectrum of an array of subwavelength-sized resonant particles (or any other resonant sub-wavelength patterned surface, including amorphous structures) as a sensor, as illustrated in Fig. 2.29. The array is positioned close to the object

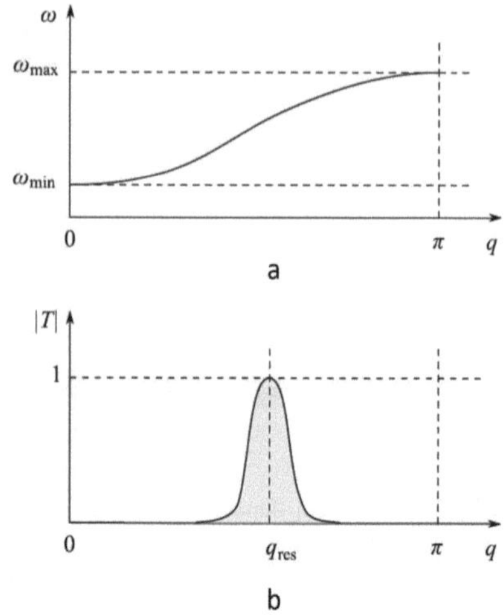

Fig. 2.30 An example dispersion curve of a periodic structure (**a**). *On the horizontal axis* $q = k_t d$ is the transverse wave number normalized to the grid period d. (**b**) Transmission through a pair of resonating arrays at a given frequency as a function of q. Reprinted with permission from [38], Copyright 2008, AIP

under test and illuminated by a frequency-sweeping probe electromagnetic wave. The array response is measured by one or two stationary probes in the vicinity of the array. The measured amplitudes and phases of the excited eigenwaves allow restoration of the spatial spectrum of the field which illuminates the array (that is, the spatial spectrum of the field scattered by the object under test). It is also possible to benefit from the effect of restoration of evanescent field components in a pair of grids of resonant particles as in the superlens based on two plasmonic surfaces (Fig. 2.30).

Typical dispersion curve of a resonant array designed for the use as a sensor shown in Fig. 2.30 corresponds to the transmission coefficient at a fixed frequency which exhibits high transmission for the resonant mode at this frequency and low transmission at all other, non-resonant, wave numbers.

Scanning the frequency from $\omega_{\min}$ up to $\omega_{\max}$ we can sequentially excite the surface modes of all possible k_t ranging from 0 to $k_{t\,\max} = \pi/d$. The amplitude of an excited mode is proportional to the amplitude of the corresponding spectral component of the incident field. We can find the amplitude of the mode from a field measurement done only at a single point in the image plane. To realize the device, it is necessary to provide strong excitation of many modes of the grid, thus, the limiting factor is the loss level in the particles. The spatial resolution is limited by the distance between the particles. Strict periodicity and planarity of the grid are not essential, because the spectrum of the grid needs to be pre-calibrated in any case. The single particles of the grid sensor should have as high quality factor as possible. As to the level of interparticle coupling (determined mainly by the average distances between the nanoparticles), we expect that there exists an optimum for the desired performance, to be optimized numerically (for very small coupling all the spheres resonate at nearly the same frequency). For optical applications, the practical challenge will be realization of (stationary) probes which can measure both amplitude and phase of light. The measurement speed is determined by the time during which the frequency scan over the resonant band of the grid can be completed.

2.5 Conclusions

This chapter has presented an overview of potential applications and potential device properties, starting from the technologically achievable nanostructures, manufactured by novel self-assembly techniques. A number of potential applications have been identified and analyzed in detail. In particular, it has been found that realizations of cloaking devices based on the field-cancellation technique, optical beaming devices, and optical superlenses become feasible with the current nanofabrication techniques used to realize amorphous metamaterial structures.

Acknowledgements Part of the work on writing this chapter was done within the framework of the EU-funded FP7 project NANOGOLD. The authors are grateful to Dr. Pekka Alitalo for providing his unpublished simulation results of the superlens based on random arrays of plasmonic nanoparticles (Figs. 2.26 and 2.27) and to Mr. Alessio Monti and Prof. Alessandro Toscano for providing their unpublished results of the cylindrical cloak (Figs. 2.3–2.6) and of the spatial filter (Fig. 2.24) based on gold nanospheres.

References

1. J.B. Pendry, D. Schurig, D.R. Smith, Controlling electromagnetic fields. Sci. **312**, 1780–1782 (2006)
2. U. Leonhardt, Optical conformal mapping. Sci. **312**, 1777–1780 (2006)
3. D. Schurig, J.J. Mock, B.J. Justice, S.A. Cummer, J.B. Pendry, A.F. Starr, D.R. Smith, Metamaterial electromagnetic cloak at microwave frequencies. Sci. **314**, 977–980 (2006)
4. W. Cai, U.K. Chettiar, A.V. Kildishev, V.M. Shalaev, Optical cloaking with metamaterials. Nat. Photonics **1**, 224–227 (2007)
5. A. Alù, N. Engheta, Achieving transparency with plasmonic and metamaterial coatings. Phys. Rev. E **72**, 016623 (2005)
6. M.G. Silveirinha, A. Alù, N. Engheta, Parallel plate metamaterials for cloaking structures. Phys. Rev. E **75**, 036603 (2007)
7. F. Bilotti, S. Tricarico, L. Vegni, Electromagnetic cloaking devices for TE and TM polarizations. New J. Phys. **10**, 115035 (2008)
8. F. Bilotti, S. Tricarico, L. Vegni, Plasmonic metamaterial cloaking at optical frequencies. IEEE Trans. Nanotechnol. **9**, 55–61 (2010)
9. S. Tricarico, F. Bilotti, A. Alù, L. Vegni, Plasmonic cloaking for irregular objects with anisotropic scattering properties. Phys. Rev. E **81**, 026602 (2010)
10. S. Tricarico, F. Bilotti, L. Vegni, Reduction of optical forces exerted on nano-particles covered by scattering cancellation based plasmonic cloaks. Phys. Rev. B **82**, 045109 (2010)
11. F. Bilotti, S. Tricarico, F. Pierini, L. Vegni, Cloaking apertureless near-field scanning optical microscopy tips. Opt. Lett. **36**, 211–213 (2011)
12. S. Mühlig, M. Farhat, C. Rockstuhl, F. Lederer, Cloaking dielectric spherical objects by a shell of metallic nanoparticles. Phys. Rev. B **83**, 195116 (2011)
13. N.A. Nicorovici, G.W. Milton, R.C. McPhedran, L.C. Botten, Quasistatic cloaking of two-dimensional polarizable discrete systems by anomalous resonance. Opt. Express **15**, 6314–6323 (2007)
14. S. Tretyakov, P. Alitalo, O. Luukkonen, C. Simovski, Broadband electromagnetic cloaking of long cylindrical objects. Phys. Rev. Lett. **103**, 103905 (2009)
15. P. Alitalo, O. Luukkonen, L. Jylha, J. Venermo, S.A. Tretyakov, Transmission-line networks cloaking objects from electromagnetic fields. IEEE Trans. Antennas Propag. **56**, 416–424 (2008)

16. V. Myroshnychenko, J. Rodríguez-Fernández, I. Pastoriza-Santos, A.M. Funston, C. Novo, P. Mulvaney, L.M. Liz-Marzán, F. Javier García de Abajo, Modelling the optical response of gold nanoparticles. Chem. Soc. Rev. **37**, 1792–1805 (2008)
17. J.C. Maxwell-Garnett, Colours in metal glasses and in metallic films. Philos. Trans. Soc. **203**, 385–420 (1904)
18. P.B. Johnson, R.W. Christy, Optical constants of the noble metals. Phys. Rev. B **6**, 4370 (1972)
19. T.G. Mackay, Lewin's homogenization formula revisited for nanocomposite materials. J. Nanophotonics **2**, 029503 (2008)
20. A. Monti, F. Bilotti, A. Toscano, L. Vegni, Possible implementation of epsilon-near-zero metamaterials working at optical frequencies. Opt. Commun. **285**(16), 3412–3418 (2012)
21. E. Prodan, C. Radloff, N.J. Halas, P. Nordlander, A hybridization model for the plasmon response of complex nanostructures. Sci. **302**, 419–422 (2003)
22. A. Monti, F. Bilotti, A. Toscano, Optical cloaking of cylindrical objects by using covers made of plasmonic nano-particles. Opt. Lett. **36**, 4479–4481 (2011)
23. R.C. Dunn, Near-field scanning optical microscopy. Chem. Rev. **99**, 2891–2928 (1999)
24. V. Sandoghdar, J. Mlynek, Prospects of apertureless SNOM with active probes. J. Opt. A, Pure Appl. Opt. **1**, 523–530 (1999)
25. F. Bilotti, S. Tricarico, F. Pierini, L. Vegni, Cloaking apertureless near-field scanning optical microscopy tips. Opt. Lett. **36**, 211–213 (2011)
26. W.G. Spitzer, D. Kleinman, D. Walsh, Infrared properties of hexagonal silicon carbide. Phys. Rev. **113**, 127–132 (1959)
27. S. Tricarico, F. Bilotti, L. Vegni, Reduction of optical forces exerted on nano-particles covered by scattering cancellation based plasmonic cloaks. Phys. Rev. B **82**, 045109 (2010)
28. A. Alù, F. Bilotti, N. Engheta, L. Vegni, Metamaterial covers over a small aperture. IEEE Trans. Antennas Propag. **54**, 1632–1643 (2006)
29. A. Alù, M.G. Silveirinha, A. Salandrino, N. Engheta, Epsilon-near-zero metamaterials and electromagnetic sources: tailoring the radiation. Phys. Rev. B **75**, 155410 (2007)
30. S. Tricarico, F. Bilotti, L. Vegni, Multi-functional dipole antennas based on artificial magnetic metamaterials. IET Microw. Antennas Propag. **4**, 1026–1038 (2010)
31. J.B. Pendry, Negative refraction make a perfect lens. Phys. Rev. Lett. **85**, 3966–3969 (2000)
32. S. Maslovski, S. Tretyakov, P. Alitalo, Near-field enhancement and imaging in double planar polariton-resonant structures. J. Appl. Phys. **96**, 1293–1300 (2004)
33. P. Alitalo, C. Simovski, A. Viitanen, S. Tretyakov, Near-field enhancement and subwavelength imaging in the optical region using a pair of two-dimensional arrays of metal nanospheres. Phys. Rev. B **74**, 235425 (2006)
34. C.C. Yan, D.H. Zhang, D.D. Li, Spherical metallic nanoparticle arrays for super-resolution imaging. J. Appl. Phys. **109**, 063105 (2011)
35. V. Yannopapas, Subwavelength imaging of light by arrays of metal-coated semiconductor nanoparticles: a theoretical study. J. Phys. Condens. Matter **20**, 255201 (2008)
36. X. Huang, S. Xiao, D. Ye, J. Huangfu, Z. Wang, L. Ran, L. Zhou, Fractal plasmonic metamaterials for subwavelength imaging. Opt. Express **18**, 10377 (2010)
37. S. Steshenko, F. Capolino, P. Alitalo, S. Tretyakov, Effective model and investigation of the near-field enhancement and subwavelength imaging properties of multilayer arrays of plasmonic nanospheres. Phys. Rev. E **84**, 016607 (2011)
38. S. Maslovski, P. Alitalo, S. Tretyakov, Subwavelength imaging based on frequency scanning. J. Appl. Phys. **104**, 103109 (2008)

Chapter 3
Homogenization of Amorphous Media

Ari Sihvola and Henrik Wallén

Abstract This chapter focuses on the homogenization of two-phase mixtures. The materials that form the mixture are assumed to be isotropic but they can be plasmonic (having negative real part of the permittivity) and/or lossy (complex-valued). The geometrical arrangements that are analyzed can be regular, disordered, or totally random morphologies. Classical mixing rules are presented and applied to various type of mixtures, and their predictions are tested and compared against computational simulations. Among the phenomena that are discussed in detail are the emergence of resonances and complex-valued responses from lossless components, as well as the effect of mixing process on the dispersive behavior of materials. Such homogenization of mixtures is a valuable means to assign effective properties to amorphous nanophotonic materials with sufficiently small unit cells, since it helps in analyzing emerging physical phenomena, and makes such materials available in the design process of a functional device.

3.1 Introduction

Interaction of electromagnetic fields with nanostructured materials is a complicated process. The essential factors determining the character of this interaction, in addition to the morphology of the medium structure, are the frequency of the temporal variation of the field and the electromagnetic response of the microscopic elements composing the mixture.

Nanotechnological products have very small spatial scales. The physical size of the elements and their interdistances is typically of the order of tens of nanometers. This means that for waves in the optical and infrared spectral range, the structural inhomogeneity of the medium is often much smaller than the spatial wavelength

A. Sihvola (✉) · H. Wallén
Department of Radio Science and Engineering, Aalto University School of Electrical Engineering, PO Box 1300, 00076 Aalto, Finland
e-mail: ari.sihvola@aalto.fi

H. Wallén
e-mail: henrik.wallen@aalto.fi

C. Rockstuhl, T. Scharf (eds.), *Amorphous Nanophotonics*,
Nano-Optics and Nanophotonics, DOI 10.1007/978-3-642-32475-8_3,

© Springer-Verlag Berlin Heidelberg 2013

of the waves that illuminate the medium. Therefore the electromagnetic response of the medium can be averaged over a volume that covers a large amount of polarizable inclusions in the medium. This homogenization, however, has to be performed in a manner that respects the field interaction between the neighboring scatterers.

This is the domain of mixing rules. Mixing rules are formulas that predict effective macroscopic parameters for heterogeneous materials. Mixing rules have been treated extensively in the classical materials modeling literature and they have also been shown, in many cases, to match experimentally with the measured macroscopic properties of mixtures. In the present chapter, the applicability of mixing and homogenization principles will be evaluated in the context of regularly-ordered and amorphous nanostructured materials.

In connection with material characterization of complex plasmonic composites, there are certain important aspects that affect the predictive power of mixing principles which are transferred from the domain of positive-permittivity and low-loss dielectric mixtures. First of all, the response of materials used in composites, like noble metals or various semiconductors, is utterly different from their behavior at other spectral ranges. The free-electron plasma model for conductors leads to negative real parts for the permittivity, and the loss factor may be surprisingly high for certain materials at particular frequencies. Such material responses create additional challenges for a stable modeling procedure.

Another issue which has a very strong effect on the macroscopic behavior of material mixtures is the regularity of the structure. If the medium has a deterministic lattice structure, its response to electromagnetic excitation is in principle exactly calculable, at least numerically to any desired accuracy. Of course also any given sample of inhomogeneous media can be exposed to a computational analysis and its macroscopic response functions can be enumerated. However, the calculated result only gives one "measurement" of the collective macroscopic property for the mixture with random internal structure. For geophysical media like snow and sand, the internal geometry cannot be described in other than probabilistic terms, and yet there are mixing rules that manage to predict the effective dielectric properties of such materials. Are these formulas generalizable to amorphous nanostructures that display plasmonic behavior?

This chapter treats mixtures and inhomogeneous structures that can be complex, not only in geometrical shape and regularity, but also in terms of the material response of the constituent components of the mixture. When inclusions with strongly dispersive dielectric response are dispersed in another dielectric medium to form an amorphous mixture, the macroscopic continuum very often displays a qualitatively different dispersive character. In this sense a nanostructured medium deserves the label "metamaterial" [1]: its material parameters are different from those of its constituent phases.

As to the notation, the convention of $\exp(\mathrm{j}\omega t)$ for the time-harmonic signal with angular frequency ω is followed. The sign of the imaginary part of the permittivity of lossy materials corresponds to power balance, and the choice leads to negative values for the imaginary part of the permittivity for dissipative media. The notation $\varepsilon_{\text{eff}} = \varepsilon'_{\text{eff}} - \mathrm{j}\varepsilon''_{\text{eff}}$ is therefore followed in this chapter. Furthermore, ε denotes *relative* permittivity and is a dimensionless (possibly complex) number.

3.2 Mixing Rules

A long history exists for mixing rules, in other words formulas with which the macroscopic effective permittivity (or another material parameter) can be computed from the knowledge of the characteristic parameters of the mixing constituents and geometrical and structural information of a given inhomogeneous sample. The oldest mixing rules for dielectric and optical characterization of materials date from the mid-19th century [2].

In the following section, various mixing principles are introduced. The permittivities of the components are taken to be constants with respect to external parameters like the frequency of the excitation. Hence the dependence of the dielectric response on frequency is not directly visible in the formulas. Later, in Sect. 3.6, the explicit effect of the frequency dependence of the constituent permittivities on the dispersion of the mixture is discussed in more detail.

3.2.1 Maxwell Garnett Formula

One of the simplest mixing rules is the following relation between the parameters of the mixture and the effective permittivity ε_{eff}:

$$\varepsilon_{\text{eff}} = \varepsilon_{\text{e}} + 3\varepsilon_{\text{e}} p \frac{\varepsilon_{\text{i}} - \varepsilon_{\text{e}}}{\varepsilon_{\text{i}} + 2\varepsilon_{\text{e}} - p(\varepsilon_{\text{i}} - \varepsilon_{\text{e}})} \tag{3.1}$$

which applies for a mixture where spherical inclusions with permittivity ε_{i} occupy a volume fraction p in a host matrix that is characterized by permittivity ε_{e}.

This relation is very often called Maxwell Garnett (MG) formula [3] in the literature and because of its simplicity combined with its sometimes astonishing accuracy, it is very widely applied to predict the macroscopic permittivity of a mixture where the inclusion size is considerably smaller than the wavelength.

Since the environment and inclusions are treated non-symmetrically in the MG mixing formula (3.1), an obvious generalization is to switch their roles as host and guest in the manner

$$\varepsilon_{\text{i}} \leftrightarrow \varepsilon_{\text{e}}, \qquad p \rightarrow 1 - p \tag{3.2}$$

This leads to the so-called complementary Maxwell Garnett formula. In this case the effective permittivity is given by the equation

$$\begin{aligned} \varepsilon_{\text{eff}} &= \varepsilon_{\text{i}} + 3\varepsilon_{\text{i}}(1-p)\frac{\varepsilon_{\text{e}} - \varepsilon_{\text{i}}}{\varepsilon_{\text{e}} + 2\varepsilon_{\text{i}} - (1-p)(\varepsilon_{\text{e}} - \varepsilon_{\text{i}})} \\ &= \varepsilon_{\text{e}} + p(\varepsilon_{\text{i}} - \varepsilon_{\text{e}})\frac{2\varepsilon_{\text{i}} + \varepsilon_{\text{e}}}{3\varepsilon_{\text{i}} - p(\varepsilon_{\text{i}} - \varepsilon_{\text{e}})} \end{aligned} \tag{3.3}$$

which predicts different values for the effective permittivity for the same mixture as (3.1).

3.2.2 *Hashin–Shtrikman Bounds*

The two mixing rules (3.1) and (3.3) are fundamental in the sense that they are very often given as upper and lower bounds for a mixture with given structural values. These bounds apply for statistically homogeneous and isotropic multiphase materials. They are often called Hashin–Shtrikman bounds [4]. A more extensive treatment of theories for bounds of effective parameters of composites is given in [5].

An essential terminological classification of heterogeneities is the distinction between so-called raisin-pudding and Swiss-cheese mixtures. In the raisin-pudding case, the inclusions generate positive dipole moments into the environment, in other words $\varepsilon_{\mathrm{i}} > \varepsilon_{\mathrm{e}}$. Correspondingly, Swiss-cheese means the complementary case such that the inclusions are like holes in a solid environment ($\varepsilon_{\mathrm{i}} < \varepsilon_{\mathrm{e}}$), causing negative dipole moments.

In connection with these two mixture types, the roles of the two Hashin–Shtrikman bounds have to be inverted. For raisin-pudding mixtures, the MG formula (3.1) is the *lower* bound for $\varepsilon_{\mathrm{eff}}$ and the complementary MG rule (3.3) is the *upper* limit. For the Swiss-cheese case, the situation is the opposite.

3.2.3 *Bruggeman Formula*

Another widely-used mixing formula is the Bruggeman mixing rule [6], especially popular in applications where materials to be modeled are random in structure, like in the theory of disordered composite materials and modeling of natural and geophysical media. The Bruggeman relation is the following, here again written for three-dimensionally spherical mixing geometry:

$$(1-p)\frac{\varepsilon_{\mathrm{e}}-\varepsilon_{\mathrm{eff}}}{\varepsilon_{\mathrm{e}}+2\varepsilon_{\mathrm{eff}}}+p\frac{\varepsilon_{\mathrm{i}}-\varepsilon_{\mathrm{eff}}}{\varepsilon_{\mathrm{i}}+2\varepsilon_{\mathrm{eff}}}=0 \tag{3.4}$$

In this implicit form for $\varepsilon_{\mathrm{eff}}$, the Bruggeman relation emphasizes the symmetry between the two constituents of the mixture, ε_{i} and ε_{e}. The normalized polarizations against the effective medium are balanced by weighing with the volume fractions of the component phases, which is an apparent interpretation of the formula (3.4). Hence the Bruggeman mixing formula is self-complementary: the operation $\varepsilon_{\mathrm{i}} \leftrightarrow \varepsilon_{\mathrm{e}}, p \to 1-p$ on (3.4) retains the same mixing relation. Due to this equality between the two phases, either one can be seen to be a guest phase embedded into the other's host environment, and subsequently a generalization of the Bruggeman mixing rule into multiphase mixtures is straightforward.

3.2.4 Lord Rayleigh Formula and Extensions

It is a well-accepted fact that Maxwell Garnett formula (3.1) seems to work reasonably well for isotropic mixtures with regularly latticed inclusion spheres that do not occupy a very large volume of the total space. Furthermore, if the dielectric contrast between the components forming the mixture is moderate, the accuracy is even better.

For non-dilute mixtures where the volume fraction of inclusions increases, more accurate estimates than the Maxwell Garnett prediction are needed. For the regular (simple cubic) lattice, a series of improved expansions for $\varepsilon_{\mathrm{eff}}$ have been presented in the literature. Of these, the Lord Rayleigh mixing formula [7] dates already from the 19th century:

$$\varepsilon_{\mathrm{eff}} = \varepsilon_{\mathrm{e}} + p\frac{3\varepsilon_{\mathrm{e}}}{\frac{\varepsilon_{\mathrm{i}}+2\varepsilon_{\mathrm{e}}}{\varepsilon_{\mathrm{i}}-\varepsilon_{\mathrm{e}}} - p - 1.305\frac{\varepsilon_{\mathrm{i}}-\varepsilon_{\mathrm{e}}}{\varepsilon_{\mathrm{i}}+4\varepsilon_{\mathrm{e}}/3}p^{10/3}} \tag{3.5}$$

This equation extends the predictive power of MG rule into higher volume fractions in p. However, due to the fixed positions of the uniform spheres there is a maximum packing density which gives the upper limit $\pi/6 \approx 52.4$ % for the volume fraction of the inclusion spheres.

An even more advanced generalization of the mixing principles for simple-cubic lattice of spheres has been presented by McPhedran and McKenzie [8]. The result is written for the effective conductivity of the mixture but due to the duality of the conductivity and permittivity problems, the result is directly applicable for the dielectric mixture that has the same geometry.

3.2.5 Examples

To illustrate how the predictions of the different mixing rules differ from each other, Fig. 3.1 shows the effective permittivity of a mixture with a dielectric parameters $\varepsilon_{\mathrm{i}} = 20$ and $\varepsilon_{\mathrm{e}} = 1$. The dielectric contrast is high enough to display the particular features of the four mixing models.

Maxwell Garnett mixing rule gives the lowest prediction for $\varepsilon_{\mathrm{eff}}$. Lord Rayleigh formula follows it to surprisingly high volume fractions but at around $p = 0.4$, the predictions start to deviate strongly. It is worth noting that the maximum volume fraction for a three-dimensional simple cubic lattice (52.4 %) is deliberately overtaken in the figure to show a warning example what may happen if the assumptions of a given mixing formula are violated. The Lord Rayleigh curve not only disregards the generous Hashin–Shtrikman bounds but also leaves them behind with increasing slope.

Also the Bruggeman prediction approaches asymptotically Maxwell Garnett for low volume fractions but the difference is conspicuous for large p values. However,

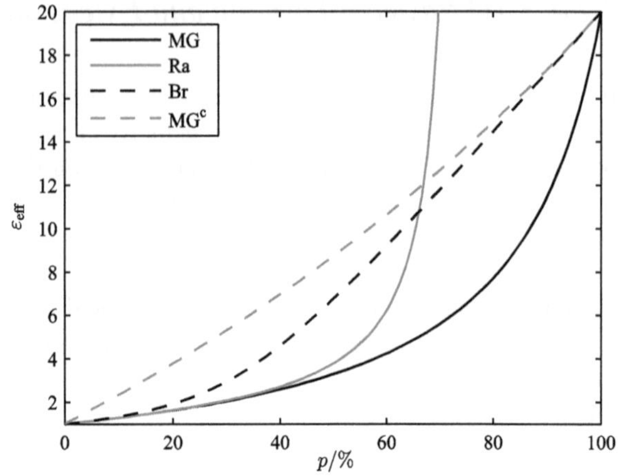

Fig. 3.1 The effective permittivity for an ordinary positive-permittivity mixture with $\varepsilon_{\mathrm{i}} = 20$ and $\varepsilon_{\mathrm{e}} = 1$, varying volume fraction p, and spherical inclusions, according to various mixing models: Maxwell Garnett, Rayleigh, Bruggeman, and complementary Maxwell Garnett. Notice that the Rayleigh formula is also plotted using larger values for p than applicable (52.4 %)

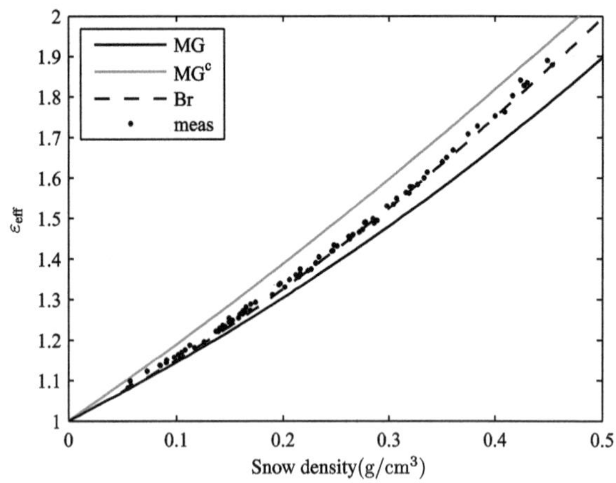

Fig. 3.2 The effective permittivity of dry snow as function of its density. Measurements [10] fall between the Maxwell Garnett and complementary Maxwell Garnett bounds, and match quite closely the Bruggeman estimate

both Bruggeman and Lord Rayleigh (below the touching-sphere limit) fall everywhere between the MG and complementary MG results, which are the Hashin–Shtrikman bounds for this raisin-pudding mixture.

Figure 3.2 shows the application of the Hashin–Shtrikman bounds to the permittivity of a real-world material, dry snow. Snow is treated as a two-component mixture of air and ice. The volume fraction of ice is linearly proportional to snow density ϱ_{snow} as $p = \varrho_{\mathrm{snow}}/\varrho_{\mathrm{ice}}$ with the density of ice being $\varrho_{\mathrm{ice}} = 0.917\,\mathrm{g/cm^3}$. The dielectric response of ice is astonishingly low-loss and dispersionless over a very broad frequency band from the megahertz range up to tens of gigahertz [9], and in the results of Fig. 3.2, the value $\varepsilon_{\mathrm{i}} = 3.17$, typical in the microwave range, is used.

The set of measurements at around 1 GHz by Christian Mätzler (University of Bern, Switzerland) are also plotted [10]. Even if the microstructure of dry snow is irregular and certainly not consisting of spherical inclusions, the experimental

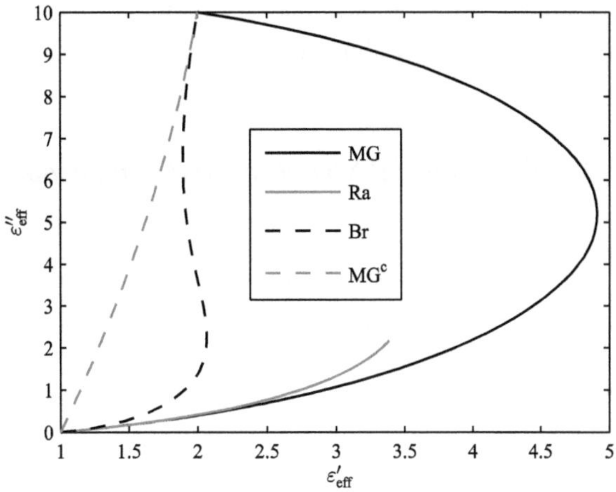

Fig. 3.3 Effective permittivity in the complex plane using different mixing formulas for the case of highly lossy inclusions: $\varepsilon_{\mathrm{i}} = 2 - \mathrm{j}\,10$ and $\varepsilon_{\mathrm{e}} = 1$. The volume fraction is $0 < p < 1$, except for the Rayleigh formula where $0 < p < \pi/6$

values can be seen to fall close to the average curve of the two bounds (3.1) and (3.3). The Bruggeman prediction seems to be very accurate for the permittivity of dry snow.

Although the example with snow may give an impression that the statistical variation of the permittivity of random samples of mixtures is quite tightly controlled by the bounds, the situation looks very different for heterogeneous media with negative-permittivity components. In fact, also the case of highly lossy dielectric mixtures displays very peculiar behavior as far as these limits are concerned.

For the lossless mixtures in Figs. 3.1 and 3.2, the upper bound at a given volume fraction $0 < p < 1$ is clearly smaller than the inclusion permittivity ε_{i} and the lower bound larger than the permittivity of the environment ε_{e} (in the Swiss-cheese case, the bounds would of course be inverted). It would be counter-intuitive to expect a mixture display larger (or smaller) permittivity than either of its components.

However, this is in a certain sense what happens for mixtures where one of the components is sufficiently lossy. Similarly to the convention of including dielectric losses into the imaginary part of bulk permittivities, also the homogenized, effective permittivities can be treated as complex-valued numbers computable from the same mixing rules that were written for real-valued material parameteres. Figure 3.3 shows the prediction of the various mixing rules for the case $\varepsilon_{\mathrm{i}} = 2 - \mathrm{j}\,10$, $\varepsilon_{\mathrm{e}} = 1$. The curves show the path of $\varepsilon_{\mathrm{eff}}$ in the complex plane as p increases from 0 to 1. There we can observe not only the astonishingly different behaviors of the four mixing principles but also the fact that the real part of the effective permittivity can exceed by a large margin the real part of the inclusion permittivity ($\varepsilon'_{\mathrm{eff,max}} \approx 4.91$ and $\varepsilon'_{\mathrm{i}} = 2$). This prediction is from the MG formula; however, also the Bruggeman result breaks the upper limit of $\max\{\varepsilon'_{\mathrm{i}}, \varepsilon'_{\mathrm{e}}\}$ with $\varepsilon'_{\mathrm{eff,max}} \approx 2.07$.

Nevertheless, the characteristics of MG and complementary MG formulas as boundaries for complex $\varepsilon_{\mathrm{eff}}$ remain: indeed, these Hashin–Shtrikman bounds separate a region in the complex plane in which the effective permittivity has to be located of various mixtures determined by given complex values ε_{e} and ε_{i} and the volume fraction p. As the figure shows, both Bruggeman and Lord Rayleigh curves restrict themselves within this region.

3.2.6 Geometrical Structure and Anisotropy

The previous mixing relations were based on the assumption that the inclusions are spherical. The sphere is an important canonical shape which is also a natural form in small scales, for example in connection with liquid particles for which the surface area is being minimized by tensional forces. However, often one also encounters natural and artificial materials for which the internal geometry is more complicated.

An accurate solution for the effective permittivity of a mixture with complicated microstructure inevitably requires a computational approach. However, to some extent one can retain the simplicity of closed-form mixing rules and still relax the assumption of spherical inclusions. For example, many of the mixing formulas can be written also for cases where the inclusions are ellipsoidal in shape [11]. To describe the shape of an ellipsoid, two additional geometrical parameters are required. This means that the dielectric response needs to be described with additional parameters, so-called depolarization factors. These depolarization factors appear consequently in the mixing formulas. Obvious special cases of ellipsoids are a needle and a disk.

The question of disorder in mixtures becomes more intricate in connection with non-symmetrical inclusions, like ellipsoids. In totally random mixtures, the inclusions' locations are not fixed to a lattice position and also their orientations do not have any preferred alignment. This leads to isotropy and the dielectric response of the mixture does not depend on the direction of the electric field vector. However, if the ellipsoids are all aligned, the macroscopic response depends on the direction of the excitation.

This type of homogenized anisotropy where the inclusion may be randomly positioned but they are orientationally ordered can be termed geometry-induced anisotropy. Of course, there is another potential reason, too, for anisotropy in mixtures: anisotropy can be intrinsically contained in the phases.

Some generalizations exist of basic mixing principles, like MG and Bruggeman, for anisotropic phases. Often the appearance of these anisotropic mixing rules is very similar to their isotropic counterparts, only the material parameters have to be replaced by dyadic quantities [11].

Into an even more general material response, electromagnetics literature contains several models and rules to homogenize mixtures that contain chiral [12], nonreciprocal [13], other magnetoelectric, and bianisotropic components phases [14].

3.2.7 Two-Dimensional (2D) Mixtures

A special case involving ellipsoidal inclusions is the case of two-dimensional mixtures. Assume that the inclusions are needle-shaped in such a limit that the circular cross-section becomes negligible in comparison with the length of the needle. If such circular cylinders are all aligned, embedded in a dielectric environment, the mixture is certainly anisotropic. However, in the transversal plane perpendicular to the axes, the medium can be considered a two-dimensional mixture, and in this plane, the homogenized mixture is again isotropic.

The previous three-dimensional mixing rules can be rewritten for this two-dimensional case, the differences mostly being that the depolarization factor of a sphere (1/3) is replaced by that of a circle (1/2).

For two-dimensional mixtures, the Maxwell Garnett formula reads

$$\varepsilon_{\mathrm{eff}} = \varepsilon_{\mathrm{e}} + 2\varepsilon_{\mathrm{e}} p \frac{\varepsilon_{\mathrm{i}} - \varepsilon_{\mathrm{e}}}{\varepsilon_{\mathrm{i}} + \varepsilon_{\mathrm{e}} - p(\varepsilon_{\mathrm{i}} - \varepsilon_{\mathrm{e}})} \tag{3.6}$$

and the complementary Maxwell Garnett rule is the following

$$\begin{aligned} \varepsilon_{\mathrm{eff}} &= \varepsilon_{\mathrm{i}} + 2\varepsilon_{\mathrm{i}}(1-p) \frac{\varepsilon_{\mathrm{e}} - \varepsilon_{\mathrm{i}}}{\varepsilon_{\mathrm{e}} + \varepsilon_{\mathrm{i}} - (1-p)(\varepsilon_{\mathrm{e}} - \varepsilon_{\mathrm{i}})} \\ &= \varepsilon_{\mathrm{e}} + p \frac{\varepsilon_{\mathrm{i}}^2 - \varepsilon_{\mathrm{e}}^2}{2\varepsilon_{\mathrm{i}} - p(\varepsilon_{\mathrm{i}} - \varepsilon_{\mathrm{e}})} \end{aligned} \tag{3.7}$$

These relations are the Hashin–Shtrikman bounds in two dimensions.

The Bruggeman rule in this case looks like

$$(1-p)\frac{\varepsilon_{\mathrm{e}} - \varepsilon_{\mathrm{eff}}}{\varepsilon_{\mathrm{e}} + \varepsilon_{\mathrm{eff}}} + p \frac{\varepsilon_{\mathrm{i}} - \varepsilon_{\mathrm{eff}}}{\varepsilon_{\mathrm{i}} + \varepsilon_{\mathrm{eff}}} = 0 \tag{3.8}$$

and the 2D Lord Rayleigh formula is the following:

$$\varepsilon_{\mathrm{eff}} = \varepsilon_{\mathrm{e}} + \frac{2p\varepsilon_{\mathrm{e}}}{\frac{\varepsilon_{\mathrm{i}}+\varepsilon_{\mathrm{e}}}{\varepsilon_{\mathrm{i}}-\varepsilon_{\mathrm{e}}} - p - \frac{\varepsilon_{\mathrm{i}}-\varepsilon_{\mathrm{e}}}{\varepsilon_{\mathrm{i}}+\varepsilon_{\mathrm{e}}}(0.3058p^4 + 0.0134p^8)} \tag{3.9}$$

For the Lord Rayleigh case which focuses on the regular square lattice, the maximum filling ratio before the circles overlap is $p_{\max} = \pi/4 \approx 78.5\,\%$. This is apparently higher than the corresponding upper limit in the 3D case (52.4 %).

Furthermore, to improve the accuracy of the 2D Lord Rayleigh formula, an exact solution for the square lattice case in a series expansion form has been presented in [15].

3.3 Limitations and Applicability

It is evident that mixing rules with such simplicity in appearance as presented above suffer from limitations which need to be kept in mind when they are applied to different homogenization problems. In particular in connection with metamaterials, this is utterly important.

For positive-permittivity mixtures, the various mixing rules were shown above to agree reasonably well for certain ranges of structural and material parameters. On the other hand, for high-loss mixtures the predictions could go greatly apart. However, it is in connection with homogenization problems of mixtures where the components have permittivities of opposite signs where various mixing models result in really qualitative differences.

A look at the MG formulas in their three- (3.1) or two-dimensional (3.6) forms reveals that if ε_i and ε_e are of different signs, the denominator may vanish. This means that the effective permittivity grows without limit. The exact relations when such singularities happen are

$$\varepsilon_\mathrm{i}/\varepsilon_\mathrm{e} = -\frac{2+p}{1-p} \quad \text{(3D)} \tag{3.10}$$

$$\varepsilon_\mathrm{i}/\varepsilon_\mathrm{e} = -\frac{1+p}{1-p} \quad \text{(2D)} \tag{3.11}$$

In addition, the complementary MG formulas (3.3) and (3.7) can grow towards infinity. This takes place at a different condition:

$$\varepsilon_\mathrm{i}/\varepsilon_\mathrm{e} = -\frac{p}{3-p} \quad \text{(3D)} \tag{3.12}$$

$$\varepsilon_\mathrm{i}/\varepsilon_\mathrm{e} = -\frac{p}{2-p} \quad \text{(2D)} \tag{3.13}$$

As to the Lord Rayleigh formulas (3.5) and (3.9), there are two singularities for the effective permittivity that correspond to certain (negative) values of the ratio $\varepsilon_\mathrm{i}/\varepsilon_\mathrm{e}$.

However, the situation for the Bruggeman formulas is radically different. To compute ε_eff from (3.4) and (3.8) leads to solving a second-order equation which involves a square root. The sign of the root is obvious in the case of positive values for the real parts of ε_i and ε_e. However, when they are of opposite signs, the branch of the square root needs to be chosen such that the resulting ε_eff has non-positive imaginary part ($\varepsilon''_\mathrm{eff} \geq 0$) in order to avoid active character for the homogenized mixture.

Indeed this is a very peculiar prediction by the Bruggeman formula. For certain combinations of negative and positive values for ε_i and ε_e, the solution for ε_eff becomes a complex number, even if the two component permittivities are purely real. In other words, according to Bruggeman model, lossless ingredients can make a lossy end product.

To illustrate these two qualitatively new phenomena that mixing of complex inclusions brings about (either infinities in the Maxwell Garnett and Lord Rayleigh case, or complex-valued permittivities in the Bruggeman case), Figs. 3.4 and 3.5 display the required conditions as functions of the parameters of the mixture. The singularity locations (3.10) and (3.12) are shown as curves in the plots, whereas the shaded region corresponds to mixture parameters for which the Bruggeman rule predicts complex-valued ε_eff.

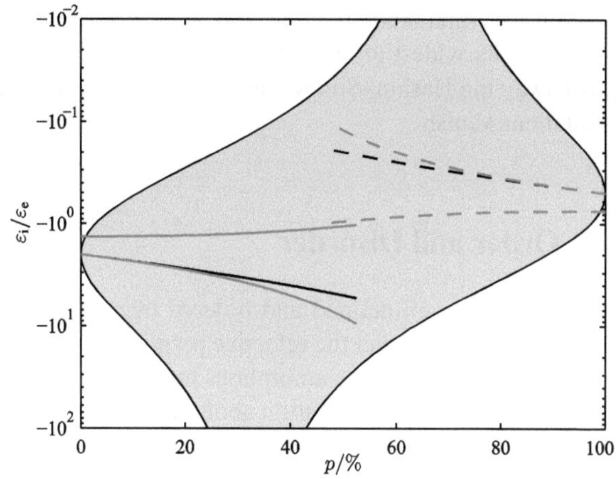

Fig. 3.4 Singularity locations for the 3D Maxwell Garnett formula (*black line*) and 3D Rayleigh formula (*gray line*), and their complementary versions (*dashed lines*). Inside the shaded region, the Bruggeman formula gives complex $\varepsilon_{\mathrm{eff}}$, although both ε_{i} and ε_{e} are assumed real

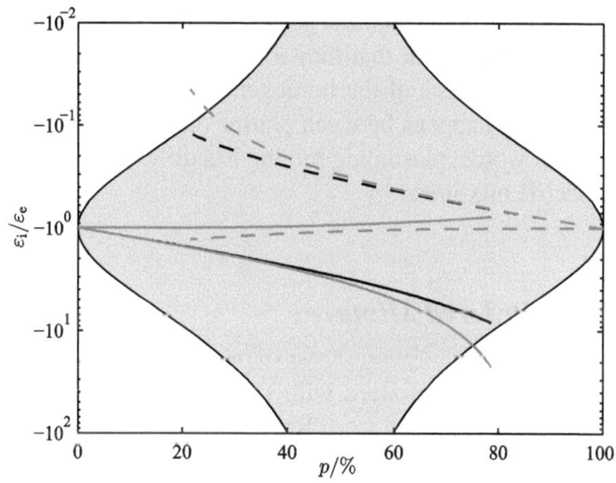

Fig. 3.5 Similarly as Fig. 3.4 using the 2D versions of the mixing formulas. Note the symmetric appearance of the shape of the curves compared to the 3D case, Fig. 3.4

The two figures show the situation in three- and two-dimensional mixing cases. Even though the general appearance is similar, it is noteworthy that the 2D figure is clearly more symmetric than the 3D figure, especially in this case when the vertical axis for the permittivity ratio is logarithmic. This is consistent with a result of a particular duality principle that is only valid for two-dimensional mixtures [16]. When one of the component permittivities is changed to its reciprocal value, the same thing happens to the effective permittivity. More exactly, the following is easily seen to hold:

$$\varepsilon_{\mathrm{eff}}(\varepsilon_{\mathrm{i}}, \varepsilon_{\mathrm{e}}, p)\, \varepsilon_{\mathrm{eff}}\big(\varepsilon_{\mathrm{e}}^2/\varepsilon_{\mathrm{i}}, \varepsilon_{\mathrm{e}}, p\big) = \varepsilon_{\mathrm{e}}^2 \tag{3.14}$$

when a two-dimensional mixing rule (3.6) or (3.7) is applied for the function $\varepsilon_{\mathrm{eff}}$.

The singularities of the Maxwell Garnett and complementary Maxwell Garnett predictions can also be interpreted in light of their role as Hashin–Shtrikman bounds. If an upper bound becomes infinite, its restrictive power vanishes. In this sense it is symptomatic that the gray area showing the problematic Bruggeman region is at its widest at around the same parameter region where the singularities push away the Hashin–Shtrikman bounds. When there are no finite boundaries, the limitations vanish.

3.4 Order and Disorder

Mixtures with well-defined and ordered microstructure can be modeled—at least numerically—to predict the effective permittivity up to a very good accuracy. However, the description of an amorphous heterogeneity requires many more degrees of freedom. Therefore the question about the exact value for the effective permittivity of a sample with given component permittivities and volume fractions cannot be answered exactly if the distribution of the inclusions in the mixture is random. The predictions for the effective permittivity are very much dependent on the magnitude of the randomness, in other words whether the inclusions are allowed to change positions to the extent that they touch their neighbors and form clusters. In addition, the characteristics of the homogenization are very strongly dependent on whether the mixing happens between phases of permittivities of the same or different sign. In other words, plasmonic mixing has different characteristics compared with pure-dielectric mixtures.

3.4.1 Ordered Mixtures

For the ordered mixture with simple cubic lattice structure in three dimensions, Maxwell Garnett and Lord Rayleigh formulas and their generalizations provide an increasingly accurate estimate for the effective permittivity. The situation is similar for two-dimensional mixtures where the square-lattice mixture is computable either approximately from (3.6), (3.9), or to any desired accuracy from the series solution [15].

Figure 3.1 showed the monotonously increasing curves of $\varepsilon_{\mathrm{eff}}$ of a positive-permittivity mixture as functions of the volume fraction p. For sufficiently large values of p, their predictions start to differ from each other. However, more marked differences between these models become visible when plasmonic mixtures are analyzed.

Figure 3.6 displays the effective permittivity of a regular 2D mixture according to various models. The volume fraction is kept constant ($p = 60$ %) but the inclusion permittivity ε_{i} varies. As discussed above, there are singularities for certain values for negative ε_{i} (here the environment is assumed free space, $\varepsilon_{\mathrm{e}} = 1$). The curves show one resonance for MG and two for Lord Rayleigh, quantitatively in agreement with the results in Fig. 3.5.

For this rather high volume fraction of 60 %, the prediction of the Maxwell Garnett formula is clearly inaccurate. Lord Rayleigh gives a much better estimate for

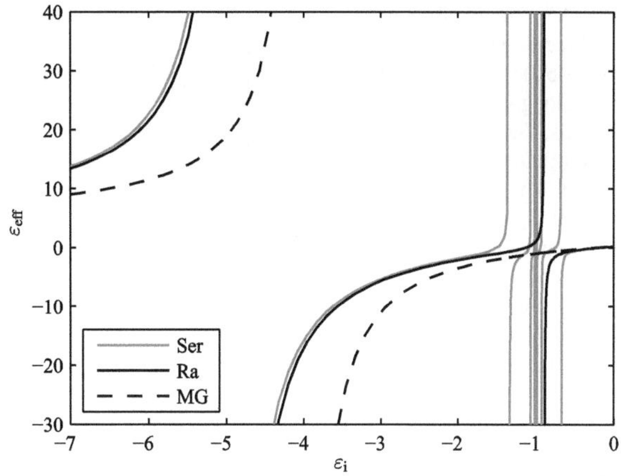

Fig. 3.6 Effective permittivity for the 2D case: negative-permittivity circular cylinders in free-space environment $\varepsilon_e = 1$. Varying ε_i, fixed $p = 60\ \%$. The three curves refer to the series solution (Ser), Rayleigh formula (Ra), and Maxwell Garnett formula (MG)

this resonance around $\varepsilon_i \approx -5$, although the slight deviation from the exact series solution is distinguishable.

The neighboring inclusions are rather close to each other in this structure (the minimum distance between neighboring boundaries is only approximately one-eighth of the diameter of the inclusion). This fact gives rise to plasmonic interaction effects which are the reason for the very rich singularity region in the series solution around the second resonance at $\varepsilon_i \approx -1$.

The success of Bruggeman mixing rule to explain the plasmonic behavior of this regular mixture is poor, due to its prediction of complex-valued effective permittivity (see Fig. 3.5), and hence it is not shown in Fig. 3.6.

3.4.2 Effect of Disorder

The calculation of the effective permittivity of a non-regular mixture sample requires numerical efforts. As an example of such studies, Figs. 3.7–3.9 show results of a computational effort on three-dimensional mixtures with quasi-random structures [17].

In [17], the samples were generated by randomly positioning spherical inclusions into a homogeneous environment up to a given volume fraction p. Using a finite difference method, the electrical field distribution of each sample was solved. From the field solution, the effective permittivity was determined. The contrast between the inclusion and host phases was rather high, $\varepsilon_i = 51\varepsilon_e$.

Two different type of mixtures were created. In the first case, the inclusions were not allowed to touch each other. All spheres were separate. In the other case, overlapping was permitted, in other words when a position for a new inclusion was randomly generated and it happened to be located close enough with an already existing sphere, the result was a cluster of two particles. Eventually, also more complicate clusters could arise in the high volume fraction end.

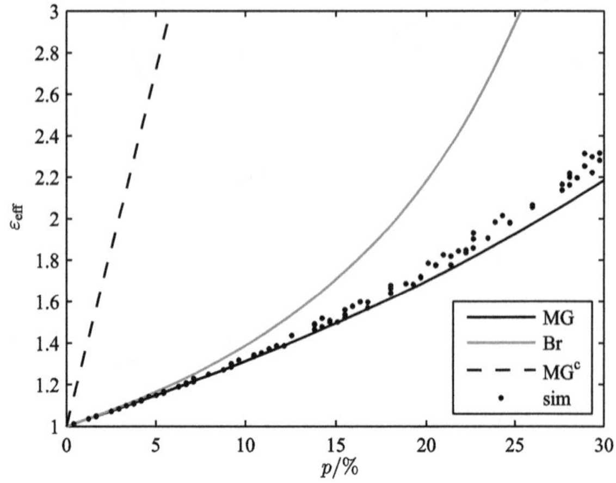

Fig. 3.7 Simulation results of three-dimensional random raisin-pudding mixtures from [17] along with basic mixing models. The relative permittivities are $\varepsilon_i = 51$ and $\varepsilon_e = 1$. The inclusions are all separate spheres which are not allowed to overlap with each other. This limits the achievable volume fractions for the inclusion phase well below the regular packing limit

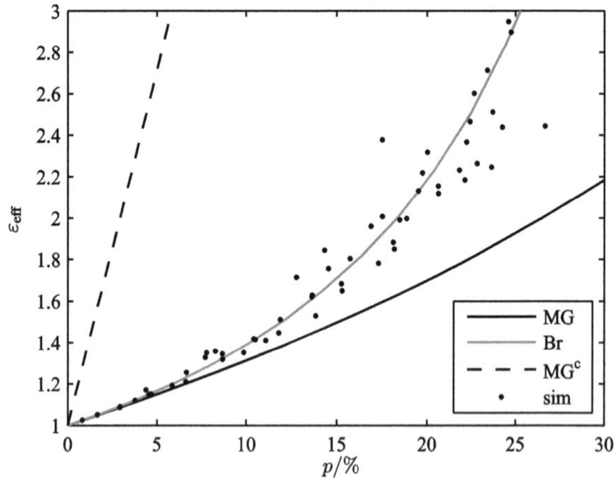

Fig. 3.8 Similarly to Fig. 3.7, but in the simulations clustering is allowed. In other words, the inclusions may touch each other

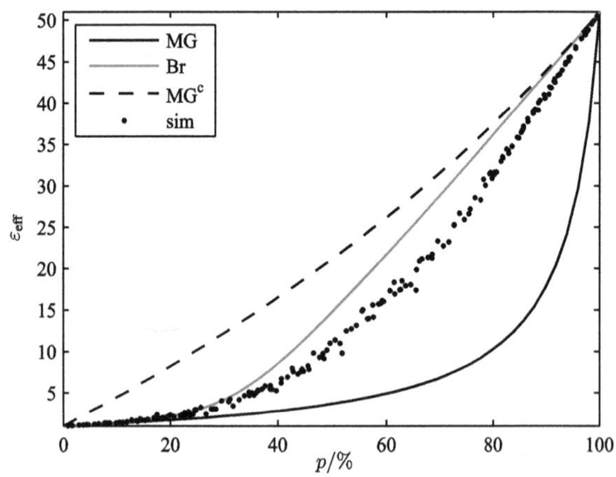

Fig. 3.9 Similarly to Fig. 3.8, but for the whole volume fraction range $0 < p < 1$. Inclusion clustering is allowed, and is in fact compulsory for high volume fractions. The simulations tend to stay between the Maxwell Garnett and Bruggeman results, see [17] for a more detailed empirical model to match the computational data

The illustrations show that the two types of disordered mixtures have clearly different macroscopic permittivity characteristics. The case when the randomly placed inclusions are separate (Fig. 3.7) follows approximately the Maxwell Garnett curve. In fact, the results in the figure are slightly higher than the MG prediction. This is to be expected because of the minimum-bound nature of MG formula for this raisin-pudding mixture ($\varepsilon_\mathrm{i} > \varepsilon_\mathrm{e}$).

However, when the inclusion phase can form clusters in the mixture (Fig. 3.8), the effective permittivity in the average is clearly higher than in the first scenario. This can be explained with the observation that since some of the inclusions are no longer spherical but rather elongated doublets or even more complicated clusters, and on the other hand non-spherical particles have always a larger polarizability in average than spheres [18], also the macroscopic permittivity will be increased. A coarse conclusion of the second scenario is that the computed results follow approximately the Bruggeman curve.

Note also that the complementary MG formula cannot explain at all the simulations. This is understandable because the simulation samples are of raisin-pudding type ($\varepsilon_\mathrm{i} > \varepsilon_\mathrm{e}$) and the complementary MG model applies to the Swiss-cheese structure.

Figure 3.9 displays the simulations and ε_eff computations over the whole range of volume fractions $0 < p < 1$. Clustering of the inclusions is allowed. (For random mixtures and equisized spheres, it is very difficult if not impossible to arrive at volume fractions over around 30 % without clustering.) Interestingly, even if the simulations seem to stay close to the Bruggeman estimate for sparse mixtures, the results tend to fall between the MG and Bruggeman predictions in the dense end of structures.

3.4.3 Disorder and Plasmonics

It is to be expected that the combination of disorder and plasmonics causes special challenges, especially concerning the issue of the effect of non-regularity on the resonance structure. Again, only a computational approach can provide reliable results for ε_eff in the case of negative-permittivity inclusions.

Figure 3.10 shows the computed macroscopic permittivity of a slightly disordered two-dimensional mixture where the inclusion disks have negative permittivity and the host environment is free space. The volume fraction is $p = 0.15$.

Using the finite element method and the COMSOL Multiphysics software, the quasistatic effective permittivity is determined as function of the negative inclusion phase permittivity ε_i. Note that due to the lacking 90° rotation symmetry, two perpendicular electric field excitations (x and y) lead to different estimates for the effective permittivity. For more details of the computational procedure, see [19].

The volume fraction 15 % is not very high, and hence the MG and Lord Rayleigh predictions predict quite similar results, although of course the second resonance at $\varepsilon_\mathrm{i} \approx -1$ is missing from the MG curve. As to the agreement with simulations, the

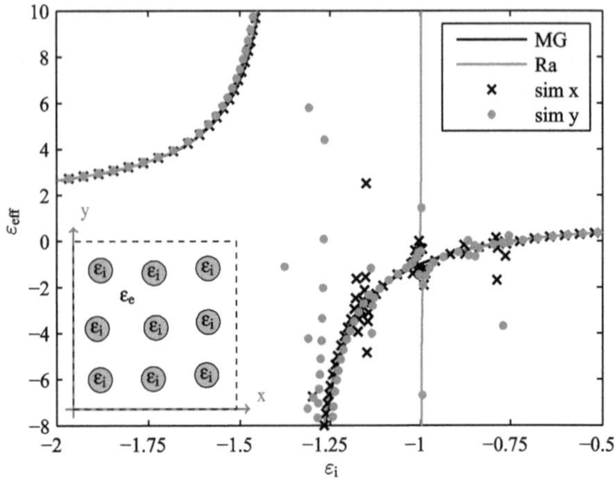

Fig. 3.10 Simulations and mixing models for a slightly disordered two-dimensional structure with plasmonic disks in free space environment ($\varepsilon_e = 1$). The volume fraction is $p = 15\ \%$. Note that due to non-regularity of the simulated structure (shown as *inset*), the numerically computed effective permittivity is not the same for x- and y-polarized field excitations

illustrations for ε_{eff} show that outside the plasmonic resonances, there is a rather good agreement with the mixing rule and the numerical computations. However, around the resonances and between them, the differences can be very large.

It is especially noteworthy that although the anisotropy of the structure is geometrically very modest (the positions of the inclusions are also shown in the figure), the two permittivity components may be strongly different within the resonance region. For corresponding results from another disordered sample where the volume fraction of the plasmonic inclusions is higher, see [19].

3.5 Inclusion Shape Effects

The previous examples were dealing with spherical inclusions (spheres for the 3D case, and disks in 2D). The singularity behavior for such mixtures is dimension-dependent, as is evident from (3.10) and (3.11) which give different resonance conditions depending on the dimension of the space. When the mixture becomes more dense, more and stronger resonances appear.

When the mixture consists of non-spherical inclusions, the plasmonic response becomes naturally more complex. For example, for ellipsoidal inclusion, the dipole moment response of an individual particle is determined by its depolarization factors as mentioned in Sect. 3.2.6. The range of singularities will be affected by these parameters.

It is interesting that there exists an exact solution for a very extreme two-dimensional system: a mixture where the inclusions are square-shaped and occupy a volume fraction $p = 1/4$ of the area in a regular square lattice [20, 21]. For this case, the effective permittivity is

$$\varepsilon_{eff} = \sqrt{\frac{\varepsilon_e + 3\varepsilon_i}{\varepsilon_i + 3\varepsilon_e}} \tag{3.15}$$

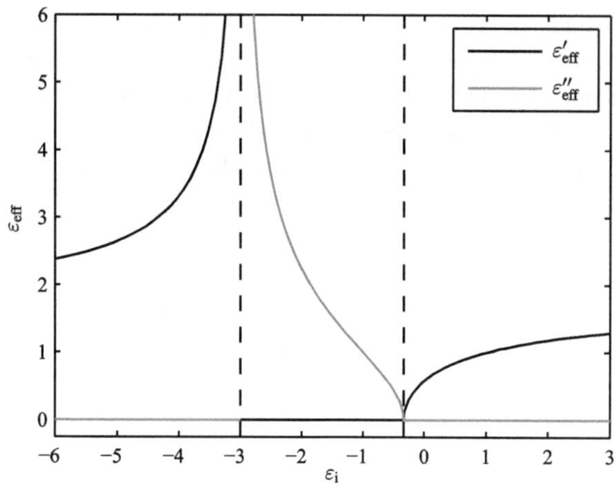

Fig. 3.11 The effective permittivity of a regular two-dimensional lattice of squares according to the result (3.15). The volume fraction of inclusions is $p = 25\ \%$ and the background permittivity $\varepsilon_e = 1$. Note the purely imaginary prediction for ε_{eff} in the range $-3 < \varepsilon_i < -1/3$

This result is displayed in Fig. 3.11.

The effective permittivity (3.15) shares the special property with Bruggeman mixing rule that within a certain range for negative values of $\varepsilon_i/\varepsilon_e$, it predicts complex values for ε_{eff}. In particular, ε_{eff} is purely imaginary for the values $-3 < \varepsilon_i/\varepsilon_e < -1/3$. Numerical computations to determine ε_{eff} for this square structure both with finite element method and in [21] the mode-matching method reconstruct very accurately the analytic result over the ranges when ε_{eff} is real. However, in the region $-3 < \varepsilon_i/\varepsilon_e < -1/3$, the numerical methods result in a wildly resonating real part for ε_{eff} instead of the smooth branch behavior of (3.15). The character of this mixture is truly peculiar within this particular region: depending on the model, either having purely imaginary permittivity, or displaying a non-convergent, strongly singular response. Both are solutions that are intuitively difficult to accept.

3.6 Dispersion

In connection with broadband applications of metamaterials, dispersion is essential. Dispersion is related to the frequency dependence of the response of the system.

The dielectric response of bulk materials often follows certain basic dispersion models, at least within a limited frequency range. Among these models the most important are Drude model (often associated with noble metals), Lorentz model (in solid state media), and Debye model (for liquids with polar molecules), each with their own characteristic frequency-dependent permittivity function.

3.6.1 Effects of Mixing on Dispersion

A look at any of the mixing rules described above is enough to convince that the frequency dependence of the macroscopic permittivity may differ strongly from the dispersive properties of the components. Mixing rules are always non-linear functions of the permittivities of the phases. It is therefore a very interesting fact that sometimes the mixture permittivity may obey a frequency function according to one of the basic dispersion models above.

For example, according to the Maxwell Garnett mixing model, a mixture where the inclusions follow the Debye frequency model and the environment is dispersionless turns out to be also a Debye-type mixture [11]. In other words, a cloud or rain (water droplets in air) follows the same frequency model as water. Of course, the model parameters are functions of the component permittivity parameters and volume fraction. One interesting feature of the Debye mixture is that the frequency range where the dispersion happens becomes pushed to over ten times higher frequencies!

Another interesting case is when mixing brings forth a dispersion model that is not present in the components. This happens—again according to the Maxwell Garnett model—when in a dispersionless environment (ε_e is constant with frequency, here assumed free space) the inclusion phase follows the Drude model with the angular frequency ω:

$$\varepsilon_i(\omega) = 1 - \frac{\omega_p^2}{\omega^2 - j\omega\nu} \tag{3.16}$$

Here ω_p is the plasma frequency, ν is the damping frequency, a measure of losses. Above the dispersive frequency band, $\varepsilon_i(\omega)$ approaches the free-space value 1.

Combination of (3.16) with (3.1) shows that the effective permittivity follows the Lorentz resonance model

$$\varepsilon_{\text{eff}}(\omega) = 1 + \frac{\omega_{\text{p,eff}}^2}{\omega_{0,\text{eff}}^2 - \omega^2 + j\omega\nu_{\text{eff}}} \tag{3.17}$$

Here, the plasma frequency of the mixture is proportional to the plasma frequency of the Drude inclusion phase: $\omega_{\text{p,eff}} = \sqrt{p}\,\omega_p$. The effective damping frequency remains the same as that in the inclusions: $\nu_{\text{eff}} = \nu$. In addition, the resonance frequency in this Lorentz model $\omega_{0,\text{eff}}$ depends on the mixture parameters in the following way:

$$\frac{\omega_{0,\text{eff}}}{\omega_p} = \sqrt{\frac{1-p}{3}} \tag{3.18}$$

Hence the resonance due to the Drude inclusions takes place at frequencies below the plasma frequency of these inclusions, and its position can be tuned with the volume fraction p. To gain more degrees of freedom in dispersion engineering of plasmonic composites, the metallic spheres can be embedded in another host medium with permittivity ε_e. Formulas for such mixtures can be found in [11].

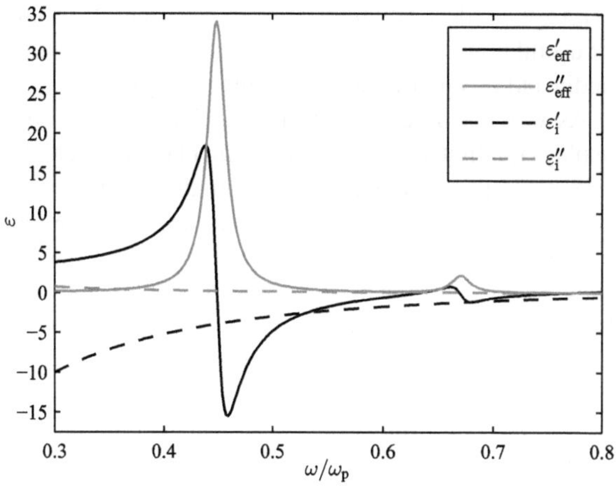

Fig. 3.12 A 3D mixture of metallic spheres in free space. The frequency dependence of the Drude spheres ε_i', ε_i'' and the Lorentzian curves for the effective permittivity $\varepsilon_{\text{eff}}'$, $\varepsilon_{\text{eff}}''$ are shown. In addition to the main Lorentzian resonance, the figure shows that Lord Rayleigh mixing rule predicts also another, smaller and more narrow-band resonance at higher frequencies

3.6.2 Example of Mixtures Containing Drude Components

As an example of the resonance-generating mixture of Drude inclusions in neutral host material, Fig. 3.12 shows the frequency dependence of the effective permittivity over the interesting frequency band. Metallic spheres occupy a volume fraction $p = 1/3$ in the host medium (3D mixture). The environment is assumed free space ($\varepsilon_e = 1$) and the inclusions follow Drude dispersion (3.16) with loss factor $\nu = \omega_p/50$.

In order to have a more realistic view of the global dispersion of the mixture, the Lord Rayleigh model is used to compute the results in Fig. 3.12. Hence, in addition to the strong resonance according to (3.18) around $\omega \approx 0.45\omega_p$, another resonance, although smaller in amplitude, appears at a somewhat higher frequency. This is due to the singularity region when the inclusion permittivity is around $\varepsilon_i \approx -1$ predicted by Lord Rayleigh formula. Figure 3.12 displays also the bulk inclusion permittivity function $\varepsilon_i(\omega)$ showing the monotonous increase from negative values of the real part of the inclusion permittivity.

The analysis above focused on the mixing case of the raisin-pudding model (it was the inclusion phase which contained the intrinsic dispersion). The complementary Swiss-cheese mixture where spherical holes are carved into bulk metal does not obey any of the simple dispersion models. However, as discussed in more detail in [19], such a mixture can offer better potential than the raisin-pudding configuration in the design of low-loss plasmonic metamaterials.

3.7 Conclusions

The calculation of the effective permittivity of heterogeneous materials is a very fundamental problem in mathematical physics in general, and in electromagnetics in particular. Electromagnetics literature contains several classical homogenization results, the oldest of which date already from the 19th century. There is still a need for useful homogenization models, and with the strong research activity on complex media and metamaterials, this demand is growing.

As was discussed in the present chapter, the effective permittivity of mixtures displays qualitatively very interesting phenonema that are not present in the component materials but rather emerge from the mixing process. This happens particularly when plasmonic mixtures are treated. And for disordered mixtures the effective permittivity response becomes furthermore complicated. Numerical computations are needed to evaluate the predictive power of classical mixing rules. The results showed that also for plasmonic and even disordered mixtures, for certain parameter ranges the easy-to-use mixing formulas were able to produce reasonable accurate estimates for the true macroscopic permittivity.

The results for the permittivity characterization of dielectric mixtures are applicable in many other fields of physics of materials, like, for example, in calculating the effective magnetic, conductivity, and elastic responses of heterogeneous media.

In summary, mixing rules definitely have a role in the analysis and design of amorphous nanomaterials. The predictions for the effective material parameters for plasmonic composites for ideal and lossless inclusions was seen to lead to singularities and other unexpected types of behavior. However, when the mixing models were taken with plasmonic material having realistic loss mechanisms (Sect. 3.6), the extremities of the variations of the effective permittivity vanished. In the modeling of amorphous nanophotonic materials with sufficiently small unit cells, mixing formulas hence make valuable tools. They help in analyzing physical phenomena that emerge from the complex microstructure, and thus provide a path towards the design of functional nanotechnological devices.

Acknowledgements The authors thank Professor Christian Mätzler for providing the dry snow data in Fig. 3.2, and Dr. Kimmo Kärkkäinen for the computational data for Figs. 3.7–3.9.

References

1. A. Sihvola, Metamaterials in electromagnetics. Metamaterials **1**(1), 2–11 (2007)
2. O.F. Mossotti, Discussione analitica sull'influenza che l'azione di un mezzo dielettrico ha sulla distribuzione dell'elettricità alla superficie di più corpi elettrici disseminati in esso. Mem. Mat. Fis. Soc. Ital. Sci. Modena **XXIV**(2), 49–74 (1850)
3. J.C. Maxwell Garnett, Colours in metal glasses and metal films. Philos. Trans. R. Soc. Lond. **203**, 385–420 (1904)
4. Z. Hashin, S. Shtrikman, A variational approach to the theory of the effective magnetic permeability of multiphase materials. J. Appl. Phys. **33**(10), 3125–3131 (1962)
5. G.W. Milton, *The Theory of Composites* (Cambridge University Press, Cambridge, 2002)
6. D.A.G. Bruggeman, Berechnung verschiedener physikalischer Konstanten von heterogenen Substanzen. I. Dielektrizitätskonstanten und Leitfähigkeiten der Mischkörper aus isotropen Substanzen. Ann. Phys. **24**(7–8), 636–679 (1935)
7. Lord Rayleigh, On the influence of obstacles arranged in rectangular order upon the properties of a medium. Philos. Mag. **34**, 481–502 (1892)
8. R.C. McPhedran, D.R. McKenzie, The conductivity of lattices of spheres. I. The simple cubic lattice. Proc. R. Soc. Lond. Ser. A, Math. Phys. Sci. **359**(1696), 45–63 (1978)

9. C. Mätzler, Dielectric properties of fresh-water ice at microwave frequencies. J. Phys. D, Appl. Phys. **20**(12), 1623–1630 (1987)
10. C. Mätzler, Microwave permittivity of dry snow. IEEE Trans. Geosci. Remote Sens. **34**(2), 573–581 (1996)
11. A. Sihvola, *Electromagnetic Mixing Formulas and Applications*. Electromagnetic Waves Series, vol. 47 (IEE Publishing, London, 1999)
12. B. Shanker, The extended Bruggeman approach for chiral-in-chiral mixtures. J. Phys. D, Appl. Phys. **29**, 281–288 (1996)
13. A.H. Sihvola, Bi-isotropic mixtures. IEEE Trans. Antennas Propag. **40**(2), 188–197 (1992)
14. T.G. Mackay, A. Lakhtakia, W.S. Weiglhofer, Strong-property-fluctuation theory for homogenization of bianisotropic composites: formulation. Phys. Rev. E **62**, 6052–6064 (2000). Erratum, Phys. Rev. E **63**, 049901(E) (2001)
15. R.C. McPhedran, D.R. McKenzie, Electrostatic and optical resonances of arrays of cylinders. Appl. Phys. A **23**(3), 223–235 (1980)
16. J.B. Keller, Conductivity of a medium containing a dense array of perfectly spheres or cylinders or nonconducting cylinders. J. Appl. Phys. **34**, 991–993 (1962)
17. K. Kärkkainen, A. Sihvola, K. Nikoskinen, Analysis of a three-dimensional dielectric mixture with finite difference method. IEEE Trans. Geosci. Remote Sens. **39**(5), 1013–1018 (2001)
18. A. Sihvola, Dielectric polarization and particle shape effects. J. Nanomater. **2007**, 45090 (2007)
19. H. Wallén, H. Kettunen, A. Sihvola, Mixing formulas and plasmonic composites, in *Metamaterials and Plasmonics: Fundamentals, Modelling, Applications, NATO Science for Peace and Security Series, B – Physics and Biophysics*, ed. by S. Zouhdi, A. Sihvola, A. Vinogradov (Springer, Dordrecht, 2009), pp. 91–102
20. Y.V. Obnosov, Periodic heterogeneous structures: New explicit solutions and effective characteristics of refraction of an imposed field. SIAM J. Appl. Math. **59**(4), 1267–1287 (1999)
21. J. Helsing, R.C. McPhedran, G.W. Milton, Spectral super-resolution in metamaterial composites. New J. Phys. **13**(11), 115005 (2011)

Chapter 4
Multipole Analysis of Self-assembled Metamaterials

Stefan Mühlig and Carsten Rockstuhl

Abstract We provide here a review on the theoretical description of the interaction of light with metamaterials fabricated by chemical self-assembling processes. Unique to the metamaterials accessible with such approaches is the amorphous arrangement of the unit-cells in space. This is in striking contrast to most of the structures previously considered, i.e. metamaterials fabricated by top-down process that usually lead to periodically arranged unit-cells. In consequence, novel concepts have to be established to describe the light interaction with their metamaterials and novel design rules have to be developed to suggest metamaterials that shall provide a desired optical response. A theoretical description based on Cartesian multipole moments is outlined in this chapter that fully satisfies these requirements. The description of the scattering response of the unit-cells is revealed as to be essential to understand amorphous metamaterials. Based on mixing rules, the propagation of light in amorphous metamaterials is properly described in terms of excited multipole moments of their unit-cells. The theoretical framework we outline here provides the methodology to discuss amorphous metamaterials and constitutes therefore an indispensable tool for the future development of optical components exploiting amorphous nanooptical materials.

4.1 Introduction

In the last decade the fusion of the fields of colloidal nanochemistry and nanooptics has been evolving to form an entirely new area of research which focuses on self-assembled nanooptical and nanophotonical structures. Chemical self-assembly processes are exploited to fabricate nanostructures required for nanooptical applications. A primary goal of these bottom-up processes is the fabrication of materials tailored at the nanoscale to affect optical properties in the visible and near infrared

S. Mühlig (✉) · C. Rockstuhl
Abbe Center of Photonics, Institute of Condensed Matter Theory and Solid State Optics, Friedrich-Schiller-Universität Jena, Max-Wien-Platz 1, 07743 Jena, Germany
e-mail: stefan.muehlig@uni-jena.de

C. Rockstuhl
e-mail: Carsten.Rockstuhl@uni-jena.de

C. Rockstuhl, T. Scharf (eds.), *Amorphous Nanophotonics*,
Nano-Optics and Nanophotonics, DOI 10.1007/978-3-642-32475-8_4,
© Springer-Verlag Berlin Heidelberg 2013

regime. These self-assembled metamaterials (MMs) are usually made of resonant unit-cells smaller than the wavelength of interest. In an obvious analogy to real materials that are made of atoms, these resonant unit-cells are termed meta-atoms. The excitation of localized surface plasmon polariton resonances (LSPRs) in plasmonic nanoparticles (NPs) is one way to promote the desired resonances. Individual but also strongly coupled ensembles of NPs are considered to cause a desired scattering response. Most notably, these meta-atoms should sustain electric as well as magnetic resonances to cause eventually materials with an electric or magnetic dispersion. Therewith, optical properties are in reach which cannot be found in nature. The final materials are then achieved by arranging the resonant meta-atoms in either a periodic or amorphous manner to form the MM.

Pioneering work by Alivisatos et al. [1] and Feldheim et al. [2] demonstrated the assembly of two plasmonic NPs, often termed a dimer, with a controlled distance to each other using DNA and molecular linkers, respectively. Such a plasmonic dimer could be seen as a possible meta-atom for a self-assembled MM. Therewith, the fabrication of MMs exploiting self-assembly processes based on colloidal nanochemistry came into reach. The field of self-assembled MMs evolved during the last ten years and today there exists a multitude of chemical processes to assemble plasmonic NPs [3–6] into well-ordered nanooptical structures. Some of them are reviewed in devoted chapters in this book, e.g. in the chapter from Bürgi and co-worker.

Concerning the optical properties of the fabricated self-assembled MMs and their theoretical description one should distinguish between two distinct groups. The first group includes self-assembled MMs offering long-range order, i.e. there exists a specific spatial arrangement between a huge number of meta-atoms. It has to be stressed that this group is not restricted to a periodic or a quasiperiodic arrangement of the meta-atoms as would be suspected from crystallography. As long as there exists any ordering between the meta-atoms the fabricated MM shall belong to this group. For example, if the meta-atoms are arranged along chains, but these chains do not offer any specific geometry, the MMs constitutes to the long-range order group. Furthermore, the optical response of these MMs should be affected by the resonances of the single meta-atom and from a collective interaction caused by the specific spatial order. The second group includes all self-assembled MMs that exhibit a specific spatial arrangement only on the short-range scale, which is essential to built up meta-atoms possessing a response deviating from that of an isolated NP. The spatial arrangement of the meta-atoms in space is amorphous. In this group the optical response of the entire MM is, as will be proven later on, primarily dominated by the resonances of the meta-atoms itself. For both introduced groups, MMs can be found that are fully three-dimensional as well as MMs offering just two-dimensional ordering existing on a substrate.

Regarding the group of long-range order self-assembled MMs, manifold chemical processes were introduced in literature to fabricate them. Block-copolymer self-assembly [7–11] is one prominent candidate. The block-copolymers consist of two or more distinct polymer blocks which are covalently bound to each other. They self-assemble into different geometries depending (in the most simple case) on one geometrical parameter like the length of one polymer chain [8]. Loading this block-copolymer host with resonant NPs forms the MM. In most of the cases the self-assembled structures are two-dimensional and therefore the fabricated MMs appear planar. Incorporating plasmonic NPs into liquid crystals presents another and effective way of achieving MMs offering long-range order [12]. The tunable anisotropy of liquid crystals can be exploited to precisely control the resonances of the incorporated NPs. Here, fully three-dimensional samples [13] as well as stacked layers of liquid crystal lamellae [14] have been demonstrated. Applying DNA to self-assemble MMs is usually related to the short-range group, since the DNA is normally used to connect two NPs with a well defined distance to each other. This fabrication technique is hard to adapt to long-range ordering. However, it has been recently demonstrated that due to hybridization interactions of DNA strands spherical NPs can be assembled into highly periodic three-dimensional lattices [15] like, e.g., face-centered cubic and body-centered cubic ones. Furthermore, by combining different manufacturing techniques like evaporation [16] or substrate patterning [17] with DNA controlled assembly the formation of two-dimensional long-range order MMs has been demonstrated. Apart from this, the use of polymer-ligands in combination with a drying-mediated entropy-driven process on an interface can be exploited to assemble spherical NPs [18] or nanorods [19] into two-dimensional long-range order structures. The process of evaporating a solution including the NPs can be seen as another commonly used technique to self-assemble long-range order MMs. Ultra large-scale arrays could be fabricated by the evaporation of NPs in a binary solvent mixture onto a substrate [20]. An additional pattering of the substrates enables a more sophisticated geometry of the entire structure [21]. Two-dimensional arrays of cylinders [22] or stripes [26] made of densely packed metallic NPs offering a magnetic resonance in the visible or even more complex structures like arrays of rings of NPs [23] were fabricated. Another way to self assemble three-dimensional MMs with long-range order is induced by the geometry of the single NPs. Recently, it has been demonstrated that polyhedral silver NPs can assemble into densest packings and exotic lattices by a sedimentation process [24]. Surprisingly, even quasicrystalline structures could be fabricated by chemical self-assembling processes [25]. Therefore, binary NP mixtures (consisting of Au and metal oxide NPs) were investigated and assembled due to the interplay of entropy-driven crystallization and interparticle interactions.

Concerning these manifold long-range order self-assembled MMs one has to mention the following points. At first, most of the MMs can be solely fabricated in a planar fashion on a substrate. Therefore, mostly two-dimensional MMs can be achieved and only a minor part of chemical self-assembly processes allows the fabrication of fully three-dimensional MMs offering long-range order [15, 24]. At second, in most of the cases, the meta-atoms of these MMs are made of single NPs. The arrangement of complex meta-atoms consisting of a desired arrangement of a few numbers of NPs, is still a challenging task. Most notably the latter restriction hampers the design of MMs whose optical properties are different from the isolated NPs. A few examples have been demonstrated, where the self-assembled structure really offers different optical properties compared to an diluted solution of the NPs, e.g., the demonstration of a magnetic response of an array of metal stripes [26].

Fig. 4.1 Self-assembled MMs offering long-range order. (**a**) Scanning force micrograph of self-assembled TiO_2 NPs using block-copolymers. Reprinted with permission from [11], ©2011 by American Chemical Society (further ACS). (**b**) Scanning electron micrograph (SEM) of self-assembled Ag truncated octahedra NPs into their densest lattice by sedimentation. Reprinted with permission from [24], ©2011 by Macmillan Publishers Limited. (**c**) Transmission electron micrograph (TEM) of a self-assembled dodecagonal quasicrystal made of binary mixture of Au and Fe_2O_3 NPs. Reprinted with permission from [25], ©2009 by Macmillan Publishers Limited

Contrary to the group of long-range order self-assembled MMs, chemical processes are exploited to assemble the NPs within a single meta-atom into a desired spatial arrangement. The chemical processes of the group of short-range order self-assembled MMs are outlined in the following. The evaporation of a solution including the NPs can be applied to assemble them onto a substrate into a desired geometry of the meta-atom. Therefore, the NPs are often coated with a specific polymer defining the inter-particle distances and therewith the optical coupling strength [27]. Magnetic resonances in trimers [28] and Fano-resonances in heptamers [27] and quadrumer clusters [29] fabricated by evaporation have been demonstrated. Another prominent short-range order self-assembling process is the arrangement of NPs using DNA [30]. First principle demonstrations by Alivisatos et al. [1] and Mirkin et al. [31] have shown the arrangement of two NPs and the fabrication of an aggregate of NPs with DNA, respectively. This concept has been developed further [32] allowing the assembly of more complex meta-atoms like NP trimer and tetramer complexes [33], the arrangement of binary mixtures of NPs [34], dimers made of asymmetric NPs [35], heteropentamer NP complexes to observe Fano resonances [36] and chains of NPs [37]. These meta-atoms are all planar fashioned onto a substrate but of course it is possible to built-up fully three-dimensional meta-atoms, though it is more challenging. For example, the assembling of asymmetric meta-atoms using three-dimensional Janus-particles as a starting point was demonstrated [38]. Furthermore, a DNA guided crystallization allows fabricating three-dimensional structures of gold NPs [39]. The fabrication of DNA scaffolds yields three-dimensional meta-atoms, e.g., pyramids of gold NPs can be assembled [40]. Apart from the experimental demonstration of DNA assembled NPs some theoretical considerations are reported [41], too. The application of organic molecular linkers to assemble the NPs offers another possibility for short-range ordering. Depending on the symmetry of the molecular linker, the NPs can be ordered to dimers [2], trimers, pyramidal structures [42], or to larger aggregates with a well-defined outer shape [43]. Another way to assemble the NPs are electrostatic forces [44]. These forces can be evoked between charged plasmonic NPs and oppositely charged NPs or other chargeable

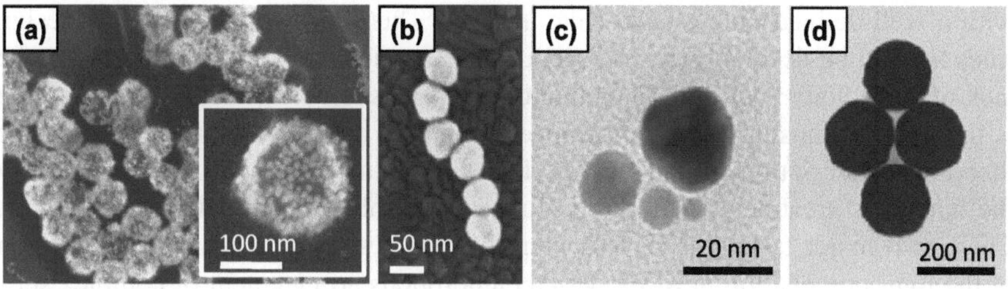

Fig. 4.2 Self-assembled MMs offering short-range order. (**a**) Plasmonic core-shell clusters offering a magnetic dipole response made of Au NPs attached to a dielectric core by electrostatic forces. Reprinted with permission from [49], ©2011 by ACS. (**b**) SEM image of a self-assembled Au NP chain by DNA. Reprinted with permission from [37], ©2011 by ACS. (**c**) TEM image of a self-assembled Au NPs into a chiral pyramidal structure by DNA. Reprinted with permission from [40], ©2009 by ACS. (**d**) TEM image of a self-assembled quadrumer cluster (offering Fano resonances) made of Au NPs by evaporation. Reprinted with permission from [29], ©2010 by ACS

substances, e.g. polymer layers [45]. Various two-dimensional meta-atoms can be fabricated like, e.g., dimers [46] or asymmetric dimers [47] made of plasmonic NPs. The advantage of this technique is its simple adaptability for all chargeable particles like, e.g., nanorods [48], independently of most of the other chemical and physical properties. Details thereof are given in other chapters in that book. Furthermore, the chemical process can be adapted to fabricate fully three-dimensional meta-atoms, like silica cores covered with tiny plasmonic nanospheres offering an isotropic magnetic response in the visible [49] or huge local field enhancements suitable for surface enhanced Raman scattering [50, 51]. Apart from that, by controlling the directional solidification of eutectics, Pawlak et al. have demonstrated the self-assembling of split-ring resonator (SRR) like nanostructures [52] which are well-known from lithographical top-down fabrication processes and are one promising candidate offering a magnetic response in the visible. Recently, the formation of meta-atoms using liquid crystals (which is normally exploited for long-range ordering) was demonstrated [53]. Therefore, defect states were introduced into the liquid crystal host to place the NPs at defined positions.

To sum up the group of short-range order self-assembled MMs, one has to conclude the following points. At one hand, the exploited chemical processes allow the assembling of complex meta-atoms in either two or three dimensions. This allows to observe complex optical responses, e.g., Fano resonances or a magnetic response of the meta-atom. This is in contrast to the long-range order group, where normally the optical response does not differ from a material consisting of diluted NPs. At the other hand, unfortunately, these self-assembling processes enable no control for long-range ordering which is the spatial arrangement of the meta-atoms. Therefore, the meta-atoms appear in most cases in an amorphous fashion on a substrate or in solution.

Both groups of long-range and short-range order self-assembled MMs require a completely different theoretical approach to describe their optical properties when compared to MMs fabricated by top-down techniques. The meta-atoms of self-assembled MMs are made of several separated NPs instead of one bulk material with a well-defined outer shape (like a SRR). Therefore, completely new design rules have to be developed to achieve a desired optical response. Furthermore, the lack of periodicity between the meta-atoms for the short-range order group (which is the more promising group to achieve a complex optical response) hampers the theoretical description of these MMs. Periodic boundary conditions that simplify the numerical treatment, e.g., are no longer applicable to simulate the structures. These two major tasks, the development of new design rules and the description of the light propagation in MMs with an amorphous arrangement of complex self-assembled meta-atoms, are at the focus of this book chapter. The chapter is arranged as follows.

In Sect. 4.2 a short review of top-down MMs is given and the advantages and disadvantages are shortly summarized in comparison to self-assembled MMs. This certainly helps to appreciate the advantages of bottom-up MMs. Afterwards, the focus lies on the theoretical description of self-assembled amorphous MMs. The framework of a multipole analysis of meta-atoms is introduced in Sect. 4.3.2; the optical response of representative examples of meta-atoms is discussed within this framework and new design rules are developed in Sect. 4.3.3. The accessibility of an isotropic magnetic response in the visible is outlined. In Sect. 4.3.4 the light propagation in self-assembled MMs with amorphously arranged meta-atoms is discussed and two different approaches are compared to each other.

4.2 Periodic Metamaterials

To motivate research on amorphous MMs as well as to understand basic guidelines for the design of meta-atoms, it is useful to reflect for a moment on traditional MMs made from deterministic unit-cells. They are usually fabricated with top-down fabrication processes and are arranged in a periodic lattice [54]. Some selected structures are shown in Fig 4.3. Such top-down MMs were essential to evolve the field of MMs towards a mature discipline after the first applications on how to use MMs were suggested [55, 56]. They were the key component to implement materials that sustain the desired response on a first glance. But they also possess limitations bottom-up, self-assembled MMs promise to lift.

To achieve properties that strongly deviate from the mere spatial average of the meta-atom constituents, resonances have to be incorporated such that light strongly interacts with matter. Whereas in self-assembled MMs these initial resonances are usually contributed by LSPRs of metallic NPs, in top-down MMs made from metals the resonances are equally contributed by LSPRs but sustained by NPs with a much more complicated, fully deterministic geometry [57]. Elongated nanowires with a rectangular cross section or thin metallic discs might serve as examples for canonical geometries [58]. Their scattering response is dominated by an electric dipole; hence leading to a dispersion in the effective permittivity. The advantage of elongated nanowires when compared to nanopsheres is the tunability of the spectral position of the resonances across a large spectral domain by modifying geometrical parameters, e.g. the aspect ratio of the naowires [59]. The disadvantage is that with a strongly modified aspect ratio, the individual meta-atom no longer possesses an

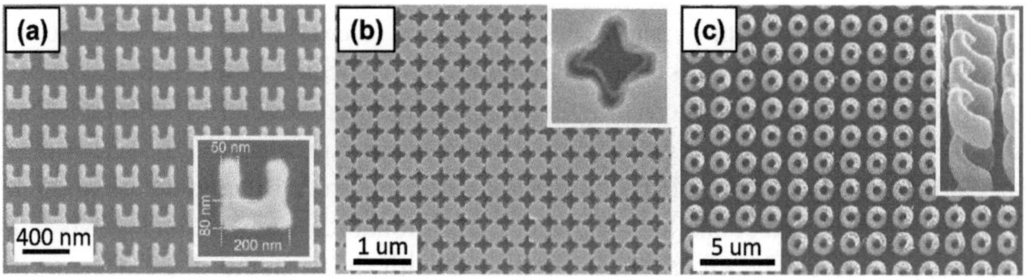

Fig. 4.3 Electron micrographs of top-down MMs periodically arranged on a substrate. (**a**) Split ring resonator array offering a magnetic resonance at 200 THz. Reprinted with permission from [68], ©2005 by American Physical Society. (**b**) Swiss cross metamaterial offering polarization-independent negative-index at 1.4 μm. Reprinted with permission from [60], ©2008 by Optical Society of America. (**c**) Periodically arranged gold helix structure that exhibits a broadband chiral response in the near infrared. Reprinted with permission from [69], ©2009 by American Association for the Advancement of Science

isotropic response. Instead, at least a biaxial anisotropic response is observed, i.e. the polarizability of the meta-atoms is different for all three main axes. The geometry can be tuned to be polarization independent at normal incidence, i.e. for a unit cell with a C_4 symmetry, but it will lead at most to a uniaxial anisotropic response [60].

To achieve meta-atoms that possess a response beyond an electric dipole, most MMs working in the visible rely on strongly coupled two NPs that sustain an electric dipolar resonance at the same frequency. The theory of plasmon hybridization suggests that in the strong coupling regime the individual modes at the same frequency are splited into two modes at different frequencies if the dipoles are oriented perpendicular to the connection line of the NPs [61]. The low (high) energy mode is called anti-symmetric (symmetric) since the currents in the two NPs oscillate π out of (in) phase. The anti-symmetric character, as will be detailed below, is associated with a scattering response that is dominated by a magnetic dipole and/or an electric quadrupole. The symmetric mode where the currents in the coupled NPs oscillate in phase, possesses a scattering response that is dominated by an electric dipole since contribution from the two particles reinforce. With that basic operational principle, various top-down metamaterials have been suggested across a wide spectral domain; ranging from the GHz down to the blue end of the visible. The basic understanding that two currents in the unit cell that flow π out of phase induce a scattering response dominated by a magnetic dipole can also be applied to explain the scattering response of other meta-atoms, e.g. the SRR, as detailed later in this chapter.

The success of top-down MMs can be found in their conceptual simple way of fabrication and their ability to allow for basic investigations of many effects discussed in the context of MMs. Examples are the evidence of a negative index [62], of negative refraction [63], negative group velocity [64], and on their imaging performance [72]. To this end, the application of laser beam writing, electron-beam lithography, or focused ion beam milling has been reported for their fabrication [54]. All these methods, however, are limited to the fabrication of planar structures with a thickness of only a few tens of nanometers for each functional layers. Possibilities exist, however, to extend this thickness, though fully bulk materials are out of reach. Referential examples for techniques to extend the thickness are the stacking of functional layers that are individually fabricated or the structuring of a complicated sequence of alternating layers of metals and dielectrics with an identical lateral geometry with a given top-down fabrication process [65, 66]. However, this led in the past only to MMs comprising a few hundreds of nanometers in thickness; and technological challenges render it questionable to achieve bulk MMs with these approaches. Therefore, MMs fabricated with these methods are necessarily meta-surfaces. Conceptually fundamentally new approaches are required to overcome these limitations.

Moreover, the proximity effect strongly affects the achievable resolution for fabricated meta-atoms and a great degree of sophistication has to be put in place to precompensate associated detrimental effects. This, however, is only feasible if each meta-atom is written with an identical environment. This led to the fabrication of strictly periodically arranged unit cells in space which causes additional problems. First, it is difficult, if not to say impossible to achieve isotropic MMs [67]. Although MM might possess a polarization independent response at normal incidence, e.g. by choosing a C_4 symmetry, the challenge to fabricate a meta-atom with identical geometrical details in the third spatial dimension, i.e. normal to the substrate on which the MM is fabricated, is not solved yet. Second, the periodic arrangement of the meta-atoms causes often strong spatial dispersion, a property which denies the introduction of biaxial anisotropic constitutive relations [70]. To do so, only weak spatial dispersion or a weak non-locality would have been required. This constitutes a major limitation and might be taken as the reason why many of the envisioned applications are not yet implemented.

In conclusion, many of the basic functionalities of MMs have been proven with structures fabricated with top-down processes, but the limitations indicated above have to be lifted to further advance the field. Self-assembled amorphous MMs are promising to nullify at least problems associated by the periodic arrangement of the meta-atoms as outlined in the next section. The challenge to discuss the properties of such amorphous MMs is the restriction on the scattering response of the individual meta-atom. It is the primary and sole source of information and at the heart of the following sections.

4.3 Amorphous Metamaterials

As outlined in Sect. 4.1 self-assembled MMs can offer long-range or short-range order and especially the last part leads to complex meta-atoms with desired optical properties. The arrangement and orientation of these meta-atoms is completely amorphous, forming an amorphous MM, since the self-assembling process only takes place at length scales to assemble the single meta-atoms. This evokes several advantages and disadvantages compared to periodic MMs which are in most of the cases fabricated by top-down processes (cf. Sect. 4.2). The amorphous orientation of the meta-atoms solves one major problem of the periodic MMs which is spatial dispersion. The optical response of an amorphous MM is completely isotropic since the propagating electromagnetic field would probe independent of the propagation direction and polarization state always the same material. This isotropy is one key feature which is required for many predicted applications of MMs. For example the well-known perfect lens, predicted by Pendry [71], requires a MM offering a negative refractive index independently on the spatial wavevector component that has to pass the lens. This requires a MM characterized by isotropic material parameters or it can be circumvented by using complex layered periodic MMs which provide at least an isotropic optical response, though material parameters cannot be assigned to [72]. This problem could be easily solved by amorphous MMs exhibiting the desired optical properties. In opposite to the striking advantage of the amorphous arrangement of the meta-atoms, it strongly hampers their theoretical description. In principle, every meta-atom of the MM, or at least the part that is illuminated by the incident field, has to be considered in the simulations of the optical response of the MM. This is in most of the cases impossible or requires computational resources in excess. Well established techniques such as periodic-boundary conditions for periodic MMs that reduce the simulation of the entire MM to that of a single meta-atom are no longer applicable for amorphous MMs. Therefore, other numerical techniques has to be evaluated to properly describe the interaction of light with amorphous MMs and evaluate design rules for them.

The description of light propagation in amorphous MMs presented in this chapter is based on the scattering response of the single meta-atom. Therefore, to start with, a compact outline about measuring the optical response of meta-atoms is given in Sect. 4.3.1. Afterwards, in Sect. 4.3.2 a multipole analysis of meta-atoms is introduced, allowing the numerical description of the scattering response of arbitrarily shaped meta-atoms in terms of multipole moments. The subsequent Sect. 4.3.3 discusses several meta-atoms in terms of their excited multipole moments and outlines the advantages of the introduced formalism. In the last Sect. 4.3.4 the light propagation in amorphous MMs is discussed in detail.

4.3.1 Optical Response of Single Meta-atoms

The challenge to access on experimental grounds the optical properties of individual meta-atoms is linked to their nanometric extension. This causes the absolute scattering response to be extremely weak and difficult to detect. Moreover, due to the strong localization in space, a possible scattering signal diffracts and it is difficult to reconstruct its angular distribution. Therefore, most experiments are performed while using periodic arrays of meta-atoms, i.e. the structure forms a grating. Then, transmission (reflection) into the zeroth diffraction order can be measured to quantify at least the scattering in forward (backward) direction. Detrimental is that results are occasionally exposed by artifacts due to the lattice, i.e. grating resonances such as Wood or Rayleigh anomalies, and it does not always help to disclose the multipolar character of the scattered field. Moreover, resonances which do not couple to free space, i.e. dark modes, are not probed. To circumvent some of these problems, a few experimental techniques were developed to measure in an experiment quantities of individual meta-atoms.

Probably the most direct approach to probe the optical properties of individual meta-atoms has been introduced by Husnik et al. [73]. They suggested and applied a technique to measure the extinction cross section of individual meta-atoms using a modulation technique. The extinction cross section is the sum of the scattered and the absorbed light normalized to the incident power onto the geometrical cross section of the meta-atom, i.e. the amount of incident light that is not transmitted in forward direction. Regarding the example of SRRs, they have been proven that the extinction cross sections can be eight times the geometrical cross sections at the frequency of the fundamental resonance. The experimental results were in excellent agreement with microscopic simulations of a plane wave scattered at an individual SRR; taking into account all the minuscule details of their geometry as extracted from scanning electron micrographs. The supporting simulations could also reveal that one third of the extinction cross section was associated with the absorption of light while the remaining two third were scattered. The advantage of the method is its ability to measure spectrally resolved quantities that characterize individual meta-atoms. But with these quantities, i.e. the extinction cross section, the character or the nature of the individual resonance cannot be resolved for and, moreover, only resonances can be accessed that can be also excited with free space radiation. Only the presence of such resonance is detected and its strength. Supporting simulations are required to further understand the meta-atoms.

To circumvent some of these limitations, measurements with an apertureless scanning near-field optical microscope (a-SNOM) can be performed [74]. Although aperture-SNOMs can be used as well, their finite spatial resolution usually denies an application to probe the near-field properties of meta-atoms. The a-SNOM usually measures close to an object the near-field upon plane wave excitation. In most cases only the component in longitudinal direction, i.e. normal to the substrate, is measured since the tips used in a-SNOMs only scatter that specific component into the far-field. Details of the method are discussed in another chapter of this book by Vogelgesang and co-worker. The main advantage of such techniques is to provide comparison to quantities usually used to discuss and classify the resonances of meta-atoms, e.g. SRRs. The disadvantage of being spectrally not always flexible can be circumvented by scaling the geometries of the individual elements such that resonance frequencies are supported at the frequency of operation for the a-SNOM. A further disadvantage with a-SNOMs is the illumination of the sample at oblique incidence. This causes the incident field to have a phase variation across the meta-atom. This is not a major problem and can be taken into account in supporting simulations, but the understanding of the measured field distributions is less intuitive.

To access optically the radiation patterns of individual meta-atoms the use of Fourier microscopy in a dark-field confocal setup that relies on an objective with a high numerical aperture has been suggested [75]. In that method the individual meta-atoms on top of a glass substrate are usually illuminated from the substrate with an oblique beam possessing an angle larger than the angle necessary for total internal reflection. The angular distribution of the scattered field is then measured in a $2f$-configuration. An objective with a large numerical aperture ensures that the scattered light can be measured over a large angular domain. The problem of having a spread of photons across the detector which leads to small signals can be circumvented by using a high power supercontinuum light source that is spectrally tuned by an acousto-optical filter.

A last optical method that can be used to probe for specific eigenmodes consists in using tailored light fields for the illumination and measuring the reflected and/or transmitted light [76]. Here the idea is that the eigenmodes sustained by the meta-atoms possess a well defined electric and magnetic field distribution in space. The excitation strength of the resonance strongly depends on the spatial overlap of the incident field with that of the eigenmode. The use of, e.g., tailored illumination with no electric but only a strong magnetic field can suppress the electric dipolar scattering response that otherwise would have dominated the scattered signal in favor of a much weaker magnetic dipole. Varying the incidence field, hence varying the excitation condition for the meta-atoms under consideration, allows, in principle, to entirely map the scattering matrix of individual meta-atoms. It will include information on how strong an incident electric dipolar field is scattered into an electric dipole, magnetic dipole; and in turn any other electro-magnetic multipolar contributions.

To probe for all resonances a meta-atom may sustain, i.e. also those which cannot be excited by free space radiation, other methods have to be put in place [77]. A prominent technique is electron energy loss spectroscopy (EELS) (cf. Fig. 4.4). In this technique electron beams are transmitted through the samples, that are required to be thin, and the spatially resolved electron energy loss spectra is recorded. While the electron traverses the structure, it can couple to the normal component of the electric field of the eigenmodes as sustained by the meta-atoms at discrete frequencies; and the electron hence looses some of its energy. After obtaining a three-dimensional data-set (energy and two transversal coordinates), various inspections of the data are useful. For example, the amplitude of the EELS spectra can be spatially looked at for a fixed frequency depending on the transversal coordinates to glimpse the spatial distribution of eigenmodes. Impressive results have been obtained with EELS while probing for resonances of single SSRs. Resonances of orders up to seven were witnessed, well beyond an order usually seen in far-field optical spectroscopy. These higher order modes usually do not couple well to free space and cannot be probed in consequence with far-field technique. But they can be seen using EELS due to the local excitation. Disadvantages of EELS are the only indirect measurement of quantities linked to the field distribution of the eigenmodes and the requirement of thin samples that the electrons can pass through.

To conclude this section on various experimental strategies to probe for properties of individual meta-atoms, it remains to be said that various approaches have been developed in the past for that purpose. They all possess advantages and disadvantages; but taking them together may provide quite comprehensive insights into the absolute scattering strength, the angular distribution of the scattered light, and

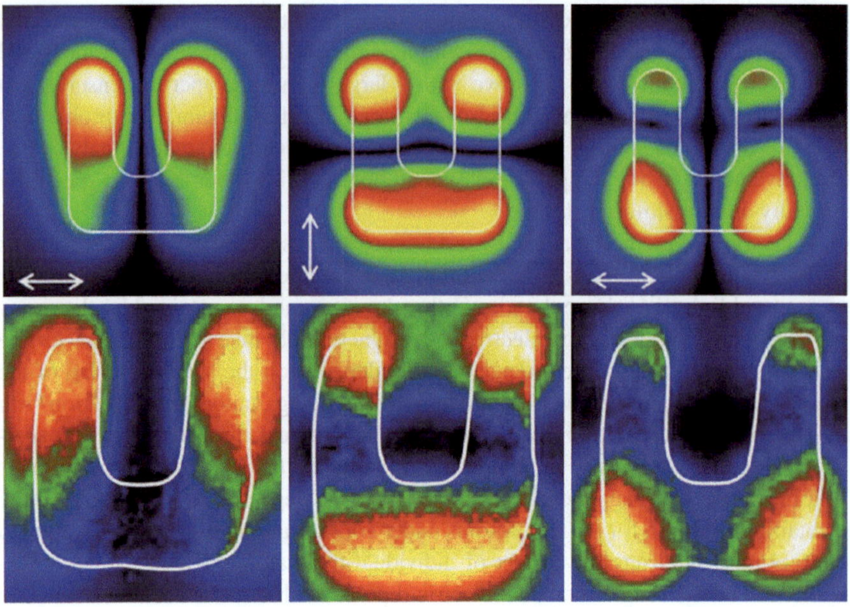

Fig. 4.4 Direct measurement of SRR resonances by EELS. *Upper row*: Calculated field distributions of the lowest three order resonances of a gold SRR with 200 nm arm and base lengths under normal incidence. Only the electric field component perpendicular to the SRR plane is shown. The *white arrows* indicate the polarization of the incident field. *Lower row*: Corresponding EELS maps of the lowest three order resonances for a fabricated SRR with identical dimensions. The EELS maps nicely correspond to the calculated field distributions of the *upper row*. Reprinted with permission from [77], ©2011 by Optical Society of America

the field distribution of individual resonances sustained by the meta-atoms. These meta-atoms constitute the heart of amorphous MMs. However, in all cases supporting theoretical considerations and simulations have to be put in place to fully understand their properties. How to explore these scattering properties on theoretical grounds is discussed in the following sections.

4.3.2 Multipole Analysis of Meta-atoms

As outlined in the previous sections the aim of this book chapter is the understanding of light propagation in amorphous MMs based on excited multipole moments of the corresponding meta-atoms. Therefore, prior the consideration of the problem how light interacts with amorphous MMs, a formalism needs to be introduced to describe how light interacts with the individual meta-atom. This is performed by accessing the scattered multipole moments of the meta-atoms which is done in this section. Furthermore, it is shown that the understanding of the scattering response of the meta-atoms plays a pivotal role to develop design rules for self-assembled

MMs and it can be used to assign material parameters to these structures; just to mention the most prominent issues. However, a theoretical description of the meta-atoms in terms of multipole moments is not just linked to the field of self-assembled amorphous MMs. Quite the contrary, it is important for the field of periodic top-down MMs to distinguish between resonances evoked by the meta-atoms itself and by their periodical arrangement.

Several theoretical models have been developed to describe the scattering response of a given meta-atom. One way is a description based on LC circuits well-known from quasistatic electrodynamics. Here the meta-atom is replaced by coupled LC circuits depending on the given geometry [78, 79]. The basic features regarding the observed resonances in the quasistatic limit can be explained by this model in a very intuitive manner. However, since the LC circuits are not based on a full electrodynamical treatment of Maxwell's equations, in some cases several problems arise, e.g., an infinite speed of light in the LC models as well as a violation of energy conservation [80, 81]. These limitations can be lifted, e.g., by treating the meta-atoms as magnetoelectric point scatterers and converting the quasistatic LC circuit model into a full electrodynamical model by considering radiation damping effects [81]. However, the parameters of the assumed LC circuit such as the impedances and the capacitances are not given by theory. Experimental results such as measured extinction spectra have to be put forward. Therewith, the free parameters of the LC circuit model are fitted to match the experimental results. Furthermore, since the theoretical description is based on dipoles, no insight is given to higher order multipole moments excited in the meta-atom that contribute to the scattering response.

Another way is the description of the meta-atom by coupled oscillators [82]. Therefore, the meta atom is assumed to consists of several plasmonic entities in which a current oscillates. A multipole expansion technique relates these coupled oscillators to an electric and magnetic dipole and higher order multipole moments. Though this description provides access to all excited multipole moments in a meta-atom, it still needs experimental or numerical results to fit the unknown parameters of the model such as the coupling strengths.

The previously mentioned theoretical descriptions of the meta-atoms are based on analytical formulas. Unfortunately, this requires in most of the cases some fitting procedure where the assumed free parameters of the analytical description (e.g. the capacity in the LC model) are assigned according to experimentally measured spectra. In contrast to the analytical treatment of the meta-atom, Maxwell's equations can be solved numerically in the frequency or time domain by well-established techniques such as the finite-difference time-domain (FDTD) method or the finite-element method (FEM). The numerical solution yields the electromagnetic field and therewith the scattering response of the meta-atom can be derived. The advantage of the numerical treatment is the prevention of any assumptions to the system (such as assuming the meta-atoms as pure dipoles) and furthermore no free parameters have to be fitted. Contrariwise, it is quite challenging to develop a deep physical understanding of the meta-atom and to explain the observed resonances. Of course it is possible by interpreting the near-field components of the meta-atom, as shown in Refs. [83, 84], but in general, the physical origin of the excited eigenmodes in the meta-atom cannot be extracted.

Therefore, it is important to reduce the complexity of the simulated field-distributions to theoretical descriptions of the problem that could be physically interpreted. The so-called T-matrix method [85–87] provides exactly such a description. It is based on a full set of eigenfunctions in spherical coordinates which are the vector spherical harmonics (VSH). The incident and the scattered field of the meta-atom are expanded into this set of eigenfunctions by an infinite series with complex expansion coefficients. The T-matrix relates the coefficients of the incident field **p** to the coefficients of the scattered field **a** (in this short notation **p** and **a** are vectors including all the coefficients)

$$\mathbf{a} = \hat{\mathbf{T}} \cdot \mathbf{p}. \tag{4.1}$$

Obviously, the T-matrix includes all information of the meta-atom; for every incident field the scattered field can be easily extracted by a matrix multiplication. In general, the entire T-matrix includes still too much information (or in other words the complexity of the problem is still too high) for most of the meta-atoms. In principle, in the experiments only a few angles of incident are important and it is enough to consider the two independent polarizations. Furthermore, it is quite challenging to calculate the entire T-matrix for a given meta-atom. Analytical formulas exist only for highly symmetric scatterers such as, e.g., spheres [88] or ellipsoids [89]. Apart form that the T-matrix has to be computed numerically [90]. However, every incident field for which the meta-atom is illuminated yields to solely one row of the T-matrix (assume that the meta-atom is illuminated by a field that can be entirely described by one VSH; this is identical to a vector **p** including predominantly zeros and one nonzero entry). Furthermore, the T-matrix is normally written down in spherical coordinates which makes it more difficult to interpret the entries and reveal a deeper physical understanding of the meta-atom compared to Cartesian coordinates.

In the following a formalism is introduced which is based on the T-matrix method and called multipole analysis of meta-atoms [91]. For an arbitrary incident field it will be shown how the Cartesian multipole moments contributing to the scattered field can be extracted from numerical simulations. In analogy to the T-matrix method all fields are expanded in VSH that consist of two orthogonal sets of eigenfunctions labeled **N** and **M** spanning the entire space in spherical coordinates. The scattered field reads as

$$\mathbf{E}_{sca}(r,\theta,\varphi) = \sum_{n=1}^{\infty} \sum_{m=-n}^{n} k^2 E_{nm} \big[a_{nm} \mathbf{N}_{nm}(r,\theta,\varphi) + b_{nm} \mathbf{M}_{nm}(r,\theta,\varphi) \big], \tag{4.2}$$

where k is the wavenumber in freespace, a_{nm} and b_{nm} are the complex expansion coefficients of the scattered field (from now on we will call them scattering coefficients) and E_{nm} is a scaling factor. The analytical formulas of E_{nm} as well as for the VSH **N** and **M** can be found in Ref. [91]. For a given incident field and highly symmetric meta-atoms, e.g. spheres or ellipsoids, analytical formulas can be derived to obtain the scattering coefficients a_{nm}, b_{nm} [92, 93]. For example the well-known

Mie coefficients connect the expansion coefficients of the incident field [for a decomposition into VSH similar to (4.2)] to the scattering coefficients for a sphere of arbitrary permittivity and permeabilty [94]. If the meta-atom is more complex shaped, the scattering coefficients have to be provided by a numerical treatment [95]. Therefore the scattering response is simulated by a suitable numerical technique (e.g. the FDTD method). Afterwards, the electromagnetic fields are projected onto the VSH by the following means [91]

$$\begin{aligned} a_{nm} &= \frac{\int_0^{2\pi}\int_0^{\pi} \mathbf{E}(r=a)\mathbf{N}_{nm}^*(r=a)\sin\theta d\theta d\varphi}{\int_0^{2\pi}\int_0^{\pi} |\mathbf{N}_{nm}(r=a)|^2 \sin\theta d\theta d\varphi}, \\ b_{nm} &= \frac{\int_0^{2\pi}\int_0^{\pi} \mathbf{E}(r=a)\mathbf{M}_{nm}^*(r=a)\sin\theta d\theta d\varphi}{\int_0^{2\pi}\int_0^{\pi} |\mathbf{M}_{nm}(r=a)|^2 \sin\theta d\theta d\varphi}. \end{aligned} \tag{4.3}$$

The asterisk indicates a complex conjugation and the integral is calculated on a surface of a virtual sphere with radius a enclosing the entire meta-atom.

The expansion of the scattered field into the VSH with complex amplitudes in (4.2) is identical, except some prefactors, to a multipole expansion in spherical coordinates. Hence, by having the scattering coefficients from (4.3) at hand the multipole moments generating the scattered field of the meta-atom are known. In the following these multipole moments are transformed into Cartesian coordinates since it appears the natural choice to discuss most of the meta-atoms. The transformation from spherical to Cartesian coordinates can be obtained by comparing the far-field expansion of the scattered fields in both coordinate systems, as demonstrated in Ref. [91]. After some cumbersome but straightforward calculations, the following transformation rules for the lowest two multipole moments (the electric dipole $\mathbf{p}$ and the magnetic dipole $\mathbf{m}$) can be found

$$\mathbf{p} = \begin{pmatrix} p_x \\ p_y \\ p_z \end{pmatrix} = C_0 \begin{pmatrix} (a_{11} - a_{1-1}) \\ i(a_{11} + a_{1-1}) \\ -\sqrt{2}a_{10} \end{pmatrix}, \qquad \mathbf{m} = cC_0 \begin{pmatrix} (b_{11} - b_{1-1}) \\ i(b_{11} + b_{1-1}) \\ -\sqrt{2}b_{10} \end{pmatrix}, \tag{4.4}$$

with $C_0 = \frac{\sqrt{6\pi i}}{cZ_0 k}$ where c is the velocity of light in vacuum and Z_0 the free space impedance. As expected, only the lowest order of multipole moments in spherical coordinates [$n = 1$ in (4.2)] contribute to the electric and magnetic dipole in Cartesian coordinates. Furthermore, as can be seen by the transformation of higher order multipole moments (such as the electric quadrupole, which is not presented here) the VSH $\mathbf{N}$ and $\mathbf{M}$ address the electric and magnetic multipole moments [91], respectively. This is indicated already by (4.4) where solely the scattering coefficients a_{1m} contribute to the electric dipole moment $\mathbf{p}$ the b_{1m} to the magnetic dipole $\mathbf{m}$.

The understanding of the scattering response of the meta-atom in terms of Cartesian multipole moments requires a formalism where the influence of different multipole moments at certain frequencies can be extracted. One possible way is to consider the scattering cross section C_{sca} of the meta-atom in the frequency range of interest and to identify resonances thereof with resonant multipole moments. Therefore, the Cartesian multipole moments from (4.4) have to be linked to C_{sca}. This can be easily done by inverting the transformation rules from (4.4) and applying the analytical formula of the C_{sca} in terms of the scattering coefficients [91]. Considering only contributions of the electric dipole moments yields

$$C_{sca} = \frac{2k^2}{C_0^2}\left(\frac{1}{2}|p_x + ip_y|^2 + |p_z|^2 + \frac{1}{2}|p_x - ip_y|^2\right) + \cdots \tag{4.5}$$

It is worth mentioning that, of course, the contribution of every higher order multipole moment to C_{sca} can be extracted by exactly the same formalism; though in the case of most of the meta-atoms no higher order contributions than electric or magnetic quadrupoles play an important role.

To sum up, a formalism was introduced allowing the description of the scattering response of arbitrary meta-atoms in terms of Cartesian multipole moments. The scattered field of the meta-atom has to be simulated by a related numerical technique and afterwards it is projected to the VSH yielding the scattering coefficients. These scattering coefficients can be transformed to Cartesian multipole moments and their contribution to C_{sca} can be derived. Therewith, it is possible to relate resonances in the scattering response of single meta-atoms to excited Cartesian multipole moments. The applications and advantages of this multipole analysis of meta-atoms are revealed by some exclusive examples in the next section.

4.3.3 Exclusive Meta-atom Geometries

The aim of this section is to demonstrate the advantages of the multipole analysis of meta-atoms by presenting an exclusive set of different meta-atom geometries. It is shown that a physical understanding of the optical response of arbitrarily shaped meta-atoms can be derived by the formalism introduced in Sect. 4.3.2. The importance of revealing the contribution of the multipole moments to the scattering response for developing design rules for self-assembled MMs is outlined. In advantage, this section serves as a primary step for the description of light propagation in self-assembled amorphous MMs which is discussed in detail in Sect. 4.3.4.

As outlined in Sect. 4.1 a big portion of self-assembled MMs is made of NP offering a spherical shape (cf. Figs. 4.1 and 4.2). Therefore, the first part of this section is devoted to meta-atoms made of properly arranged resonant nanospheres. This implicates a huge numerical advantage. The scattering coefficients of these meta-atoms can be simulated by an extension of Mie-theory [92, 93] which is a quasi-analytical solution of Maxwell's equations. Therefore, the overlap integrals of (4.3) have not to be computed explicitly. The second part of this section deals with two exclusive examples of meta-atoms, namely the cut-plate pair and the split-ring resonator (SRR), that are commonly fabricated by top-down fabrication processes. Here, the scattering response of the meta-atoms was computed by FDTD simulations and the Cartesian multipole moments are derived by the formalism described in Sect. 4.3.2.

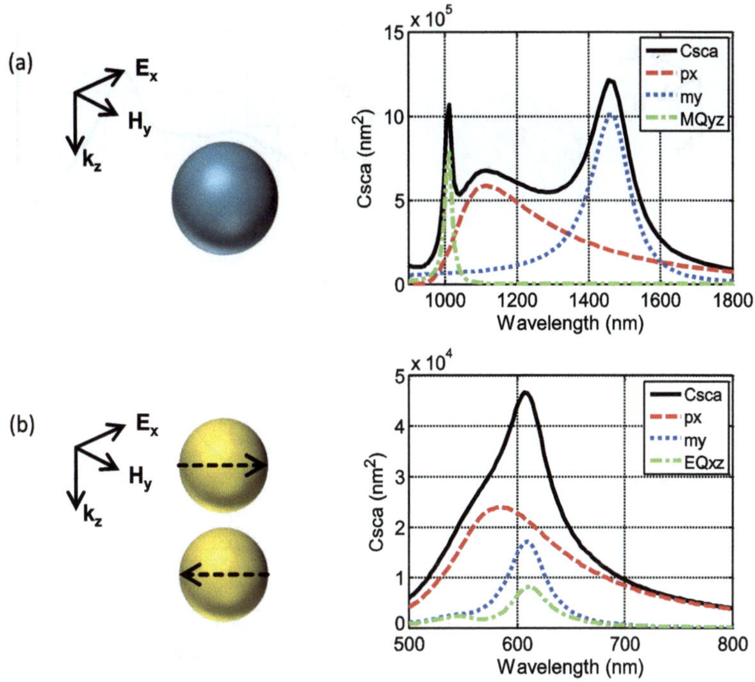

Fig. 4.5 Cartesian multipole moments (as they contribute to the scattering cross section) of meta-atoms made of resonant nanopsheres as a function of the wavelength; p is electric dipole, m is magnetic dipole, EQ is electric quadrupole, MQ is magnetic quadrupole. (**a**) A silicon sphere of 200 nm radius is illuminated by a plane wave as sketched. The introduced coordinate system and illumination scenario are identical for (**b**). (**b**) Cartesian multipole moments of a meta-atom made of two gold spheres (40 nm radii and 5 nm separation) embedded in a usual dielectric ($\varepsilon = 2.25$)

The multipole moments of meta-atoms made of nanospheres are presented in Figs. 4.5 and 4.6. All multipole moments are sketched as they contribute to the scattering cross section [cf. (4.5)], in other words the sum of all shown multipole moments for a specific meta-atom yields the entire scattering cross section. Exclusive examples are discussed where the focus lies on meta-atoms fabricated by self-assembly offering a magnetic dipole response. The magnetic dipole response of these meta-atoms is important for manifold predicted applications, e.g., on the field of transformation optics [96]. Various meta-atom designs are presented and advantages and disadvantages are outlined. Especially, it is shown that the designs of meta-atoms of self-assembled MMs are completely different compared to commonly top-down MMs.

In Fig. 4.5 (a) the Cartesian multipole moments of a silicon sphere with 200 nm radius in the near infrared are shown. In this wavelength regime the permittivity of silicon can be assumed to be a real valued constant of $\approx$12 [97]. The expected separation of the multipole resonances of a high permittivity dielectric sphere can be observed. The first order resonance (which appears at the lowest frequencies or

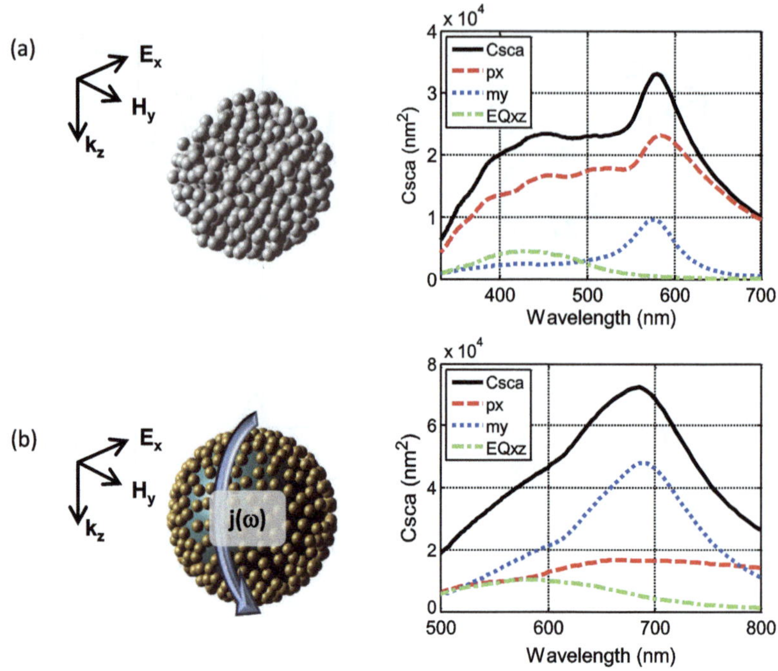

Fig. 4.6 Cartesian multipole moments (as they contribute to the scattering cross section) of meta-atoms made of resonant nanopsheres as a function of the wavelength; p is electric dipole, m is magnetic dipole, EQ is electric quadrupole. (**a**) The meta-atom made of amorphously arranged silver nanospheres (6 nm radii) arranged into a spherical shape (75 nm radius) surrounded by a dielectric with permittivity $\varepsilon = 2.6$ is illuminated by a plane wave as sketched. The introduced coordinate system and illumination scenario are identical for (**b**). (**b**) Meta-atom made of a silica core sphere (130 nm radius) where approximately 350 gold nanospheres (10 nm radii) are forming a shell; the meta-atom is dissolved in water

longest wavelengths) is a pure magnetic dipole resonance [98, 99] at about 1500 nm in Fig. 4.5 (a). The next order resonance (when going to shorter wavelengths) is dominated by an electric dipole at approximately 1100 nm. Both dipolar resonances are broad compared to the magnetic quadrupole resonance at about 1000 nm. In principle, the magnetic dipole resonance at 1500 nm should be sufficient to fabricate a MM out of somehow in space arranged silicon spheres that exhibits a significant response to the magnetic field. The magnetic dipole resonances are exited when the wavelength inside the nanosphere is comparable to its diameter [100]. Therefore, the nanospheres have to be made of a material offering a huge real part of the permittivity [100–106]. Unfortunately, such materials do not exist for the visible regime. Materials like the discussed silicon nanosphere or silicon carbide nanospheres [107, 108] offer a magnetic resonance only in the near infrared regime. Therefore, novel concepts needs to be developed for the visible domain.

One possibility to achieve a magnetic dipole resonance in the visible is shown in Fig. 4.5 (b). Here, the meta-atom consists of two strongly coupled gold spheres, a plasmonic dimer. The LSPR of the single gold sphere is dominated by an electric dipole contribution and can be excited in the visible. By properly illuminating the coupled spheres both electric dipole moments of the single spheres can be exited to oscillate 180° out of phase [as sketched in Fig. 4.5 (b)] causing a magnetic dipole resonance of the entire structure [109, 110]. This resonance can be observed at around 600 nm in Fig. 4.5 (b). Unfortunately, the contribution of the electric quadrupole at this wavelength is not negligible. This may cause problems to describe a MM made of these meta-atoms by simple constitutive relations relying on homogeneous material parameters [70]. The impact of such a quadrupole contribution is discussed in the next section. Furthermore, a meta-atom consisting of two coupled spheres offers an anisotropic optical response and only for the presented illumination scenario the magnetic dipole resonance can be excited [91]. Of course an isotropic magnetic resonance could be achieved by a random arrangement of the dimers in space forming an amorphous MM. Anyhow, the influence of the magnetic dipole resonance on the scattering of the amorphous MM gets lowered. The reason is that the magnetic dipole resonance is evoked only for one of the three independent orientations of the dimers to the incident field.

The two meta-atoms presented in Fig. 4.6 (a) and (b) circumvent the problem of an anisotropic response. They offer a spherical shape by their own yielding an isotropic response of the meta-atom. In Fig. 4.6 (a) the meta-atom consists of an amorphous arrangement of silver nanospheres offering a spherical outer shape [43, 111, 112]. The observed magnetic resonance at about 550 nm can be understood as follows. A material made of densely packed silver nanospheres can be described by an effective permittivity that offers a Lorentzian line shaped resonance which is dictated by the LSPR of the single nanosphere. Forming a sphere out of such a material is somehow identical to the situation in Fig. 4.5 (a). The resulting meta-atom can be understand as a homogeneous sphere which exhibits (due to the Lorentzian line shaped permittivity) wavelengths ranges where the real part of the permittivity is much larger than one. Therefore, as for the silicon sphere, a magnetic dipole resonance of the structure is possible [113]. The advantages compared to the previously discussed meta-atoms are the following. At first the magnetic dipole resonance appears in the visible compared to the silicon sphere. At second the response is fully isotropic due to the spherical shape of the meta-atom avoiding disadvantages of an anisotropic response, like for the plasmonic dimer. However, the excited magnetic dipole resonance does not dominate the scattering spectra, there is still a dominating influence of the electric dipole of the meta-atom [cf. Fig. 4.6 (a)].

One way to enhance the influence of the magnetic dipole resonance is shown for the presented meta-atom in Fig. 4.6 (b). The meta-atom is a core-shell cluster where the silica core sphere is decorated by a huge number of gold nanospheres [49]. The appearance of a magnetic dipole resonance at around 700 nm can be understood as follows. The optical response of the single gold nanospheres is dominated by the excitation of an electric dipole oscillation at the LSPR. By illuminating the core-shell cluster in Fig. 4.6 (b) by an external plane wave all these dipoles can oscillate in phase around the core sphere causing a current [sketched in Fig. 4.6 (b) by the arrow] that can be associated with a magnetic dipole [114]. In contrast to the meta-atom in

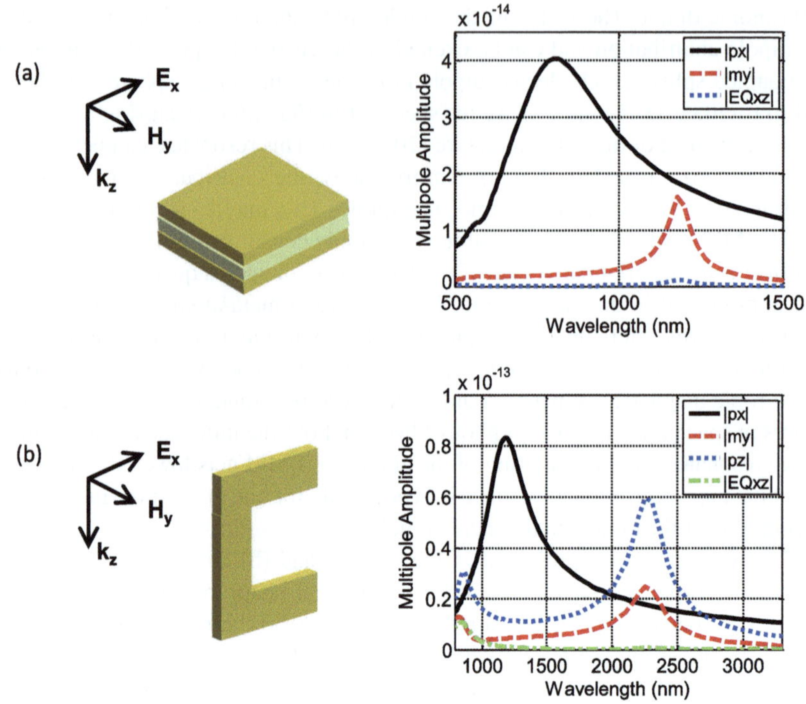

Fig. 4.7 Cartesian multipole amplitudes (p is electric dipole, m is magnetic dipole, EQ is electric quadrupole) of two selected meta-atoms commonly fabricated by top-down fabrication processes as a function of the wavelength. (**a**) The shown meta-atom (a cut-plate pair) consists of two strongly coupled gold plates (180 nm lateral dimension and 30 nm thickness) with a MgO layer in between (same lateral dimensions and 45 nm thickness). The cut-plate pair is illuminated by a plane wave as sketched. The introduced coordinate system and illumination scenario is identical for (**b**) where the multipole amplitudes for a SRR made of gold (300 nm arm and base length; 40 × 40 nm geometrical cross section) are shown

Fig. 4.6 (a) the magnetic dipole resonance of the core-shell cluster dominates the scattering response of the meta-atom. One reason is the reduced amount of metal in the meta-atom design yielding more pronounced resonances due to the lowered absorption. Therefore, the presented core-shell cluster can be seen as one prominent candidate offering an isotropic magnetic response in the visible domain.

In the second part of this section meta-atoms are in the focus which are commonly fabricated by top-down fabrication processes. The multipole amplitudes of the so-called cut-plate pair [95] can be seen in Fig. 4.7 (a). By comparing the multipole response with that of the previously discussed plasmonic dimer [cf. Fig. 4.5 (b)] one can draw the following conclusion. The excited multipole moments and their relative resonance positions to each other are similar for both meta-atoms. At short wavelengths a broad electric dipole resonance is excited. The magnetic dipole resonance and a contribution of an electric quadrupole can be observed at longer wavelengths in Fig. 4.7 (a) and Fig. 4.5 (b). This outlines one of the advantages of the multipole analysis. Though both considered geometries are totally different, their multipole response looks similar and therefore the physical origin of the observed multipole resonances should be the same, which is indeed the case. As for the plasmonic dimer structure electric dipoles are excited in both metal plates of the cut-plate pair. Their out of phase oscillation at a specific wavelength can be associated with a magnetic dipole. Furthermore, the multipole analysis allows for a comparison of both meta-atoms. For example the quadrupole contributions at the magnetic resonance are different for both meta-atoms and it is weaker for the cut-plate pair. Therefore, the cut-plate pair serves as a more prominent candidate to describe the propagation of light in a MM by simple constitutive relations since the quadrupole is negligible small [95]. Problems arising from the quadrupole moment like spatial dispersion should not dominate the optical response as shown in the next section.

The second geometry presented in Fig. 4.7 (b) shows the multipole moments of a split-ring resonator (SRR) under illumination parallel to its base [91, 95]. Clearly, the first order resonance at around 2200 nm can be identified with a contribution of the desired magnetic dipole moment. Contrary to most of the observations in literature, a second dipole moment contributes at this wavelength range which is the electric dipole parallel to the wavevector of the incident field. This electric dipole can be associated with a current oscillation in the SRR base. It is caused by a conductive coupling of the excited currents in the SRR arms. If they are oscillating out of phase, which causes the observed magnetic dipole resonance, a current is generated in the base of the SRR. Normally, the electric dipole contribution due to this current is not observed in regular lattices of SRRs since it does radiate neither in forward nor in backward direction. Anyhow, it has to be taken into account for randomly arrange SRRs as they were currently investigated [52]. This causes more complicated constitutive relations to describe the light propagation in such systems. By considering extended designs (derived from the presented multipole analysis) of meta-atoms consisting of several SRRs, the contribution of the mentioned electric dipole to the first order resonance could be totally suppressed [95]. The second order resonance at about 1200 nm is identified with an electric dipole parallel to the incident field. Here the current in both SRR arms are oscillating in phase. Therefore, no current is generated in the base and the electric dipole parallel to the base does not exist. The third order resonance of the SRR at 900 nm exhibits contributions from several dipole and quadrupole moments. This hampers the description of the SRR at this wavelengths by simple constitutive relations.

To sum up, in this section the multipole analysis of meta-atoms has been intensively discussed by several exemplarily geometries. The first part was devoted to magnetic dipole resonances in self-assembled MMs. By starting from a single dielectric sphere more challenging geometries were developed offering an isotropic magnetic dipole response in the visible domain. The second part of this section presented the multipole response of the cut-plate pair and the SRR that are commonly fabricated by top-down fabrication process. Here, the advantages of the multipole analysis for comparing different meta-atom geometries as well as identifying unexpected multipole contributions to resonances were presented.

4.3.4 Light Propagation in Amorphous Metamaterials

The introduction of self-assembled MMs in Sect. 4.1 clearly reveals that most of these MMs offer solely short-range ordering which is required to fabricate the single meta-atoms. The arrangement of these meta-atoms in space forming the entire MM is in most of the cases fully amorphous. The advantage of this amorphous arrangement is its isotropic optical response. Therefore, the light propagation through this amorphous MMs does not depend on the polarization state and the angle of incidence of the illuminating field. In contrary, such amorphous ordering of the meta-atoms hampers the theoretical description on how light propagates in these MMs. In the case of periodically arranged meta-atoms, as they usually appear for top-down fabrication processes, periodic boundary conditions can be introduced. In principle, this reduces the complexity of the problem to the question on how light interacts with the single meta-atom. The aim of this section is to introduce exactly such an reduction of the highly complex problem of light propagation in amorphous self-assembled MMs by exploiting the previously introduced multipole analysis of meta-atoms in Sects. 4.3.2 and 4.3.3.

One common way to describe the light propagation in MMs is to assume them acting like a homogeneous material. As long as this assumption holds, i.e. the meta-atoms are small enough to offer solely electric and magnetic dipole responses, one could introduce so-called effective parameters and replace the MM by a homogeneous material assigning these parameters. To retrieve the effective parameters one way is to consider reflection R and transmission T of a slab made of the MM. The important point is that R and T are required in amplitude and phase. As long as the assumption of a homogeneous material holds, the inversion of the analytical Fresnel formulas for reflection and transmission on a homogeneous slab yields the effective parameters [115, 116]. This procedure is of course not restricted to periodic MMs since the only requirement is the availability of the complex reflection and transmission. Anyhow, measuring the phase of R and T for a MMs slab remains still a challenging experimental task, and only a few approaches on solutions can be found in literature [64, 117–119]. Therefore, in most of the cases the amplitudes of R and T are measured and the phase information is taken from simulations of the samples. For periodic MMs these simulations, as mentioned previously, can be done by exploiting the periodicity of the system. This allows, e.g., to treat the light propagation in terms of Bloch-waves in periodic materials and the effective parameters can be directly related to the fundamental Bloch eigenmode of the system [120–122]. This treatment, in principle, reduces the complexity of the problem to the solution of the scattering problem of the single unit-cell which is the meta-atom. If the arrangement of these meta-atoms is not periodic, meaning amorphous in general, then one possibility to rigorously simulate the light propagation in such an amorphous MM slab are large scale simulations with a so-called supercell ansatz. Here the amorphous arrangement of the meta-atoms is taken fully into account in simulations until some fixed lateral extension of the MM slab. This so-called supercell is then periodically continued along the lateral dimensions of the slab. If the lateral extension of the

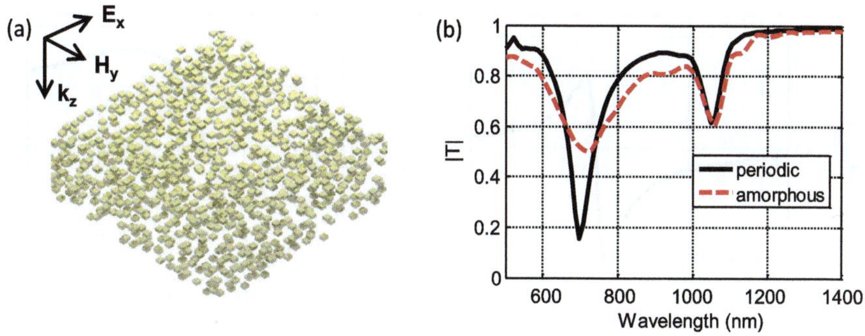

Fig. 4.8 (**a**) Supercell geometry and illumination scenario for amorphously arranged cut-plate pairs. The geometrical parameters of the cut-plate pairs are identical to Fig. 4.7 (a). (**b**) Respective amplitude of the transmission for the supercell from (**a**) and for periodically arranged cut-plate pairs

supercell is large enough (compared to the wavelength) the periodicity of the system will not influence the optical response. An example for a supercell including cut-plate pairs amorphously arranged in 4 layers can be seen in Fig. 4.8 (a). The amorphous arrangement of the cut-plate pairs is as follows. They are all orientated in the same direction concerning the incident field identical to Fig. 4.7 (a). This allows the excitation of the magnetic resonance. Furthermore, the supercell consists of four layers of amorphously arranged cut-plate pairs with a fixed distance between adjacent layers. In other words, the amorphous arrangement takes only place inside the four layers. More details can be found in Ref. [123].

The respective amplitudes of the transmission spectra received from large-scale FDTD simulations considering the presented supercell [95] can be seen in Fig. 4.8 (b). For comparison the transmission is shown for a periodical arrangement of the cut-plate pairs. By comparing the shown transmission spectra with the multipole amplitudes of the single cut-plate pair from Fig. 4.7 (a) one can conclude the following points. The resonance at around 1100 nm in transmission should be caused by the magnetic resonance of the cut-plate pair. Furthermore, the second resonance in transmission at about 700 nm can be clearly related to the electric resonance of the cut-plate pair. By comparing the resulting transmission spectra for a periodic and an amorphous arrangement of the meta-atoms in Fig. 4.8 (b) one can see that the transmission spectra of the magnetic resonance is nearly identical whereas there are remarkable differences for the electric resonance. Here the amplitude is lowered and the width is increased for the amorphous arrangement. This was later explained by the ratio of the average distance of particles to the wavelength which is close to unity for the electric resonance and with the much smaller radiation loss of the magnetic resonance. This causes a strongly reduced interaction among all the particles at this wavelength since the magnetic resonance is rather dominated by absorption.

With the complex transmission and reflection (not shown here) spectra from large-scale simulations for the amorphous MM at hand, it is possible to assign ef-

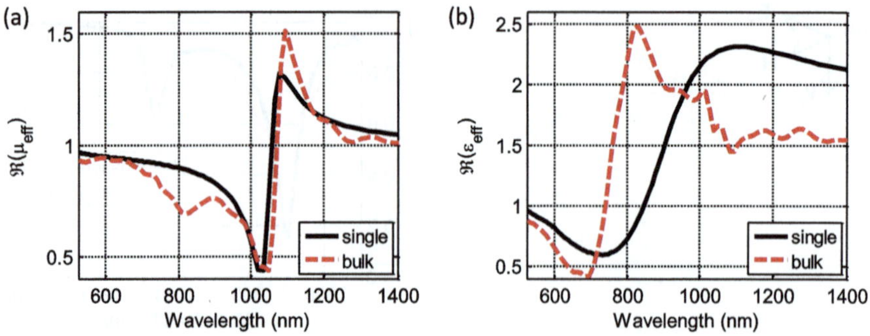

Fig. 4.9 Real part of the assigned effective permeability (**a**) and permittivity (**b**) to the amorphous MM slab consisting of cut-plate pairs from Fig. 4.8 (a). The *solid lines* refer to results obtained by the Clausius-Mossotti equation and the *dashed lines* to the inversion of complex reflection and transmission

fective parameters as described previously. However, a reduction of the simulation issue to considering solely the single cut-plate pair like for periodic MMs would be highly desirable. One possible way of doing this is to exploit mixing rules as they are discussed in detail in other chapters of this book, e.g. by Sihvola and co-workers. The most common ansatz is to use the Clausius-Mossotti equation which relates the polarizability of a molecule (in the introduced nomenclature this is the meta-atom) to an effective permittivity and permeability of the MM in the following way

$$\varepsilon_{\text{eff}} = \varepsilon_0 \frac{3 + 2N\alpha^e}{3 - N\alpha^e}, \qquad \mu_{\text{eff}} = \mu_0 \frac{3 + 2N\alpha^m}{3 - N\alpha^m}, \tag{4.6}$$

where ε_0 and μ_0 are the permittivity and permeability of the surrounding, N is the volume filling fraction of the meta-atoms, and α^e, α^m are the electric and magnetic polarizability, respectively. The electric polarizability is linked to the electric dipole of the meta-atoms by the electric field at the origin $\alpha^e = p/E(r = 0)$. The same holds for the magnetic polarizability by applying the magnetic field. In other words, the Clausius-Mossotti equation relates the dipolar response of single meta-atoms to effective parameters of the corresponding amorphous MM. Obviously, one restriction to apply this equation is that the meta-atom should exhibit only electric and magnetic dipole amplitudes at the considered wavelength domain. This could be easily verified by the introduced multipole analysis of meta-atoms and it was indeed the case for the cut-plate pair, cf. Fig. 4.7 (a).

The comparison of both introduced methods to assign effective parameters to amorphous MMs [via complex R and T or by (4.6)] can be seen in Fig. 4.9.

An almost perfect agreement between both methods can be observed for the effective permeability presented in Fig. 4.9 (a). Here the resonance positions, their strengths and the widths nicely fell together. Of course some minor wiggling of the effective permeability at off-resonance wavelengths for inverting R and T is not reproduced by the Clausius-Mossotti equation. This is obvious since for Clausius-Mossotti no fine details of the ordering of the cut-plate pairs is taken into account.

Anyhow, this small wiggling should only weakly influence the light propagation in the amorphous MM. For the assigned effective permittivity in Fig. 4.9 (b) almost the same trend could be observed for both applied methods. The resonance positions of the effective permittivity are slightly different but the resonance strengths as well the widths appear to be comparable. As for the effective permeability there exist some fine details for the effective permittivity by inverting R and T which are not reproduced by Clausius-Mossotti. The presented effective parameters in Fig. 4.9 demonstrate that the light propagation through a slab of an amorphous MM can be properly described by exploiting the Clausius-Mossotti equation instead of performing large-scale simulations to obtain R and T.

To sum up, a method was introduced to describe the light propagation in amorphous self-assembled MMs. It is based on the assumption that the MM could be described by homogeneous material parameters. Based on the introduced multipole analysis of meta-atoms in Sects. 4.3.2 and 4.3.3 effective parameters were assigned using the Clausius-Mossotti equation. The validity of this approach was counterchecked by rigorous large-scale simulations of the transmission and reflection of a MM slab consisting of amorphously arranged cut-plate pairs. With the introduced approach the complexity of the description of light propagation through amorphous MMs could be reduced to simulating the scattering response of the single meta-atoms and performing a multipole analysis.

4.4 Conclusion

The theoretical description of the interaction of light with amorphous MMs fabricated by self-assembly processes was outlined. The optical response of the single meta-atoms was identified as the most important information to understand and provide deep physical insights into the optical response of amorphous MMs. A formalism, called the multipole analysis of meta-atoms, was introduced. It allows the identification of contributions of Cartesian multipole moments to resonances of the scattering response of arbitrarily shaped meta-atoms. Advantages of this formalism were outlined by discussing exclusive examples of MMs fabricated by bottom-up as well as top-down processes. Based on the multipole analysis of meta-atoms one possible way of describing the light propagation in amorphous MMs was introduced. Therefore, mixing rules, as also intensively discussed in other chapters of this book were exploited to assign effective parameters to amorphous MMs.

Acknowledgements The research leading to these results has received funding from the European Union's Seven Framework Programme (FP7/2007-2013) under Grant Agreement No. 228455-NANOGOLD (Self-organized nanomaterials for tailored optical and electrical properties). We would also like to thank especially the Federal Ministry of Education and Research that supported this work within the project PhoNa and the Thuringian State Government for support in the MeMa project.

References

1. A.P. Alivisatos, K.P. Johnsson, X. Peng, T.E. Wilson, C.J. Loweth, M.P. Bruchez, P.G. Schultz, Nature **382**, 609–611 (1996)
2. L.C. Brousseau III, J.P. Novak, S.M. Marinakos, D.L. Feldheim, Adv. Mater. **11**, 447–449 (1999)
3. K.J.M. Bishop, C.E. Wilmer, S. Soh, B.A. Grzybowski, Small **5**, 1600–1630 (2009)
4. C.L. Choi, A.P. Alivisatos, Annu. Rev. Phys. Chem. **61**, 369–389 (2010)
5. J.M. Romo-Herrera, R.A. Alvarez-Puebla, L.M. Liz-Marzan, Nanoscale **3**, 1304–1315 (2011)
6. A. Guerrero-Martínez, J.L. Alonso-Gómez, B. Auguié, M.M. Cid, L.M. Liz-Marzán, Nano Today **6**, 381–400 (2011)
7. J.Y. Cheng, C.A. Ross, E.L. Thomas, H.I. Smith, G.J. Vancso, Appl. Phys. Lett. **81**, 3657–3659 (2002)
8. S. Darling, Prog. Polym. Sci. **32**, 1152–1204 (2007)
9. T. Smart, H. Lomas, M. Massignani, M.V. Flores-Merino, L.R. Perez, G. Battaglia, Nano Today **3**, 38–46 (2008)
10. J. Bang, U. Jeong, D.Y. Ryu, T.P. Russell, C.J. Hawker, Adv. Mater. **21**, 4769–4792 (2009)
11. J. Polleux, M. Rasp, I. Louban, N. Plath, A. Feldhoff, J.P. Spatz, ACS Nano **5**, 6355–6364 (2011)
12. X. Wang, D.-H. Kwon, D.H. Werner, I.-C. Khoo, A.V. Kildishev, V.M. Shalaev, Appl. Phys. Lett. **91**, 143122 (2007)
13. R. Pratibha, K. Park, I.I. Smalyukh, W. Park, Opt. Express **17**, 19459–19469 (2009)
14. R. Pratibha, W. Park, I.I. Smalyukh, J. Appl. Phys. **107**, 063511 (2010)
15. R.J. Macfarlane, B. Lee, M.R. Jones, N. Harris, G.C. Schatz, C.A. Mirkin, Science **334**, 204–208 (2011)
16. W. Cheng, M.J. Campolongo, J.J. Cha, S.J. Tan, C.C. Umbach, D.A. Muller, D. Luo, Nat. Mater. **8**, 519–525 (2009)
17. A.M. Hung, C.M. Micheel, L.D. Bozano, L.W. Osterbur, G.M. Wallraff, J.N. Cha, Nat. Nanotechnol. **5**, 121–126 (2010)
18. Y. Chen, J. Fu, K.C. Ng, Y. Tang, W. Cheng, Cryst. Growth Des. **11**, 4742–4746 (2011)
19. K.C. Ng, I.B. Udagedara, I.D. Rukhlenko, Y. Chen, Y. Tang, M. Premaratne, W. Cheng, ACS Nano **6**, 925–934 (2012)
20. T. Wen, S.A. Majetich, ACS Nano **5**, 8868–8876 (2011)
21. K.J. Stebe, E. Lewandowski, M. Ghosh, Science **325**, 159–160 (2009)
22. J.H. Lee, Q. Wu, W. Park, Opt. Lett. **34**, 443–445 (2009)
23. T. Lerond, J. Proust, H. Yockell-Lelièvre, D. Gérard, J. Plain, Appl. Phys. Lett. **99**, 123110 (2011)
24. J. Henzie, M. Grünwald, A. Widmer-Cooper, P.L. Geissler, P. Yang, Nat. Mater. **11**, 131–137 (2011)
25. D.V. Talapin, E.V. Shevchenko, M.I. Bodnarchuk, X. Ye, J. Chen, C.B. Murray, Nature **461**, 964–967 (2009)
26. V.A. Tamma, J.-H. Lee, Q. Wu, W. Park, Appl. Opt. **49**, A11–A17 (2009)
27. J.A. Fan, C. Wu, K. Bao, J. Bao, R. Bardhan, N.J. Halas, V.N. Manoharan, P. Nordlander, G. Shvets, F. Capasso, Science **328**, 1135–1138 (2010)
28. L. Chuntonov, G. Haran, Nano Lett. **11**, 2440–2445 (2011)
29. J.A. Fan, K. Bao, C. Wu, J. Bao, R. Bardhan, N.J. Halas, V.N. Manoharan, G. Shvets, P. Nordlander, F. Capasso, Nano Lett. **10**, 4680–4685 (2010)
30. S.J. Tan, M.J. Campolongo, D. Luo, W. Cheng, Nat. Nanotechnol. **6**, 268–276 (2011)
31. C.A. Mirkin, R.L. Letsinger, R.C. Mucic, J.J. Storhoff, Nature **382**, 607–609 (1996)
32. O.I. Wilner, I. Willner, Chem. Rev. **112**, 2528–2556 (2012)
33. S.A. Claridge, S.L. Goh, J.M.J. Fréchet, S.C. Williams, C.M. Micheel, A.P. Alivisatos, Chem. Mater. **17**, 1628–1635 (2005)
34. C.M. Soto, A. Srinivasan, B.R. Ratna, J. Am. Chem. Soc. **124**, 8508–8509 (2002)

35. S. Sheikholeslami, Y.-w. Jun, P.K. Jain, A.P. Alivisatos, Nano Lett. **10**, 2655–2660 (2010)
36. J.A. Fan, Y. He, K. Bao, C. Wu, J. Bao, N.B. Schade, V.N. Manoharan, G. Shvets, P. Nordlander, D.R. Liu, F. Capasso, Nano Lett. **11**, 4859–4864 (2011)
37. S.J. Barrow, A.M. Funston, D.E. Gómez, T.J. Davis, P. Mulvaney, Nano Lett. **11**, 4180–4187 (2011)
38. H. Xing, Z. Wang, Z. Xu, N.Y. Wong, Y. Xiang, G.L. Liu, Y. Lu, ACS Nano **6**, 802–809 (2012)
39. D. Nykypanchuk, M.M. Maye, D. van der Lelie, O. Gang, Nature **451**, 549–552 (2008)
40. A.J. Mastroianni, S.A. Claridge, A.P. Alivisatos, J. Am. Chem. Soc. **131**, 8455–8459 (2009)
41. O.-S. Lee, T.R. Prytkova, G.C. Schatz, J. Phys. Chem. Lett. **1**, 1781–1788 (2010)
42. J.P. Novak, D.L. Feldheim, J. Am. Chem. Soc. **122**, 3979–3980 (2000)
43. S. Mühlig, C. Rockstuhl, V. Yannopapas, T. Bürgi, N. Shalkevich, F. Lederer, Opt. Express **19**, 9607–9616 (2011)
44. D.A. Walker, B. Kowalczyk, M.O. De la Cruz, B.A. Grzybowski, Nanoscale **3**, 1316–1344 (2011)
45. G. Decher, Science **277**, 1232–1237 (1997)
46. A. Cunningham, S. Mühlig, C. Rockstuhl, T. Bürgi, J. Phys. Chem. C **115**, 8955–8960 (2011)
47. L.V. Brown, H. Sobhani, J.B. Lassiter, P. Nordlander, N.J. Halas, ACS Nano **4**, 819–832 (2010)
48. M.N. Hyder, S.W. Lee, F.C. Cebeci, D.J. Schmidt, Y. Shao-Horn, P.T. Hammond, ACS Nano **5**, 8552–8561 (2011)
49. S. Mühlig, A. Cunningham, S. Scheeler, C. Pacholski, T. Bürgi, C. Rockstuhl, F. Lederer, ACS Nano **5**, 6586–6592 (2011)
50. M. Gellner, S. Niebling, H.Y. Kuschelmeister, C. Schmuck, S. Schlücker, Chem. Commun. **47**, 12762–12764 (2011)
51. M. Gellner, D. Steinigeweg, S. Ichilmann, M. Salehi, M. Schütz, K. Kömpe, M. Haase, S. Schlücker, Small **7**, 3445–3451 (2011)
52. D.A. Pawlak, S. Turczynski, M. Gajc, K. Kolodziejak, R. Diduszko, K. Rozniatowski, J. Smalc, I. Vendik, Adv. Funct. Mater. **20**, 1116–1124 (2010)
53. B. Senyuk, J.S. Evans, P.J. Ackerman, T. Lee, P. Manna, L. Vigderman, E.R. Zubarev, J.v.d. Lagemaat, I.I. Smalyukh, Nano Lett. **12**, 955–963 (2012)
54. H.O. Moser, C. Rockstuhl, Laser Photonics Rev. **6**, 219–244 (2012)
55. D.R. Smith, J.B. Pendry, M.C.K. Wiltshire, Science **305**, 788–792 (2004)
56. V.M. Shalaev, Nat. Photonics **1**, 41–48 (2007)
57. C.M. Soukoulis, M. Wegener, Nat. Photonics **5**, 523–530 (2011)
58. T. Okamoto, *Near-Field Optics and Surface Plasmon Polaritons* (Springer, Berlin, 2001)
59. Y.-Y. Yu, S.-S. Chang, C.-L. Lee, C.R.C. Wang, J. Phys. Chem. B **101**, 6661–6664 (1997)
60. C. Helgert, C. Menzel, C. Rockstuhl, E. Pshenay Severin, E. B. Kley, A. Chipouline, A. Tünnermann, F. Lederer, T. Pertsch, Opt. Lett. **34**, 704–706 (2009)
61. E. Prodan, C. Radloff, N.J. Halas, P. Nordlander, Science **302**, 419–422 (2003)
62. S. Zhang, W. Fan, N.C. Panoiu, K.J. Malloy, R.M. Osgood, S.R.J. Brueck, Phys. Rev. Lett. **95**, 137404 (2005)
63. J. Yao, Z. Liu, Y. Liu, Y. Wang, C. Sun, G. Bartal, A.M. Stacy, X. Zhang, Science **321**, 930 (2008)
64. G. Dolling, C. Enkrich, M. Wegener, C.M. Soukoulis, S. Linden, Science **312**, 892–894 (2006)
65. N. Liu, H. Guo, L. Fu, S. Kaiser, H. Schweizer, H. Giessen, Nat. Mater. **7**, 31–37 (2008)
66. J. Valentine, S. Zhang, T. Zentgraf, E. Ulin-Avila, D.A. Genov, G. Bartal, X. Zhang, Nature **455**, 376–379 (2008)
67. C. Menzel, A. Andryieuski, C. Rockstuhl, R. Iliew, R. Malureanu, F. Lederer, A. Lavrinenko, Phys. Rev. B **81**, 195123 (2010)
68. C. Enkrich, M. Wegener, S. Linden, S. Burger, L. Zschiedrich, F. Schmidt, J.F. Zhou, Th. Koschny, C.M. Soukoulis, Phys. Rev. Lett. **95**, 203901 (2005)

69. J.K. Gansel, M. Thiel, M.S. Rill, M. Decker, K. Bade, V. Saile, G. von Freymann, S. Linden, M. Wegerner, Science **325**, 1513–1515 (2009)
70. C. Menzel, T. Paul, C. Rockstuhl, T. Pertsch, S. Tretyakov, F. Lederer, Phys. Rev. B **81**, 035320 (2010)
71. J.B. Pendry, Phys. Rev. Lett. **85**, 3966–3969 (2000)
72. T. Paul, C. Menzel, C. Rockstuhl, F. Lederer, Adv. Mater. **22**, 2354–2357 (2010)
73. M. Husnik, M.W. Klein, N. Feth, M. König, J. Niegemann, K. Busch, S. Linden, M. Wegener, Nat. Photonics **2**, 614–617 (2008)
74. T. Zentgraf, J. Dorfmüller, C. Rockstuhl, C. Etrich, R. Vogelgesang, K. Kern, T. Pertsch, F. Lederer, H. Giessen, Opt. Lett. **33**, 848–850 (2008)
75. I. Sersic, C. Tuambilangana, A.F. Koenderink, New J. Phys. **13**, 083019 (2011)
76. P. Banzer, U. Peschel, S. Quabis, G. Leuchs, Opt. Express **18**, 10905–10923 (2010)
77. F. von Cube, S. Irsen, J. Niegemann, C. Matyssek, W. Hergert, K. Busch, S. Linden, Opt. Mater. Express **1**, 1009–1018 (2011)
78. J.B. Pendry, A.J. Holden, D.J. Robbins, W.J. Stewart, IEEE Trans. Microw. Theory Tech. **47**, 2075–2084 (1999)
79. S. Linden, C. Enkrich, M. Wegener, J. Zhou, T. Koschny, C.M. Soukoulis, Science **306**, 1351–1353 (2004)
80. I. Sersic, M. Frimmer, E. Verhagen, A.F. Koenderink, Phys. Rev. Lett. **103**, 213902 (2009)
81. I. Sersic, C. Tuambilangana, T. Kampfrath, A.F. Koenderink, Phys. Rev. B **83**, 245102 (2011)
82. J. Petschulat, C. Menzel, A. Chipouline, C. Rockstuhl, A. Tünnermann, F. Lederer, T. Pertsch, Phys. Rev. A **78**, 043811 (2008)
83. C. Rockstuhl, F. Lederer, C. Etrich, T. Zentgraf, J. Kuhl, H. Giessen, Opt. Express **14**, 8827–8836 (2006)
84. T.D. Corrigan, P.W. Kolb, A.B. Sushkov, H.D. Drew, D.C. Schmadel, R.J. Phaneuf, Opt. Express **16**, 19850 (2008)
85. P.C. Waterman, Phys. Rev. D **3**, 825–839 (1971)
86. M.I. Mishchenko, L.D. Travis, D.W. Mackowski, J. Quant. Spectrosc. Radiat. Transf. **55**, 535–575 (1996)
87. T.A. Nieminen, N.R. Heckenberg, H. Rubinsztein-Dunlop, in *SPIE Conference Series*, vol. 5514, ed. by K. Dholakia, G.C. Spalding (2004), pp. 514–523
88. D.W. Mackowski, M.I. Mishchenko, J. Opt. Soc. Am. A **13**, 2266–2278 (1996)
89. J.B. Schneider, I.C. Peden, IEEE Trans. Antennas Propag. **36**, 1317–1321 (1988)
90. T.A. Nieminen, V.L.Y. Loke, A.B. Stilgoe, G. Knöner, A.M. Brańczyk, N.R. Heckenberg, H. Rubinsztein-Dunlop, J. Opt. A, Pure Appl. Opt. **9**, 196–203 (2007)
91. S. Mühlig, C. Menzel, C. Rockstuhl, F. Lederer, Metamaterials **5**, 64–73 (2011)
92. Y.-l. Xu, Appl. Opt. **34**, 4573–4588 (1995)
93. S. Mühlig, C. Rockstuhl, J. Pniewski, C.R. Simovski, S.A. Tretyakov, F. Lederer, Phys. Rev. B **81**, 075317 (2010)
94. C.F. Bohren, D.R. Huffman, *Absorption and Scattering of Light by Small Particles*, (1998)
95. C. Rockstuhl, C. Menzel, S. Mühlig, J. Petschulat, C. Helgert, C. Etrich, A. Chipouline, T. Pertsch, F. Lederer, Phys. Rev. B **83**, 245119 (2011)
96. U. Leonhardt, T.G. Philbin, in *Progress in Optics*, vol. 53 (Elsevier, Amsterdam, 2009), pp. 69–152
97. E.D. Palik, *Handbook of Optical Constants of Solids*, (1985)
98. A.B. Evlyukhin, C. Reinhardt, A. Seidel, B.S. Luk'Yanchuk, B.N. Chichkov, Phys. Rev. B **82**, 045404 (2010)
99. A. García-Etxarri, R. Gómez-Medina, L.S. Froufe-Pérez, C. López, L. Chantada, F. Scheffold, J. Aizpurua, M. Nieto-Vesperinas, J.J. Sáenz, Opt. Express **19**, 4815–4826 (2011)
100. Q. Zhao, J. Zhou, F. Zhang, D. Lippens, Mater. Today **12**, 60–69 (2009)
101. V. Yannopapas, A. Moroz, J. Phys. Condens. Matter **17**, 3717–3734 (2005)
102. M.S. Wheeler, J.S. Aitchison, M. Mojahedi, Phys. Rev. B **72**, 193103 (2005)
103. B.-J. Seo, T. Ueda, T. Itoh, H. Fetterman, Appl. Phys. Lett. **88**, 161122 (2006)
104. L. Jylhä, I. Kolmakov, S. Maslovski, S. Tretyakov, J. Appl. Phys. **99**, 043102 (2006)

105. V. Yannopapas, Phys. Rev. B **75**, 035112 (2007)
106. V. Yannopapas, Appl. Phys. A **87**, 259–264 (2007)
107. J.A. Schuller, R. Zia, T. Taubner, M.L. Brongersma, Phys. Rev. Lett. **99**, 107401 (2007)
108. M.S. Wheeler, J.S. Aitchison, J.I.L. Chen, G.A. Ozin, M. Mojahedi, Phys. Rev. B **79**, 073103 (2009)
109. P. Nordlander, C. Oubre, E. Prodan, K. Li, M.I. Stockman, Nano Lett. **4**, 899–903 (2004)
110. S. Riikonen, I. Romero, F.J. Garcia de Abajo, Phys. Rev. B **71**, 235104 (2005)
111. N. Shalkevich, A. Shalkevich, L. Si-Ahmed, T. Bürgi, Phys. Chem. Chem. Phys. **11**, 10175 (2009)
112. J. Dintinger, S. Mühlig, C. Rockstuhl, T. Scharf, Opt. Mater. Express **2**, 269–278 (2012)
113. C. Rockstuhl, F. Lederer, C. Etrich, T. Pertsch, T. Scharf, Phys. Rev. Lett. **99**, 017401 (2007)
114. C.R. Simovski, S.A. Tretyakov, Phys. Rev. B **79**, 045111 (2009)
115. D.R. Smith, S. Schultz, P. Markoš, C.M. Soukoulis, Phys. Rev. B **65**, 195104 (2002)
116. C. Menzel, C. Rockstuhl, T. Paul, F. Lederer, T. Pertsch, Phys. Rev. B **77**, 195328 (2008)
117. S. Zhang, W. Fan, N.C. Panoiu, K.J. Malloy, R.M. Osgood, S.R.J. Brueck, Phys. Rev. Lett. **95**, 137404 (2005)
118. V.P. Drachev, W. Cai, U. Chettiar, H.-K. Yuan, A.K. Sarychev, A.V. Kildishev, G. Klimeck, V.M. Shalaev, Laser Phys. Lett. **3**, 49–55 (2006)
119. E. Pshenay-Severin, F. Setzpfandt, C. Helgert, U. Hübner, C. Menzel, A. Chipouline, C. Rockstuhl, A. Tünnermann, F. Lederer, T. Pertsch, J. Opt. Soc. Am. B **27**, 660–666 (2010)
120. C. Rockstuhl, C. Menzel, T. Paul, T. Pertsch, F. Lederer, Phys. Rev. B **78**, 155102 (2008)
121. T. Paul, C. Rockstuhl, C. Menzel, F. Lederer, Phys. Rev. B **79**, 115430 (2009)
122. T. Paul, C. Menzel, W. Śmigaj, C. Rockstuhl, P. Lalanne, F. Lederer, Phys. Rev. B **84**, 115142 (2011)
123. C. Helgert, C. Rockstuhl, C. Etrich, C. Menzel, E.-B. Kley, A. Tünnermann, F. Lederer, T. Pertsch, Phys. Rev. B **79**, 233107 (2009)

Chapter 5
Electrodynamic Theory of Three-Dimensional Metamaterials of Hierarchically Organized Nanoparticles

Vassilios Yannopapas, Alexandros G. Vanakaras, and Demetri J. Photinos

Abstract To date, the most promising candidate structures for exhibiting photo-induced magnetism and negative refractive index in the optical regime are the so called Mie resonance-based metamaterials which consist of scatterers of simple geometrical shape, e.g., spherical or cylindrical, and are made of a high-index material. When such a structure is illuminated by an electromagnetic wave of frequency around the Mie resonance of a single scatterer, strong polarization currents are generated within the surface of the scatterers resulting in a macroscopic magnetization of the metamaterial. Due to the lack of naturally occurring materials with high refractive index in the optical regime, one can envisage a metamaterial which consists of meta-atoms that are clusters of metallic nanoparticles wherein strong polarization currents can also be induced under illumination. These type of metamaterials are hierarchically organized as they possess two length scales: the inter-particle distance within the cluster and the inter-cluster separation within the metamaterial. The nanoparticle clusters can be formed by direct or template-assisted self-organization and are generally amorphous due to the random positioning of the nanoparticles in air or within a cavity. The amorphous arrangement of such strongly scattering objects constitutes a major challenge for the field of theoretical and computational nanophotonics. In order to tackle this computational problem in the framework of metamaterials, we adopt a hierarchical theoretical strategy in proportion to the hierarchical organization of such structures. To this end, we develop a layer-multiple-scattering formalism for electromagnetic waves in order to model the optical response of metamaterials formed as collections of cavities filled by amorphous clusters of hierarchically organized spherical nanoparticles. It is based on a three-stage process where we take fully into account all the multiple-scattering processes experienced by photons: (a) among the particles of the cluster inside the cavity, (b) be-

V. Yannopapas (✉) · A.G. Vanakaras · D.J. Photinos
Department of Materials Science, University of Patras, 26504 Patras, Greece
e-mail: vyannop@upatras.gr

A.G. Vanakaras
e-mail: a.g.vanakaras@upatras.gr

D.J. Photinos
e-mail: photinos@upatras.gr

C. Rockstuhl, T. Scharf (eds.), *Amorphous Nanophotonics*, Nano-Optics and Nanophotonics, DOI 10.1007/978-3-642-32475-8_5,
© Springer-Verlag Berlin Heidelberg 2013

tween the cluster and the cavity and (c) among the cavities (containing the clusters) within the metamaterial. We demonstrate the applicability of the method to the case of a silica-inverted opal whose voids contain clusters of gold nanoparticles. We find, in particular, such a metamaterial acts as a super absorber over a wide frequency range, from 2–4 eV.

5.1 Introduction

Metamaterials are man-made structures with electromagnetic (EM) properties which are not met in naturally occurring materials, such as artificial magnetism [1], negative refractive index (NRI) [2], near-field amplification [3], cloaking [4], and perfect absorption [5]. The magnetic response stems from the induction of strong currents within the metamaterials when illuminated externally and can lead to strong paramagnetic (permeability $\mu > 1$) and diamagnetic response (permeability $\mu < 1$ or even $\mu < 0$) in frequency regions where such a response is not met in naturally occurring materials, like the near-infrared and optical regions. Magnetic activity in these regions of the EM spectrum is of great technological importance since it allows for the realization of devices such as compact cavities, adaptive selective lenses, tunable mirrors, isolators, converters, optical polarizers, filters, and phase shifters [6, 7].

Ever since the concept of NRI metamaterials was introduced [2], the holy grail of research in this discipline has been the realization of an artificial material with negative refractive index in the visible regime. To this end, top-down technologies and lithographic techniques have been employed for the realization of optical metamaterials [8–15]. However, fabrication restrictions in the top-down approaches have limited the operating wavelength of the produced metamaterials up to the near infrared regime. Bottom-up approaches based on self-assembly technology allow for the realization of true optical metamaterials with the promise of lower cost, high throughput, and small sensitivity to damage or fabrication errors.

Optical metamaterials based on lithographic design are usually miniatures of their microwave counterparts being periodic arrays of LC circuits such as metallic split-ring resonators [16–18] and their variations. A viable alternative to the mainstream LC circuit-based metamaterials are the so-called Mie resonance-based metamaterials: periodic lattices consisting of scatterers of simple geometry such as two-dimensional (2D) arrays of cylinders [19, 20] or three-dimensional (3D) arrays of spherical particles [21–30] exhibiting strong magnetic activity as well as NRI in the microwave and infrared regimes (for a recent review see [31]). The emergence of magnetic response in these structures relies on a different mechanism from the excitation of an LC resonance. Namely, the materials which the particles are made of must exhibit a certain type of internal resonance around which the electric permittivity (dielectric function) of the material assumes very high values. Such a resonance might be a phonon-polariton or an exciton-polariton resonance and the corresponding materials can be ionic solids or polar semiconductors. When an array of particles made of such a resonant material is illuminated by an EM wave of fre-

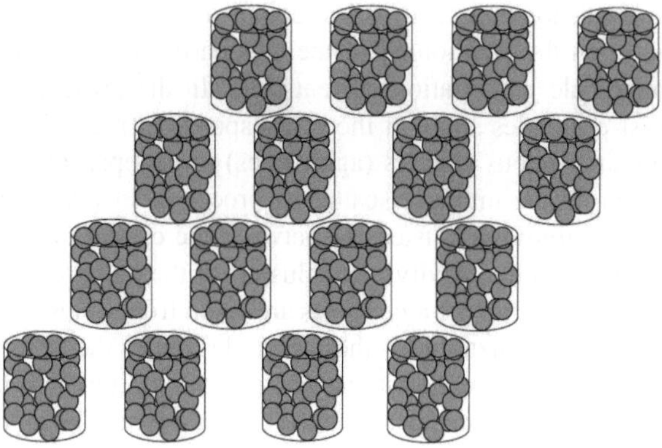

Fig. 5.1 Schematic of a 2D array of cylindrical cavities in a dielectric material containing amorphous clusters of metallic nanoparticles

quency around the Mie resonance of the particles, strong polarization currents are generated within the particles resulting in a macroscopic magnetization of the array of particles.

Proof-of-concept experiments demonstrating magnetic activity and/or NRI in Mie resonance-based metamaterials have been conducted in the millimeter [32–38] and infrared regimes [39]. Scaling down the above designs to the near infrared and optical regimes requires materials with high permittivity in these regimes which, however, are particularly rare.

Recently, a bottom-up approach for realizing Mie-resonance based metamaterials in the optical regime has been proposed [40–42] wherein the required high-dielectric material is formed by the self-organization of metallic nanoparticles of a few nm in radius. Such metamaterials have been realized by template-assisted colloidal self-organization wherein clusters of metallic nanoparticles reside within the voids of a 2D periodically perforated dielectric slab [43] (see Fig. 5.1) or within the trenches of a 1D grating [44]. The resulting structure does not demonstrate significant magnetic response due to the large size of the gold nanoclusters (390–420 nm). Further shrinking of the cluster size is not feasible due to the restrictions imposed by the lithographic patterning of the template which hosts the nanoparticles [43]. Thus, the original bottom-up approach of [40] which is based solely on self-organization of the nanoparticles via their functionalization with liquid-crystalline molecules (e.g., dendrimers, mesogens) appears to be the most viable solution for implementing the idea of a periodic metamaterial consisting of nanoclusters. To this end, arrays of nanostrings of gold nanoparticles coated with nematogenic ligands have already been reported [45].

The theoretical modeling of the above metamaterials relies on treating the EM response of the amorphous cluster of nanoparticles as a homogeneous body 'cropped' out of an infinite metamaterial [40–44]. This approach is potentially valid only when the metallic nanoparticles are small enough (few nm) and the corresponding aggregates are sufficiently large (several hundreds of nm) so that surface effects are negligible. For small aggregates containing tens or few hundreds of metallic nanoparticles, most particles reside at the surface of the cluster rendering an effective-medium approach inadequate. Furthermore, this approach does not account for several phenomena such as the actual dependence of the effective parameters (ϵ_{eff}, μ_{eff}) on the positioning of the nanoparticles within the cluster, surface roughness, nanoscale polarization currents, etc. In the present study we aim at presenting a first-principles study of the EM response of periodic structures of cavities containing amorphous clusters (aggregates) of nanoparticles by accurately taking into account all the multiple-scattering processes involved: (a) among the gold nanoparticles *within* the cluster, (b) between the cluster and the cavity, and (c) among the repeating units (cavity+nanocluster) of the metamaterial. The positions of the gold particles within the aggregates are taken from a rigorous Monte-Carlo simulation of the self-organization of these particles under the spherically confining potential of the cavity. The multiple-scattering theory of light within a single nanocluster+cavity provides a *total* scattering matrix. Having found the latter matrix, we embed it within the existing layer-multiple-scattering (LMS) formalism [46–48] which provides us with the transmittance, reflectance and absorbance of light incident on one or many two-dimensional lattices (planes) of clusters. Alternatively, the EM response can be studied by means of the complex frequency band structure of an infinitely periodic metamaterial of clusters of particles. Besides the apparent application of the present work in metamaterials realized by template-assisted self-assembly, the formalism presented here constitutes a significant extension of the existing LMS formulations [47–49] enabling the study of periodic structures with more than one scatterers in the unit cell.

In case where randomness is evident in the spatial arrangement of the clusters *within* the metamaterial as well, e.g., we may have amorphous clusters of NPs occupying *randomly* the positions of a three-dimensional lattice, one can extend the above cluster-LMS formalism to non-periodic metamaterials of NP clusters by employing the average T-matrix approximation.

5.2 Theory

5.2.1 Multipole Expansion of the EM Field

Consider a harmonic EM wave, of angular frequency ω which is described by its electric-field component

$$\mathbf{E}(\mathbf{r},t) = \text{Re}\big[\mathbf{E}(\mathbf{r})\exp(-\mathrm{i}\omega t)\big]. \tag{5.1}$$

In a homogeneous medium characterized by a dielectric function $\epsilon(\omega)\epsilon_0$ and a magnetic permeability $\mu(\omega)\mu_0$, where ϵ_0, μ_0 are the electric permittivity and magnetic permeability of vacuum, Maxwell's equations imply that $\mathbf{E}(\mathbf{r})$ satisfies a vector Helmholtz equation, subject to the condition $\nabla\cdot\mathbf{E}=0$, with wave number $q=\omega/c$, where $c = 1/\sqrt{\mu\epsilon\mu_0\epsilon_0} = c_0/\sqrt{\mu\epsilon}$ is the velocity of light in the medium. The spherical-wave expansion of $\mathbf{E}(\mathbf{r})$ is given by [50]

$$\mathbf{E}(\mathbf{r}) = \sum_{l=1}^{\infty}\sum_{m=-l}^{l}\left\{a_{lm}^{H} f_l(qr)\mathbf{X}_{lm}(\hat{\mathbf{r}}) + a_{lm}^{E}\frac{\mathrm{i}}{q}\nabla\times\big[f_l(qr)\mathbf{X}_{lm}(\hat{\mathbf{r}})\big]\right\}, \tag{5.2}$$

where a_{lm}^{P} ($P = E, H$) are coefficients to be determined. l is the total angular momentum number and m the azimuthal number. $\mathbf{X}_{lm}(\hat{\mathbf{r}})$ are the so-called vector spherical harmonics [50] and f_l may be any linear combination of the spherical Bessel function, j_l, and the spherical Hankel function, h_l^+. The corresponding magnetic induction, $\mathbf{B}(\mathbf{r})$, can be readily obtained from $\mathbf{E}(\mathbf{r})$ using Maxwell's equations,

$$\mathbf{B}(\mathbf{r}) = \frac{\sqrt{\epsilon\mu}}{c_0} \sum_{l=1}^{\infty} \sum_{m=-l}^{l} \left\{ a_{lm}^{E} f_l(qr)\mathbf{X}_{lm}(\hat{\mathbf{r}}) - a_{lm}^{H} \frac{\mathrm{i}}{q} \nabla \times \left[f_l(qr)\mathbf{X}_{lm}(\hat{\mathbf{r}}) \right] \right\}, \tag{5.3}$$

and we shall not write it down explicitly in what follows.

5.2.2 Scattering by a Single Scatterer

In this subsection we summarize the solution to the problem of EM scattering from a single sphere (Mie scattering theory [50, 51]) along with an extension to the case of irregular (nonphysical) solutions which are necessary for a wave emitted by the center of the sphere. We will make use of the compact notation of Ref. [52] for the eigenfunctions and the angular-momentum indices which allows for easier computer coding.

We consider a sphere of radius S, with its center at the origin of coordinates, and assume that its electric permittivity ϵ_s and/or magnetic permeability μ_s are different from those of the surrounding homogeneous medium, ϵ_h, μ_h. An EM plane wave incident on this scatterer is described, respectively, by (5.2) with $f_l = j_l$ (since the plane wave is finite everywhere) and appropriate coefficients a_L^0, where L denotes collectively the indices Plm. That is,

$$\mathbf{E}^0(\mathbf{r}) = \sum_L a_L^0 \mathbf{J}_L(\mathbf{r}) \tag{5.4}$$

where

$$\mathbf{J}_{Elm}(\mathbf{r}) = \frac{\mathrm{i}}{q_h} \nabla \times j_l(q_h r)\mathbf{X}_{lm}(\hat{\mathbf{r}}), \qquad \mathbf{J}_{Hlm}(\mathbf{r}) = j_l(q_h r)\mathbf{X}_{lm}(\hat{\mathbf{r}}) \tag{5.5}$$

and $q_h = \sqrt{\epsilon_h \mu_h}\,\omega/c_0$. The coefficients a_L^0 depend on the amplitude, polarization and propagation direction of the incident EM plane wave [46–48, 50].

Similarly, the wave that is scattered from the sphere is described by (5.2) with $f_l = h_l^+$, which has the asymptotic form appropriate to an outgoing spherical wave: $h_l^+ \approx (-\mathrm{i})^l \exp(\mathrm{i}q_h r)/\mathrm{i}q_h r$ as $r \to \infty$, and appropriate expansion coefficients a_L^+. Namely,

$$\mathbf{E}^+(\mathbf{r}) = \sum_L a_L^+ \mathbf{H}_L(\mathbf{r}) \tag{5.6}$$

where

$$\mathbf{H}_{Elm}(\mathbf{r}) = \frac{\mathrm{i}}{q_h} \nabla \times h_l^+(q_h r)\mathbf{X}_{lm}(\hat{\mathbf{r}}), \qquad \mathbf{H}_{Hlm}(\mathbf{r}) = h_l^+(q_h r)\mathbf{X}_{lm}(\hat{\mathbf{r}}). \tag{5.7}$$

The wavefield for $r > S$ is the sum of the incident and scattered waves, i.e., $\mathbf{E}^{out} = \mathbf{E}^0 + \mathbf{E}^+$. The spherical-wave expansion of the field $\mathbf{E}^I$ for $r < R$ (inside the sphere) is obtained in a similar manner by the requirement that it be finite at the origin ($\mathbf{r}=\mathbf{0}$), i.e.,

$$\mathbf{E}^{I}(\mathbf{r})=\sum_{L} a_{L}^{I}\mathbf{J}_{L}^{s}(\mathbf{r}) \tag{5.8}$$

where $\mathbf{J}_{L}^{s}(\mathbf{r})$ are given from (5.5) by replacing q_h with $q_s=\sqrt{\epsilon_s\mu_s}\omega/c_0$.

By applying the requirement that the tangential components of $\mathbf{E}$ and $\mathbf{H}$ be continuous at the surface of the scatterer, we obtain a relation between the expansion coefficients of the incident and the scattered field, as follows:

$$a_{L}^{+}=\sum_{L'} T_{LL'}a_{L'}^{0}, \tag{5.9}$$

where $T_{LL'}$ are the elements of the so-called scattering transition T-matrix [51]. Equation (5.9) is valid for any shape of scatterer; for spherically symmetric scatterers each spherical wave scatters independently of all others, which leads to a transition T-matrix that does not depend on m and is diagonal in l, i.e., $T_{LL'}=T_{L}\delta_{LL'}$; it is given by

$$T_{El}(\omega)=\left[\frac{j_l(q_s r)\frac{\partial}{\partial r}(rj_l(q_h r))\epsilon_s-j_l(q_h r)\frac{\partial}{\partial r}(rj_l(q_s r))\epsilon_h}{h_l^{+}(q_h r)\frac{\partial}{\partial r}(rj_l(q_s r))\epsilon_h-j_l(q_s r)\frac{\partial}{\partial r}(rh_l^{+}(q_h r))\epsilon_s}\right]_{r=S} \tag{5.10}$$

$$T_{Hl}(\omega)=\left[\frac{j_l(q_s r)\frac{\partial}{\partial r}(rj_l(q_h r))\mu_s-j_l(q_h r)\frac{\partial}{\partial r}(rj_l(q_s r))\mu_h}{h_l^{+}(q_h r)\frac{\partial}{\partial r}(rj_l(q_s r))\mu_h-j_l(q_s r)\frac{\partial}{\partial r}(rh_l^{+}(q_h r))\mu_s}\right]_{r=S}. \tag{5.11}$$

The expansion coefficients a_{L}^{I} of the field inside the spheres are given in terms of a_{L}^{0} by a similar expression

$$a_{L}^{I}=\sum_{L'} C_{LL'}a_{L'}^{0}. \tag{5.12}$$

The C-matrix is also independent of m and diagonal in l, i.e., $C_{LL'}=C_{L}\delta_{LL'}$; it is given in terms of $T_{LL'}$ by

$$C_{El}(\omega)=\sqrt{\frac{\epsilon_h\mu_s}{\epsilon_s\mu_h}}\left[\frac{j_l(q_h S)}{j_l(q_s S)}+\frac{h_l^{+}(q_h S)}{j_l(q_s S)}T_{El}\right] \tag{5.13}$$

$$C_{Hl}(\omega)=\frac{j_l(q_h S)}{j_l(q_s S)}+\frac{h_l^{+}(q_h S)}{j_l(q_s S)}T_{Hl} \tag{5.14}$$

In a similar manner one can define a wave which is infinite at the origin and matches continuously an outgoing spherical wave of given L outside the sphere. This definition is needed in order to solve the scattering problem from a source radiating from the center of the sphere. For this purpose, the electric field outside the sphere is written as

$$\mathbf{E}^{0}(\mathbf{r})=\sum_{L} b_{L}^{0}\mathbf{H}_{L}(\mathbf{r}) \tag{5.15}$$

and inside the sphere as

$$\mathbf{E}^I(\mathbf{r}) = \sum_L \left[b_L^I \mathbf{J}_L^s(\mathbf{r}) + b_L^{I+} \mathbf{H}_L^s(\mathbf{r}) \right] \tag{5.16}$$

where $\mathbf{H}_L^s(\mathbf{r})$ is given by (5.7) with q_s instead of q_h. Again, by requiring that the tangential components of the electric and magnetic fields be continuous on the surface of the sphere, one can express b_L^I and b_L^{I+} in terms of b_L^0 as follows

$$b_L^{I+} = \sum_{L'} Q_{LL'} b_{L'}^0, \qquad b_L^I = \sum_{L'} P_{LL'} b_{L'}^0, \tag{5.17}$$

where, similar to the T-matrix, $Q_{LL'}$ and $P_{LL'}$ are diagonal in l and do not depend on m. They are provided by

$$Q_{El}(\omega) = \sqrt{\frac{\epsilon_h \mu_s}{\epsilon_s \mu_h}} \left[\frac{h_l^+(q_h S)}{h_l^+(q_s S) + j_l(q_s S) V_{El}} \right] \tag{5.18}$$

$$Q_{El}(\omega) = \frac{h_l^+(q_h S)}{h_l^+(q_s S) + j_l(q_s S) V_{Hl}} \tag{5.19}$$

$$P_{El}(\omega) = V_{El} Q_{El} \tag{5.20}$$

$$P_{Hl}(\omega) = V_{Hl} Q_{Hl} \tag{5.21}$$

and

$$V_{El}(\omega) = \left[\frac{h_l^+(q_s r)\frac{\partial}{\partial r}(r h_l^+(q_h r))\epsilon_s - h_l^+(q_h r)\frac{\partial}{\partial r}(r h_l^+(q_s r))\epsilon_h}{h_l^+(q_h r)\frac{\partial}{\partial r}(r j_l(q_s r))\epsilon_h - j_l(q_s r)\frac{\partial}{\partial r}(r h_l^+(q_h r))\epsilon_s} \right]_{r=S} \tag{5.22}$$

$$V_{Hl}(\omega) = \left[\frac{h_l^+(q_s r)\frac{\partial}{\partial r}(r h_l^+(q_h r))\mu_s - h_l^+(q_h r)\frac{\partial}{\partial r}(r h_l^+(q_s r))\mu_h}{h_l^+(q_h r)\frac{\partial}{\partial r}(r j_l(q_s r))\mu_h - j_l(q_s r)\frac{\partial}{\partial r}(r h_l^+(q_h r))\mu_s} \right]_{r=S}. \tag{5.23}$$

5.2.3 Multiple-Scattering by a Collection of Spheres

Next we consider a collection of N_s nonoverlapping spherical scatterers centered at sites $\mathbf{R}_n$ in a homogeneous host medium. An outgoing vector spherical wave about $\mathbf{R}_{n'}$ can be expanded in a series of incoming vector spherical waves around $\mathbf{R}_n$ as follows

$$\mathbf{H}_{L'}(\mathbf{r} - \mathbf{R}_{n'}) = \sum_L \Omega_{LL'}^{nn'} \mathbf{J}_L(\mathbf{r} - \mathbf{R}_n). \tag{5.24}$$

An outgoing vector spherical wave about $\mathbf{R}_{n'}$ can be expanded in a series of outgoing vector spherical waves around $\mathbf{R}_n$ as follows

$$\mathbf{H}_{L'}(\mathbf{r} - \mathbf{R}_{n'}) = \sum_L \Xi_{LL'}^{nn'} \mathbf{H}_L(\mathbf{r} - \mathbf{R}_n) \tag{5.25}$$

and similarly for incoming vector spherical waves

$$\mathbf{J}_{L'}(\mathbf{r} - \mathbf{R}_{n'}) = \sum_L \Xi_{LL'}^{nn'} \mathbf{J}_L(\mathbf{r} - \mathbf{R}_n). \tag{5.26}$$

Explicit formulae for the matrices Ω and Ξ are given in the appendix. These matrices do not depend on the material properties of the scatterers but on their particular arrangement in space. From (5.24) we can express an outgoing EM wave about $\mathbf{R}_{n'}$, $\sum_{L'} b_{L'}^{+n'} \mathbf{H}_{L'}(\mathbf{r}-\mathbf{R}_{n'})$, as an incoming EM wave about $\mathbf{R}_n$, $\sum_L b_L^{'n} \mathbf{J}_L(\mathbf{r}-\mathbf{R}_n)$, as follows

$$b_L^{'n}(n') = \sum_{L'} \Omega_{LL'}^{nn'} b_{L'}^{+n'}. \tag{5.27}$$

The wave scattered from the sphere at $\mathbf{R}_n$ is determined by the total incident wave on that sphere, i.e.,

$$b_L^{+n} = \sum_{L'} T_{LL'}^{n} \left[a_{L'}^{0n} + \sum_{n' \neq n} b_{L'}^{'n}(n') \right], \tag{5.28}$$

where $T_{LL'}^{n} = T_L^n \delta_{LL'}$ is the T-matrix for the sphere at $\mathbf{R}_n$ and a_L^{0n} are the spherical-wave expansion coefficients of an externally incident wave. Equation (5.28) can be written as

$$\sum_{n'L'} \left[\delta_{nn'} \delta_{LL'} - \sum_{L''} T_{LL''}^{n} \Omega_{L''L'}^{nn'} \right] b_{L'}^{+n'} = \sum_{L'} T_{LL'}^{n} a_{L'}^{0n}. \tag{5.29}$$

The above equation is the basic equation of multiple scattering and can be solved either by standard linear-system numerical solvers or iteratively [53]. The solution provides the scattering wave b_L^{+n} outgoing from each sphere of the collection for a given externally incident wave a_L^{0n}. Having calculated b_L^{+n} from (5.29) one can readily find the coefficients $b_L^{'n}(n')$ from (5.27) and therefore the total incident wave to each sphere of the collection given by the square brackets of (5.28). Using the total incident wave as input to (5.12) one can determine the multipole coefficients a_L^{In} within each sphere of the collection. The corresponding electric field is given similarly to (5.16).

The electric field outside the spheres, $\mathbf{E}^{out}$, is written as the sum of the scattered field from all spheres plus the incident wave field, i.e.,

$$\mathbf{E}^{out}(\mathbf{r}) = \mathbf{E}^{sc}(\mathbf{r}) + \mathbf{E}^{0}(\mathbf{r}) \tag{5.30}$$

where the incident field $\mathbf{E}^0$ is given by (5.4) and $\mathbf{E}^{sc}$ is given as follows

$$\mathbf{E}^{sc}(\mathbf{r}) = \sum_{n=1}^{N_s} \sum_{L} b_L^{+n} \mathbf{H}_L(\mathbf{r}-\mathbf{R}_n). \tag{5.31}$$

For studying a metamaterial of clusters of NPs, we need to extend the existing LMS with the theory presented so far. The LMS method is an efficient computational method for the study of the EM response of three-dimensional photonic structures consisting of nonoverlapping spheres [46–48] and axisymmetric nonspherical particles [49]. It applies equally well to non-absorbing systems and to absorbing ones. Its chief advantage over the other existing numerical methods lies in its efficient and reliable treatment of systems containing strongly dispersive materials such as Drude-like and polaritonic materials.

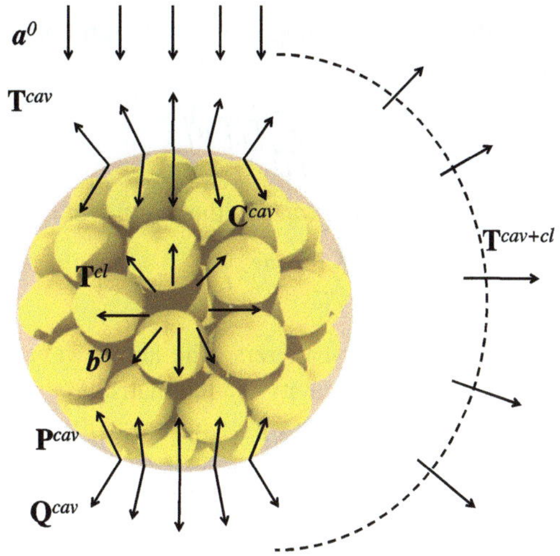

Fig. 5.2 Matrices involved in the calculation of the total scattering matrix $\mathbf{T}^{cav+cl}$ of a cluster of particles within a cavity. A plane wave (corresponding to multipole coefficients a^0) incident on the cavity (without the cluster) is either scattered off the cavity via $\mathbf{T}^{cav}$ or enters the cavity via $\mathbf{C}^{cav}$. A spherical wave (corresponding to multipole coefficients b^0) outgoing from the center of the cavity is either scattered back at the boundary of the cavity via $\mathbf{P}^{cav}$ or escapes the cavity via $\mathbf{Q}^{cav}$. The scattering matrix of the cluster is $\mathbf{T}^{cl}$. (Reprinted from V. Yannopapas and A. G. Vanakaras, Phys. Rev. B **84**, 085119 (2011). With permission)

In order to incorporate a cluster of spherical scatterers within the existing LMS code as single scattering entity we need to calculate the scattering T-matrix $T^{cl}_{LL'}$ of the entire cluster. It can be shown that the scattering matrix $T^{cl}_{LL'}$ assumes the form [52]

$$T^{cl}_{LL'} = \sum_{nn'} \sum_{L''L'''} \Xi^{0n}_{LL''} \left[[\mathbf{I} - \mathbf{T}\Omega]^{-1} \mathbf{T} \right]^{nn'}_{L''L'''} \Xi^{n'0}_{L'''L'}, \tag{5.32}$$

where the matrix $[\mathbf{I} - \mathbf{T}\Omega]$ is the one appearing in the left-hand side of (5.29). $T^{cl}_{LL'}$ contains nondiagonal elements in general. We note that alternative formulations of the EM scattering by a finite number of scatterers have been developed in the past [54–60]. However, the formalism presented above is suitable for embedding the T-matrix of (5.32) in the existing LMS formalism.

5.2.4 Multiple-Scattering in a Spherical Cavity

We consider a cluster of scatterers described by a T-matrix T^{cl} embedded within a spherical cavity (see Fig. 5.2) which is associated with a scattering matrix T^{cav}. The system of the cavity containing the cluster of scatterers is illuminated by a wave of the form of (5.4) with multipole coefficients a^0_L. This wave can be directly scattered off the cavity producing an outgoing wave of the form of (5.6) with multipole coefficients a_L^+ given by

$$a_L^+ = \sum_{LL'} T_{LL'}^{cav} a_{L'}^0 \tag{5.33}$$

or, it can enter the cavity producing a wave in the manner of (5.12),

$$\sum_{LL'} C_{LL'}^{cav} a_{L'}^0, \tag{5.34}$$

which is incident on the cluster of scatterers. The above wave is scattered off the cluster via $\mathbf{T}^{cl}$ and then escapes the cavity via the matrix $\mathbf{Q}^{cav}$ [first of (5.17)]. So, the second contribution to the scattering waves of the cavity+clusters amounts to

$$\sum_{L'} [\mathbf{Q}^{cav}\mathbf{T}^{cl}\mathbf{C}^{cav}]_{LL'} a_{L'}^0. \tag{5.35}$$

However, the wave outgoing from the cluster $\sum_{L'}[\mathbf{T}^{cl}\mathbf{C}^{cav}]_{LL'} a_{L'}^0$ can be scattered at the inner surface of the cavity and return back to the cluster via the matrix $\mathbf{P}^{cav}$ [second of (5.17)] producing this way a new incident field on the cluster. The latter is again scattered off the cluster via $\mathbf{T}^{cl}$ and escape the cavity via $\mathbf{Q}^{cav}$ producing a wave of the form

$$\sum_{L'} [\mathbf{Q}^{cav}\mathbf{T}^{cl}\mathbf{P}^{cav}\mathbf{T}^{cl}\mathbf{C}^{cav}]_{LL'} a_{L'}^0. \tag{5.36}$$

It can be easily understood that this process can be repeated infinite times giving rise to a series of multiple-scattering events

$$\begin{aligned}
&\sum_{L'} [\mathbf{Q}^{cav}\mathbf{T}^{cl}\mathbf{C}^{cav} + \mathbf{Q}^{cav}\mathbf{T}^{cl}\mathbf{P}^{cav}\mathbf{T}^{cl}\mathbf{C}^{cav} \\
&\quad + \mathbf{Q}^{cav}\mathbf{T}^{cl}\mathbf{P}^{cav}\mathbf{T}^{cl}\mathbf{P}^{cav}\mathbf{T}^{cl}\mathbf{C}^{cav} + \cdots]_{LL'} a_{L'}^0 \\
&\quad = \sum_{L'} [\mathbf{Q}^{cav}\mathbf{T}^{cl}[\mathbf{I} + \mathbf{P}^{cav}\mathbf{T}^{cl} + \mathbf{P}^{cav}\mathbf{T}^{cl}\mathbf{P}^{cav}\mathbf{T}^{cl} + \cdots]\mathbf{C}^{cav}]_{LL'} a_{L'}^0 \\
&\quad = \sum_{L'} [\mathbf{Q}^{cav}\mathbf{T}^{cl}[\mathbf{I} - \mathbf{P}^{cav}\mathbf{T}^{cl}]^{-1}\mathbf{C}^{cav}]_{LL'} a_{L'}^0.
\end{aligned} \tag{5.37}$$

The scattering matrix of the system cavity+clusters $\mathbf{T}^{cav+cl}$ is the sum of (5.33) and (5.37), i.e.,

$$\mathbf{T}^{cav+cl} = \mathbf{T}^{cav} + \mathbf{Q}^{cav}\mathbf{T}^{cl}[\mathbf{I} - \mathbf{P}^{cav}\mathbf{T}^{cl}]^{-1}\mathbf{C}^{cav}. \tag{5.38}$$

5.2.5 Multiple-Scattering Within a 2D Plane of Scatterers

The structures we are interested in usually contain more than one plane of scatterers but, to begin with, we consider just one plane, at $z = 0$, in which case the scatterers, which do not overlap with each other, are centered on the sites $\mathbf{R}_n$ of a given 2D lattice. We define the 2D reciprocal vectors $\mathbf{g}$, and the surface Brillouin zone (SBZ) corresponding to this lattice in the usual manner [47, 48].

Let the plane wave, described by (5.1), be incident on this plane of scatterers. We can always write the component of its wavevector parallel to the plane of scatterers, as follows

$$\mathbf{q}_{\parallel} = \mathbf{k}_{\parallel} + \mathbf{g}' \tag{5.39}$$

where the reduced wavevector $\mathbf{k}_{\parallel}$ lies in the SBZ and $\mathbf{g}'$ is a certain reciprocal vector. In what follows we shall write the wavevector of a plane wave of given $q = \sqrt{\mu\epsilon}\,\omega/c$ and given $\mathbf{q}_{\parallel} = \mathbf{k}_{\parallel} + \mathbf{g}$ as follows

$$\mathbf{K}_{\mathbf{g}}^{\pm} = \left(\mathbf{k}_{\parallel} + \mathbf{g},\ \pm\left[q^2 - (\mathbf{k}_{\parallel} + \mathbf{g})^2\right]^{1/2}\right) \tag{5.40}$$

where the $+, -$ sign defines the sign of the z component of the wavevector. We note that when $q^2 < (\mathbf{k}_{\parallel} + \mathbf{g})^2$, the above defines a decaying wave; the positive sign in (5.40) describes a wave propagating or decaying to the right and the negative sign describes a wave propagating or decaying to the left.

We write the electric field of the incident wave in the form

$$\mathbf{E}_{in}^{s'}(\mathbf{r}) = \sum_{i'=1}^{2} [E_{in}]_{\mathbf{g}'i'}^{s'} \exp\left(\mathrm{i}\mathbf{K}_{\mathbf{g}'}^{s'} \cdot \mathbf{r}\right)\hat{\mathbf{e}}_{i'} \tag{5.41}$$

where $s' = +(-)$ corresponds to a propagating or decaying wave incident on the plane of spheres from the left (right), and $\hat{\mathbf{e}}_1$, $\hat{\mathbf{e}}_2$ are the polar and azimuthal unit vectors, respectively, which are perpendicular to $\mathbf{K}_{\mathbf{g}'}^{s'}$. In the same manner [according to (5.40)] we define, for given $\mathbf{k}_{\parallel}$ and q, a wavevector $\mathbf{K}_{\mathbf{g}}^{s}$ and the corresponding $\hat{\mathbf{e}}_i$ for any $\mathbf{g}$ and $s = \pm$. In this way we can expand the electric-field component of an EM wave into p- and s-polarized transverse plane waves, i.e. polarized along $\hat{\mathbf{e}}_1$ and $\hat{\mathbf{e}}_2$, respectively. We note that, in the case of a decaying wave, the unit vectors $\hat{\mathbf{e}}_1$ and $\hat{\mathbf{e}}_2$ are complex but they are still orthonormal ($\hat{\mathbf{e}}_i \cdot \hat{\mathbf{e}}_j = \delta_{ij},\ i, j = 1, 2$). The coefficients a_L^0 in the expansion (5.4) of the plane wave (5.41) can be written in the following form

$$\mathrm{a}_L^0 = \sum_{i'=1}^{2} A_{L;i'}^0(\mathbf{K}_{\mathbf{g}'}^{s'})[E_{in}]_{\mathbf{g}'i'}^{s'}, \quad \text{for } P = E, H \tag{5.42}$$

where the coefficients $\mathbf{A}_L^0$ are provided by

$$\begin{aligned}\mathbf{A}_{Elm}^0(\hat{\mathbf{K}}_{\mathbf{g}'}^{s'}) = {} & \frac{4\pi \mathrm{i}^l(-1)^{m+1}}{\sqrt{l(l+1)}}\left\{\mathrm{i}\left[\alpha_l^m\, e^{\mathrm{i}\phi}\, Y_l^{-m-1}(\hat{\mathbf{K}}_{\mathbf{g}'}^{s'}) - \alpha_l^{-m}\, e^{-\mathrm{i}\phi}\, Y_l^{-m+1}(\hat{\mathbf{K}}_{\mathbf{g}'}^{s'})\right]\hat{\mathbf{e}}_1\right. \\ & - \left[\alpha_l^m \cos\theta\, e^{\mathrm{i}\phi}\, Y_l^{-m-1}(\hat{\mathbf{K}}_{\mathbf{g}'}^{s'}) + m\sin\theta\, Y_l^{-m}(\hat{\mathbf{K}}_{\mathbf{g}'}^{s'})\right. \\ & \left.\left. + \alpha_l^{-m}\cos\theta\, e^{-\mathrm{i}\phi}\, Y_l^{-m+1}(\hat{\mathbf{K}}_{\mathbf{g}'}^{s'})\right]\hat{\mathbf{e}}_2\right\},\end{aligned} \tag{5.43}$$

and

$$\mathbf{A}^0_{Hlm}(\hat{\mathbf{K}}^{s'}_{\mathbf{g}'}) = \frac{4\pi \mathrm{i}^l(-1)^{m+1}}{\sqrt{l(l+1)}} \{ [\alpha_l^m \cos\theta\, e^{\mathrm{i}\phi}\, Y_l^{-m-1}(\hat{\mathbf{K}}^{s'}_{\mathbf{g}'}) + m \sin\theta\, Y_l^{-m}(\hat{\mathbf{K}}^{s'}_{\mathbf{g}'}) + \alpha_l^{-m} \cos\theta\, e^{-\mathrm{i}\phi}\, Y_l^{-m+1}(\hat{\mathbf{K}}^{s'}_{\mathbf{g}'})]\hat{\mathbf{e}}_1 + \mathrm{i}[\alpha_l^m\, e^{\mathrm{i}\phi}\, Y_l^{-m-1}(\hat{\mathbf{K}}^{s'}_{\mathbf{g}'}) - \alpha_l^{-m}\, e^{-\mathrm{i}\phi}\, Y_l^{-m+1}(\hat{\mathbf{K}}^{s'}_{\mathbf{g}'})]\hat{\mathbf{e}}_2 \}, \quad (5.44)$$

where θ, ϕ are the angular variables $(\hat{\mathbf{K}}^{s'}_{\mathbf{g}'})$ of $\mathbf{K}^{s'}_{\mathbf{g}'}$.

Because of the 2D periodicity of the plane of scatterers, the wave scattered from it, when the wave (5.41) is incident upon it, has the following form

$$\mathbf{E}_{sc}(\mathbf{r}) = \sum_{\mathbf{R}_n} \exp(\mathrm{i}\mathbf{k}_\parallel \cdot \mathbf{R}_n) \sum_L b_L^+ \mathbf{H}_L(\mathbf{r}_n) \quad (5.45)$$

where $\mathbf{r}_n = \mathbf{r} - \mathbf{R}_n$. The coefficients b_L^+, which depend linearly on the amplitude of the incident wave, can be written as follows

$$\mathrm{b}_L^+ = \sum_{i'=1}^{2} B^+_{L;i'}(\mathbf{K}^{s'}_{\mathbf{g}'})[E_{in}]^{s'}_{\mathbf{g}'i'}. \quad (5.46)$$

We obtain $\mathbf{B}_L^+$ in terms of the coefficients $\mathbf{A}_L^0$ of (5.43) and (5.44), by solving the following system of linear equations [47, 48]

$$\sum_{L'} \Big[\delta_{LL'} - \sum_{L''} T_{LL''} \Omega_{L''L'} \Big] B^+_{L';i'}(\mathbf{K}^{s'}_{\mathbf{g}'}) = \sum_{L'} T_{LL'} \mathrm{A}^0_{L';i'}(\mathbf{K}^{s'}_{\mathbf{g}'}). \quad (5.47)$$

The matrix elements $\Omega_{LL'}$ depend on the geometry of the plane, on the reduced wavevector $\mathbf{k}_\parallel$ and on the frequency ω of the incident wave. They are the Fourier transform of the $\Omega^{nn'}_{LL'}$ introduced in Sect. 5.2.3 (explicit relations of which are provided in the Appendix). The scattering matrix $T_{LL'}$ for a single sphere is provided by (5.10) and (5.11). For a cluster of spheres, $T_{LL'}$ is provided by (5.32) or (5.38) depending on whether the cluster is placed within a cavity or not.

Finally, the scattered wave (5.45) is expressed as a sum of plane waves as follows

$$\mathbf{E}^s_{sc}(\mathbf{r}) = \sum_{i=1}^{2} \sum_{\mathbf{g}} [E_{sc}]^s_{\mathbf{g}i} \exp(\mathrm{i}\mathbf{K}^s_{\mathbf{g}} \cdot \mathbf{r})\hat{\mathbf{e}}_i \quad (5.48)$$

where the superscript $s = +(-)$ holds for $z > 0$ $(z < 0)$. Though the scattered wave consists, in general, of a number of diffracted beams corresponding to different $\mathbf{g}$ vectors, only beams for which $K^s_{\mathbf{g}z}$ is real constitute propagating waves. The coefficients in (5.48) are given by

$$[E_{sc}]^s_{\mathbf{g}i} = \sum_L \Delta_{L;i}(\mathbf{K}^s_{\mathbf{g}}) B^+_{L;i}(\mathbf{K}^{s'}_{\mathbf{g}'}) \quad (5.49)$$

where $\Delta_{L;i}(\mathbf{K}^s_{\mathbf{g}})$ are provided by

$$\Delta_{Elm}(\mathbf{K}_{\mathbf{g}}^{s}) = \frac{2\pi(-\mathrm{i})^{l}}{qA_0K_{\mathbf{g}z}^{+}\sqrt{l(l+1)}}\{\mathrm{i}[\alpha_l^{-m}\,e^{\mathrm{i}\phi}\,Y_l^{m-1}(\hat{\mathbf{K}}_{\mathbf{g}}^{s}) - \alpha_l^{m}\,e^{-\mathrm{i}\phi}\,Y_l^{m+1}(\hat{\mathbf{K}}_{\mathbf{g}}^{s})]\hat{\mathbf{e}}_1$$
$$-\big[\alpha_l^{-m}\cos\theta\,e^{\mathrm{i}\phi}\,Y_l^{m-1}(\hat{\mathbf{K}}_{\mathbf{g}}^{s}) - m\sin\theta\,Y_l^{m}(\hat{\mathbf{K}}_{\mathbf{g}}^{s})$$
$$+\alpha_l^{m}\cos\theta\,e^{-\mathrm{i}\phi}\,Y_l^{m+1}(\hat{\mathbf{K}}_{\mathbf{g}}^{s})\big]\hat{\mathbf{e}}_2\}, \tag{5.50}$$

$$\Delta_{Hlm}(\mathbf{K}_{\mathbf{g}}^{s}) = \frac{2\pi(-\mathrm{i})^{l}}{qA_0K_{\mathbf{g}z}^{+}\sqrt{l(l+1)}}\{\big[\alpha_l^{-m}\cos\theta\,e^{\mathrm{i}\phi}\,Y_l^{m-1}(\hat{\mathbf{K}}_{\mathbf{g}}^{s})$$
$$-m\sin\theta\,Y_l^{m}(\hat{\mathbf{K}}_{\mathbf{g}}^{s}) + \alpha_l^{m}\cos\theta\,e^{-\mathrm{i}\phi}\,Y_l^{m+1}(\hat{\mathbf{K}}_{\mathbf{g}}^{s})\big]\hat{\mathbf{e}}_1$$
$$+\mathrm{i}\big[\alpha_l^{-m}\,e^{\mathrm{i}\phi}\,Y_l^{m-1}(\hat{\mathbf{K}}_{\mathbf{g}}^{s}) - \alpha_l^{m}\,e^{-\mathrm{i}\phi}\,Y_l^{m+1}(\hat{\mathbf{K}}_{\mathbf{g}}^{s})\big]\hat{\mathbf{e}}_2\}, \tag{5.51}$$

where θ, ϕ denote the angular variables $(\hat{\mathbf{K}}_{\mathbf{g}}^{s})$ of $\mathbf{K}_{\mathbf{g}}^{s}$. We note that the z component of $\mathbf{K}_{\mathbf{g}}^{s}$ (denoted by $K_{\mathbf{g}z}^{s}$) can be real or imaginary. We point out that, according to (5.49), $[E_{sc}]_{\mathbf{g}i}^{s}$ depend on the incident plane wave through the coefficients $B_{L;i'}^{+}(\mathbf{K}_{\mathbf{g}'}^{s'})$. These coefficients are to be evaluated for an incident plane wave with parallel wavevector $\mathbf{k}_{\|} + \mathbf{g}'$, incident from the left (right) corresponding to $s' = +(-)$, with an electric field, along the i'th direction, of magnitude equal to unity. In other words, $B_{L;i'}^{+}(\mathbf{K}_{\mathbf{g}'}^{s'})$ are calculated from (5.47) by substituting in the right-hand side of this equation $A_{L;i'}^{0}(\mathbf{K}_{\mathbf{g}'}^{s'})$ from (5.43) and (5.44).

For example, when a plane wave (5.41) is incident on the plane of scatterers from the left, the transmitted wave (incident+scattered) on the right of the plane of scatterers is given by

$$\mathbf{E}_{tr}^{+}(\mathbf{r}) = \sum_{i=1}^{2}\sum_{\mathbf{g}}[E_{tr}]_{\mathbf{g}i}^{+}\exp(\mathrm{i}\mathbf{K}_{\mathbf{g}}^{+}\cdot\mathbf{r})\hat{\mathbf{e}}_i, \quad z>0 \tag{5.52}$$

with

$$[E_{tr}]_{\mathbf{g}i}^{+} = [E_{in}]_{\mathbf{g}'i}^{+}\delta_{\mathbf{g}\mathbf{g}'} + [E_{sc}]_{\mathbf{g}i}^{+} = \sum_{i'} M_{\mathbf{g}i;\mathbf{g}'i'}^{++}[E_{in}]_{\mathbf{g}'i'}^{+} \tag{5.53}$$

and the reflected wave by

$$\mathbf{E}_{rf}^{-}(\mathbf{r}) = \sum_{i=1}^{2}\sum_{\mathbf{g}}[E_{rf}]_{\mathbf{g}i}^{-}\exp(\mathrm{i}\mathbf{K}_{\mathbf{g}}^{-}\cdot\mathbf{r})\hat{\mathbf{e}}_i, \quad z<0 \tag{5.54}$$

with

$$[E_{rf}]_{\mathbf{g}i}^{-} = [E_{sc}]_{\mathbf{g}i}^{-} = \sum_{i'} M_{\mathbf{g}i;\mathbf{g}'i'}^{-+}[E_{in}]_{\mathbf{g}'i'}^{+}. \tag{5.55}$$

Similarly, we can define the transmission matrix elements $M_{\mathbf{g}i;\mathbf{g}'i'}^{--}$ and the reflection matrix elements $M_{\mathbf{g}i;\mathbf{g}'i'}^{+-}$ for a plane wave incident on the plane of scatterers from the right. Using (5.49) we obtain

$$M_{\mathbf{g}i;\mathbf{g}'i'}^{ss'} = \delta_{ss'}\delta_{\mathbf{g}\mathbf{g}'}\delta_{ii'} + \sum_{L}\Delta_{L;i}(\mathbf{K}_{\mathbf{g}}^{s})B_{L;i'}^{+}(\mathbf{K}_{\mathbf{g}'}^{s'}). \tag{5.56}$$

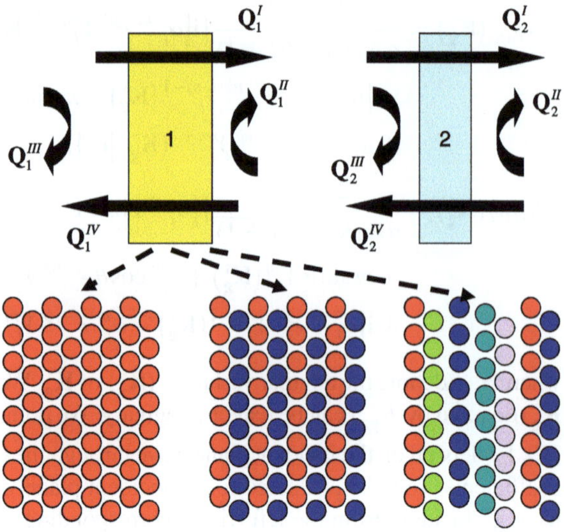

Fig. 5.3 Schematic definition of the scattering **Q**-matrices of two finite slabs. For the calculation of the scattering matrix of the pair of slabs one takes into account all the multiple-scattering paths between the two constituent slabs

The matrix elements $M^{ss'}_{\mathbf{g}i;\mathbf{g}'i'}$ obey the following symmetry relation [47, 48]

$$M^{-\mathrm{s}-\mathrm{s}'}_{\mathbf{g}i;\mathbf{g}'i'} = (-1)^{i+i'} M^{\mathrm{ss}'}_{\mathbf{g}i;\mathbf{g}'i'}. \tag{5.57}$$

The transmission/reflection matrices of a stack of planes of scatterers with the same 2D periodicity parallel to the xy plane are obtained from the transmission/reflection matrices of the individual planes [47, 48]. We express the waves on the left of a given plane of scatterers with respect to an origin, $\mathbf{A}_l$, on the left of the plane at $-\mathbf{d}_l$ from its center and the waves on the right of this plane with respect to an origin, $\mathbf{A}_r$, on the right of the plane at $\mathbf{d}_r$ from its center. With the above choice of origins the transmission/reflection matrix elements of a plane of scatterer become

$$\begin{aligned}
Q^{\mathrm{I}}_{\mathbf{g}i;\mathbf{g}'i'} &= M^{++}_{\mathbf{g}i;\mathbf{g}'i'} \exp\big(\mathrm{i}\big(\mathbf{K}^+_{\mathbf{g}} \cdot \mathbf{d}_r + \mathbf{K}^+_{\mathbf{g}'} \cdot \mathbf{d}_l\big)\big) \\
Q^{\mathrm{II}}_{\mathbf{g}i;\mathbf{g}'i'} &= M^{+-}_{\mathbf{g}i;\mathbf{g}'i'} \exp\big(\mathrm{i}\big(\mathbf{K}^+_{\mathbf{g}} \cdot \mathbf{d}_r - \mathbf{K}^-_{\mathbf{g}'} \cdot \mathbf{d}_r\big)\big) \\
Q^{\mathrm{III}}_{\mathbf{g}i;\mathbf{g}'i'} &= M^{-+}_{\mathbf{g}i;\mathbf{g}'i'} \exp\big(-\mathrm{i}\big(\mathbf{K}^-_{\mathbf{g}} \cdot \mathbf{d}_l - \mathbf{K}^+_{\mathbf{g}'} \cdot \mathbf{d}_l\big)\big) \\
Q^{\mathrm{IV}}_{\mathbf{g}i;\mathbf{g}'i'} &= M^{--}_{\mathbf{g}i;\mathbf{g}'i'} \exp\big(-\mathrm{i}\big(\mathbf{K}^-_{\mathbf{g}} \cdot \mathbf{d}_l + \mathbf{K}^-_{\mathbf{g}'} \cdot \mathbf{d}_r\big)\big)
\end{aligned} \tag{5.58}$$

In what follows we shall write the above matrix elements in compact form as follows: $\mathbf{Q}^{\mathrm{I}}$, $\mathbf{Q}^{\mathrm{II}}$, $\mathbf{Q}^{\mathrm{III}}$, and $\mathbf{Q}^{\mathrm{IV}}$ which implies a definite sequence in the ordering of the indices: $\mathbf{g}_1 1, \mathbf{g}_1 2, \mathbf{g}_2 1, \mathbf{g}_2 2, \ldots$.

5.2.6 Multiple-Scattering Within a Slab of Many 2D Planes of Scatterers

We obtain the transmission and reflection matrices of two successive 2D planes of scatterers, 1 and 2, by combining the matrices of the two elements, as shown schematically in Fig. 5.3. It is worth noting that each element need not be exclusively a single plane of scatterers but it may comprise many 2D planes of identical or different scatterers wherein each scatterer may be a single sphere or a cluster of spheres. One can easily prove that the transmission and reflection matrices for the pair of elements, denoted by $\mathbf{Q}(1,2)$ are

$$\begin{aligned}
\mathbf{Q}^{\mathrm{I}}(1,2) &= \mathbf{Q}^{\mathrm{I}}(2)\left[\mathbf{I}-\mathbf{Q}^{\mathrm{II}}(1)\mathbf{Q}^{\mathrm{III}}(2)\right]^{-1}\mathbf{Q}^{\mathrm{I}}(1)\\
\mathbf{Q}^{\mathrm{II}}(1,2) &= \mathbf{Q}^{\mathrm{II}}(2)+\mathbf{Q}^{\mathrm{I}}(2)\mathbf{Q}^{\mathrm{II}}(1)\left[\mathbf{I}-\mathbf{Q}^{\mathrm{III}}(2)\mathbf{Q}^{\mathrm{II}}(1)\right]^{-1}\mathbf{Q}^{\mathrm{IV}}(2)\\
\mathbf{Q}^{\mathrm{III}}(1,2) &= \mathbf{Q}^{\mathrm{III}}(1)+\mathbf{Q}^{\mathrm{IV}}(1)\mathbf{Q}^{\mathrm{III}}(2)\left[\mathbf{I}-\mathbf{Q}^{\mathrm{II}}(1)\mathbf{Q}^{\mathrm{III}}(2)\right]^{-1}\mathbf{Q}^{\mathrm{I}}(1)\\
\mathbf{Q}^{\mathrm{IV}}(1,2) &= \mathbf{Q}^{\mathrm{IV}}(1)\left[\mathbf{I}-\mathbf{Q}^{\mathrm{III}}(2)\mathbf{Q}^{\mathrm{II}}(1)\right]^{-1}\mathbf{Q}^{\mathrm{IV}}(2)
\end{aligned} \tag{5.59}$$

All matrices refer of course to the same ω and $\mathbf{k}_{\|}$. We note in particular that the waves on the left (right) of the pair of elements are referred to an origin at $-\mathbf{d}_l(1)$ ($+\mathbf{d}_r(2)$) from the center of the 1st (2nd) element.

It is obvious that we can repeat the process to obtain the transmission and reflection matrices of three elements, by combining those of the pair of elements with those of the third element; and that we can in similar fashion repeat the process to obtain the scattering matrices of a slab consisting of any finite number of elements. However, a slab may consist of a number of identical elements stacked together along the z-axis (normal to the surface of the slab). We assume that the slab consists of 2^N elements, where $N = 0, 1, 2, \ldots$.

Having calculated the $\mathbf{Q}$-matrix elements of a single element, we obtain those of the slab, by a doubling-layer scheme as follows: we first obtain the $\mathbf{Q}$-matrix of two consecutive elements in the manner described above; then, using as units the $\mathbf{Q}$-matrix of a pair of elements, we obtain those of four consecutive elements, and in this way, by doubling the number of elements at each stage we finally obtain the $\mathbf{Q}$-matrix elements of the 2^N elements of the slab.

In summary, for a plane wave $\sum_i [E_{in}]^{+}_{\mathbf{g}'i} \exp(\mathrm{i}\mathbf{K}^{+}_{\mathbf{g}'} \cdot (\mathbf{r}-\mathbf{A}_L))\hat{e}_i$, incident on the slab from the left, we finally obtain a reflected wave $\sum_{\mathbf{g}i} [E_{rf}]^{-}_{\mathbf{g}i} \exp(\mathrm{i}\mathbf{K}^{-}_{\mathbf{g}} \cdot (\mathbf{r} - \mathbf{A}_L))\hat{e}_i$ on the left of the slab and a transmitted wave $\sum_{\mathbf{g}i} [E_{tr}]^{+}_{\mathbf{g}i} \exp(\mathrm{i}\mathbf{K}^{+}_{\mathbf{g}} \cdot (\mathbf{r} - \mathbf{A}_R))\hat{e}_i$ on the right of the slab, where $\mathbf{A}_L$ ($\mathbf{A}_R$) is the appropriate origin on the left (right) of the slab. We have

$$[E_{tr}]^{+}_{\mathbf{g}i} = \sum_{i'} Q^{\mathrm{I}}_{\mathbf{g}i;\mathbf{g}'i'} [E_{in}]^{+}_{\mathbf{g}'i'} \tag{5.60}$$

$$[E_{rf}]^{-}_{\mathbf{g}i} = \sum_{i'} Q^{\mathrm{III}}_{\mathbf{g}i;\mathbf{g}'i'} [E_{in}]^{+}_{\mathbf{g}'i'} \tag{5.61}$$

where the $\mathbf{Q}$-matrix elements are those of the whole slab.

When we have calculated the transmitted wave (5.60) and the reflected wave (5.61), corresponding to the given incident wave, we can obtain the transmittance $\mathcal{T}$ and the reflectivity $\mathcal{R}$ of the slab. $\mathcal{T}(\mathcal{R})$ is defined as the ratio of the flux of the transmitted (reflected) wave to the flux of the incident wave. Integrating the Poynting vector over the xy-plane, on the appropriate side of the slab, and taking the time average over a period $T = 2\pi/\omega$, we obtain

$$\mathcal{T} = \frac{\sum_{\mathbf{g},i} [E_{tr}]^{+}_{\mathbf{g}i} ([E_{tr}]^{+}_{\mathbf{g}i})^{*} K^{+}_{\mathbf{g}z}}{\sum_{i} [E_{in}]^{+}_{\mathbf{g}'i} ([E_{in}]^{+}_{\mathbf{g}'i})^{*} K^{+}_{\mathbf{g}'z}}$$
$$\mathcal{R} = \frac{\sum_{\mathbf{g},i} [E_{rf}]^{-}_{\mathbf{g}i} ([E_{rf}]^{-}_{\mathbf{g}i})^{*} K^{+}_{\mathbf{g}z}}{\sum_{i} [E_{in}]^{+}_{\mathbf{g}'i} ([E_{in}]^{+}_{\mathbf{g}'i})^{*} K^{+}_{\mathbf{g}'z}} \tag{5.62}$$

where the $*$ sign denotes complex conjugation as usual. If the structure contains light-absorbing materials, the requirement of energy conservation implies that the absorbance $\mathcal{A}$ of the slab is

$$\mathcal{A} = 1 - \mathcal{T} - \mathcal{R} \tag{5.63}$$

5.2.7 Amorphous Metamaterial of Clusters of Nanoparticles

We next consider the case where the metamaterial is amorphous in the manner of substitutional disorder, i.e. the metamaterial is a periodic lattice whose sites, however, are randomly occupied by scatterers of different shape, material and/or size. In order to study the EM response of such an amorphous metamaterial, the average T-matrix approximation (ATA) can be employed within the cluster-LMS framework. The ATA method which has been originally introduced for the study of the electron properties of disordered metals and surfaces [61, 62], has been also employed for the study of the optical properties of random 2D [63, 64] and 3D [65, 66] photonic alloys of spherical scatterers. In the ATA approach, the actual disordered crystal is replaced by an effectively periodic one whose lattice sites are occupied by same scatterers of an average T-matrix. More specifically, one assumes that actual random system consists of scatterers of type A with concentration C_A and of scatterers B with concentration C_B. If the EM response of the scatterers A is described by a scattering T-matrix $T^{A}_{LL'}$ and the response of scatterers B by $T^{B}_{LL'}$, then the T-matrix $T^{ATA}_{LL'}$ of the effective scatterers of the ATA periodic crystal is given by

$$T^{ATA}_{LL'} = C_A T^{A}_{LL'} + C_B T^{B}_{LL'}. \tag{5.64}$$

Obviously, $C_A + C_B = 1$. ATA provides the correct configurational average of all scattering events off the same site, i.e. the average over all possible configurations of a substitutional alloy [62]. As such, the ATA c-LMS provides reasonably good results for moderate scattering regimes such as the subwavelength regime we consider here.

5.3 An Example: 3D Metamaterial of Air Cavities Containing Clusters of Gold Nanoparticles

We consider a metamaterial made of air cavities in silica (SiO_2) where each cavity contains a cluster of 100 nonoverlapping gold (Au) nanoparticles (see Fig. 5.4).

Fig. 5.4 3D orthorhombic metamaterial made of air cavities in silica containing clusters of gold nanoparticles. Each cluster consists of 100 nonoverlapping gold nanoparticles of radius $S = 8.8$ nm in a nearly close-packed arrangement, with cluster radius 42.67 nm. Each cluster is placed at a center of a cavity of radius 44 nm. The metamaterial is viewed as a succession of (001) planes (*square lattices*) of clusters of gold NPs, parallel to the xy-plane. The lattice constant of each square lattice is $a_x = a_y = 85.22$ nm whilst the lattice constant in the z-direction is $a_z = 87.86$ nm. (Reprinted from V. Yannopapas and A. G. Vanakaras, Phys. Rev. B **84**, 085119 (2011). With permission)

The positions of the gold particles within a cluster are taken from a Monte-Carlo simulation of the self-organization of these particles under a spherically confining potential. All particles have the same radius $S = 8.8$ nm; the average radius of the cluster is $S_{cl} = 42.67$ nm whilst the cavity radius $S_{cav} = 44$ nm. The average interparticle gap is about 1.2 nm. The dielectric function of a single gold nanoparticle is taken from experiment [67] with corrections accounting for the electron scattering at the boundaries of a nanoparticle [see Eqs. (3)–(4) of [68]]. The metamaterial is a slightly elongated cubic (orthorhombic) lattice viewed as a succession of 2D square lattices of the above clusters of gold NPs, parallel to the xy-plane [see Fig. 5.4]. The lattice constant of the 2D square lattice is $a_x = a_y = 85.22$ nm whilst the lattice constant in the z-direction is $a_z = 87.86$ nm. Note in passing that we have deliberately assumed an orthorhombic crystal in order to demonstrate the applicability of the presented formalism which can treat on equal footing the effects of inhomogeneity, dispersion, anisotropy and disorder in the metamaterials under study without any approximations.

In Fig. 5.5 we show the transmittance, reflectance and absorbance spectra for light incident normally on 8 planes of cavities of gold clusters. The nanoparticles

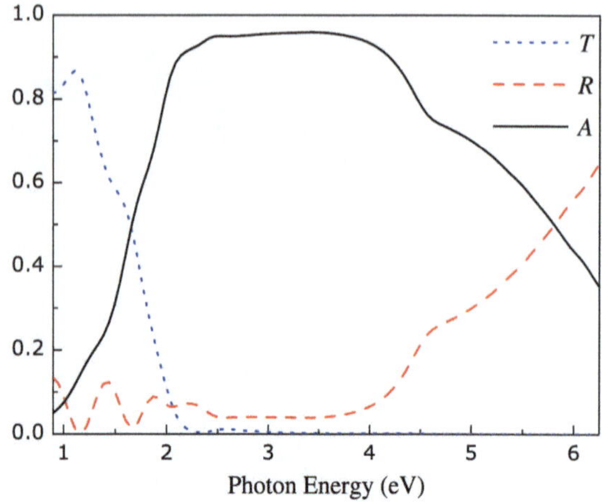

Fig. 5.5 Transmittance (T), reflectance (R), and absorbance (A) spectra for light incident normally on a slab of the metamaterial of Fig. 5.4 consisting of 8 unit planes

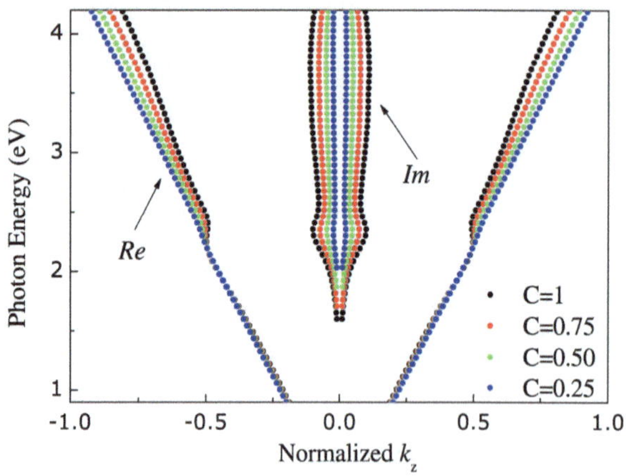

Fig. 5.6 Photon dispersion lines for a 3D nonperiodic metamaterial made of air cavities in silica containing clusters of gold NPs of 8 nm radius with different amounts of disorder C, as calculated by the ATA c-LMS method. The NP clusters are the same as those of Fig. 5.3

are small enough (8.8 nm radius) so that taking into account only dipole terms ($l_{max} = 1$) in the multipole expansion of the EM field [see (5.2) and (5.3)] suffices for achieving convergence in the absorption spectra [69]. In the plane-wave expansion of the EM field [see (5.48)] we have considered 21 reciprocal-lattice vectors. As it is evident from Fig. 5.5, there appears a distinct local plateau from 2–4 eV wherein the absorbance is about 95 %. The enhanced absorbance is due to the excitation of surface plasmons on each single gold nanoparticle; the interaction of the surface plasmons of each nanoparticle leads to a surface-plasmon frequency band which gives rise to the broad plateau of the absorbance appearing in Fig. 5.5. We note that, due to the absorption performance depicted in Fig. 5.5, the metamaterial under study behaves as super absorber [70] promising important application in thermophotovoltaic cells [68, 71–75].

In Fig. 5.6 we show the dispersion lines for infinite (bulk) amorphous metamaterials of clusters of NPs, i.e., the cavities+clusters are randomly positioned in space following the substitutional model of disorder described in Sect. 5.2.7 (type-A scatterer is a cavity+NP cluster and type-B is just the silica host—unoccupied lattice site). With C we denote the % fraction of the sites being occupied quantifying this way the amount of disorder. Obviously, for $C = 1$ we have a periodic metamaterial. With *Re* (*Im*) we depict the real (imaginary) lines, i.e., frequency (in eV units) as a function of the $\Re k_z$ ($\Im k_z$), respectively. For frequencies below 1.6 eV, $\Im k_z$ is smaller than 10^{-3} and this is why the imaginary dispersion lines are not shown below this frequency. It is obvious that for frequencies below 2.2 eV the real frequency lines, which describe the wave propagation within the metamaterial, are independent of the amount of disorder. Also, above 2.2 eV, $\Re k_z$ assumes different values for different C but, the slopes (describing the phase velocity of light within the metamaterial) are practically the same. These two effects reflect the subwavelength nature of the metamaterials under study. On the other hand, the imaginary frequency lines, which describe the wave attenuation within the metamaterial, dependent relatively more on the amount of disorder.

Finally, a note on the possible realization of the metamaterial under study. As stated in the introduction, 1D and 2D lattices decorated by clusters of metallic NPs have already been realized by template-assisted colloidal self-organization. Namely, clusters of metallic NPs have been deposited within the voids of a 2D periodically perforated dielectric slab [43] or within the trenches of a 1D grating [44]. Here, we have a 3D metamaterial made of air cavities containing clusters of metallic nanoparticles. In this case, a template-assisted colloidal self-assembly should be applied to 3D dielectrics with spherical voids such as inverted opals. The latter are fcc photonic crystals of air holes in a dielectric host such as SiO_2, TiO_2 or Si [76]. By infiltrating the air holes with gold NPs one can realize the metamaterial of Fig. 5.4. Alternatively, a bottom-up self-assembly technique without the need of a (lithograpically fabricated) template can be employed [77, 78]. Namely, highly stable gold NPs are synthesized by the Turkey-Frens method [80] which are dispersed within an aqueous solution. By adding a ligand molecule within the solution, the NPs start to agglomerate into supramolecular clusters [77].

5.4 Conclusions

The development of a rigorous electrodynamic theory for the theoretical study of amorphous electromagnetic structures such as disordered metamaterials wherein the randomness lies in multiple length scales, constitutes a major challenge for the field of theoretical and computational nanophotonics. In order to tackle efficiently this computational problem without resorting to any kind of approximations, we have employed a hierarchical theoretical strategy wherein we employ an EM multiple-scattering formalism for each length scale. The connection between the different scales has been realized via the concept of the scattering T-matrix. The above hierarchical strategy has been focused on metamaterials of amorphous clusters of nanoparticles embedded or not within a cavity. Namely, we have taken into account all the multiple-scattering events experienced by the EM waves: scattering within the cluster of nanoparticles, between the cluster and the cavity boundaries, and among the clusters of nanoparticles within the metamaterial. As a case study, we have applied the presented formalism to the case of a lattice of air voids in silica containing clusters of gold nanoparticles. Thick slabs of the above metamaterial act as super absorbers exhibiting a very broad plateau of 95 % light absorption promising application in thermophotovoltaic cells.

Acknowledgements The research leading to these results has received funding from the European Union's Seven Framework Programme (FP7/2007-2013) under Grant Agreement No. 228455-NANOGOLD (Self-organized nanomaterials for tailored optical and electrical properties).

Appendix

The matrix Ω for a vector field is given by [52, 79]

$$\begin{aligned}\Omega^{nn'}_{Elm;El'm'} = \Omega^{nn'}_{Hlm;Hl'm'} = (\psi_l \psi_{l'})^{-1}\big[&2\alpha_l^{-m}\alpha_{l'}^{-m'} G_{l'm'-1;lm-1}(\mathbf{R}_{nn'}; q_h)\\ &+ mm' G_{l'm';lm}(\mathbf{R}_{nn'}; q_h)\\ &+ 2\alpha_l^{m}\alpha_{l'}^{m'} G_{l'm'+1;lm+1}(\mathbf{R}_{nn'}; q_h)\big]\end{aligned} \tag{5.65}$$

$$\begin{aligned}\Omega^{nn'}_{Hlm;El'm'} = -\Omega^{nn'}_{Elm;Hl'm'} = (2l+1)&(\psi_l \psi_{l'})^{-1}\\ &\times \big[-2\alpha_{l'}^{-m'}\gamma_l^{m} G_{l'm'-1;l-1m-1}(\mathbf{R}_{nn'}; q_h)\\ &+ m'\zeta_l^m G_{l'm';l-1m}(\mathbf{R}_{nn'}; q_h)\\ &+ 2\alpha_{l'}^{m'}\gamma_l^{-m} G_{l'm'+1;l-1m+1}(\mathbf{R}_{nn'}; q_h)\big]\end{aligned} \tag{5.66}$$

where

$$\psi_l = \sqrt{l(l+1)} \tag{5.67}$$

$$\alpha_l^m = \frac{1}{2}\big[(l-m)(l+m+1)\big]^{1/2} \tag{5.68}$$

$$\gamma_l^m = \frac{1}{2}\big[(l+m)(l+m-1)\big]^{1/2}/\big[(2l-1)(2l+1)\big]^{1/2} \tag{5.69}$$

$$\zeta_l^m = \big[(l+m)(l-m)\big]^{1/2}/\big[(2l-1)(2l+1)\big]^{1/2}. \tag{5.70}$$

$G_{LL'}(\mathbf{R}_{nn'}; q_h)$ transforms an outgoing scalar spherical wave about $\mathbf{R}_{n'}$ to a series of incoming scalar spherical waves around $\mathbf{R}_n$. It is given by

$$\begin{aligned}G_{lm;l'm'}(\mathbf{R}_{nn'}; q_h) = 4\pi &\sum_{l''m''} (-1)^{(l-l'-l'')/2}(-1)^{m'+m''}\\ &\times B_{lm}(l''m''; l'm') h^+_{l''}(q_h R_{nn'}) Y_{l''-m''}(\hat{\mathbf{R}}_{nn'})\end{aligned} \tag{5.71}$$

with

$$B_{lm}(l''m''; l'm') = \int d\Omega\, Y_{lm}(\hat{\mathbf{r}}) Y_{l'-m'}(\hat{\mathbf{r}}) Y_{l''m''}(\hat{\mathbf{r}}). \tag{5.72}$$

$Y_{lm}(\hat{\mathbf{r}})$ are the usual scalar spherical harmonics [50].

The matrix Ξ for a vector field is given by [52, 79]

$$\begin{aligned}\Xi^{nn'}_{Elm;El'm'} = \Xi^{nn'}_{Hlm;Hl'm'} &= (\psi_l \psi_{l'})^{-1}\big[2\alpha_l^{-m}\alpha_{l'}^{-m'}\xi_{l'm'-1;lm-1}(\mathbf{R}_{nn'};q_h)\\ &\quad + mm'\xi_{l'm';lm}(\mathbf{R}_{nn'};q_h)\\ &\quad + 2\alpha_l^{m}\alpha_{l'}^{m'}\xi_{l'm'+1;lm+1}(\mathbf{R}_{nn'};q_h)\big] \qquad (5.73)\\ \Xi^{nn'}_{Hlm;El'm'} = -\Xi^{nn'}_{Elm;Hl'm'} &= (2l+1)(\psi_l\psi_{l'})^{-1}\\ &\quad \times\big[-2\alpha_{l'}^{-m'}\gamma_l^{m}\xi_{l'm'-1;l-1m-1}(\mathbf{R}_{nn'};q_h)\\ &\quad + m'\zeta_l^{m}\xi_{l'm';l-1m}(\mathbf{R}_{nn'};q_h)\\ &\quad + 2\alpha_{l'}^{m'}\gamma_l^{-m}\xi_{l'm'+1;l-1m+1}(\mathbf{R}_{nn'};q_h)\big]. \qquad (5.74)\end{aligned}$$

$\xi_{LL'}(\mathbf{R}_{nn'};q_h)$ transforms an outgoing (incoming) scalar spherical wave about $\mathbf{R}_{n'}$ to a series of outgoing (incoming) scalar spherical waves around $\mathbf{R}_n$ [see (5.25) and (5.26)]. It is given by

$$\begin{aligned}\xi_{lm;l'm'}(\mathbf{R}_{nn'};q_h) = 4\pi \sum_{l''m''}(-1)^{(-l+l'+l'')/2}(-1)^{m'+m''}\\ \times B_{lm}\big(l''m'';l'm'\big)\, j_{l''}(q_h R_{nn'})Y_{l''-m''}(\hat{\mathbf{R}}_{nn'}). \qquad (5.75)\end{aligned}$$

References

1. J.B. Pendry, A.J. Holden, D.J. Robbins, W.J. Stewart, IEEE Trans. Microw. Theory **47**, 2075 (1999)
2. J.B. Pendry, Phys. Rev. Lett. **85**, 3966 (2000)
3. M.C.K. Wiltshire, J.B. Pendry, I.R. Young, D.J. Larkman, D.J. Gilderdale, J.V. Hajnal, Science **291**, 849 (2001)
4. D. Schurig, J.J. Mock, B.J. Justice, S.A. Cummer, J.B. Pendry, A.F. Starr, D.R. Smith, Science **314**, 5801 (2006)
5. N.I. Landy, S. Sajuyigbe, J.J. Mock, D.R. Smith, W.J. Padilla, Phys. Rev. Lett. **100**, 207402 (2008)
6. L.V. Panina, A.N. Grigorenko, D.P. Makhnovskiy, Phys. Rev. B **66**, 155411 (2002)
7. T.J. Yen et al., Science **303**, 1494 (2004)
8. T.J. Yen et al., Science **303**, 1494 (2004)
9. V.M. Shalaev et al., Opt. Lett. **30**, 3356 (2005)
10. S. Linden et al., Science **306**, 1351 (2004)
11. C.M. Soukoulis, S. Linden, M. Wegener, Science **315**, 788 (2007)
12. V.M. Shalaev, Nat. Photonics **1**, 41 (2007)
13. N. Liu et al., Adv. Mater. **19**, 3628 (2007)
14. N. Liu et al., Nat. Mater. **7**, 31 (2008)
15. N. Liu, H. Liu, S. Zhu, H. Giessen, Nat. Photonics **3**, 157 (2009)
16. R.A. Shelby, D.R. Smith, S. Schultz, Science **292**, 77 (2001)
17. C.G. Parazzoli, R.B. Greegor, K. Li, B.E.C. Koltenbah, M. Tanielian, Phys. Rev. Lett. **90**, 107401 (2003)
18. A.A. Houck, J.B. Brock, I.L. Chuang, Phys. Rev. Lett. **90**, 137401 (2003)
19. S. O'Brien, J.B. Pendry, J. Phys. Condens. Matter **14**, 4035 (2002)
20. K.C. Huang, M.L. Povinelli, J.D. Joannopoulos, Appl. Phys. Lett. **85**, 543 (2004)

21. C.L. Holloway, E.F. Kuester, J. Baker-Jarvis, P. Kabos, IEEE Trans. Antennas Propag. **51**, 2596 (2003)
22. V. Yannopapas, A. Moroz, J. Phys. Condens. Matter **17**, 3717 (2005)
23. M.S. Wheeler, J.S. Aitchison, M. Mojahedi, Phys. Rev. B **72**, 193103 (2005)
24. M.S. Wheeler, J.S. Aitchison, M. Mojahedi, Phys. Rev. B **73**, 045105 (2006)
25. L. Jylhä, I. Kolmakov, S. Maslovski, S. Tretyakov, J. Appl. Phys. **99**, 043102 (2006)
26. T.G. MacKay, A. Lakhtakia, J. Appl. Phys. **100**, 063533 (2006)
27. V. Yannopapas, N.V. Vitanov, Phys. Rev. B **74**, 193304 (2006)
28. V. Yannopapas, Phys. Rev. B **75**, 035112 (2007)
29. A.G. Kussow, A. Akyurtlu, A. Semichaevsky, N. Angkawisittpan, Phys. Rev. B **76**, 195123 (2007)
30. C.W. Qiu, L. Gao, J. Opt. Soc. Am. B **25**, 1728 (2008)
31. Q. Zhao, J. Zhou, F. Zhang, D. Lippens, Mater. Today **12**, 60 (2009)
32. L. Peng, L. Ran, H. Chen, H. Zhang, J.A. Kong, T.M. Grzegorczyk, Phys. Rev. Lett. **98**, 157403 (2007)
33. B.I. Popa, S.A. Cummer, Phys. Rev. Lett. **100**, 207401 (2008)
34. Q. Zhao, L. Kang, B. Du, H. Zhao, Q. Xie, X. Huang, B. Li, J. Zhou, L. Li, Phys. Rev. Lett. **101**, 027402 (2008)
35. Q. Zhao, B. Du, L. Kang, H. Zhao, Q. Xie, B. Li, X. Zhang, J. Zhou, L. Li, Y. Meng, Appl. Phys. Lett. **92**, 051106 (2008)
36. O. Acher, M. Ledieu, A. Bardaine, F. Levassort, Appl. Phys. Lett. **93**, 032501 (2008)
37. O. Acher, J.H. Le Gallou, M. Ledieu, Metamaterials **2**, 18 (2008)
38. X.Q. Lin et al., Appl. Phys. Lett. **92**, 131904 (2008)
39. J.A. Schuller, R. Zia, T. Taubner, M.L. Brongersma, Phys. Rev. Lett. **99**, 107401 (2007)
40. C. Rockstuhl, F. Lederer, C. Etrich, T. Pertsch, T. Scharf, Phys. Rev. Lett. **99**, 017401 (2007)
41. Q. Wu, W. Park, Appl. Phys. Lett. **92**, 153114 (2008)
42. W. Park, Q. Wu, Solid State Commun. **146**, 221 (2008)
43. H.J. Lee, Q. Wu, W. Park, Opt. Lett. **34**, 443 (2009)
44. V.A. Tamma, J.H. Lee, Q. Wu, W. Park, Appl. Opt. **49**, A11 (2010)
45. X. Zeng et al., Adv. Mater. **21**, 1746 (2009)
46. N. Stefanou, V. Karathanos, A. Modinos, J. Phys. Condens. Matter **4**, 7389 (1992)
47. N. Stefanou, V. Yannopapas, A. Modinos, Comput. Phys. Commun. **113**, 49 (1998)
48. N. Stefanou, V. Yannopapas, A. Modinos, Comput. Phys. Commun. **132**, 189 (2000)
49. G. Gantzounis, N. Stefanou, Phys. Rev. B **73**, 035115 (2006)
50. J.D. Jackson, *Classical Electrodynamics* (Wiley, New York, 1975)
51. C.F. Bohren, D.R. Huffman, *Absorption and Scattering of Light by Small Particles* (Wiley, New York, 1983)
52. R. Sainidou, N. Stefanou, A. Modinos, Phys. Rev. B **69**, 064301 (2004)
53. V. Yannopapas, N.V. Vitanov, Phys. Rev. B **75**, 115124 (2007)
54. M. Inoue, K. Ohtaka, J. Phys. Soc. Jpn. **52**, 3853 (1983)
55. H. Xu, E.J. Bjerneld, M. Käll, L. Börjesson, Phys. Rev. Lett. **83**, 4357 (1999)
56. H. Xu, J. Aizpurua, M. Käll, P. Apell, Phys. Rev. E **62**, 4318 (2000)
57. H. Xu, M. Käll, Phys. Rev. Lett. **89**, 246802 (2002)
58. H. Xu, J. Opt. Soc. Am. A **21**, 804 (2004)
59. K. Zhao, H. Xu, B. Gu, Z. Zhang, J. Chem. Phys. **125**, 081102 (2006)
60. Z. Li, H. Xu, J. Quant. Spectrosc. Radiat. Transf. **103**, 394 (2007)
61. J.L. Beeby, J. Phys. C **1**, 82 (1968)
62. A. Gonis, *Green Functions for Ordered and Disordered Systems* (North-Holland, Amsterdam, 1992)
63. N. Stefanou, A. Modinos, J. Phys. Condens. Matter **3**, 8135 (1991)
64. N. Stefanou, A. Modinos, J. Phys. Condens. Matter **3**, 8149 (1991)
65. A. Modinos, V. Yannopapas, N. Stefanou, Phys. Rev. B **61**, 8099 (2000)
66. V. Yannopapas, Phys. Rev. B **75**, 035112 (2007)
67. R.B. Johnson, R.W. Christy, Phys. Rev. B **6**, 4370 (1972)

68. V. Yannopapas, Phys. Rev. B **73**, 113108 (2006)
69. V.M. Shalaev, *Nonlinear Optics of Random Media: Fractal Composites and Metal-Dielectric Films* (Springer, Heidelberg, 2000)
70. K. Aydin, V.E. Ferry, R.M. Briggs, H.A. Atwater, Nat. Commun. (2011). doi:10.1038/ncomms1528
71. J.G. Fleming, S.Y. Lin, I. El-Kady, R. Biswas, K.M. Ho, Nature (London) **417**, 52 (2002)
72. M.U. Pralle, N. Moelders, M.P. McNeal, I. Puscasu, A.C. Greenwald, J.T. Daly, E.A. Johnson, T. George, D.S. Choi, I. El-Kady, R. Biswas, Appl. Phys. Lett. **81**, 4685 (2002)
73. I. Celanovic, F.O. Sullivan, M. Ilak, J. Kassakian, D. Perreault, Opt. Lett. **29**, 863 (2004)
74. A. Narayanaswamy, G. Chen, Phys. Rev. B **70**, 125101 (2004)
75. S. Enoch, J.-J. Simon, L. Escoubas, Z. Elalmy, F. Lemarquis, P. Torchio, G. Albrand, Appl. Phys. Lett. **86**, 261101 (2005)
76. A. Blanco, E. Chomski, S. Grabtchak, M. Ibisate, S. John, S.W. Leonard, C. Lopez, F. Meseguer, H. Miguez, J.P. Mondia, G.A. Ozin, O. Toader, H.M. van Driel, Nature (London) **405**, 437 (2000)
77. N. Shalkevich, A. Shalkevich, L. Si-Ahmed, T. Bürgi, Phys. Chem. Chem. Phys. **11**, 10175 (2009)
78. S. Mühlig, C. Rockstuhl, V. Yannopapas, T. Bürgi, F. Lederer, Opt. Express **19**, 9607 (2011)
79. A. Modinos, Physica A **141**, 575 (1987)
80. G. Frens, Nat. Phys. Sci. **241**, 20 (1973)

Chapter 6
Deterministic Amorphous Metamaterials and Their Optical Far-Field Response

Christian Helgert and Thomas Pertsch

Abstract We review the optical far-field response of artificially structured media—metamaterials—and the impact of deterministic amorphization thereon. The interest on such deterministic amorphous metamaterials recently emerged because they may bridge a conceptional gap between bottom-up and top-down fabricated metamaterials. Artificial amorphization can be implemented by means of mathematical algorithms governed by randomization or disorder parameters that assign an individual position to each single building-block, i.e. each meta-atom, of a large assembly. These meta-atoms sustain characteristic and shape-dependent optical resonances which, as consistently confirmed in experiments, show different sensitivities to deterministic amorphization. In consequence, the widely alleged necessity of a periodic lattice can be relaxed in many cases. This newly emerging direction of nanophotonic research offers novel interdisciplinary approaches to tailor the electromagnetic properties of artificial composites with elevated structural complexities.

6.1 Introduction

The development of artificial materials which can control the flow of electromagnetic waves in unprecedented ways has evoked a great interest in modern nanophotonics. The fabrication of nano-sized structures has provided us means to create artificial materials that have no natural counterparts. The subclass of these materials which probably attracted the highest attention are the so-called metamaterials [1–4]. These man-made media are composed of nanostructured elements in the same sense as natural matter consists of atoms. Therefore and throughout this chapter these structural elements will be termed meta-atoms [5]. Ideally, they are supposed to be considerably smaller than the wavelength of the interacting electromagnetic fields.

C. Helgert (✉) · T. Pertsch
Abbe Center of Photonics, Institute of Applied Physics, Friedrich-Schiller-Universität Jena, Max-Wien-Platz 1, 07743 Jena, Germany
e-mail: christian.helgert@uni-jena.de

T. Pertsch
e-mail: thomas.pertsch@uni-jena.de

C. Rockstuhl, T. Scharf (eds.), *Amorphous Nanophotonics*,
Nano-Optics and Nanophotonics, DOI 10.1007/978-3-642-32475-8_6,

© Springer-Verlag Berlin Heidelberg 2013

The meta-atoms themselves are made of conventional materials such as noble metals, but their specifically tailored shapes and arrangements promote their constituting ensembles to a superior level of structural organization of matter. Most meta-atoms are made of metallo-dielectric composites in order to support resonant surface plasmon polariton eigenmodes. Since nanofabrication technology practically facilitates to create meta-atoms according to predefined designs [6–8], their electromagnetic response can be tailored on purpose. The characterization and understanding of such nano-optical systems classically rely on the interpretation of their far-field spectral properties. Specifically, distinctive dips and/or peaks in the transmittance and reflectance spectra of metamaterials are mostly caused by increased (or resonant) scattering or absorptive responses. Such resonances lend themselves to be connected to the characteristic electromagnetic eigenmodes which are sustained by the meta-atoms [9]. Since far-field characterization tools are well established, commercially available and thus relatively easy to apply for measurement purposes, this approach allows for a straightforward understanding of the working principle of many metamaterials.

This chapter is devoted to the far-field response of a particularly fascinating class of metamaterials, which meet the characteristics of an amorphous, rather than periodic or crystalline, arrangement of their meta-atoms. By definition, "amorphous" refers to a conglomerate whose constituents inhibit a limited degree of short-range order, while long-range order is completely absent. The most prominent example of a classical amorphous material is glass, which is known as an amorphous solid that transforms into a liquid upon heating it through the glass transition. Thus, the term "amorphous metamaterials" refers to artificial media whose meta-atoms are arranged in a disordered or randomized way, without any mutual long-range order. Such amorphous arrangements are quite naturally achieved when a metamaterial fabrication technique is based on a so-called bottom-up process [10–12]. Bottom-up nanofabrication approaches employ or may be complemented by self-organization, functionalization and external control parameters like heat, mechanical stress, chemical additives or prepatterned surfaces to create and assemble meta-atoms. This field of nanostructuring techniques has attracted a continuously increasing interest because of the rich assembling behavior and the collective properties that can potentially be engineered [13–29]. The further usage of bottom-up approaches is very likely to expand with regard to amorphous metamaterials for nanooptics in the near future [30–43]. Selected examples of bottom-up metamaterials are depicted in Fig. 6.1. Many bottom-up processes generate useful nanopatterns but are not directly applicable to optical metamaterial fabrication. However, they can be upgraded by exploiting complementary nanopattern transfer steps including noble metal deposition [44–48]. This selection of studies devoted to amorphous metamaterials, which is by no means exhaustive, illustrates the emerging research focus on this topic and is accompanied by former efforts to understand localization, transport and wave scattering phenomena in randomized media [49–52].

Nevertheless, one considerable shortcoming is the generally irreproducible, i.e. *non-deterministic* nature of bottom-up fabrication processes, in the sense that a number of uncontrollable (or even unconceivable) side conditions and process param-

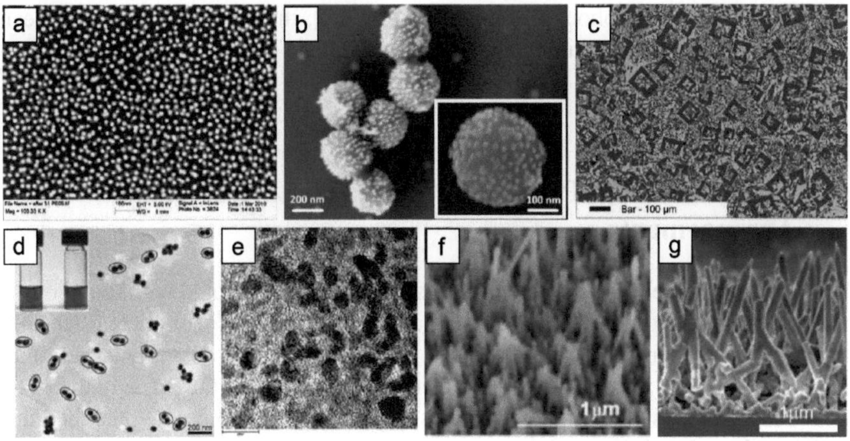

Fig. 6.1 Selected examples of bottom-up fabricated metamaterials with their constituting meta-atoms in amorphous arrangements. (**a**) Two distinctively separated layers made of gold nanoparticles with radii of 10 nm. Reprinted with permission from [42], Copyright 2011, American Chemical Society. (**b**) Core–shell (dielectric-gold) clusters fabricated by electrostatic assembly. Reprinted with permission from [41], Copyright 2011, American Chemical Society. (**c**) $SrTiO_3$-TiO_2 eutectic microstructures. Reprinted with permission from [36], Copyright 2010, Wiley-VCH. (**d**) Assembly of gold nanosphere dimers formed by amphiphilic polymer brushes, along with a low fraction of monomers and other multimers. Reprinted with permission from [43], Copyright 2011, American Chemical Society. (**e**) Dispersed silver nanoparticles obtained by the reduction of a silver salt. Reprinted with permission from [40], Copyright 2010, Wiley-VCH. (**f**) Gold nanowires grown out of an alumina nanotemplate. Reprinted with permission from [53], Copyright 2008, American Institute of Physics. (**g**) Upright gallium phosphide nanowires fabricated by metal-organic vapor phase-epitaxy. Reprinted with permission from [32], Copyright 2008, Wiley-VCH

eters may affect the overall outcome. We will show in the following that amorphous metamaterials can be investigated comprehensively by experiments based on simple mathematical design guidelines, or with other words, based on a *deterministic* background. Complementary to the "classical", bottom-up route towards amorphous metamaterials, this chapter will give an introduction to the more recent developments of *deterministic* amorphous metamaterials. A method, algorithm or model is said to be deterministic when its result is entirely determined by its initial state and inputs, and is neither random nor stochastic. Consequently, a deterministic amorphous metamaterial is a structure, whose state and degree of amorphization can be tailored intentionally during fabrication. In this sense, when fabricating a deterministic amorphous metamaterial, the final result including the size, shape and the very arrangement of each single meta-atom should be predictable and controllable within very narrow fabrication tolerances, an aspect which regrettably rarely applies to bottom-up nanofabrication methods. Recent studies performed by the Dal Negro group exemplified by highly involved aperiodic patterns [54–56], how deterministic approaches offer a superior level of control and information on the optical far-field properties of nanoplasmonic systems. In this discussion, we will mostly focus on the deterministic transition from periodic towards amorphous metamaterials [57–60], thus bridging a gap between the historically well separated areas of bottom-up and top-down fabrication techniques.

6.2 Transition from Periodic to Amorphous Meta-atom Arrangements

To date, most contemporary metamaterials have been investigated with their constituting meta-atoms ordered on a periodic lattice. It remained rather unclear whether their optical properties could be retained if this lattice was broken and the meta-atoms were put in an amorphous arrangement. This state of the art is mainly due to two reasons. First, the limited resources of contemporary computational facilities call for adapted and time-efficient numerical algorithms. In a periodic structure, it is often sufficient to compute the response of a single unit cell and employ periodic boundary conditions to mimic periodicity. Second, since many contemporary metamaterial fabrication techniques rely on lithographic methods, the lithographic data preparation is considerably facilitated if a unit cell can be repeated on a periodic lattice.

Thus the approaches described here inhibit a fundamental difference to previous investigations of strictly periodic optical metamaterials. An amorphous metamaterial cannot be treated straightforward by standardized numerical methods which would strictly rely on a periodic arrangement of the meta-atoms. Rather, the description of the interaction of light with an amorphous metamaterial exhibits a many-body problem which is often impractical to treat numerically, though not impossible [61, 62]. As a result, direct comparisons between experimentally available disordered systems containing a large number of constituents and their rigorous numerical treatment are comparably rare in the present literature [57, 61–64]. In the studies summarized here, we present different realizations of deterministic amorphous metamaterials and correlate the scattering response of the individual meta-atoms to the averaged far-field response of the ensemble. For one particular case, effective optical properties for amorphous metamaterials based on rigorous numerical simulations are introduced.

6.2.1 A Deterministic Algorithm for Amorphous Structures

First we give the technical details along with the assessment of an algorithm that allows for the deterministic and quasi-continuous amorphization of two-dimensional metamaterials whose meta-atoms are assembled on a plane area. While we restrict ourselves to the mathematical notation as implemented in [57, 58, 63], we note that similar approaches using slightly different parametrization sets have been utilized in other optical systems, including photonic crystals [52, 65–67], nanoapertures [68–70], nanoantennas [71], and plasmonic metamaterials [55, 59, 60, 72, 73]. In any case, the output of the algorithm is a data set containing the intended position of each single meta-atom, whose actual shape and composition must not be specified at this stage. This data set can be used to lithographically define an amorphous metamaterial. For instance, using state-of-the-art electron-beam lithography for the fabrication of the metamaterial samples, it is possible to precisely control the positions of the meta-atoms within a typical accuracy of 1 nm. Hence deterministic amorphization algorithms along with an advanced nanofabrication technology provide powerful and highly versatile tools for introducing a well-defined degree of artificial disorder and amorphization into a metamaterial.

In the following, the xy-plane is supposed to be the main lateral plane where the meta-atoms are assembled, while z denotes the principal propagation direction of light. In general, positional disorder in a planar metamaterial can be introduced by adding a random displacement to the position of each meta-atom in each lateral direction independently. Here we choose the displacements Δx and Δy such that they are uniformly distributed within the intervals

$$\Delta x, \Delta y \in \left[-\frac{Dp}{2}, +\frac{Dp}{2} \right], \tag{6.1}$$

where p is the lattice constant of the referential periodic system in either lateral direction ($p = p_x = p_y$). D is a dimensionless disorder parameter to control and quantize the degree of disorder. With other words, $D = 0$ corresponds to a periodic lattice arrangement while for $D = 1$ the center position of each meta-atom can randomly extend between the boundaries of the former unit cell. If D is larger than one, each meta-atom can even be placed beyond those boundaries. For practical reasons, a minimum interparticle separation d_{min} between adjacent meta-atoms has to be enforced. The reason for this is twofold: First, the undesired proximity effect in lithographic methods [74] may lead to the deformation and merging of adjacent structures during fabrication, if they are too closely spaced. Second, the plasmonic resonance properties of the individual meta-atoms would be predominantly affected by evanescent coupling. For the studies reported here, we want to focus on the scattering properties of the meta-atoms, explicitly excluding the regime of evanescent coupling. However, the practical requirement for d_{min} does not pose a strong restriction on the general applicability of the algorithm, because d_{min} appropriately reflects the physical situation of hard spheres in standard routines to fabricate self-organized metamaterials, given that conglomeration is prohibited [46].

In practice, the algorithm works in the following way: Starting from a periodic lattice, each meta-atom is given a number. Then, the order of all positions (x, y) of all meta-atoms is randomized, resulting in a randomized renumbering. This randomization of the order of placement itself ensures that the positions of all meta-atoms are not correlated [65]. At next, a displacement pair Δx and Δy is generated following (6.1) and added to the first (x, y) pair. The corresponding meta-atom has now been given a new position, which is memorized and kept fixed. Thus and due to the minimum interparticle distance d_{min}, a "forbidden zone" around this meta-atom is created, which can no more be occupied by another meta-atom. After that, the positions of all other meta-atoms on the renumbered list are subsequentially

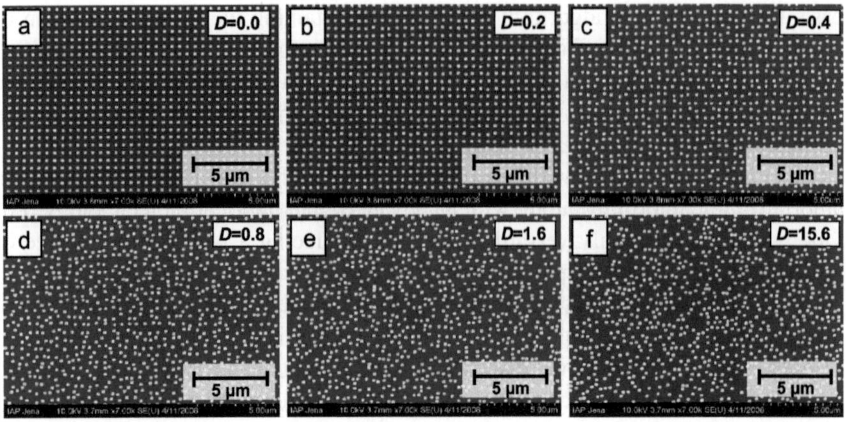

Fig. 6.2 Top view scanning electron microscope images of meta-atoms arranged with implementations of different values of disorder D

randomized. If the placement of one meta-atom violates the restriction enforced on $d_{\min}$ with respect to any other meta-atom already fixed, this specific pair of Δx and Δy is rejected and a new pair within $[-D/2p, D/2p]$ can be generated. The final result is a rearrangement of all meta-atoms governed by D and $d_{\min}$. We performed numerous test runs of this algorithm, confirming that it converged quickly as long as $d_{\min}$ was not chosen too large.

After implementing the algorithm in a numerical code, two-dimensional coordinate sets of disordered meta-atoms corresponding to different values of D can be generated. Figure 6.2 shows six scanning electron microscope images of fabricated metamaterials where each bright dot represents a single meta-atom, whose specific shape will be specified later in Sect. 6.3.3. While D varies, $d_{\min}$ was kept constant to about one third of the lateral size of a single meta-atom [57, 63]. The choice of this specific value turned out to be a good compromise between two counteracting aspects: On the one hand, smaller values of $d_{\min}$ result in the unintentional merging of a rather big number of adjacent meta-atoms during fabrication. On the other hand, if $d_{\min}$ was too large the disorder algorithm fails to converge. In addition to that, the larger $d_{\min}$ the stronger is the enforcement of the periodicity which is actually to be perturbed.

6.2.2 Statistical Assessment of an Amorphous System

Now we will classify the "randomness" in our metamaterial and verify that really amorphous structures are created with this algorithm. As defined above, the constituents of an amorphous medium may exhibit a finite degree of short-range order and practically no long-range order. Among various mathematical means, the most appropriate one to classify positional disorder is the radial distribution function, also called *pair correlation function* $g(r)$ [75, 76]. This dimensionless quantity can be calculated for any system containing a large number of particles at fixed positions, just like our meta-atoms. The parameter r denotes the radius of a circle which is normalized to the interparticle separation in a referential periodic system. The pair correlation function $g(r)$ then describes how the particle density varies as a function of the distance from an arbitrary meta-atom. Hence $g(r)$ offers a statistical mapping of the distances between the centers of pairs of particles [77]. Applied to the situation of two dimensions, it is evaluated by counting the numbers of particles that lie within a circular shell, dr, of radius r from an arbitrary origin in the plane. The absolute value of dr must be chosen sufficiently small with respect to r to ensure a reasonable spatial discretization. In the present case $r/dr \gg 10$ was required. The counting of particles within a circular shell area $da = 2\pi r dr$ is repeated for a range of radii and a large number of arbitrary points selected as the origin. The statistical average of these numbers, normalized with respect to the average particle number density $\langle\rho\rangle$ and the sampling area da for a particular radial distance r, gives the pair correlation function

$$g(r) = \frac{1}{\langle\rho\rangle} \frac{dn(r, r+dr)}{da(r, r+dr)}. \tag{6.2}$$

This formula is a valuable means for our purpose in that its characteristic features allow to distinguish between, e.g., periodic, randomized and amorphous particle assemblies. In periodic structures $g(r)$ shows not only one peak at $r = 1$ (corresponding to short-range order) but also higher order peaks at higher values of r (corresponding to long-range order). On the other hand, long-range order should be completely absent in a truly amorphous structure. Therefore, if our metamaterial is really amorphous, we expect only one discernible peak around $r = 1$ and a smooth evolution for large r converging to $g(r \to \infty) = 1$. For completeness it is noteworthy that $g(r)$ of a fully randomized structure will not show any features at all indicating the lack of both short- and long-range order [76]. Such a structure could be implemented in our case by setting $d_{\min}$ equal to zero while D is sufficiently large. However, this consideration is practically not relevant due to the finite extension of any physical meta-atom.

The pair correlation function $g(r)$ was calculated for a sufficiently large section of the arrangements depicted in Fig. 6.2. The results are shown in Fig. 6.3. As expected, $g(r)$ displays discrete peaks for the periodic arrangement (Fig. 6.3a). With increasing D, those peaks are gradually reduced and smeared out (note that the y-axes scale differently in the subfigures of Fig. 6.3). Though there is no distinct threshold between a partially disordered and an amorphous system, it can be clearly seen in Fig. 6.3e and f that for large values of D only the first peak around $r = 1$ survives while $g(r > 1)$ gives a noisy signal with a mean value around one. Note that the principal behavior of $g(r)$ will always support this statement, although its actual values depend on the individual implementation of disorder which can vary for a given D. Short-range correlation is present because the particles were not allowed to take positions in the ultimate vicinity of other particles. Consequently, $g(r)$ is identical to zero for small values of r, i.e. $g(r) = 0$ for all r smaller than the sum

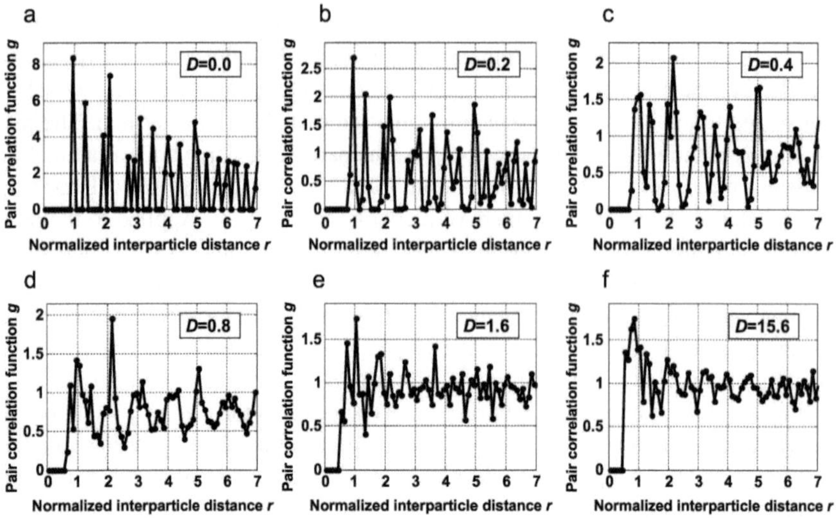

Fig. 6.3 Pair correlation function $g(r)$ for a two-dimensional assembly of meta-atoms with different degrees of disorder D as outlined in the text. (**a**) Periodic, (**b**)–(**d**) partially disordered and (**e**)–(**f**) amorphous system. Note that the y-axis scales differently in (**a**)–(**f**)

of $d_{\min}$ and the lateral size of a meta-atom. Correlation functions of that type are a clear indication of amorphous structures [78]. It is therefore proven that (6.1) together with the side condition $d_{\min} > 0$ can mimic the transition of a metamaterial from a periodic (or crystalline) to an amorphous state.

6.3 Far-Field Properties of Deterministic Amorphous Metamaterials

The optical properties of metamaterials are predominantly determined by the meta-atoms they are composed of. These meta-atoms act as strongly scattering entities sustaining eigenmodes for interacting electromagnetic waves. In order to maximize the efficiency of the scattering mechanism, noble metals are often incorporated into the meta-atoms to provide loosely bound or free charge carriers which may resonantly interact with external electromagnetic fields, giving rise to localized surface plasmon polaritons [79]. Accordingly, a meta-atom may be represented by an arbitrarily complex charge density distribution. By means of semi-analytical models it was shown how the characteristic near-field distributions of the eigenmodes of meta-atoms can be used for the classification and identification of molecular multipole moments [80–84]. Multipolar near-fields can be rigorously decomposed into a series of tensorial electromagnetic moments, e.g. contributions of first order (electric dipole), second order (magnetic dipole and electric quadrupole), third order (magnetic quadrupole and electric octupole), et cetera [85]. For detailed properties of the associated electromagnetic near-fields of these multipoles, the reader may conveniently resort to Chap. 7 written by Worawut Khunsin and Ralf Vogelgesang within this book. In the optical far-field spectra however, the footprints of the electromagnetic multipolar contributions are detectable as distinctive peaks and dips. It is then the task of the observer to attribute these spectral characteristics to the respective multipolar moments, for which complementary information like numerical simulations or optical near-field measurements are usually helpful if not necessary. In the following we will classify three deterministic amorphous metamaterials according to their wavelength operation regimes and relate their respective sensitivities against positional disorder to the coupling mechanisms of the excited electromagnetic multipoles.

6.3.1 Amorphous Metamaterials for Microwaves

In a study from Papasimakis et al., the concept of electromagnetically coupled and uncoupled ensembles of meta-atoms was introduced under the terminology "coherent" and "incoherent" metamaterials. The approach was underpinned by microwave experiments on specifically selected meta-atoms, forming either periodic or amorphous metamaterials [58]. These two kinds of structures, namely asymmetrically split rings and double concentric rings, represent antipodes for coherent and incoherent metamaterials and are schematically depicted in Fig. 6.4. In this context, a coherent metamaterial is defined to consist of meta-atoms exhibiting a *collective* far-field response, which is supported by the mutual interaction of scattered waves upon resonant excitation. Naturally, such a coherent response is sensitive to a rearrangement of the meta-atoms. Contrariwise, the meta-atoms of an incoherent metamaterial show an *individual* response, support only marginal short- and long-range interactions with their neighbors, and thus are virtually insensitive to positional disorder and amorphization.

The deterministic approach to introduce disorder into these microwave metamaterials is governed by a disorder control parameter D analogous to the one defined in Sect. 6.2.1. The experimental proofs of different sensitivities of concentric rings and asymmetrically split rings to such deterministic disorder are summarized in Fig. 6.5. In both cases, a resonant dip in the spectra denotes the position of the characteristic resonance of the respective meta-atoms on a periodic grid ($D = 0$). For the metamaterial composed of double concentric rings this resonant dip remains unaltered and robust with increasing disorder, including also the fully amorphous state corresponding to a value of D larger than one (Fig. 6.5a). This observation does not apply for the asymmetrically split ring metamaterial, where the respective dip smoothens out rapidly with an increasing value of D and finally vanishes almost completely for the amorphous modification of the metamaterial (Fig. 6.5b).

These completely oppositional characteristics can be explained on the basis of the dominant electromagnetic multipole of the main resonances sustained by the meta-atoms [58]. In both cases, magnetic dipoles oriented perpendicular to the plane of

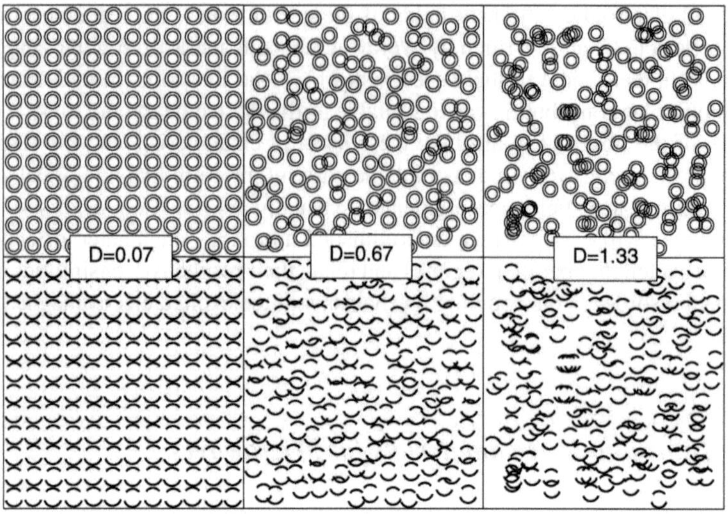

Fig. 6.4 Schematics of metamaterials composed of concentric rings (*top row*) and asymmetrically split rings (*bottom row*), with a increasing degree of positional disorder *D from the left to the right*. The definition of the disorder parameter *D* in [58] is identical to the one given in Sect. 6.2.1. The *two subfigures on the right* represent deterministic amorphous metamaterials. Reprinted with permission from [58], Copyright 2009, American Physical Society

the metamaterials are excited. However, while for the concentric rings these magnetic dipoles stem from antisymmetric electron currents in the two rings and point in opposite directions, they are oscillating perfectly in phase for regularly arranged asymmetric split rings (see insets in Fig. 6.5). Thus in the first case the interactions between the meta-atoms (both of electric dipole and magnetic dipole type) are negligible. The resonance shows the footprint of the individual structure and the collective response of the metamaterial is merely the sum of these individual meta-atom contributions (regardless of the actual arrangement), giving rise to the interpretation as an incoherent response. In the latter case, the electron currents induced in the upper and lower arcs of the rings create in-phase, coherent magnetic dipole moments. Increasing disorder leads to an increase of scattering losses, a decrease of the excitation strength of the collective resonance and ultimately cancels out the magnetic response when a complete amorphization is reached. These observations are in full agreement with the results of an earlier study on disorder in microwave metamaterials composed of double concentric rings and double split-ring resonators by Aydin et al. [72]. In this work, however, the authors restricted themselves to the case of comparably small perturbations of a periodic lattice mimicking typical fabrication tolerances and effectively excluded completely amorphous metamaterials.

We would like to point out that the terms "coherent" and "incoherent" as introduced by Papasimakis et al. refer rather to an abstract concept of metamaterials in general than to a distinctive class of physical structures. In Sect. 6.3.3 it will be shown that a particular metamaterial may optically respond either in a coherent or an incoherent way at different excitation wavelengths, depending on the actually

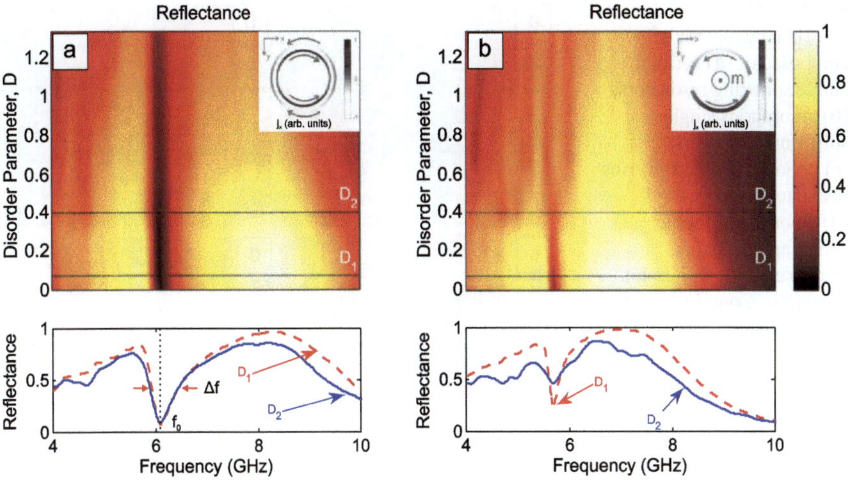

Fig. 6.5 Measured reflectances as functions of the frequency and the disorder parameter D for (**a**) arrays of concentric rings and (**b**) asymmetrically split rings. Large values of D correspond to amorphous metamaterials. The *two marked cross sections* show reflectance profiles at $D_1 = 0.07$ (*dashed line*) and $D_2 = 0.40$ (*solid line*), respectively. The current distributions inside the according meta-atoms are shown in the *insets*. Reprinted with permission from [58], Copyright 2009, American Physical Society

excited eigenmode. Hence, one may denote a specific resonance of a metamaterial (under differentiated excitation conditions) to be of "coherent" or "incoherent" nature rather than a whole subclass of metamaterials.

6.3.2 Amorphous Metamaterials for Terahertz Waves

The so-called "split-ring resonator" is considered as one of the most prototypical meta-atoms ever, since it has been the primary structure to study artificial magnetism at any range of the electromagnetic spectrum down to wavelengths of about 800 nm [86–89]. The resonances of split-ring resonators at normal light incidence are commonly understood as their odd and even plasmon-polaritonic eigenmodes, whose excitation conditions depend on the orientation of the incident electric field [9].

In their study, Singh et al. performed far-field measurements on periodic and randomized (or amorphous) single and double split-ring resonators and experimentally verified different sensitivities on positional disorder of their respective eigenmodes [59]. The authors fabricated four sets of samples as shown in Fig. 6.6, differentiating the cases of single and double split-ring resonators positioned on periodic or amorphous grids, respectively. Upon illuminating these metamaterials under normal incidence and depending on the polarization state, either the odd or even eigenmodes of the structures were excited. The key findings of this work can be

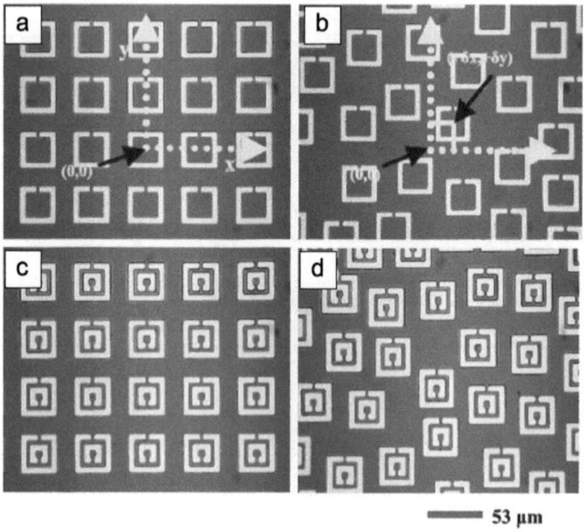

Fig. 6.6 Microscope images of THz metamaterials composed of (**a**) periodic single split-ring resonators, (**b**) amorphous single split-ring resonators, (**c**) periodic double split-ring resonators, and (**d**) amorphous double split-ring resonators. Reprinted with permission from [59], Copyright 2010, IOP Publishing

summarized as follows: The lowest order odd eigenmode of an assembly of split-ring resonators (historically often referred to as an *LC* resonance [87, 88]), which is excited when the incoming electric field is polarized parallel to the split-ring gap, is nearly independent of the very arrangement of the meta-atoms. Differently, when the electric field of the incoming radiation is oriented perpendicular to the split-ring gaps, the next order odd eigenmode and the lowest order even eigenmode are excited. These modes, which are characterized by in-plane electric dipole oscillations, turned out to be seriously affected if disorder is introduced into the arrangement of the split-ring resonators, causing the strength of the observable far-field spectra to decrease considerably [59].

The authors interpreted the persistence of the lowest order odd eigenmode on the basis of an *LC* circuit model, whose detailed analysis however is beyond the scope of this chapter. In addition, the vulnerability of an ensemble of electric dipole oscillations to positional disorder is physically intuitive, since the strong dipolar coupling between adjacent split-ring resonators is diminished in a complicated way. We note that only a limited number of samples has been considered in the work by Singh et al., leaving room for further studies on split-ring resonators with a deterministic and more continuous tuning of the degree of disorder. However, in the next section we will consider this full transition from a periodic to an amorphous metamaterial for the case of another meta-atom supporting similar mode features as split-ring resonators. In this context we give a unifying approach in showing how the general sensitivity of distinct eigenmodes of arbitrary meta-atoms on disorder can be consistently explained by an analysis of the respective scattering response and the absorption levels.

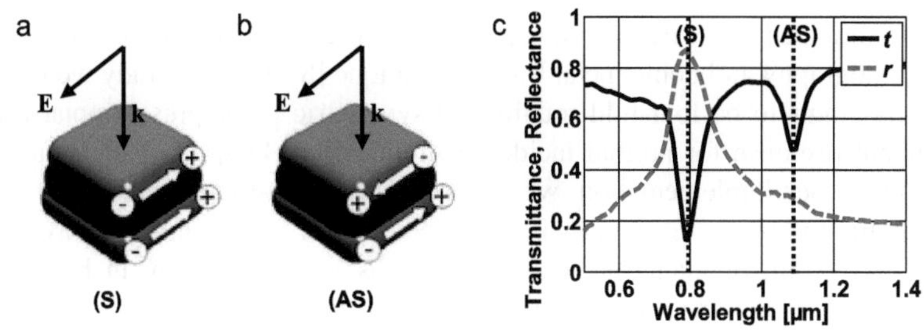

Fig. 6.7 Schematic of a cut-wire pair meta-atom excited at (**a**) the symmetric (S) and (**b**) the antisymmetric (AS) plasmonic eigenmode. The wave vector **k** denotes the light propagation direction and **E** the polarization of the electric field. The *arrows* represent the (**a**) in-phase and (**b**) out-of-phase oscillations of the current densities in the metal wires. (**c**) Simulated prototypical spectrum showing the two resonances corresponding to the symmetric (S) and antisymmetric (AS) mode in transmittance t and reflectance r

6.3.3 Amorphous Metamaterials for Optical Wavelengths

The previous sections were devoted to a series of studies on disordered and amorphous metamaterials in each of which only one single electromagnetic multipole was excitable and detectable under fixed illumination conditions such as incoming polarization and angle of incidence. The focus was put on asymmetrically split rings, double concentric rings and split-ring resonators [58, 59, 72]. It is imperative to extend those studies toward more complex meta-atoms that support also higher order multipoles for a given light input scheme (polarization and angle of incidence) to develop a comprehensive understanding of disordered metamaterials. The simplest structure that fulfills this goal is the cut-wire pair [90]. This meta-atom received an enormous attention from the community since both an electric quadrupole and a magnetic dipole can be excited at normal incidence. The cut-wire pair is therefore regarded as one of the key building blocks to obtain a magnetic response at optical frequencies [91–93].

In the following we consider a cut-wire pair as a structure consisting of two nanoscale metallic wires which are separated by a dielectric spacer. The plasmonic properties of this meta-atom can be schematically understood by a plasmon hybridization model [94], as illustrated in Fig. 6.7. Each metallic cut-wire supports a localized plasmon polariton. In a dimer system (two cut-wires in close proximity), the strong coupling of plasmonic modes causes a hybrid splitting of the resonance into a symmetric and an antisymmetric mode [80, 94]. Since for the high-frequency symmetric mode the current densities in both arms are in phase, it effectively acts as an electric dipole. By contrast, the current densities in both arms oscillate π out-of-phase for the low-frequency antisymmetric mode evoking an electric quadrupole and a magnetic dipole moment [80, 81]. In an ensemble of meta-atoms, the two modes translate into the far-field response by dips in the transmittance and peaks in the reflectance, although the latter is rather weak for the antisymmetric eigenmode (Fig. 6.7).

We fabricated cut-wire pair metamaterials on a silica substrate by electron-beam lithography and noble metal lift-off technique [8]. For this study we required the meta-atoms to be four-fold rotationally symmetric to suppress a polarization-dependent response for normal incidence illumination. Despite of their quadratic footprint in our implementation, we will consistently use the expression *cut-wire pairs* in the following. Each cut-wire pair consisted of two 30 nm gold layers separated by 45 nm of magnesia with side lengths $s_{x,y}$ of 180 nm in both lateral directions. For reference, a perfectly periodic sample ($D = 0$) with lattice constants $p_{x,y} = 512$ nm was fabricated. The corresponding area filling fraction of $f = s_{x,y}^2/p_{x,y}^2 \approx 12.4\,\%$ remained unaltered on average for all samples by keeping the number of meta-atoms and hence their average density constant. Eleven different samples with different values of D ranging from 0 to 1000 were fabricated on the same wafer. Each sample had an area footprint of 3 mm × 3 mm, from which a selection of scanning electron microscopy images was shown in Fig. 6.2. As we outlined in Sect. 6.2.2, this set represents a gradual transition from a periodic to an amorphous planar metamaterial.

The optical properties of all samples were characterized by transmission and reflection spectroscopy. At normal incidence virtually identical spectra have been obtained for both linear polarization states parallel to either lattice vector, which is in agreement with the symmetry of quadratic cut-wire pair meta-atoms. The investigated spectral domain spanned wavelengths between 0.5 μm and 1.2 μm, comprising the two principal resonances of the metamaterial. The spectral results are shown in Fig. 6.8. We note that the low signal-to-noise ratio in the wavelength range from 0.9 μm to 1.0 μm in the reflection spectra is due to low sensitivity of the detector and does not bear any physical meaning. For the periodic arrangement ($D = 0$) two dips appear in the transmission spectrum (Fig. 6.8a) situated at wavelengths of 0.8 μm and 1.05 μm which are related to the symmetric and antisymmetric plasmon polariton eigenmodes of the meta-atom, respectively. The former is also confirmed by a noticeable peak in reflection (Fig. 6.8b) at 0.8 μm. A considerably different sensitivity of the two resonances was observed for an increasing degree of disorder. While the antisymmetric resonance at $\lambda = 1.05$ μm almost perfectly sustains in width and magnitude even for $D = 1000$, the symmetric resonance blue shifts by about 30 nm, broadens and its magnitude decays already at a low level of disorder $D = 1$.

These astoundingly different characteristics can be explained by considering the absorption levels of the respective cut-wire pair eigenmodes and the chosen average interparticle distance. In a model developed by Albooyeh et al. [95], the reflective, transmissive and absorptive properties of meta-atoms both on periodic and amorphous grids were analyzed. The theory provides a unifying answer to the observed strong differences in resonance broadening and decay at electrically and/or magnetically resonant responses, such as the cut-wire pairs with symmetric and antisymmetric resonances as observed in Figs. 6.7 and 6.8. With the assumptions of an optically dense array of meta-atoms, arranged with an average interparticle distance r, the negligence of bi-anisotropic coupling, and the assumption that the meta-atoms sustain both electric and magnetic resonances with Lorentzian line-shapes at

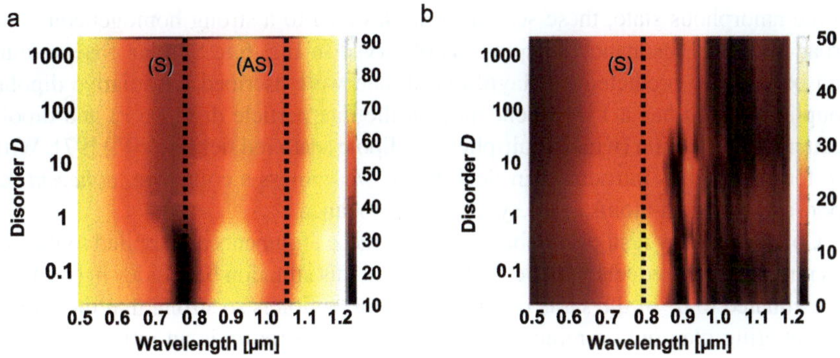

Fig. 6.8 Measured (**a**) transmittance and (**b**) reflectance as functions of the wavelength and the disorder parameter D (logarithmic scale). Both spectra are recorded for discrete values of D and interpolated to guide the eye. The *black dotted lines* indicate the positions of the symmetric (S) and antisymmetric (AS) resonance. Reprinted with permission from [57], Copyright 2009, American Physical Society

distinct (spectrally well separated) wavelengths, the following inequality which has to hold close to the respective resonance frequency can be derived [95]:

$$\omega_{0e,m}\frac{\Gamma_{e,m}}{A_{e,m}} + \frac{k_0 r}{2} \gg \frac{k_0^3 r^3}{6\pi}. \tag{6.3}$$

Herein, $\omega_{0e,m}$, $\Gamma_{e,m}$, $A_{e,m}$ and $k_0 = \omega\sqrt{\epsilon_0\mu_0}$ denote, respectively, the corresponding resonance frequencies, the dissipation losses in the meta-atoms, the resonance amplitudes and the wave number in the surrounding space. The subscripts distinguish between the electric (e) and magnetic (m) resonance case. While r corresponds to the lattice period p for the situation of a periodic arrangement of the meta-atoms, it can be considered an averaged particle distance in the amorphous case. If the condition formulated in (6.3) is satisfied, the effect of the specific arrangement of the meta-atoms on the resonance lineshape would be negligible, and accordingly the response of periodic and amorphous metamaterials would be similar. Physically, this condition is fulfilled if absorption (the first term of the left-hand side) and coherent plane-wave reflection (created by surface-averaged currents [96] and corresponding to the second term of the left-hand side) dominate over scattering (the right-hand side term). This is particularly true for weak resonances with high dissipative losses, and small interparticle distances r.

Applying this analysis to the cut-wire pairs considered here, the interparticle distances r and thus surface-induced plane waves are nearly identical for both resonances, i.e. the wavelengths were they are supported do not differ too significantly. However, we find a pronounced difference in the ratio of the absorption and the scattering losses at the respective resonance frequencies. At the symmetric resonance, associated with dominating electric dipole scattering, it was found that (6.3) is not satisfied, i.e. scattering effects are strong and clearly dominating [57, 95]. Note that electric dipolar resonances in general do couple well to the free space radiation, hence a large scattered field is generated. Upon the transition from the periodic to the amorphous state, these scattering losses lead to a strong homogeneous line broadening and resonance damping, as observed in Fig. 6.8. Similar observations had been firstly predicted by Haynes et al. and were ascribed to radiative dipolar coupling (which has a $1/r$ dependence on the interparticle distance r) and dipole scattering retardation (which multiplies the dipole scattered field by e^{ikr}) [97]. With the terminology as introduced in Sect. 6.3.1, we witness a collective, *coherent* response of the meta-atoms at the symmetric resonance.

Complementary results are obtained when these arguments are applied to the antisymmetric low frequency mode (Fig. 6.7b). In this situation both sides in (6.3) can be estimated to be of the same order, i.e. the absorption is so strong that the effects of scattering play only a minor role. The higher absorption is an intrinsic feature of higher order electromagnetic multipolar resonances. They couple less efficiently to free space radiation which causes non-radiative losses usually to dominate over radiative losses, i.e. these resonances are dominated by absorption. Therefore, the meta-atoms scatter considerably more weakly at the anti-symmetric resonance than at the symmetric resonance. This of course strongly reduces any interference effect of the external field and the scattered field from all the other meta-atoms which might have driven the meta-atoms all with a slightly different phase. In consequence, the spectral response is independent of the arrangement, namely periodic or amorphous. In other words, at the antisymmetric resonance the meta-atoms respond in an individual, *non-coherent* manner to the external excitation [58].

It is noteworthy that the physical phenomena leading to resonance damping are the same for electrically or magnetically resonant meta-atoms. In principal, although not achieved to date, it should be possible to design arrays of electric scatterers which give nearly the same response in periodic and amorphous arrangements, provided that their scattering losses are designed to be much smaller than the dissipation losses. Alternatively, the average interparticle distance r could be minimized to satisfy condition (6.3).

6.3.4 Large-Scale Numerical Simulations of Amorphous Systems

To compare these spectra from Fig. 6.8 with theoretical predictions, numerical simulations of the system for no ($D = 0.0$), moderate ($D = 0.3$), and high ($D = 3.0$) positional disorder were performed [57, 63]. In the case of non-periodic optical systems, the method of choice to solve Maxwell's equations rigorously is the finite-difference time-domain (FDTD) method [98]. In brief, for this technique the Maxwell's equations in the temporal domain and relations connecting polarization and conduction with the electric field are directly discretized in space and time on the so-called Yee-grid.

For the system described in this section, we discretized comparably large supercells (7 μm × 7 μm), each containing 196 meta-atoms, with a spatial resolution of 5 nm. These supercells were truncated by periodic boundaries in the lateral dimensions whereas perfectly matched layers were used along the light propagation direction. A Drude model with adjusted plasma frequency and damping constant was

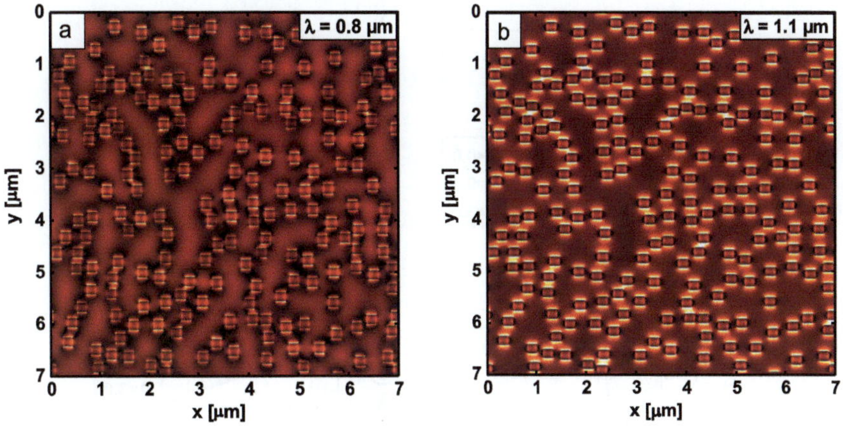

Fig. 6.9 Optical near-fields of amorphous cut-wire pair metamaterials obtained by supercell FDTD simulations. The figures show the modulus of the electric field 10 nm above the meta-atoms for a y-polarized plane wave illumination at normal incidence. (**a**) Amorphous structure with disorder parameter $D = 3.0$ at the symmetric resonance ($\lambda = 0.8$ μm). (**b**) Amorphous structure with disorder parameter $D = 3.0$ at the anti-symmetric resonance ($\lambda = 1.1$ μm). Note that the color scale is differently chosen in the two subfigures. Dark and bright colors mark spots of low and high electric fields, respectively. Reprinted with permission from [63], Copyright 2011, Springer

implemented to simulate induced currents. Then, this adapted FDTD method solves the wave equations in the time domain, meaning that each wavelength has to be treated separately. The temporal evolution of a system must be computed sufficiently long to obtain the steady-state field. After imposing a spatial Fourier-transform on this field and extracting the transmitted zeroth order diffraction efficiency from the scattering background, the complex transmission coefficient of the system under consideration is obtained for one single wavelength. An equivalent treatment is applied to the reflected field, where in addition the incident field must be subtracted. In order to obtain a spectrally resolved response, this procedure is repeated for a finite number of discrete wavelengths. Besides the transmission and reflection spectra, this treatment provides also the spatially resolved optical near-fields of the system under consideration, as exemplarily shown in Fig. 6.9. A detailed discussion of this numerical technique is beyond the scope of this section, but can be found in [61, 62].

The two plasmonic resonances of interest are confirmed to appear at the wavelengths 0.8 μm and 1.1 μm for $D = 0.0$. Their identification as the symmetric and antisymmetric eigenmodes of the cut-wire pair structure is supported by a close investigation of the near-field at the resonant wavelengths (Fig. 6.9). The full spectral evolution is summarized for no ($D = 0.0$), moderate ($D = 0.3$), and high disorder ($D = 3.0$) in Fig. 6.10. First we note that the antisymmetric resonance appears at slightly longer wavelengths in comparison to the experiment, which is most likely due to small deviations from the topography of the meta-atoms. Owing to the huge computational effort required by the FDTD method it is usually possible to perform geometry parameter scans on such large spatial domains. Nevertheless, all experimental observations in transmission and reflection that occur for an increasing de-

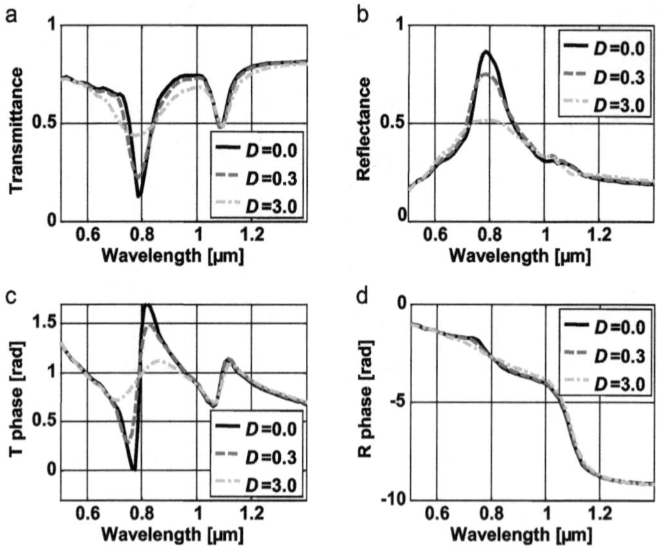

Fig. 6.10 Simulated (**a**) transmittance, (**b**) reflectance, (**c**) transmission phase delay and (**d**) reflection phase delay of the cut-wire pair metamaterial as functions of the wavelength and D. Reprinted with permission from [57], Copyright 2009, American Physical Society

gree of disorder are fully reproduced. With D increasing, the symmetric resonance at 0.8 μm broadens, blue shifts, decays, and nearly vanishes already for $D = 3.0$. This evolution does not apply for the antisymmetric eigenmode at 1.1 μm. While D increases, the dip in the transmitted amplitude and its corresponding phase evolution retain their strength and width almost unaffected. Thus it is confirmed that the antisymmetric eigenmode is nearly invulnerable to deterministic amorphization for this particular implementation [57] and supports all findings of the analytical model [95]. Similar conclusions were drawn by Guida et al. for magnetic eigenmodes in disordered metamaterials composed of silver square nanospirals, however the authors did not monitor the whole transition into an amorphous arrangement of meta-atoms but restricted their study to comparably small degrees of positional disorder [60].

6.3.5 Effective Optical Properties of Amorphous Metamaterials

Thanks to the fact that the FDTD delivers not only amplitude (Figs. 6.10a and b) but also phase information of the optical far-field (Figs. 6.10c and d) and based on the good agreement between measurements (Fig. 6.8) and simulations (Fig. 6.10), the computed complex spectra allowed us to determine effective material properties for the periodic, the weakly disordered, and the amorphous metamaterial. This complete set of data allows for calculating effective optical far-field properties of an amorphous metamaterial by using the standard retrieval algorithm [99, 100]. Concerning this technique, a grain of salt is due. Effective optical properties such as $\epsilon_{\mathrm{eff}}(\omega)$ and $\mu_{\mathrm{eff}}(\omega)$ have been extensively applied to quantify the performance of numerous metamaterials. As a first order approximation, they serve as a convenient means to describe the light-matter interaction in a simplified way in order to gain preliminary insight into the underlying physics in such complex and heterogeneous nanostructures. However, it is now well established that the concept of effective optical properties may not accurately reflect the complicated constitutive relations of mesoscopic and complexly shaped nanostructures [101, 102]. Particularly, the implications of strong spatial dispersion [103] in optical metamaterials have been confirmed by numerous experimental observations [104–109]. Nowadays the assignment of effective optical properties is applied with increasing reluctance [110–113] while alternative optical target functions will be essential for a physically meaningful description [114, 115]. Within a historical context we will show and discuss in the following the effective optical properties of the amorphous cut-wire pair metamaterial, implying that the limits of their validity must always be carefully borne in mind.

In Fig. 6.11 we attribute effective properties to the metamaterial along its transition from a periodic to an amorphous state. From those properties similar conclusions can be drawn as for the transmission and reflection spectra. Firstly, the high frequency resonance at $\lambda = 0.8$ μm that induces a Lorentzian dispersion in the effective permittivity $\epsilon_{\mathrm{eff}}(\lambda)$ strongly decreases with increasing disorder D (Figs. 6.11a and b). Since the electric dipoles excited in the meta-atoms cease to oscillate in phase, the resonance is homogeneously broadened, causing a degradation of the induced dispersion. Our second important finding concerns the persistence of the antisymmetric resonance of $\mu_{\mathrm{eff}}(\lambda)$ at $\lambda = 1.1$ μm, whose strength, line shape, and width are preserved practically independently of D. Since this resonance is usually employed as a "magnetic resonance" in metamaterial designs [91–93], one may conclude that the effective magneto-optical properties of many current left-handed metamaterials are only marginally affected by the very arrangement of the meta-atoms, provided that condition (6.3) is fulfilled.

6.3.6 Top-Down and Bottom-Up Amorphous Metamaterials

A "top-down" fabrication scheme accounts for a method in which the entire structure is created starting from a predefined arrangement, e.g. a stack of layers, that is sequentially processed, typically by electron, photon or ion beam lithography. On the other hand, "bottom-up" approaches refer to schemes which start with entities at a smaller scale than the final structure, where regularly ordered lattices cannot be simply realized without additional efforts [20, 36, 44–46]. Hence, the meta-atoms in bottom-up fabricated metamaterials are almost inevitably put in an amorphous arrangement, as exemplarily illustrated in Fig. 6.1.

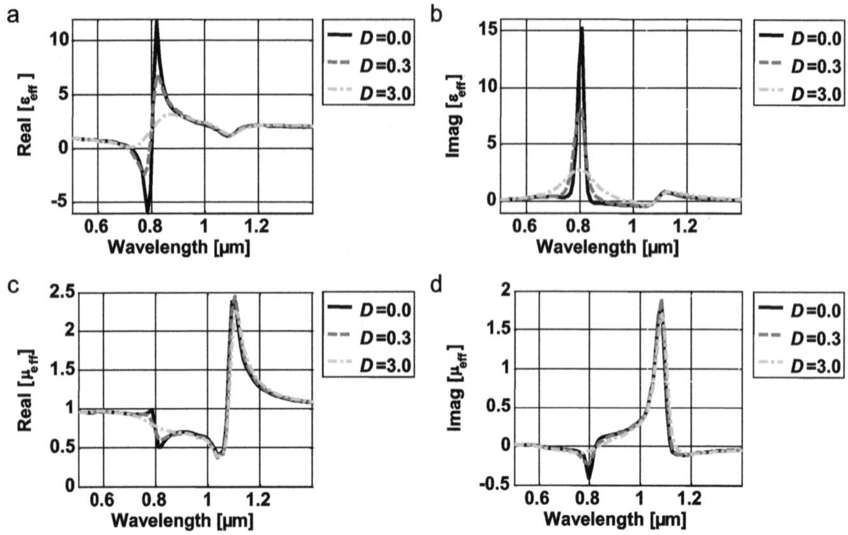

Fig. 6.11 Wavelength-dependent effective optical properties of the periodic ($D = 0.0$), the weakly disordered ($D = 0.3$), and the amorphous cut-wire pair metamaterial ($D = 3.0$). (**a**) Real and (**b**) imaginary parts of the effective electric permittivity $\epsilon_{\mathrm{eff}}(\lambda)$. (**c**) Real and (**d**) imaginary parts of the effective magnetic permeability $\mu_{\mathrm{eff}}(\lambda)$. Reprinted with permission from [57], Copyright 2009, American Physical Society

In the preceding sections it was numerically, analytically and experimentally shown that the optical resonances of some specific types of meta-atoms may become invulnerable to positional disorder and deterministic amorphization [57–60]. Among these resonances is the so-called magnetic resonance in planar metamaterials, which has been frequently implemented in the design for left-handedness [91–93]. This finding, though experimentally realized by deterministic top-down fabrication schemes, establishes a previously missing link to metamaterials fabricated by bottom-up approaches, like chemical syntheses or self-assembling nanofabrication methods. Nowadays the vast majority of optical metamaterials is fabricated using top-down approaches, which often inhibit costly and time consuming serial writing processes such as electron-beam lithography. Moreover, it seems elusive to fabricate true bulk metamaterials by a sequential stacking of single functional layers [116–119]. This situation hinders the transfer of fundamental concepts from an academic into an industrial environment where true applications shall be implemented.

Accordingly, quick and reliable fabrication schemes based on self-assembling or chemically randomized processes will ease the applicability of large-scale metamaterials. From our findings we can conclude that a magnetic resonance will most likely be retained if the top-down fabrication of metamaterials was substituted by bottom-up approaches. Moreover, in order to incorporate a potentially left-handed metamaterial into an imaging application, it ought to show a weak spatial dispersion [120]. However, the mesoscopic arrangement of meta-atoms in most of the present metamaterials causes inevitably a high degree of anisotropy and strong spatial dispersion [101, 121–123]. These constraints can be possibly lifted by employing amorphous metamaterials, provided that the meta-atoms obey to the quasi-static limit [124]. Accordingly, true optical isotropy is anticipated for sufficiently small meta-atoms that are randomly arranged in space [125, 126]. Thus the results presented here are especially valuable for the so far elusive prediction of optical properties of bottom-up fabricated metamaterials.

6.4 Conclusions

One of the multiple fascinating branches of the rapidly evolving field of optical metamaterial fabrication has spread into the field of chemically synthesized mixtures and amorphous conglomerates of artificial nanocomposites. Along with this evolution, a number of *deterministic* experiments relying on far-field spectroscopy in various wavelength regimes have consistently confirmed astoundingly different sensitivities of the characteristic resonances of optical metamaterials against positional disorder. We have reviewed how deterministic amorphous metamaterials can offer information on the transition from a periodic or crystalline optical medium to its amorphous counterpart. In particular, the key to understand the coupling of a resonantly excited meta-atom with its nearest neighbors is an analytical model of the optical scattering response and the level of dissipation loss at the particular electromagnetic eigenmode. With this conceptual approach it can be clearly predicted whether a particular metamaterial is affected by a disordered or amorphous arrangement of its building-blocks or not. These exciting findings establish a long-term aspired link between bottom-up and top-down fabricated metamaterials and offer a new platform to investigate a large variety of amorphous composites in terms of their electromagnetic properties.

Acknowledgements Financial support by the Federal Ministry of Education and Research (PhoNa and MetaMat) as well as from the Thuringian State Government (MeMa) is gratefully acknowledged. The authors wish to thank M. Albooyeh and S. Tretyakov for fruitful discussions on the scattering response of amorphous magnetic dipoles.

References

1. D. Smith, J. Pendry, M. Wiltshire, Science **305**, 788 (2004)
2. V. Shalaev, Nat. Photonics **1**, 41 (2007)
3. C. Caloz, Mater. Today **12**, 12 (2009)
4. C. Soukoulis, M. Wegener, Nat. Photonics **5**, 523 (2011)
5. C. Rockstuhl, C. Menzel, S. Mühlig, J. Petschulat, C. Helgert, C. Etrich, A. Chipouline, T. Pertsch, F. Lederer, Phys. Rev. B **83**, 245119 (2011)
6. A. Boltasseva, V. Shalaev, Metamaterials **2**, 1 (2008)
7. H. Moser, C. Rockstuhl, Laser Photonics Rev. **6**, 219 (2012)
8. C. Helgert, K. Dietrich, D. Lehr, T. Käsebier, T. Pertsch, E.B. Kley, Microelectron. Eng. **97**, 181 (2012)

9. C. Rockstuhl, F. Lederer, C. Etrich, T. Zentgraf, J. Kuhl, H. Giessen, Opt. Express **14**, 8827 (2006)
10. S. Zhang, Nat. Biotechnol. **21**, 1171 (2003)
11. S. Glotzer, M. Solomon, Nat. Mater. **6**, 557 (2007)
12. T. Sau, A. Rogach, Adv. Mater. **22**, 1781 (2010)
13. J. Aizenberg, P. Braun, P. Wiltzius, Phys. Rev. Lett. **84**, 2997 (2000)
14. I. Hamley, Nanotechnology **14**, R39 (2003)
15. J. Cheng, A. Mayes, C. Ross, Nat. Mater. **3**, 823 (2004)
16. M. Stoykovich, M. Müller, S. Kim, H. Solak, E. Edwards, J. de Pablo, P. Nealey, Science **308**, 1442 (2005)
17. M. Li, R. Bhiladvala, T. Morrow, J. Sioss, K. Lew, J. Redwing, C. Keating, T. Mayer, Nat. Nanotechnol. **3**, 88 (2008)
18. T. Ming, X. Kou, H. Chen, T. Wang, H. Tam, K. Cheah, J. Chen, J. Wang, Angew. Chem., Int. Ed. Engl. **47**, 9685 (2008)
19. M. Rycenga, J. McLellan, Y. Xia, Adv. Mater. **20**, 2416 (2008)
20. N. Shalkevich, A. Shalkevich, L. Si-Ahmed, T. Bürgi, Phys. Chem. Chem. Phys. **11**, 10175 (2009)
21. J. Lee, Q. Wu, W. Park, Opt. Lett. **34**, 443 (2009)
22. S. Park, D. Lee, J. Xu, B. Kim, S. Hong, U. Jeong, T. Xu, T. Russell, Science **323**, 1030 (2009)
23. M. Byun, N. Bowden, Z. Lin, Nano Lett. **10**, 3111 (2010)
24. A. Hung, C. Micheel, L. Bozano, L. Osterbur, G. Wallraff, J. Cha, Nat. Nanotechnol. **5**, 121 (2010)
25. J. Fan, K. Bao, C. Wu, J. Bao, R. Bardhan, N. Halas, V. Manoharan, G. Shvets, P. Nordlander, F. Capasso, Nano Lett. **10**, 4680 (2010)
26. J. Fan, C. Wu, K. Bao, J. Bao, R. Bardhan, N. Halas, V. Manoharan, P. Nordlander, G. Shvets, F. Capasso, Science **328**, 1135 (2010)
27. M. Personick, M. Langille, J. Zhang, C. Mirkin, Nano Lett. **11**, 3394 (2011)
28. S. Tan, M. Campolongo, D. Luo, W. Cheng, Nat. Nanotechnol. **6**, 268 (2011)
29. T. Lerond, J. Proust, H. Yockell-Lelièvre, D. Gérard, J. Plain, Appl. Phys. Lett. **99**, 123110 (2011)
30. Q. Wu, W. Park, Appl. Phys. Lett. **92**, 153114 (2008)
31. D. Pawlak, K. Kolodziejak, K. Rozniatowski, R. Diduszko, M. Kaczkan, M. Malinowski, M. Piersa, J. Kisielewski, T. Lukasiewicz, Cryst. Growth Des. **8**, 1243 (2008)
32. O. Muskens, S. Diedenhofen, M. van Weert, M. Borgström, E. Bakkers, J. Rivas, Adv. Funct. Mater. **18**, 1039 (2008)
33. D. Aydin, M. Schwieder, I. Louban, S. Knoppe, J. Ulmer, T. Haas, H. Walczak, J. Spatz, Small **5**, 1014 (2009)
34. J. Cho, D. Gracias, Nano Lett. **9**, 4049 (2009)
35. K. Stebe, E. Lewandowski, M. Ghosh, Science **325**, 159 (2009)
36. D. Pawlak, S. Turczynski, M. Gajc, K. Kolodziejak, R. Diduszko, K. Rozniatowski, J. Smalc, I. Vendik, Adv. Funct. Mater. **20**, 1116 (2010)
37. B. Casse, W. Lu, Y. Huang, E. Gultepe, L. Menon, S. Sridhar, Appl. Phys. Lett. **96**, 023114 (2010)
38. V. Tamma, J. Lee, Q. Wu, W. Park, Appl. Opt. **49**, A11 (2010)
39. C. Bohley, M. Yau, C. Erk, Y. Wang, R. Wehrspohn, S. Schlecht, M. Steinhart, Phys. Status Solidi (b) **10**, 2470 (2010)
40. J.J. Park, X. Bulliard, J. Lee, J. Hur, K. Im, J.M. Kim, P. Prabhakaran, N. Cho, K.S. Lee, S.Y. Min, T.W. Lee, S. Yong, D.Y. Yang, Adv. Funct. Mater. **20**, 2296 (2010)
41. S. Mühlig, A. Cunningham, S. Scheeler, C. Pacholski, C. Pacholski, T. Bürgi, C. Rockstuhl, F. Lederer, ACS Nano **5**, 6586 (2011)
42. A. Cunningham, S. Mühlig, C. Rockstuhl, T. Bürgi, Phys. Chem. C **115**, 8955 (2011)
43. L. Cheng, J. Song, J. Yin, H. Duan, J. Phys. Chem. Lett. **2**, 2258 (2011)
44. R. Glass, M. Möller, J. Spatz, Nanotechnology **14**, 1153 (2003)

45. H. Fredriksson, Y. Alaverdyan, A. Dmitriev, C. Langhammer, D. Sutherland, M. Zäch, B. Kasemo, Adv. Mater. **19**, 4297 (2007)
46. A. Dmitriev, T. Pakizeh, M. Käll, D. Sutherland, Small **3**, 294 (2007)
47. W. Lee, S. Lee, R. Briber, O. Rabin, Adv. Funct. Mater. **21**, 3424 (2011)
48. K. Lodewijks, N. Verellen, W. Van Roy, V. Moshchalkov, G. Borghs, P. Van Dorpe, Appl. Phys. Lett. **98**, 091101 (2011)
49. P. Sheng, *Introduction to Wave Scattering, Localization and Mesoscopic Phenomena*, vol. 88 (Springer, Berlin, 2006)
50. M. Albada, A. Lagendijk, Phys. Rev. Lett. **55**, 2692 (1985)
51. T. Pertsch, U. Peschel, J. Kobelke, K. Schuster, H. Bartelt, S. Nolte, A. Tünnermann, F. Lederer, Phys. Rev. Lett. **93**, 53901 (2004)
52. M. Rechtsman, A. Szameit, F. Dreisow, M. Heinrich, R. Keil, S. Nolte, M. Segev, Phys. Rev. Lett. **106**, 193904 (2011)
53. L. Menon, W. Lu, A. Friedman, S. Bennett, D. Heiman, S. Sridhar, Appl. Phys. Lett. **93**, 123117 (2008)
54. A. Gopinath, S. Boriskina, N. Feng, B. Reinhard, L. Dal Negro, Nano Lett. **8**, 2423 (2008)
55. C. Forestiere, G. Miano, S. Boriskina, L. Dal Negro, Opt. Express **17**, 9648 (2009)
56. L. Dal Negro, S. Boriskina, Laser Photonics Rev. **6**(2), 178 (2011)
57. C. Helgert, C. Rockstuhl, C. Etrich, C. Menzel, E.B. Kley, A. Tünnermann, F. Lederer, T. Pertsch, Phys. Rev. B **79**, 233107 (2009)
58. N. Papasimakis, V. Fedotov, Y. Fu, D. Tsai, N. Zheludev, Phys. Rev. B **80**, 041102(R) (2009)
59. R. Singh, X. Lu, Z. Tian, W. Zhang, J. Opt. **12**, 015101 (2010)
60. G. Guida, B. Gallas, R. Abdeddaim, A. Priou, J. Rivory, K. Robbie, Appl. Phys. A **103**, 783 (2011)
61. R. Esteban, R. Vogelgesang, J. Dorfmüller, A. Dmitriev, C. Rockstuhl, C. Etrich, K. Kern, Nano Lett. **8**, 3155 (2008)
62. C. Rockstuhl, C. Etrich, C. Helgert, C. Menzel, T. Paul, S. Fahr, T. Pertsch, J. Dorfmüller, R. Esteban, W. Khunsin, R. Vogelgesang, K. Kern, A. Dmitriev, K. Bittkau, T. Beckers, R. Carius, F. Lederer, Proc. SPIE **7604**, 76040D (2010)
63. C. Helgert, C. Rockstuhl, C. Etrich, E.B. Kley, A. Tünnermann, F. Lederer, T. Pertsch, Appl. Phys. A **103**, 591 (2011)
64. W. Khunsin, B. Brian, J. Dorfmüller, M. Esslinger, R. Vogelgesang, C. Etrich, C. Rockstuhl, A. Dmitriev, K. Kern, Nano Lett. **11**, 2765 (2011)
65. D. Nau, A. Schönhardt, C. Bauer, A. Christ, T. Zentgraf, J. Kuhl, M. Klein, H. Giessen, Phys. Rev. Lett. **98**, 133902 (2007)
66. J. Khurgin, G. Sun, Appl. Phys. Lett. **94**, 221111 (2009)
67. S. Takeda, S. Hamada, R. Peretti, P. Viktorovitch, M. Obara, Appl. Phys. B **106**, 95 (2011)
68. C. Rockstuhl, F. Lederer, T. Zentgraf, H. Giessen, Appl. Phys. Lett. **91**, 151109 (2007)
69. A. Yanik, X. Wang, S. Erramilli, M. Hong, H. Altug, Appl. Phys. Lett. **93**, 081104 (2008)
70. S. Orbons, M. Haftel, C. Schlockermann, D. Freeman, M. Milicevic, T. Davis, B. Luther-Davies, D. Jamieson, A. Roberts, Opt. Lett. **33**, 821 (2008)
71. R. Adato, A. Yanik, C. Wu, G. Shvets, H. Altug, Opt. Express **18**, 4526 (2010)
72. K. Aydin, K. Guven, N. Katsarakis, C. Soukoulis, E. Ozbay, Opt. Express **12**, 5896 (2004)
73. B. Auguié, W. Barnes, Opt. Lett. **34**, 401 (2009)
74. E. Kley, Microelectron. Eng. **34**, 261 (1997)
75. J. Van Leeuwen, J. Groeneveld, J. De Boer, Physica **25**, 792 (1959)
76. S. Allen, E. Thomas, R. Jones, *The Structure of Materials* (Wiley, New York, 1999)
77. R. Rengarajan, D. Mittleman, C. Rich, V. Colvin, Phys. Rev. E **71**, 016615 (2005)
78. A. Clarke, J. Wiley, Phys. Rev. B **35**, 7350 (1987)
79. A. Henry, J. Bingham, E. Ringe, L. Marks, G. Schatz, R. Van Duyne, J. Phys. Chem. C **115**, 9291 (2011)
80. J. Petschulat, C. Menzel, A. Chipouline, C. Rockstuhl, A. Tünnermann, F. Lederer, T. Pertsch, Phys. Rev. A **78**, 043811 (2008)
81. D. Cho, F. Wang, X. Zhang, R. Shen, Phys. Rev. B **78**, 121101(R) (2008)

82. T. Davis, D. Gomez, K. Vernon, Nano Lett. **10**, 2618 (2010)
83. J. Petschulat, J. Yang, C. Menzel, C. Rockstuhl, A. Chipouline, P. Lalanne, A. Tünnermann, F. Lederer, T. Pertsch, Opt. Express **18**, 14454 (2010)
84. E. Pshenay-Severin, A. Chipouline, J. Petschulat, U. Hübner, A. Tünnerman, T. Pertsch, Opt. Express **19**, 6269 (2011)
85. J. Jackson, *Classical Electrodynamics* (Wiley, New York, 1999)
86. J. Pendry, A. Holden, D. Robbins, W. Stewart, IEEE Trans. Microw. Theory Tech. **47**, 2075 (1999)
87. D. Smith, W. Padilla, D. Vier, S. Nemat-Nasser, S. Schultz, Phys. Rev. Lett. **84**, 4184 (2000)
88. S. Linden, C. Enkrich, M. Wegener, J. Zhou, T. Koschny, C. Soukoulis, Science **306**, 1351 (2004)
89. C. Enkrich, M. Wegener, S. Linden, S. Burger, L. Zschiedrich, F. Schmidt, J. Zhou, T. Koschny, C. Soukoulis, Phys. Rev. Lett. **95**, 203901 (2005)
90. V. Podolskiy, A. Sarychev, V. Shalaev, J. Nonlinear Opt. Phys. Mater. **11**, 65 (2002)
91. V. Shalaev, W. Cai, U. Chettiar, H. Yuan, A. Sarychev, V. Drachev, A. Kildishev, Opt. Lett. **30**, 3356 (2005)
92. G. Dolling, C. Enkrich, M. Wegener, J. Zhou, C. Soukoulis, S. Linden, Opt. Lett. **30**, 3198 (2005)
93. C. Helgert, C. Menzel, C. Rockstuhl, E. Pshenay-Severin, E.B. Kley, A. Chipouline, A. Tünnermann, F. Lederer, T. Pertsch, Opt. Lett. **34**, 704 (2009)
94. E. Prodan, C. Radloff, N. Halas, P. Nordlander, Science **302**, 419 (2003)
95. M. Albooyeh, D. Morits, S. Tretyakov, Phys. Rev. B **85**, 205110 (2012)
96. S. Tretyakov, *Analytical Modeling in Applied Electromagnetics* (Artech House, Norwood, 2003)
97. C. Haynes, A. McFarland, L. Zhao, R. Van Duyne, G. Schatz, L. Gunnarsson, J. Prikulis, B. Kasemo, M. Käll, J. Phys. Chem. B **107**, 7337 (2003)
98. A. Taflove, S. Hagness, *Computational Electrodynamics: The Finite-Difference Time-Domain Method*, 3rd edn. (Artech House, Norwood, 2005)
99. D. Smith, S. Schultz, P. Markoš, C. Soukoulis, Phys. Rev. B **65**, 195104 (2002)
100. C. Menzel, C. Rockstuhl, T. Paul, F. Lederer, Appl. Phys. Lett. **93**, 233106 (2008)
101. C. Menzel, T. Paul, C. Rockstuhl, T. Pertsch, S. Tretyakov, F. Lederer, Phys. Rev. B **81**(3), 035320 (2010)
102. C. Simovski, S. Tretyakov, Photonics Nanostruct. Fundam. Appl. **8**, 254 (2010)
103. C. Simovski, S. Tretyakov, Phys. Rev. B **75**, 195111 (2007)
104. C. Menzel, C. Helgert, J. Üpping, C. Rockstuhl, E.B. Kley, R. Wehrspohn, T. Pertsch, F. Lederer, Appl. Phys. Lett. **95**, 131104 (2009)
105. A. Minovich, D. Neshev, D. Powell, I. Shadrivov, M. Lapine, I. McKerracher, H. Hattori, H. Tan, C. Jagadish, Y. Kivshar, Phys. Rev. B **81**, 115109 (2010)
106. P. Kolb, T. Corrigan, H. Drew, A. Sushkov, R. Phaneuf, A. Khanikaev, S. Mousavi, G. Shvets, Opt. Express **18**, 24025 (2010)
107. Z. Liu, K.P. Chen, X. Ni, V. Drachev, V. Shalaev, A. Kildishev, J. Opt. Soc. Am. B **27**, 2465 (2010)
108. L. Lin, A. Roberts, Appl. Phys. Lett. **97**, 061109 (2010)
109. M. Burresi, D. Diessel, D. Oosten, S. Linden, M. Wegener, L. Kuipers, Nano Lett. **10**, 2480 (2010)
110. A. Minovich, D. Neshev, D. Powell, I. Shadrivov, Y. Kivshar, Appl. Phys. Lett. **96**, 193103 (2010)
111. N. Mortensen, M. Yan, O. Sigmund, O. Breinbjerg, J. Eur. Opt. Soc., Rapid Publ. **5**, 10010 (2010)
112. S. Raghunathan, N. Budko, Phys. Rev. B **81**, 054206 (2010)
113. L. Jelinek, R. Marques, J. Machac, Opt. Express **18**, 17940 (2010)
114. T. Paul, C. Menzel, C. Rockstuhl, F. Lederer, Adv. Mater. **22**, 2354 (2010)
115. C. Helgert, E. Pshenay-Severin, M. Falkner, C. Menzel, C. Rockstuhl, E.B. Kley, A. Tünnermann, F. Lederer, T. Pertsch, Nano Lett. **11**, 4400 (2011)

116. J. Valentine, S. Zhang, T. Zentgraf, E. Ulin-Avila, D. Genov, G. Bartal, X. Zhang, Nature **455**, 376 (2008)
117. N. Liu, L. Fu, S. Kaiser, H. Schweizer, H. Giessen, Adv. Mater. **20**, 3859 (2008)
118. N. Liu, H. Guo, L. Fu, S. Kaiser, H. Schweizer, H. Giessen, Nat. Mater. **7**, 31 (2008)
119. C. Menzel, C. Helgert, C. Rockstuhl, E.B. Kley, A. Tünnermann, T. Pertsch, F. Lederer, Phys. Rev. Lett. **104**, 253902 (2010)
120. T. Paul, C. Rockstuhl, C. Menzel, F. Lederer, Phys. Rev. B **79**, 115430 (2009)
121. C. Menzel, A. Andryieuski, C. Rockstuhl, R. Iliew, R. Malureanu, F. Lederer, A. Lavrinenko, Phys. Rev. B **85**, 195123 (2010)
122. C. Simovski, Metamaterials **2**, 169 (2008)
123. C. Simovski, Opt. Spectrosc. **107**, 726 (2009)
124. C. Bohren, D. Huffman, *Absorption and Scattering of Light by Small Particles* (Wiley, New York, 1983)
125. B. Seo, T. Ueda, T. Itoh, H. Fetterman, Appl. Phys. Lett. **88**, 161122 (2006)
126. A. Kussow, A. Akyurtlu, N. Angkawisittpan, Phys. Status Solidi (b) **245**, 992 (2008)

Chapter 7
Recent Advances in Nearfield Optical Analysis and Description of Amorphous Metamaterials

Worawut Khunsin and Ralf Vogelgesang

Amorphous metamaterials, a system of conglomerated plasmonic atoms arranged in a nondeterministic way, is an emerging class of nanostructured plasmonic material that is easily fabricated in large quantities and over macroscopic areas. It consists of individual, nominally identical, plasmonic resonators, whose spatial arrangement is random. These elements have been studied intensely over the last decade on a single particle level. Investigations of dimers and more complex plasmonic molecules created from such elements have lead to a thorough understanding of their nearfield-mediated interactions. Adding more plasmonic particles augments the degree of control over the ensemble response. When plasmonic particles are arranged in an array, 1D, 2D or 3D, additional optical features arising from periodic arrangement of the particles appear, adding further complexity to the system. The behavior of these periodically arranged particles is relatively easily captured in exact theoretical calculations using periodic boundary condition. On the contrary, the study of amorphous plasmonics has received much less attention, at least partially due to high computational costs that severely limit an exact treatment. At the same time, successful approximate descriptions of the ensemble behavior are largely missing. In this chapter, we review studies on plasmonic near-field interaction, starting from basic system of dimers, to periodic, and to completely disordered arrangement. We briefly review various fabrication techniques, and discuss in depth the use of Apertureless Scanning Nearfield Microscopy in amorphous plasmonics. Finally, we propose a metaglass theory that provides an explicit analytical model description of a homogenized effective medium response of amorphous plasmonic systems.

W. Khunsin · R. Vogelgesang (✉)
Max Planck Institut für Festkörperforschung, Heisenbergstraße 1, 70569 Stuttgart, Germany
e-mail: r.vogelgesang@nanoopt.org

W. Khunsin
e-mail: worawut.khunsin@icn.cat

C. Rockstuhl, T. Scharf (eds.), *Amorphous Nanophotonics*,
Nano-Optics and Nanophotonics, DOI 10.1007/978-3-642-32475-8_7,

© Springer-Verlag Berlin Heidelberg 2013

7.1 Introduction

Plasmonics deals with the interaction of light and free electrons in metals. At the interface between a dielectric and a metal this results in wave-like phenomena along the interface, called Surface Plasmon Polaritons (SPP). At extended interfaces SPPs are traveling waves, whereas in confined structures like metallic colloids they appear as standing waves. The resulting plasmonic particle resonances are referred to as Localized Surface Plasmon Resonances (LSPR). At extended planar interfaces the parallel wavevector of traveling SPPs is usually different from that of photonic modes in the dielectric medium, which prevents their direct coupling [1]. Correspondingly, their fields are bound to the interface, decaying evanescently in the normal direction.

In order to excite SPPs with photons the momentum mismatch has to be compensated. Many schemes have been proposed for this purpose; most notable are the Kretschmann [2] or Otto [3] configurations and the grating approach. The former employ back-side illumination through a material of higher index of refraction (n), or coupling of evanescent fields available in total internal reflection. The latter method relies on augmentation of the wavevector of the exciting radiation with a grating or nanoscale discontinuity fabricated on top of the metal film to be excited. The excitation of LSPRs, on the other hand, is not as strictly governed by wavevector matching conditions as SPPs, because the finite particle size ensures the presence of a broad range of field components in reciprocal space. Coupling of photonic modes—i.e. excitation by and emission into photons—generally is possible. However, the coupling strength depends on various factors, such as material, size, shape, environment, frequency, and angle of incidence. Each of these gives rise to a continuously tunable response with characteristic resonant behavior.

One of the first to study plasmonic excitations of metallic structures was R.W. Wood. Already in 1902 he uncovered not only the celebrated Wood anomaly in metal gratings [4], but also the much less known is his attribution of the brilliant colors of metallic particles observed in transmission to LSPRs [5, 6]. The theoretical description was put on firm ground in 1908 by G. Mie's seminal explanation of light scattering by spherical particles [7].

Before the 1980ies, controlled fabrication of metallic nanostructures had long been a difficult issue. Important discoveries have been made with colloidal particles, whose optical properties Kreibig and Zacharia explicitly attributed to surface plasmons in 1970 [8]. Another important route to subwavelength scale structured surfaces was statistical roughening by non-uniform etching processes. The phenomenon of surface-enhanced Raman scattering (SERS), discovered by Fleischman et al. in 1974 [9], continues to drive research to this day. Attempts to reproduce the scattering efficiency found in large, statistically patterned metallic surface areas with man-made structures still fall short by orders of magnitude. Evidently, this random nature of such metal surfaces holds a large potential for scientific discoveries.

The advances in nanofabrication techniques in recent decades have resulted in single nanoparticles now being available in tremendously diverse sizes and shapes. The possibility of deliberately placing them on different substrates offers unprecedented degrees of freedom to engineer plasmonic responses of individual as well as extended arrays of nanostructures. From a close analogy to chemical materials and made possible by advanced nanofabrication technology, the science of metamaterials emerged in the last decade. In such materials the role of an atom is played by individual metallic particles, whose particle plasmons may hybridize with that of other meta-atoms, if in close proximity. The aggregation to larger meta-molecules follows much the same rules as the construction of chemical molecules [10]. Similarly, extended periodic arrangements of such individual units form what might be called metacrystals, which exhibit Bloch-type modes and band structures, just like crystalline solids [11].

A class of metamaterials that has not received much attention so far is the equivalent of glassy, amorphous solids. As we have alluded to above, large areas (or volumes) of randomly varied plasmonic structures are at the heart of fascinating effects like SERS. One may expect to find qualitatively different behavior in such metaglass on the macro- and microscopic level. In this chapter we report on recent progress in the study of nearfield optical properties of plasmonic nanostructured ensembles consisting of deterministic (regular) and nondeterministic (amorphous) arrangements of meta-atoms. In the last part we will also outline a possible theoretical description of the global or average behavior of a random metamaterial based on an analogy to the treatment of x-ray diffraction in amorphous solids.

7.2 Deterministic vs. Amorphous/Glassy Plasmonic Systems

At the single particle level, LSPRs have been engineered through variation in size, shape and composition of the nanoparticles [12–15]. For aggregate ensembles, additional degrees of control are afforded by the deliberate placement of particles relative to each other. By varying coupling distance [16–25] and spatial arrangement [26, 27] rich optical feature variations are obtained, which are not possible with single particles [28]. In particular, clusters of plasmonic nanoparticles have been exploited as a platform to demonstrate plasmonic Fano resonance [29, 30], electromagnetically induced transparency [31] and metamolecular structures like plasmonic oligomers [32–35].

A natural extension of meta-molecules is their repeated arrangement in one, two, and three dimensions. As in solid state theory, the placement can be in the form of a deterministic arrangement—periodic like crystals or aperiodic like quasi-crystals. Or it may be fully non-deterministically disordered. We refer to the latter class as amorphous metamaterials or metaglass, again in close analogy to solid state theory.

One immediate distinction from atomic crystals emerges, when we consider the dominant interaction mechanisms: in the solid state we are dealing with short-ranged overlaps between orbitals of neighboring atoms or molecules. Typically, a short ranged, hopping-type interaction is sufficient and long-range interactions are efficiently screened. In contrast, optical fields are usually not efficiently screened in metamaterials, and we should expect having to consider both near- and farfield interactions.

Plasmonic clusters consisting of a small number of nanoparticles are successfully described in terms of plasmon mode hybridization [10, 30, 36, 37], which are strongly mediated by nearfield interactions, very similar to the aforementioned bonding of chemical molecular orbitals. However, in metamaterials the optical farfields give rise to non-negligible long-range interactions. Thus, we can expect the properties of extended plasmonic [38] arrays to be greatly enriched by scattering contributions from distant particles [38–41].

7.2.1 Plasmonic Arrays

Plasmonic arrays studied so far include linear [42–49] and two dimensional [47, 50–54] arrays of metal nanoparticles. The third dimension is considerably harder to investigate with current manufacturing technology [55–58]. Using an array of silver nanoparticles, Zhao et al., showed that for polarization parallel to the plane of the array the plasmon resonance first blue and then red shifts for decreasing lattice spacing, in contrast to the monotonous red shifts observed in isolated dimers [52]. The authors assigned the blue shift to long-range radiative dipolar coupling $\propto 1/r$, whereas the red shift is due to short-range dipolar coupling which varies as $\propto 1/r^3$. The authors further showed that the coherent interaction between the LSPR of the nanoparticles and the photonic mode of the array lead to extremely narrow line shapes in extinction spectra of the system. Based on a similar system, Zou and Schatz demonstrated theoretically that electromagnetic field enhancement of about seven orders of magnitude is possible [59]. In 2008, Kravets and co-workers [53] demonstrated experimentally using an array of gold dots that these sharp resonances are observed at Rayleigh's cutoff wavelengths for Wood anomalies and are due to diffraction coupling of LSPRs. Almost simultaneously, similar experimental work was reported for the case of nanorod arrays made of gold by Auguie and Barnes [60].

Exploiting the coupling between LSPRs via diffraction resonances of the array, Giannini et al. demonstrated the existence of hybrid lattice surface mode (LSM) and showed that the radiative LSM can enhance spontaneous emission of a fluorescence dye embedded in the PVB (Polyvinyl butyral) layer that covers the nanoantennas [54]. Recently, Zhou and Odom reported a new mechanism to localized light using subradient out-of-plane LSM sustained by a two-dimensional array of large nanoparticles [61].

7.2.2 Metaglass

In contrast to regular plasmonic arrays, metaglass is still largely unexplored. One of the first studies on amorphous plasmonic structures was the work reported by Rindzevicius and co-workers [62], in which the authors studied long-range refractive index sensing using plasmonic nanostructures consisting of gold nanodisks arranged in an amorphous manner. The authors observed an oscillatory shift of the plasmon resonance as a function of dielectric layer coating thickness, the magnitude of which is larger than shifts calculated using simple image-dipole coupling models. Although the authors did not consider coherent interactions of neighboring particles, the finding represents an important step in understanding the optical behavior of metaglass.

Among the first contributions that touched upon the coherent interaction of scattered fields from surrounding metal nanoparticles were those by Helgert et al. [63], and Papasimakis et al. from 2009 [64]. The former concentrated on the influence of structural disorder on electric and magnetic resonances of a system consisting of meta-atoms made of circular cut-wire pairs [65]. Interestingly, they found that the electric resonance degrades strongly as disorder increases, whereas the magnetic resonance is robust even in the case of high disorder approaching a glassy arrangement of the meta-atoms. This observation is explained based on the coherent superposition of direct excitation with scattered fields from all meta-atoms. As disorder increases, scattered fields from individual meta-atoms loose their fixed relative phase relationship, which results in homogeneous line broadening and resonance damping. Papasimakis's work on the influence of structural disorder on optical properties of coherent and incoherent metamaterials confirms the finding. In brief, these two seminal contributions strongly suggest the importance of coherent and collective interaction between meta-atoms spatially distributed in space [66].

In addition to coherent interaction of neighboring particles, the importance of the phase retardation associated with excitation at oblique incidence [38, 67] and of interference effects between direct excitation and coherently scattered fields [68] offer additional degrees of freedom to the control of ensemble properties. The tuning in the former is achieved by varying the angle of incidence, while the latter can be achieved, for instance, by changing the path length of the direct excitation beam [69].

In Sect. 7.4 of this chapter we discuss recent experimental results in amorphous nanoplasmonic system [38], where it is demonstrated how collective long-rage coherent interactions in metaglass extend their influence far beyond the average nearest-neighbor distance. This suggests a route to controlling the electromagnetic nearfield structure by using an excitation beam of specific field distribution profiles. Such ability to engineer nearfield structure through farfield means could be advantageous in the deterministic localization of plasmonic fields for application in SERS, for example.

In terms of theoretical investigations, metaglass still presents considerable challenges. Due to the nondeterministic arrangement of its constituent particles, numerical simulations often require extensive computational resources beyond currently available capacities. This greatly limits the development in the field and calls for analytical approaches to the problem. In Sect. 7.5, we propose an averaging procedure that may be useful to describe the ensemble behavior of metaglass. We note that the term "metaglass" underlines this effective medium approach, which largely follows that of the theory of x-ray diffraction in amorphous media.

7.3 Experimental Approach

7.3.1 Fabrication of Nanostructures

We briefly outline the variety of synthesis methods available for fabricating metal nanostructures on supporting substrates; comprehensive reviews are available elsewhere [12, 70]. The advance in nanofabrication techniques greatly facilitates crafting plasmonic nanostructures to study the effects of size, shape, material, and environment on their plasmonic response. For the study of coupling and plasmonic mode hybridization, particularly in more complex plasmonic structures, the most commonly employed methods are electron beam lithography (EBL) and focus ion beam (FIB) milling. In recent years two other techniques have received increasing attention: soft lithography (SL) [71] and nanosphere lithography (NSL) [72–76].

In soft lithography, plasmonic structures are generated through the use of an elastomeric stamp as the pattern transfer element. Nanosphere lithography, on the other hand, utilizes self-assembled nanospheres as a mask for metal deposition. The nanosphere mask can easily be removed, for example, by sonicating the sample in an appropriate solvent. The advantage of this technique over SL is that no master stamp is required, which is usually made by a lithography process followed by etching. A variation of nanosphere lithography that is able to fabricate various plasmonic nanostructures not directly possible by NSL is the so-called Hole-Mask Colloidal Lithography (HCL). In HCL, nanospheres or colloidal nanoparticles are used as the primary mask for the creation of a second, holey mask. The technique has recently been used to produce metaglass and metamaterial structures consisting of gold dipolar disks [74] and split ring resonators [76]. The major advantages of SL and NSL techniques lie in their ability to fabricate large area metaglass at low cost, as they use only simple and fast steps. Indeed, amorphous plasmonics nanostructures can be produced easily over areas as large as cm^2 or more, in contrast to the serial fabrication of positionally controlled regular arrays over sub-mm^2 area.

As such, metaglass is also very attractive for economically viable applications that can even be made on a flexible substrate using the SL and NSL techniques mentioned above. Potential uses of metaglass have been demonstrated in long-range refractive index sensing [62, 77], in broadband and omni-directional antireflection coating with prospective use in solar cell [78, 79]. Recently, its potential in ultrahigh nanorods biosensing has been demonstrated by Kabashin and co-workers [80].

7.3.2 Experimental Investigation Techniques

The easiest and straight forward technique to characterize optical property of plasmonic structures is farfield spectroscopy, such as extinction or reflection spectroscopy. However, much of what takes place in the nearfield zone does not have a signature in farfield spectra. Hence, there has been increasing interest in the nearfield interactions of amorphous materials [81]. This applies particularly to the study of

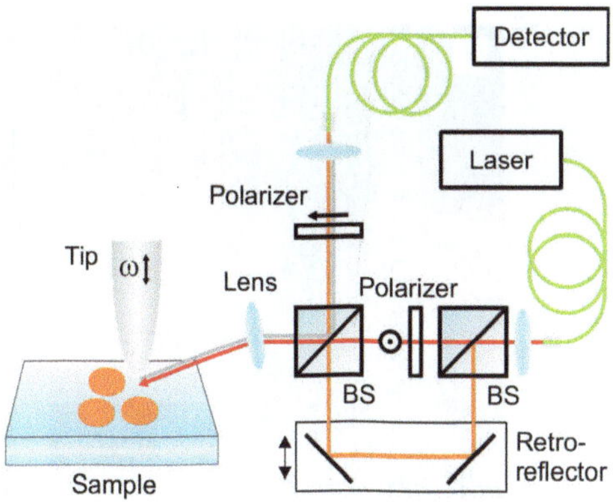

Fig. 7.1 Schematic of the aSNOM setup. Reprinted with permission from [86]. Copyright (2012), American Institute of Physics

mode hybridization, where charge distribution in coupled plasmonic nanostructures is of primary interest [82].

Scanning Nearfield Optical Microscopy (SNOM) [83] comes in handy for probing the localized fields around the particles. The two main variations are aperture-based SNOM and apertureless or scattering SNOM. In the former, metal coated tapered optical fibers are typically used to collect the nearfield signal, whereas in the latter case an AFM tip acts as a local scatterer, which converts bound nearfield into the propagating farfield that can be detected by usual farfield means. The primary drawback of the fiber SNOM is a necessary trade-off between the amount of light collected through the aperture and the optical resolution obtainable. Usually, the resolution obtained in fiber SNOM is between 50 nm and 100 nm. In contrast, in apertureless SNOM or aSNOM the resolution is only limited by the size of the local scatterer, which is easily sharper than 10 nm in commercially available AFM tips [84].

Over the last years many important advances have been made in the understanding and implementation of aSNOM [84–86], including the important considerations of geometry of the tip scatterer [87], anharmonicity of tip oscillation [85], tip-substrate interaction [88], interferometric amplification [89], data analysis of detected signal [90], and the parasitic background suppression [86]. Here, we give a brief review of the crossed-polarization aSNOM setup. We note here that although the description given below pertains to a specific type of experimental setup, the basic principles and techniques are quite general and are applicable to other setups, as well, particularly those that require high signal-to-noise fidelity.

The cross-polarization aSNOM setup is based on a commercial atomic force microscope fitted with a dielectric silicon tip of nominal apex diameter of less than 10 nm. The polarization of the excitation source is set by the polarizer to excite the sample with s-polarized light. The signal scattered back by the tip is mixed with a reference beam to achieve interferometric amplification. A Mach-Zehnder interferometry configuration provides independent control of the two optical paths, the signal and the reference arm, respectively (see Fig. 7.1). A homodyne scheme is used in our setup to allow measurement of both the amplitude and the phase difference between the signal (S_0) and reference (S_{90}) beam via two measurements with

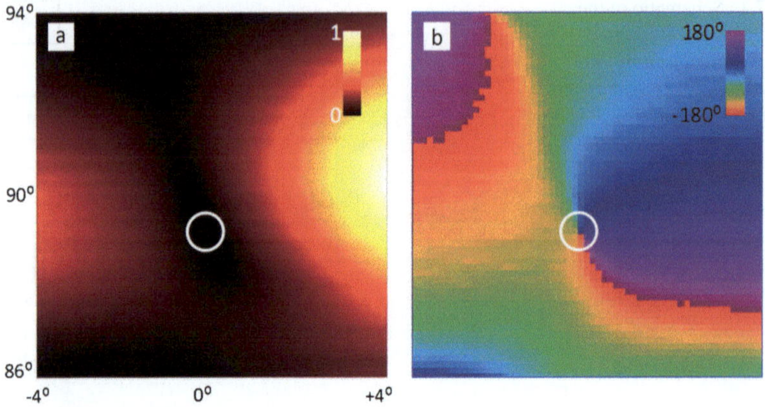

Fig. 7.2 Typical polarization scan (x-axis: polarizer, y-axis: analyzer): (**a**) optical amplitude and (**b**) optical phase according to (7.1) obtained with the sample retracted from the tip. Reprinted with permission from [86]. Copyright (2012), American Institute of Physics

90 degrees out-of-phase of the reference beam. The total optical amplitude (S) and phases (ϕ) can be obtained from the complex measured signal S_0 and S_{90} as follows:

$$S = \sqrt{S_0^2 + S_{90}^2}, \qquad \tan\phi = \frac{S_0}{S_{90}}. \tag{7.1}$$

Two important aspects for obtaining background-free imaging of plasmonic nearfield signals are (1) confocal arrangement for the detection path and (2) the cross polarization scheme. The confocal setup is achieved with single mode optical fibers that conveniently serve also as spatial filters leading to much reduced parasitic background radiation and improved interferometric visibility. Unfortunately, the parasitic background cannot be eliminated entirely through the confocal setup. On the one hand, it is known that a modulation-demodulation scheme is used to filter the nearfield signal from parasitic background signals, e.g., by filtering the optical signal at a suitable harmonic of the AFM vibration resonance frequency. On the other hand, we put forward the use of a cross polarization scheme in which the excitation and the detected signal are cross polarized. Such a configuration has been demonstrated to routinely yield near-perfect background suppression [38, 91–94]. The theoretical ground for the background suppression is explained below in terms of a phase singularity in the polarizer scan [86].

Figure 7.2 shows a typical experimental cross polarization scan. It is recorded on the bare probe tip, with the sample being absent. The zero background condition is obtained slightly off the ideal polarizer/analyzer angles of 0°/90° because of imperfect shape of the AFM tip, the imperfect Gaussian profile of the illumination beam, and/or other experimental imperfections.

The emergence of a phase singularity in such a cross-polarization scan can be understood with the Jones formalism. The scattered signal from the tip is modeled

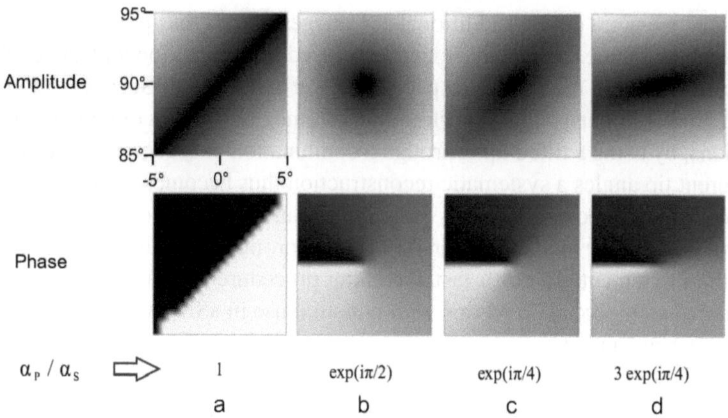

Fig. 7.3 Simulation of cross-polarization scan using (7.4). Reprinted with permission from [86]. Copyright (2012), American Institute of Physics

using a polarization ellipsoid with two orthogonal tip modes in response to the excitation beam. For simplicity, the off-diagonal elements of the Jones matrix describing the tip backscattering are zero in a symmetric idealized system:

$$T = \begin{pmatrix} \alpha_s & 0 \\ 0 & \alpha_p \end{pmatrix}, \tag{7.2}$$

where the complex-valued coefficients $\alpha\alpha$ describe the strength of the tip scattering of the two modes. The instrumental function of the linear polarizer and analyzer are described by

$$P(\theta) = \begin{pmatrix} \cos(\theta)\cos(\theta) & \cos(\theta)\sin(\theta) \\ \sin(\theta)\cos(\theta) & \sin(\theta)\sin(\theta) \end{pmatrix}. \tag{7.3}$$

The detected signal at the detector is then dependent on polarizer/analyzer angles and the characteristic of light scattered by the tip:

$$E_{\text{out}} = P(\theta_{\text{out}}) \cdot T \cdot P(\theta_{\text{in}}) \cdot E_{\text{in}}. \tag{7.4}$$

Sample images of signal strengths simulated with (7.4) are shown in Fig. 7.3, which demonstrate how the phase singularity comes about and how an imperfect tip shape may alter its appearance both in amplitude and phase.

When the polarizer and analyzer are set at the phase singularity position, faithful nearfield information free from parasitic background even from complex plasmonic structures can be obtained. The beauty of this technique is that commercial AFM tips are normally manufactured and optimized for their mechanical response. As such, they contain unavoidable shape deformation, which significantly and adversely affect the optically scattered signal. With the ability to adjust the excitation and detection beam properties to obtain the perfect cross-polarization configuration, the fidelity and reproducibility of aSNOM experiments are greatly improved.

Finally, we point out two additional advantages of the cross-polarization scheme. First, such a scheme can be envisioned to be used in conjunction with specially fabricated tips that are tilted with respect to the surface normal. Such a configuration should allow similar background-free measurements of field components in other directions, by using a tilted tip. Through a combination of successive measurements at different tip angles a systematic reconstruction thus becomes possible for the entire electric field vectors. Second, the scheme can also be used to characterize the polarizability tensor of an unknown AFM tip. A promising application of this might be in the implementation of a fast screening procedure to characterize the optical property of commercial AFM tips with potential use in aSNOM, Tip-enhanced Raman Scattering, and etc.

7.4 Evidence for Long-Range Interactions in Nearfield Optical Maps of Metaglass

Broadly speaking, the options for exciting nanoplasmonic resonances are firstly direct excitation by some farfield source, and secondly indirect nearfield excitation by neighboring plasmonic structures. In the case of large arrangements of many individual plasmonic nanostructures the interaction between neighboring plasmonic structures become quite complex. In such a disordered system, nearfield interaction varies locally and light localization as well as its propagation will exhibit strong statistical dependency on the arrangement of each individual plasmonic structure. Such a system provides a new possibility to 'engineer' interaction of light at nanoscale through, for example, spatially modulated beam profile as we discuss below. In particular, we describe below how an amorphous arrangement of plasmonic structures can support another non-negligible means for excitation, namely an indirect long-range excitation. Evidence is found for a long-range collective and coherent influence on each particle of a metaglass from its entire active neighborhood. This influence extends far beyond distances that have been reported for the interaction in isolated dimer systems [18–24, 27].

The structures under study are metaglass made of gold nanodisks with varying density or inter-particle distances. The nanodisks are fabricated using HCL [74] with a nominal diameter $D = 190$ nm and height $h = 25$ nm. These particular dimensions support a broad dipolar resonances centered at $\lambda = 780$ nm, which appears as a pronounced dip in the farfield transmission spectra (Fig. 7.4).

Additional spectral features are observed at 1570 nm and 2230 nm. They are due to dipole excitations of conglomerated dimers and larger clusters, respectively. For the low coverage the nanodisks in the array are essentially non-interacting [94]. Decreasing the inter-particle spacing, thus increasing neighbor interaction, does not show the typical red shift of the dipole resonance normally observed in dimer systems. A slight distribution in sizes among the ensemble of disks and increased nearfield interaction manifest themselves as inhomogeneous broadening of the transmission suppression at the dipole excitation.

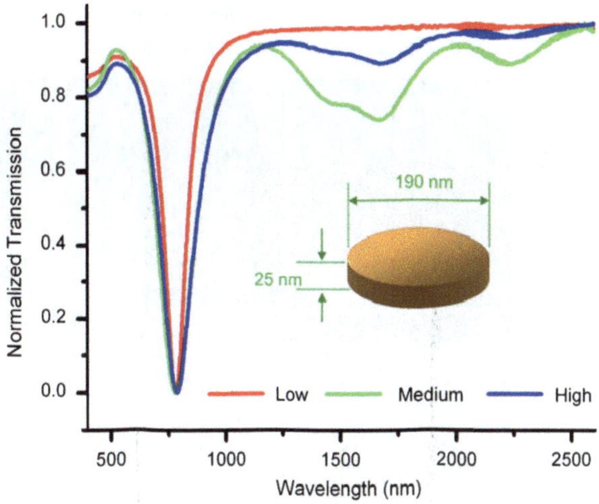

Fig. 7.4 Experimental farfield transmission spectra. The average center-to-center distance between nearest neighbor gold disks is about 388 nm, 315 nm, and 228 nm for low, medium and high-density samples, respectively. These numbers correspond to a factor of 2.05, 1.65, and 1.2 times the corresponding disk diameter D. Adapted with permission from [38]. Copyright (2011) American Chemical Society

The comparatively mundane farfield behavior is in stark contrast to what is observed in nearfield microscopic scans, such as those shown in Fig. 7.5. The local properties of this metaglass are analyzed by our cross-polarization aSNOM introduced in the previous section [86]. The results are compared with a large-scale finite-difference time-domain (FDTD) simulation [95] that takes fully into account the exact details of the samples. In order to obtain reliable statistics, aSNOM characterizations are performed on extended areas of 10 μm by 10 μm (containing about 3000 disks in total). The spatial resolution is set to 20 nm.

For ease of viewing, cut-outs of 3 μm by 3 μm in size are shown in Fig. 7.5. Three samples having different particle density were characterized as denoted on the top as low, medium and high-density arrays, respectively. The top row shows AFM topographs, whereas the second and third rows show the corresponding measured and simulated amplitudes of the E-field component of the nearfields normal to the surface. A glance at the nearfield images reveals that the in-plane dipole orientations are predominantly dictated by the polarization of the exciting radiation, but a variation in the dipole orientations exists and is higher for denser samples. The full-area FDTD simulations show a similar trend, and in both experiment and simulation a variation in dipole excitation strength is observed with features akin to plasmonic hot spots [96].

A closer look at the dipole orientation reveals a significant statistical difference between the experimental and simulated results. Noticeably, the simulated dipole orientations show a much stronger fluctuation, which is already discernible for the low density array. Here, the average nearest-neighbor distance is the range where nearfield coupling becomes less important than far-field dipole-dipole interaction [97]. The variation in local dipole orientation must therefore be attributed to a different factor, namely the total coherent contribution of fields scattered by all surrounding disks.

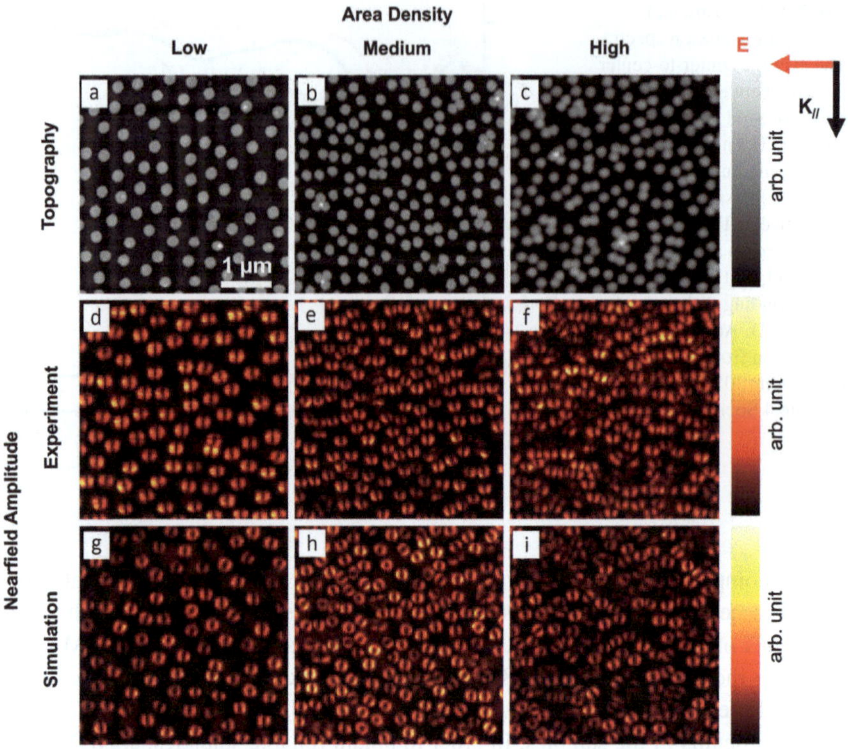

Fig. 7.5 (**a–c**) AFM topography of the three metaglasses under investigation, labeled according to nanodisk area density; (**d–f**) experimentally measured nearfield amplitudes; (**g–i**) simulated nearfield amplitude taken at 20 nm above the sample surface consisting of the exact same disk arrangement as those in panels (**a–c**). The excitation wavelength is 897.8 nm. The polarization and the in-plane wavevector of the exciting radiation are indicated on the *top-right corner* of the figure. The scale bar applies to all subfigures. Adapted with permission from [38]. Copyright (2011) American Chemical Society

To approach the properties of a metaglass system at a quantitative level, we consider in-plane dipole moments $\mathbf{P} = (p_x, p_y)$ for each nanodisk, which can be conveniently defined as a function over all pixels of a micrograph that belong to the disk area:

$$\mathbf{P} = \frac{1}{N} \sum_{i=1}^{N} \big(E_i - \langle E \rangle\big) \cdot \big(\mathbf{X}_i - \langle \mathbf{X} \rangle\big). \tag{7.5}$$

Here, $\mathbf{X}_i$ is the location of the i-th pixel and $\langle\,\rangle$ denotes an average over all N pixels. E_i is the z-component of the electric field. As the metaglass exhibits short-range order, a natural choice for a statistically meaningful variable is the nearest neighbor inter-particle distance. In the following, it is correlated with the in-plane

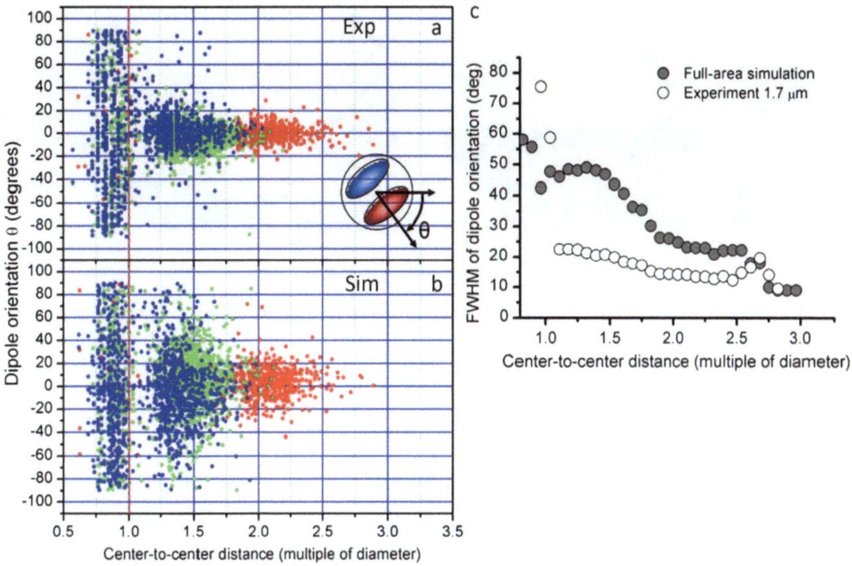

Fig. 7.6 Dipole orientation as a function of nearest-neighbor separation (multiple of diameters) for (**a**) experiment and (**b**) simulation. *Red, green, and blue colors* represent data points obtained from low, medium, and high density samples, respectively. (**c**) Standard deviation of dipole orientations as a function of nearest-neighbor separation. The simulation is undertaken with a plane wave illumination of the whole sample (*solid circles*). In the experiment the sample is excited by focused radiation (spot size of 1.7 μm, *open circles*). Adapted with permission from [38]. Copyright (2011) American Chemical Society

dipole orientation, which can be extracted from **P** with a variant of the Stokes formalism,

$$\begin{aligned} I &= |p_x|^2 + |p_y|^2, \\ Q &= |p_x|^2 - |p_y|^2, \\ U &= 2\Re(p_x \bar{p}_y), \\ V &= 2\Im(p_x \bar{p}_y). \end{aligned} \tag{7.6}$$

It allows to write the in-plane angle of the dipole as

$$\theta = \frac{1}{2} \arctan\left(\frac{U}{Q}\right). \tag{7.7}$$

The results extracted from the measured and corresponding simulated data are shown in Figs. 7.6a and 7.6b, respectively. A general trend emerges: the distribution of dipole orientations is wider for the simulated data than for the measured data.

To further condense the information, Fig. 7.6c plots the standard deviation of dipole orientations against nearest-neighbor distance. For a center-to-center nearest-neighbor distance of more than about 2.5 to 3 times the nanodisk diameter the dipole

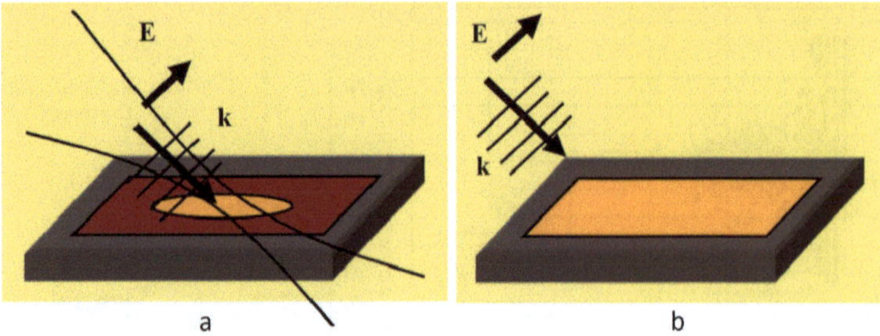

Fig. 7.7 Schematics showing illumination configuration used in the (**a**) experiment and (**b**) simulation. Polarization and wave vector of the exciting radiation are as indicated. *Orange shaded region* represents the illuminated area, *dark red*—total measured area (in the experiment), and *grey*—unprobed region. Adapted with permission from [38]. Copyright (2011) American Chemical Society

orientation is close to that of an isolated nanodisk, i.e., with $\theta = 0°$. For smaller inter-particle distances, however, the spread in dipole orientations increases until it encompasses the possible range, $-90° \leq \theta \leq +90°$, for nearly-touching and connected dimers (data points left of the red dotted line in Fig. 7.6a and b). Evidently, the disagreement between experiment and simulation becomes even more apparent when the gap between nearest-neighbors becomes nearly zero. The amorphous arrangement of the nanodisks implies a largely varied coupling of the surface charges between particle pairs. It may exhibit attractive as well as repulsive components that reduce or enhance the dipole moment of individual particles [18]. The observed behavior exhibits a trend similar to that reported in Ref. [19], and is supported by a smaller effect of repulsive interactions compared to that of attractive interactions [18], which is primarily due to a larger separation between charges forming the repulsive forces.

What is the reason for the observed discrepancy between experiment and simulation? Excitation wavelength, angle of incidence, and sample topography are the same and thus are not responsible for the observation. However, a crucial difference emerges when consider the exact geometry of excitation between experiment—which uses focused beams—and the simulation, where a plane wave excitation is used. Thus the active, directly illuminated area is much larger in the simulation. Figure 7.7 illustrates the conditions in (a) experiment and (b) simulation. An immediate consequence of this difference is the vastly different numbers of disks in experiment and simulation that contribute with (first-order) scattered fields to the indirect excitation of the central disk. A simple geometrical argument now suggests as a general rule for randomly arranged scatterers that larger illumination area implies increased variability in the local response.

As mentioned above, the behavior of any chosen particle is a result of the collective influence of scattered fields from all the neighboring active particles. To understand the magnitude of this effect, one may consider a first-order Born approximation of the total scattered field $\mathbf{E}_{\text{sca}}$. It is the convolution of the Green function $\mathbf{G}$ with the direct excitation $\mathbf{E}_{\text{exc}}$:

$$\mathbf{E}_{\text{sca}}(\mathbf{r}) = \int_V \mathbf{G}(\mathbf{r}, \mathbf{r}')\omega^2\mu_0\delta\varepsilon(\mathbf{r}')\mathbf{E}_{\text{inc}}(\mathbf{r}')\,d^2r', \tag{7.8}$$

where, $\delta\varepsilon(\mathbf{r}')$ are the scattering particles distributed in space. In terms of scattering geometry it is crucial that the free space propagator is proportional to the inverse distance $R = |\mathbf{r} - \mathbf{r}'|$. (The appropriate Green function for a substrate-air interface behaves similarly.) Therefore, $\mathbf{G}(\mathbf{r}, \mathbf{r}') \propto \exp(\iota kR)/4\pi R$ sums up to a term on the order of unity, when integrated over any ring area $A = 2\pi R\Delta r$ of uniformly distributed particles: $|\mathbf{G}|A \propto \Delta r$. However, the ring's radius R cancels out. Thus, the sum over any ring of randomly arranged scatterers generally is significant (being somewhat diminished by coherence) regardless of the size of this ring. An extended description of the theory can be found in Sect. 7.5 of this chapter.

To appreciate how this argument can explain the observed differences between experiment and simulation, the area of direct excitation may be thought as split into a sequence of concentric rings, each of the same width. Hence, each contributes to (7.8) with a similar magnitude and the total field at the center becomes less and less predictable, the larger the total area of all excited rings. To verify how the overall statistical behavior depends on the illumination size, we performed a series of experiments with different focus spot size.

Straightforwardly, different focus spot sizes can be obtained through the use of an iris aperture that controls the illumination beam width. The cross-sectional area of a well focused Gaussian beam is

$$\frac{4M^2\lambda f}{\pi D}, \tag{7.9}$$

where $M^2 \approx 1$ is the beam mode parameter, f—the focal length of the aspheric lens (10 mm), λ—the wavelength (here 898 nm) and D—the diameter of the excitation beam. The excitation intensity was kept constant while varying the beam diameter.

Figure 7.8a shows the results obtained from the experiment with varying excitation spot size: clearly, larger spot sizes result in wider spreads of in-plane dipole orientations. In the present case, the experimentally observed standard deviations approach those obtained from simulation for a spot size of about 4 µm. Besides the increased area of directly excited nanodisks that contribute the total variation of in-plane dipole orientations, the increasingly plane-wave like character of less strongly focused excitation radiation may also be a factor in the improving correspondence between experimental and simulated results.

At this point, one may wonder what results a simulation with a Gaussian focus beam as the input will give. Although this simulation will provide a direct comparison with experiment; however, this is not worth the effort for a number of reasons. Firstly, a single illumination only allows the extraction of information from a single spatial point, i.e., at the center of the Gaussian beam that is assumed to coincide with the AFM tip position. The extracted information represents one single pixel of data and for our investigation a single image similar to that shown in Fig. 7.5d, e or f

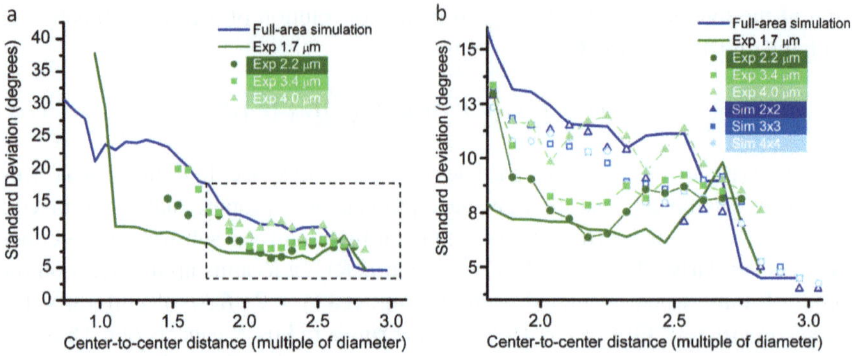

Fig. 7.8 (**a**) Standard deviation of dipole orientations as a function of center-to-center distance. The full-area FDTD simulation is represented by a *blue line*. The experimental results (*green line*) are the same as those shown in Fig. 7.6c (*open circles*); they were obtained with a focus spot size of 1.7 μm. Measured data obtained from measurements with varying focus spot sizes are shown as *scattered plot* with the corresponding focus sizes indicated. (**b**) A zoom-in of the area enclosed by the *dashed rectangle* in panel (**a**). *Blue scattered plots* represent cut-out simulation domains obtained by sub-dividing the full-area simulation into equally sized $N \times N$ sub-domains. Adapted with permission from [38]. Copyright (2011) American Chemical Society

would require one million full-area simulations; hence, three such images would require three times as much computational effort. This heavy computational demand is beyond current computational feasibility. Secondly, we would like to stress that the evolution of such systems is one of the rare examples where simulations cannot help anymore but only devoted experiments provide an answer. We also note that the effect of changing the simulation size to match the width of illumination focus by subdividing the full-area simulation into equally sized cut-outs can be investigated without much additional effort. Interestingly, however, the results thus obtained show a negligible effect (Fig. 7.8b): the statistics rarely departs from that of the full-area simulation. Most probably this is due to the fact that in such cut-out simulations the imposed periodic boundary conditions yield a strong interaction regime which is similar to the case of full-area simulation.

Metaglass thus shows significantly more varied nearfield behavior than that found in farfield spectra. It indicates strong coherent neighbor interaction well beyond distances of a typical dimer system. In addition to the average nearest neighbor separation, control of the external illumination structure has a profound influence on the optical nearfields. In particular, by increasing the focus size we demonstrate that variability of the local nearfield response also increases.

The finding is of immediate interest to the field of hot-spot and energy concentration engineering. An ability to vary the local response of plasmonic entity through specially configured farfield excitation has potential applications, for example, in enhanced Raman scattering spectroscopy. The work also indicates that the very high field enhancements achieved in SERS [98–100] might not have their origin in the very local surrounding alone, but originate also from coherent interactions of a larger region [68].

Finally, we note the direct relevance in the context of earlier work on energy localization, which have been demonstrated in the temporal domain based on polarization pulse shaping [101], and through the use of purposefully tailored structures [102]. Recently, localization control has also been achieved in the spatial domain via coherent superposition of higher-order beams [103] and spatially-tailored phase profiles [104, 105]. Very similar spatial and/or temporal excitation beam profiling should also provide good control of local responses in metaglass, which thus could serve as an inexpensive, yet versatile platform for hot spot engineering.

7.5 Metaglass Theory

Amorphous media show no long-range, but may show short-range order. In contrast to periodic systems, one does not expect to be able to formulate the eigenmode spectrum in terms of Bloch modes. Nevertheless, it is a well-known fact that sound waves, for instance, can travel very well also through amorphous media like liquids or glass. When the wavelength is significantly longer than the characteristic scale of disorder in the medium's substructure the medium should be describable by effective material parameters. This is indeed the case in typical amorphous metamaterials. Here, we sketch a procedure to derive an effective medium description for metaglass.

Our intent is to elucidate the main features of the optical response of amorphous media by an explicit analytic model. A number of simplifying assumptions are necessary, which we expect can be lifted in a more quantitative approach. We begin by introducing what is frequently referred to as the coupled-dipole model. Its exact solution is well known, but does not yield much intuitive understanding. In a second step, therefore we apply an averaging procedure to this model to arrive at a homogenized description of amorphous metamaterials.

7.5.1 Exact Eigenmodes of a Random System of Dipole Scatterers

Here we briefly review the coupled-dipole model for a random arrangement of point-like scatterers. It assumes a time-harmonic model, in which all phenomena oscillate in time with the same factor $\exp(-\iota\omega t)$. The interaction between individual meta-atoms is mediated by the Green function of the background medium, $\mathbf{G}_B(\mathbf{r}, \mathbf{r}')$, which is in general a complex function of the angular frequency ω as well as dimensionality and geometry, for example of the substrate. To keep the treatment explicit and accessible, in the following we shall assume as background medium isotropic 3-dimensional space, in which case

$$
\begin{aligned}
g_B(\mathbf{r}, \mathbf{r}'; \omega) &= \frac{\exp(\imath k|\mathbf{r} - \mathbf{r}'|)}{4\pi |\mathbf{r} - \mathbf{r}'|}, \\
\mathbf{G}_B(\mathbf{r}, \mathbf{r}'; \omega)_{ij} &= -\left\{\mathbf{1}_{ij} + \frac{\nabla_i \nabla_j}{k^2}\right\} g_B(\mathbf{r}, \mathbf{r}'; \omega), \\
k^2 &= \omega^2 \varepsilon_B \mu_B,
\end{aligned}
\tag{7.10}
$$

where ε_B and μ_B are the dielectric constant and permeability of the background medium, respectively. As meta-atoms we assume particles that may be viewed as point-like, anisotropic dipole scatterers. Their polarizability tensor is approximated by that of an oblate spheroid of height h and diameter d. In the formulation of Venermo and Sihvola [106], these determine the geometrical anisotropy factor N_z as

$$
\begin{aligned}
N_z(d, h) &= \frac{1 + s^2}{s^3}\bigl(s - \arctan(s)\bigr), \\
s &= \sqrt{\left(\frac{d}{h}\right)^2 - 1}.
\end{aligned}
\tag{7.11}
$$

The in-plane and normal polarizabilities are then

$$
\begin{aligned}
\alpha_{xy} &= 3V \varepsilon_0 \varepsilon_{\mathrm{med}} \frac{\varepsilon - \varepsilon_{\mathrm{med}}}{\varepsilon_{\mathrm{med}} + \frac{1}{2}(1 - N_z)(\varepsilon - \varepsilon_{\mathrm{med}})}, \\
\alpha_z &= 3V \varepsilon_0 \varepsilon_{\mathrm{med}} \frac{\varepsilon - \varepsilon_{\mathrm{med}}}{\varepsilon_{\mathrm{med}} + N_z(\varepsilon - \varepsilon_{\mathrm{med}})}.
\end{aligned}
\tag{7.12}
$$

An example of the spectral dependence of an individual meta-atom's polarizability is presented in Fig. 7.9. The geometric and material parameters are chosen such that they correspond well to the nanodisks discussed in Sect. 7.4. The dielectric constant is described with the values of Johnson and Christy [107], fitted by a Drude model with two critical point functions [108].

Clearly, the doubly degenerate in-plane dipole resonance at $\approx$780 nm dominates the optical response, whereas the dipole in normal direction shows only a very weak resonance around $\approx$510 nm.

In the coupled-dipole model, an arbitrary system of such meta-atoms is easily calculated. A large number of meta-atoms, $i = 1, 2, \dots, N$ are located at positions $\mathbf{r}_i$. Note that these locations may be periodic, but for our present purposes we consider them randomly distributed over the flat surface of the substrate. At this point, we also disregard any short-range ordering effects, as are frequently encountered in hard-sphere or hard-disk models.

The general oscillation amplitude $\mathbf{u}_i$ of each scattering meta-atom is the response to the total field at its location $\mathbf{E} = \mathbf{E}_{\mathrm{exc}} + \mathbf{E}_{\mathrm{sca}}$, which is the sum of the excitation and scattering response,

$$
\mathbf{u}_i = \boldsymbol{\chi}_i \cdot \mathbf{E}(\mathbf{r}_i). \tag{7.13}
$$

Here $\boldsymbol{\chi}_i = \omega^2 \mu_0 \boldsymbol{\alpha}_i$ is the susceptibility tensor of the i-th oscillator. For each excited oscillator a scattered wave emerges from it. Its travel through the background

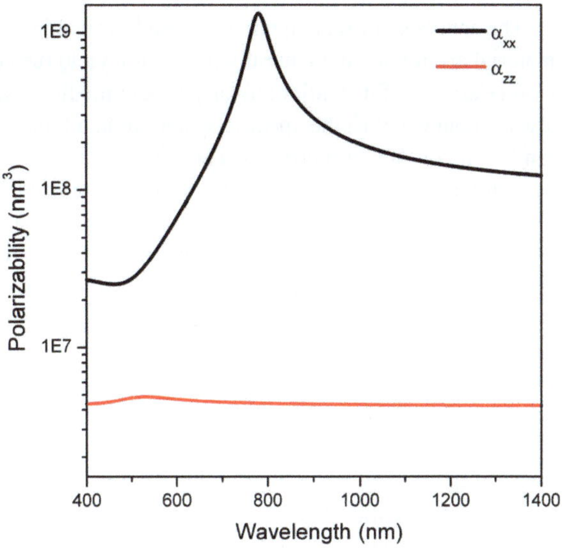

Fig. 7.9 The polarizabilities $\alpha_{xx} = \alpha_{yy}$ (*in black*) and α_{zz} (*in red*) of a gold nanodisk, modeled as an oblate spheroid of diameter $d = 190$ nm and height $h = 18$ nm in a dielectric medium $\varepsilon_{\text{med}} = (1 + 2.25)/2$

medium is described with the help of the background medium's Green propagator as

$$\mathbf{G}_B(\mathbf{r}, \mathbf{r}_i) \cdot \mathbf{u}_i\,. \tag{7.14}$$

The final step in evaluating the response of the whole system of scatterers is to consider all the mutual couplings between all scatterers. In contrast to hopping-type, short-range only interactions, this leads to a densely populated system matrix. The total field at oscillator i is thus due to the direct excitation field as well as the superposition of all other oscillators' scattered fields:

$$\mathbf{E}(\mathbf{r}_i) = \mathbf{E}_{\text{exc}}(\mathbf{r}_i) + \sum_{(j \neq i)} \mathbf{G}_B(\mathbf{r}_i, \mathbf{r}_j) \cdot \chi_j \cdot \mathbf{E}(\mathbf{r}_j). \tag{7.15}$$

By adopting the abbreviation that quantities at oscillator i shall be designated by the index, we write the self-consistent condition for the system response matrix $\mathbf{M}_{ij}$ as

$$\mathbf{E}_{\text{exc},i} = \sum_j \mathbf{M}_{ij} \cdot \mathbf{E}_j, \tag{7.16a}$$

$$\mathbf{M}_{ij} = (\delta_{ij} - 1)\mathbf{G}_{B,ij} \cdot \chi_j + \delta_{ij}. \tag{7.16b}$$

For a given excitation field and distribution of meta-atoms, we obtain first the total fields $\mathbf{E}_i$ at each scatterer with (7.16a) and (7.16b) and then the total field at arbitrary locations with (7.15). Without external excitations, i.e., vanishing left-hand-side in (7.16a), we may solve for the eigenvalues and eigenvectors of the whole scattering system. This yields the exact eigenmodes.

The analysis of exact eigenmodes is a formidable undertaking for a random metamaterial arrangement. Generally it does not yield the equivalent of a generic dispersion relation as for continuous or periodic media. Instead, the exact description is always connected to the specific system at hand and it is difficult to generalize the random medium's properties in a statistical sense.

In light of these considerations, we should tackle the problem in a different fashion, i.e., to devise an "average medium" description. Similar to the statistical approach in many-particle physics, which in the end yields a statistical, average, or effective description in terms of extensive and intensive variables, i.e., properties of the individual constituent particle and their statistical distribution parameters, like density.

As we will outline in the following, such a description may indeed be successful. The central elements of this model are a representative central scatterer and a surrounding sea of scatterers, which may be regarded as a continuous averaged medium in the long-wavelength limit. That is, on a scale much larger than the average nearest-neighbor distance, we expect this medium to support plane (as well as circular or spherical) waves just like any continuous medium.

7.5.2 Recovering a Dispersion Relation for Metaglass

In anticipation, we consider eigenmodes of the averaged random medium of the plane-wave type,

$$\mathbf{E}_i = \mathbf{E}_0 \cdot \exp(\iota \mathbf{K} \mathbf{r}_i), \tag{7.17}$$

where $\mathbf{E}_0$ is the field amplitude and $\mathbf{K}$—the wavevector. Note that in the following we will concentrate on a two-dimensional scenario. The three-dimensional variant follows an analogous treatment. Also, we cannot, at this point, make any further statements regarding the relation of $\mathbf{E}_0$ and $\mathbf{K}$. In particular, one cannot fix a priori the relative orientation, whether they are orthogonal or parallel to each other, corresponding to transverse and longitudinal waves, or if they may be skewed. To obtain a self-consistent description of such waves, we calculate the average or representative behavior of a particle in the random system, i.e., the average of (7.16a) and (7.16b) over all sites. In the absence of any excitation field this yields the average eigenmodes of the system and thus the dispersion relation of the average random medium. However, we may expect that an exact solution is not feasible.

In the spirit of an approximate plane wave being a near-resonant response to an external excitation, we consider therefore how the response according to (7.16a) and (7.16b) is to a plane wave excitation, $\mathbf{E}_{\text{exc}}(\mathbf{r}) = \mathbf{E}_{\text{exc},0} \cdot \exp(\iota \mathbf{K} \mathbf{r})$. After averaging over all meta-atoms, we obtain

$$\mathbf{E}_{\text{exc},0} = \left[\frac{1}{N} \sum_i \sum_j M_{ij} \cdot \exp\big(\iota \mathbf{K}(\mathbf{r}_j - \mathbf{r}_i)\big) \right] \cdot \mathbf{E}_0. \tag{7.18}$$

If relation (7.18) holds exactly, this establishes a direct relation for the response amplitude $\mathbf{E}_0$ to a given excitation $\mathbf{E}_{\text{exc}}$. However, a priori, there is no evidence that justifies (7.17). We may, therefore, regard (7.18) as a test for its validity. Also note that it would eliminate the need to study many specific individual scatterers and their respective individual neighborhoods by passing from $\mathbf{E}_{\text{exc}}$ to $\mathbf{E}_0$ directly.

Unfortunately, though, the exact discrete distribution of the scatterers still enters through the summations over locations. Therefore, we ought to strive to replace these summations by integrating over an appropriate average distribution. Such discrete-to-continuous substitutions have been used extensively in the past, notably in the theory of x-ray diffraction from amorphous media. They provide guidance for our current development. Specifically, we now outline how the discrete distribution of point scatterers is transformed into an average radial density description.

7.5.3 Radial Density Function (RDF)

The main distinction of glassy system is the complete absence of long-range order usually accompanied by perfect isotropy. A meaningful statistical description therefore depends only on distances, and not on orientation. In order to make the discrete summation equation (7.18) treatable, we convert it to an integral expression with the help of a distance correlation function. In effect, we construct a homogenized medium.

The concept of radial density is mainly used in the analysis of three-dimensional amorphous crystals. We therefore consider the familiar three-dimensional RDF first and translate the formulation to the two-dimensional case. The average radial number density function for a volume distribution of random scatterers is given by

$$\rho_V(R) = \lim_{dr \to 0^+} \lim_{N \to \infty} \frac{1}{N} \sum_i \left[\frac{1}{4\pi R^2 \cdot dr} \left[\sum_j \left(\left| |\mathbf{r}_i - \mathbf{r}_j| - R \right| < \frac{dr}{2} \right) \right] \right], \tag{7.19}$$

where the term $(||\mathbf{r}_i - \mathbf{r}_j| - R| < \frac{dr}{2})$ is a Boolean counter. It contributes 1, if the (center-to-center) distance between oscillators i and j is in the range $R - \frac{dr}{2} \cdots R + \frac{dr}{2}$ and 0 otherwise. $\rho_V(R)$ is a continuous function. For large arguments, $\rho_V(R)$ will approach the large scale volume number density, $\lim_{R\to\infty} \rho_V(R) = \rho_N$. For short distances, a short-range order might be present in the sample, giving rise to systematic deviations from the constant value. It should be normalized to integrate out over the long range according to

$$\int_0^{\pi} \int_0^{2\pi} \int_0^{R} \rho_V(r) \sin(\theta) r^2 \, dr \, d\phi \, d\theta = N, \tag{7.20}$$

where N is the number of scatterers contained in a large sphere of radius R.

In the following, we will be dealing with a random distribution of scatterers on a substrate, which are confined to a single plane, say $z = 0$. For this case,

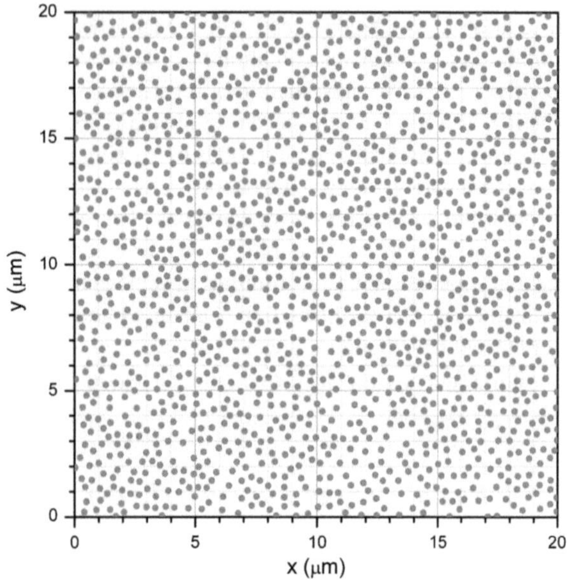

Fig. 7.10 Sample distribution of 1600 disks with an average density of $\rho_N = 4\ \mu\text{m}^{-2}$ in a square of length $L = 20\ \mu\text{m}$

we introduce the areal number density according to $\rho_A(r)\,dA\,dz = \rho_V(r)\,dV$ or $\rho_A(r)\delta z = \rho_V(r) \cdot 2r$. Alternatively, we may think of

$$\rho_A(R) = \lim_{dr\to 0^+} \lim_{N\to\infty} \frac{1}{N} \sum_i \left[\frac{1}{2\pi R \cdot dr} \left[\sum_j \left(\left| |\mathbf{r}_i - \mathbf{r}_j| - R \right| < \frac{dr}{2} \right) \right] \right]. \tag{7.21}$$

One can also deal with a random distribution of scatterers on a line in an analogous fashion.

7.5.4 Example for a Simulated and Analytical RDF

In applications of RDF's it is crucial that an analytical form is available. Before we discuss the dispersion relation of amorphous media, we illustrate here the case of the two-dimensional hard-sphere model—appropriate for disk-like scatterers. That is, their center location is random, but never closer than one diameter to the nearest neighbor disk. A reasonable simulation is straightforward to implement with Monte Carlo methods, as illustrated in Fig. 7.10.

From this discrete distribution, we obtain an approximate radial distribution function as the probability histogram H_i of all the distances between pairs of scatterers. The $N(N-1)/2$ distinct pairs are binned into intervals of width dR ($R_i = i \cdot dR$). The radial density function extracted from this histogram essentially counts the number of pairs in the interval $R_i \ldots R_{i+1}$, as shown in Fig. 7.11,

$$\rho_i = \frac{H_i N}{2\pi R_i\, dR}. \tag{7.22}$$

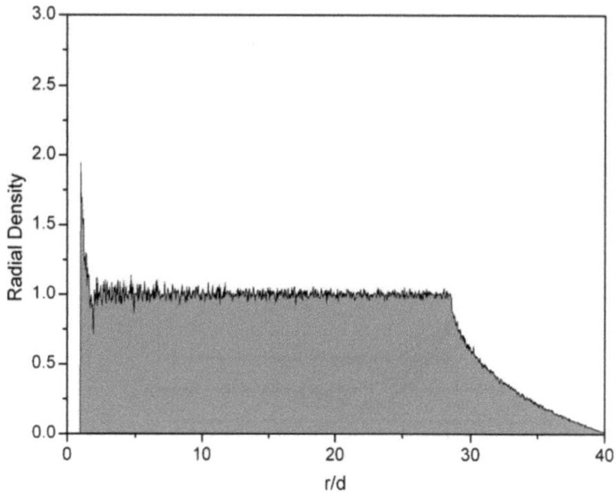

Fig. 7.11 Histogram of the simulated radial distribution function. For distances significantly larger than one disk diameter, the distribution approaches the global average number density. The dropping tail at the largest distances (here beyond ~28 diameters) stems from the finite size of the simulation area

Analytical expressions for the radial distribution function have been developed already in the early days of x-ray scattering analysis of disordered solids. Following Nijboer and van Hove [109], we assume a two dimensional hard sphere gas and set

$$g_0(r) = \begin{cases} 0 & \text{if } r < d \\ 1 & \text{otherwise.} \end{cases} \tag{7.23}$$

The general RDF is then expanded in powers of the global density as

$$g(r) = g_0(r)\big(1 + \rho_N g_1(r) + \rho_N^2 g_2(r) + \cdots\big). \tag{7.24}$$

In the hard sphere model the first correction term represents simply the overlap area of two circles, whose centers are spaced apart by r. This can be evaluated by elementary trigonometry [110]:

$$g_1(r) = 2d^2 \operatorname{Re}\left(\arccos\left(\frac{r}{2d}\right) - \sqrt{1 - \frac{r^2}{4d^2}} \cdot \frac{r}{2d}\right). \tag{7.25}$$

The higher order terms are rather more complicated to evaluate.

Figure 7.12 illustrates the adequacy of this approximation for the kinds of number densities typically encountered in realistic metaglass. Due to the hard spheres assumption, the density must be zero for center-to-center distances smaller than the disk diameter. Some of that eliminated density is re-distributed to the interval $r = 1d \ldots 2d$—the phenomenon known as short-range order. For larger distances the statistical distribution simply fluctuates—due to the finite total number of simulated disks—around the global average value. For our present purposes, the following approximation to the radial number density function is sufficient:

$$RDF(r) = g(r) \cdot \rho_N. \tag{7.26}$$

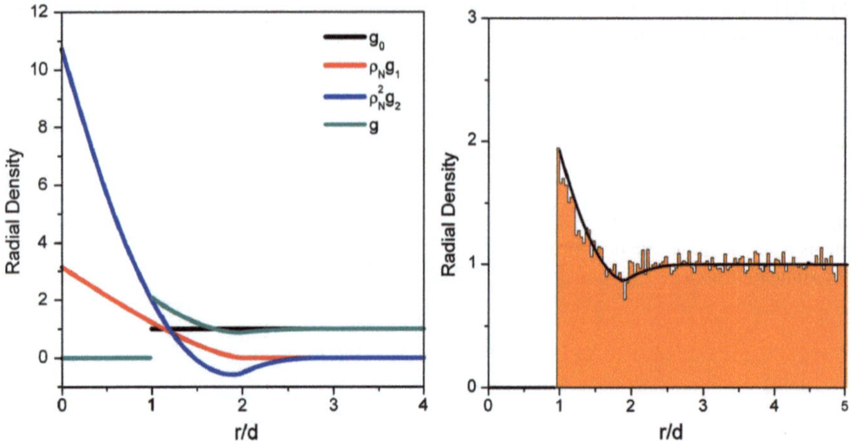

Fig. 7.12 *Left*: the three lowest order terms in the radial density function for the hard disk model (*black*, *red*, *blue*) as well as their sum (*green*). *Right*: comparison of the analytical description (*black line*) with the simulated histogram (*orange area*) for the radial density function of Fig. 7.10

7.5.5 Homogenization of the Discrete Medium

With an analytical radial density function at hand, we are now ready to replace the discrete version of (7.18):

$$\mathbf{E}_{\text{exc},0} = \left[\mathbf{1} + \frac{1}{N} \sum_i \sum_j (\delta_{ij} - 1) \mathbf{G}_{B,ij} \cdot \chi_j \cdot \exp\bigl(\imath \mathbf{K}(\mathbf{r}_j - \mathbf{r}_i)\bigr) \right] \cdot \mathbf{E}_0, \qquad (7.27)$$

with a smoothed or averaged version:

$$\mathbf{E}_{\text{exc},0} = \left[\mathbf{1} - \int_0^{2\pi} \int_0^{\infty} \mathbf{G}_B(r, \phi) \chi \exp\bigl(\imath K r \cos(\phi)\bigr) \rho_A(r) r \, dr \, d\phi \right] \cdot \mathbf{E}_0. \qquad (7.28)$$

Here, the representative scatterer is thought to reside at the center, ϕ is the angle between the wavevector $\mathbf{K}$ and the scattering element at $\mathbf{r} = (x, y, 0)$.

The expression in square brackets in (7.28) represents a key result: the closer to zero, the stronger the response $\mathbf{E}_0$. This situation may be regarded as approaching a resonant response. In fact, vanishing of this term is the condition for an eigenmode of the system. That is, in order to extract the dispersion relation for such a homogenized random medium, we need to find the wavenumber K that fulfills

$$\mathbf{0} = \mathbf{1} - \int_0^{2\pi} \int_0^{\infty} \mathbf{G}_B(r, \phi) \chi \exp\bigl(\imath K r \cos(\phi)\bigr) \rho_A(r) r \, dr \, d\phi. \qquad (7.29)$$

Note the tensorial character of this equation. As there are three vector components available to $\mathbf{E}_0$, we can expect there will be three independent branches to the dispersion relation.

7.5.6 *Analytic Dispersion Relations of Plane Waves in Two-Dimensional Metaglass*

The plane wave dispersion of a metaglass is the relation of average eigenmode wavevectors K to the frequency $\omega = c_0 k$. To find the (average) eigenmodes of a random system, we need to solve (7.29) in a self-consistent fashion:

$$\mathbf{0} = \mathbf{1} - \int_0^{2\pi} \int_0^{\infty} \mathbf{G}_B(r,\phi) \frac{k^2}{\varepsilon_0} \begin{pmatrix} \alpha_{xx} & 0 & 0 \\ 0 & \alpha_{yy} & 0 \\ 0 & 0 & \alpha_z \end{pmatrix} \exp\big(\iota K r \cos(\phi)\big) RDF(r) r \, dr \, d\phi, \tag{7.30}$$

where both the particle polarizabilities $\alpha_{xx}, \alpha_{yy}, \alpha_z$ and the Green dyadic $\mathbf{G}_B$ are functions of frequency. In the $z = 0$ plane, with $\mathbf{r} = r(\cos(\phi), \sin(\phi), 0)$, the Green dyadic equation (7.10) assumes simpler form

$$\begin{aligned} \mathbf{G}_B(r) &= \left[G_0 - G_1 \begin{pmatrix} \cos(\phi)^2 & \cos(\phi)\sin(\phi) & 0 \\ \cos(\phi)\sin(\phi) & \sin(\phi)^2 & 0 \\ 0 & 0 & 0 \end{pmatrix} \right] g_B(r), \\ G_0 &= \frac{k^2 r^2 + \iota k r - 1}{k^2 r^2}, \\ G_1 &= \frac{k^2 r^2 + 3\iota k r - 3}{k^2 r^2} \end{aligned} \tag{7.31}$$

This can be further simplified, if we express the ϕ-integrations with the help of an integral representation of Bessel functions, $\int_0^{2\pi} \exp(\iota K r \cos(\phi)) \cos(n\phi) \, d\phi = 2\pi \iota^n J_n(Kr)$. Equation (7.30) then reads

$$\begin{aligned} \mathbf{0} = \frac{\varepsilon_0}{2\pi k^2} \begin{pmatrix} \alpha_{xx} & 0 & 0 \\ 0 & \alpha_{yy} & 0 \\ 0 & 0 & \alpha_z \end{pmatrix}^{-1} \\ - \int_0^{\infty} \left[G_0 J_0 - \frac{G_1}{2} \begin{pmatrix} J_0 - J_2 & 0 & 0 \\ 0 & J_0 + J_2 & 0 \\ 0 & 0 & 0 \end{pmatrix} \right] g_B \cdot RDF \cdot r \, dr. \end{aligned} \tag{7.32}$$

We realize that the three polarizations are decoupled for this planar amorphous system. For a given excitation frequency $\omega = c_0 k$ three different wavenumbers K of the correspondingly polarized in-plane modes may be found with the help of (7.32), as illustrated in Fig. 7.13.

One way to represent the resulting dispersion at least approximately is to consider the inverse absolute value of the right-hand-side of (7.32). The in-plane trans-

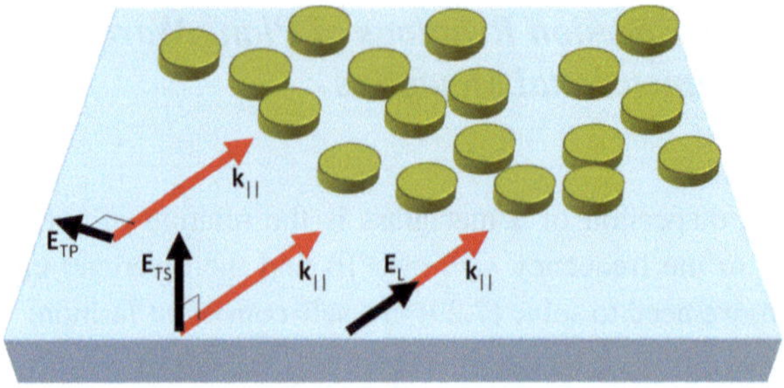

Fig. 7.13 The three linear polarization states available for plane-wave-like modes in a 2-dimensional metaglass: transverse-parallel (TP), transverse-perpendicular (TS), and longitudinal (L)

verse, out-of-plane transverse, and longitudinal branches of the dispersion are thus described, respectively, by

$$\begin{aligned} D_{TP}(\omega, K) &= \left(\frac{\varepsilon_0}{2\pi k^2 \alpha_{xx}} \right. \\ &\quad \left. - \int_0^\infty \left[G_0 J_0 - \frac{G_1}{2}(J_0 + J_2) \right] g_B \cdot RDF \cdot r\, dr \right)^{-1}, \\ D_{TS}(\omega, K) &= \left(\frac{\varepsilon_0}{2\pi k^2 \alpha_{yy}} \right. \\ &\quad \left. - \int_0^\infty \left[G_0 J_0 - \frac{G_1}{2}(J_0 - J_2) \right] g_B \cdot RDF \cdot r\, dr \right)^{-1}, \\ D_L(\omega, K) &= \left(\frac{\varepsilon_0}{2\pi k^2 \alpha_{zz}} - \int_0^\infty G_0 J_0 g_B \cdot RDF \cdot r\, dr \right)^{-1}. \end{aligned} \tag{7.33}$$

The emergence of longitudinal modes might appear surprising at first, since in a homogeneous, isotropic bulk medium they are allowed only in the case of vanishing dielectric constant. Here, however, we are dealing with a flat two-dimensional sheet-like medium, embedded in a three dimensional background medium. Only inside the sheet is the field longitudinal, in the surrounding host medium, the field lines also carry transverse components.

7.5.7 Discussion of a Typical Two-Dimensional Metaglass

According to (7.28), the terms in (7.33) represent the response amplitudes to a unit excitation of the respective polarization. That is, they are closely related to the absorption spectra of the two-dimensional metaglass. In case of a resonant divergence, they describe the excitation of an eigenmode in the dispersion.

Figure 7.14 shows a typical dispersion map obtained with (7.33) for a density of 5 meta-atoms per square micrometer. The brighter shaded areas of near-resonant solutions to (7.33) display a behavior for the whole system that resembles that of

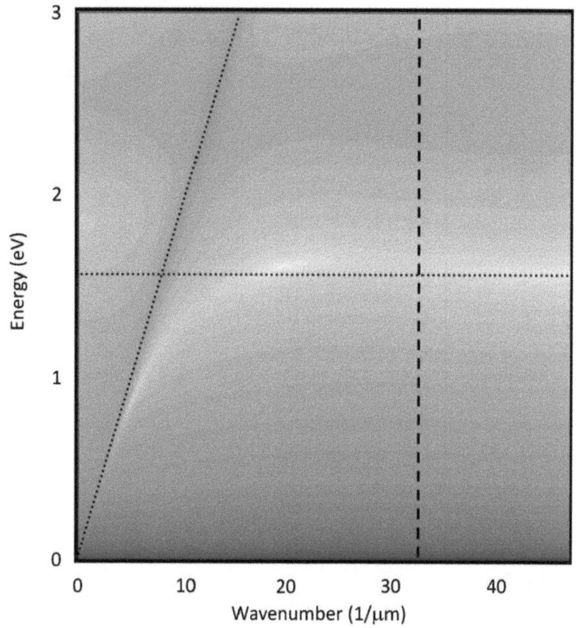

Fig. 7.14 Typical dispersion map calculated with (7.33). Here, the TP case is shown for a particle density of 5/μm^2. The grayscale is proportional to the negative logarithm of D_{TP}. *Black* indicates far off-resonant excitations, whereas *white zones* are close to resonant modes. The *slanted and horizontal dotted lines* represent the light line in vacuum and the resonant energy level of an individual meta-atom, respectively. The *vertical dashed line* is located at $K = 2\pi/d$, where d is the diameter of the meta-atoms

an anti-crossing between two independent systems: the free space photonic modes with the light line as their dispersion and the system of localized particle resonance, whose dispersion is flat in k-space.

In conventional transmission or reflection spectroscopy, the parallel wavenumbers are limited to the triangle left of the light line. The dispersion right of the light line resembles the case of surface plasmons bound to an interface between a metal and a dielectric. To excite modes in this part of the dispersion map, the same techniques can be employed as for surface plasmons, viz., Kretschmann and Otto configuration, and thin film metaglasses should be similarly beneficial for sensing applications. With the help of maps such as calculated from (7.33) the optimal operation wavenumber and energy can be found.

In Fig. 7.15 we illustrate the rich phenomena that can be encountered upon variation of the polarization state and number density.

First we notice the relatively flat dispersions for TS modes, whose polarization is perpendicular out-of-plane. This is a direct consequence of the constituent meta-atoms' weak and broad polarizability in the z-direction (see Fig. 7.9). For low number densities—at 1/μm^2 or less—we find for the TP and L polarizations a prominent band of near-resonant modes around the individual particles' in-plane resonance energy. Except for wavenumbers very close to the light line, however, not much variation is observed. Evidently, the radiative long-range interactions are not strong enough to evoke a pronounced average coherent effect in the metaglass.

As we progress to number densities of 5/μm^2 or higher, we notice more dramatic changes to the dispersion maps. At this density the average area occupied by one

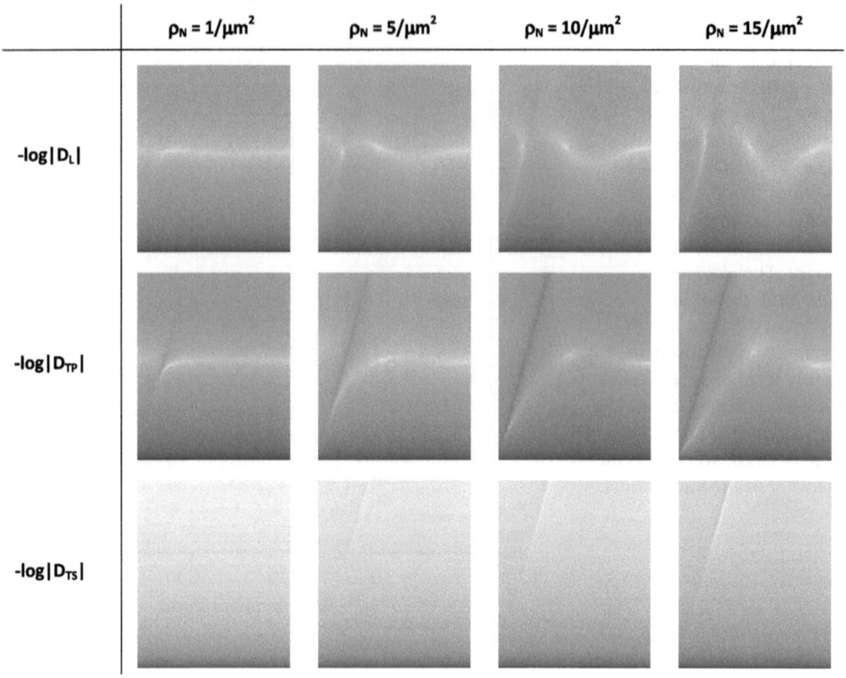

Fig. 7.15 Several dispersion maps evaluated with (7.33). The same dimensions are applied as with Fig. 7.14. *Horizontally*, the wavenumbers range from 0 to 45.6 inverse micrometer; *vertically*, the energies range from 0 to 3 eV. The *grey scales* for each map is individually normalized

meta-atom is a square with a side length of 2.4 times its diameter. This correlates well with the inter-particle distances that have been found to induce significant hybridization and resonance energy shifting in isolated dimers. Apparently, similar effects begin to play a role when the average nearest neighbor distance becomes significantly less than two or three particle diameters. Notably, the TP polarization becomes progressively forbidden on the light line and develops a continuous band of nearly resonant modes, whose spatial wavenumbers are exclusively on the right side of the light line. That is, these modes will not radiate strongly into photonic modes but progress in the plane of the metaglass. It must be underlined, though, that for wavenumbers beyond the value corresponding to an inverse particle diameter ($k > 2\pi/d$) the associated wavevectors correspond to features smaller than the particle size. This is analogous to reaching the edge of a Brillouin zone in periodic systems, where wavevectors of magnitudes outside the first Brillouin zone are shifted back by the diameter of the Brillouin zone. In the case of metaglass—without explicit periodicity—it is an interesting question how far the analogy between periodic dispersion curves and the dispersion maps introduced here can be taken.

7.6 Conclusion and Perspective

Metaglasses are a rising class of functional nanostructured materials that offer additional advantages over deterministic arrays of similar constituents. One promising application of amorphous plasmonic is in sensors and SERS. Another prospective use is in the area of dynamically engineered light localization through structured profile of farfield excitation. Furthermore, metaglass offers practical advantages over deterministic counterparts through its simple, fast, low-cost and large-scale fabrication techniques such as SL and NSL, as opposed to expensive and time-consuming EBL and FIB, and even photolithography method.

We showed in Sect. 7.4 how the sole study of farfield properties does not yield a complete picture. For the additional study of the highly variable nearfields, SNOM is perfectly suitable. In our own nearfield investigations, we uncovered direct evidence for long-range indirect interactions in metaglass. This further stresses the important role of collective phenomena in photonic nanostructures [28].

Due to the non-deterministic arrangement of individual particles, random or amorphous systems do not lend themselves to periodic boundary condition usually applied in simulation of deterministic ordered array to cut down on simulation cost by simulating only the irreducible symmetry volume. The lack of a unit cell in amorphous systems implies an enormous increase in computational resources needed for their direct simulation. Currently even a square millimeter sized area is not feasible. Hence, we put forward an analytical theory that aims to predict the average behavior of metaglass. In this theory, individual plasmonic particles are assumed to be uniformly distributed over space such that there is no long-range order. When each such meta-atom is assumed a point dipole that interacts with all others through an appropriate Green propagator, it is straightforward to derive an effective medium description of the system. From this the overall response of the metaglass including its dispersion relation is calculated.

Amorphous plasmonic metamaterials with their low-cost, fast, and large-area fabrication are excellent candidates for viable economic application, specifically in sensing and light localization applications. Further research in this direction includes, for example, extending the study of amorphous plasmonic into 3D structures, for instance, using a low-cost self assembly of core-shell metallic-dielectric particles. Another interesting, more fundamental direction of research may be deliberate, continuous transitions from deterministic periodic structures to partially disordered and finally fully amorphous structures. The corresponding control over the radial density function (and its reciprocal space equivalent) might be exploited for example in the spatial management of a broad spectrum of wavelengths, such as in efficiency enhanced solar cells.

References

1. H. Raether, *Surface-Plasmons on Smooth and Rough Surfaces and on Gratings* (Springer, New York, 1988), pp. 1–133

2. E. Kretschmann, H. Raether, Z. Naturforsch. Teil A **23**(12), 2135 (1968)
3. A. Otto, Z. Phys. **216**(4), 398 (1968)
4. R.W. Wood, Philos. Mag. **4**(19–24), 396–402 (1902)
5. R.W. Wood, Philos. Mag. **3**(13–18), 396–410 (1902)
6. R.W. Wood, Philos. Mag. **4**(19–24), 425–429 (1902)
7. G. Mie, Ann. Phys. **25**(3), 377–445 (1908)
8. U. Kreibig, P. Zacharia, Z. Phys. **231**(2), 128 (1970)
9. M. Fleischmann, P.J. Hendra, A.J. Mcquillan, Chem. Phys. Lett. **26**(2), 163–166 (1974)
10. E. Prodan, C. Radloff, N.J. Halas, P. Nordlander, Science **302**, 419–422 (2003)
11. A.A. Zharov, I.V. Shadrivov, Y.S. Kivshar, Phys. Rev. Lett. **91**(3), 037401 (2003)
12. C.J. Murphy, T.K. Sau, A.M. Gole, C.J. Orendorff, J. Gao, L. Gou, S.E. Hunyadi, T. Li, J. Phys. Chem. B **109**(29), 13857–13870 (2005)
13. C.J. Orendorff, T.K. Sau, C.J. Murphy, Small **2**(5), 636–639 (2006)
14. C. Noguez, J. Phys. Chem. C **111**(10), 3806–3819 (2007)
15. C.L. Nehl, J.H. Hafner, J. Mater. Chem. **18**(21), 2415–2419 (2008)
16. M.L. Sandrock, C.A. Foss, J. Phys. Chem. B **103**(51), 11398–11406 (1999)
17. H. Tamaru, H. Kuwata, H.T. Miyazaki, K. Miyano, Appl. Phys. Lett. **80**(10), 1826–1828 (2002)
18. W. Rechberger, A. Hohenau, A. Leitner, J.R. Krenn, B. Lamprecht, F.R. Aussenegg, Opt. Commun. **220**(1–3), 137–141 (2003)
19. K.H. Su, Q.H. Wei, X. Zhang, J.J. Mock, D.R. Smith, S. Schultz, Nano Lett. **3**(8), 1087–1090 (2003)
20. T. Atay, J.H. Song, A.V. Nurmikko, Nano Lett. **4**(9), 1627–1631 (2004)
21. P.K. Jain, S. Eustis, M.A. El-Sayed, J. Phys. Chem. B **110**(37), 18243–18253 (2006)
22. I. Romero, J. Aizpurua, G.W. Bryant, F.J. García de Abajo, Opt. Express **14**(21), 9988–9999 (2006)
23. P.K. Jain, M.A. El-Sayed, J. Phys. Chem. C **112**(13), 4954–4960 (2008)
24. C. Tabor, R. Murali, M. Mahmoud, M.A. El-Sayed, J. Phys. Chem. A **113**(10), 1946–1953 (2009)
25. N. Meinzer, M. König, M. Ruther, S. Linden, G. Khitrova, H.M. Gibbs, K. Busch, M. Wegener, Appl. Phys. Lett. **99**(11), 111104 (2011)
26. C. Tabor, D. Van Haute, M.A. El-Sayed, ACS Nano **3**(11), 3670–3678 (2009)
27. A.M. Funston, C. Novo, T.J. Davis, P. Mulvaney, Nano Lett. **9**(4), 1651–1658 (2009)
28. S.V. Boriskina, M. Povinelli, V.N. Astratov, A.V. Zayats, V.A. Podolskiy, Opt. Express **19**(22), 22024–22028 (2011)
29. N. Verellen, Y. Sonnefraud, H. Sobhani, F. Hao, V.V. Moshchalkov, P. van Dorpe, P. Nordlander, S.A. Maier, Nano Lett. **9**(4), 1663–1667 (2009)
30. J.A. Fan, K. Bao, C. Wu, J. Bao, R. Bardhan, N.J. Halas, V.N. Manoharan, G. Shvets, P. Nordlander, F. Capasso, Nano Lett. **10**(11), 4680–4685 (2010)
31. N. Liu, L. Langguth, T. Weiss, J. Kästel, M. Fleischhauer, T. Pfau, H. Giessen, Nat. Mater. **8**(9), 758–762 (2009)
32. N.A. Mirin, K. Bao, P. Nordlander, J. Phys. Chem. A **113**(16), 4028–4034 (2009)
33. J.A. Fan, C. Wu, K. Bao, J. Bao, R. Bardhan, N.J. Halas, V.N. Manoharan, P. Nordlander, G. Shvets, F. Capasso, Science **328**(5982), 1135–1138 (2010)
34. M. Hentschel, M. Saliba, R. Vogelgesang, H. Giessen, A.P. Alivisatos, N. Liu, Nano Lett. **10**(7), 2721–2726 (2010)
35. J.B. Lassiter, H. Sobhani, J.A. Fan, J. Kundu, F. Capasso, P. Nordlander, N.J. Halas, Nano Lett. **10**(8), 3184–3189 (2010)
36. P. Nordlander, C. Oubre, E. Prodan, K. Li, M.I. Stockman, Nano Lett. **4**(5), 899–903 (2004)
37. E. Prodan, P. Nordlander, J. Chem. Phys. **120**(11), 5444–5454 (2004)
38. W. Khunsin, B. Brian, J. Dorfmüller, M. Esslinger, R. Vogelgesang, C. Etrich, C. Rockstuhl, A. Dmitriev, K. Kern, Nano Lett. **11**(7), 2765–2769 (2011)
39. F.J. García de Abajo, Rev. Mod. Phys. **79**(4), 1267–1290 (2007)

40. D.M. Natarov, V.O. Byelobrov, R. Sauleau, T.M. Benson, A.I. Nosich, Opt. Express **19**(22), 22176–22190 (2011)
41. S.H. Mousavi, A.B. Khanikaev, B. Neuner, D.Y. Fozdar, T.D. Corrigan, P.W. Kolb, H.D. Drew, R.J. Phaneuf, A. Alù, G. Shvets, Opt. Express **19**(22), 22142–22155 (2011)
42. Y. Alaverdyan, E.-M. Hempe, A.N. Vamivakas, E. Haibo, S.A. Maier, M. Atatüre, Appl. Phys. Lett. **94**(2) (2009)
43. S.A. Maier, P.G. Kik, H.A. Atwater, Phys. Rev. B **67**(20), 205402 (2003)
44. S.A. Maier, P.G. Kik, H.A. Atwater, Appl. Phys. Lett. **81**(9), 1714–1716 (2002)
45. A.M. Serebrennikov, Opt. Commun. **284**(21), 5043–5054 (2011)
46. C.P. Huang, X.G. Yin, L.B. Kong, Y.Y. Zhu, J. Phys. Chem. C **114**(49), 21123–21131 (2010)
47. S.L. Zou, G.C. Schatz, J. Chem. Phys. **121**(24), 12606–12612 (2004)
48. E.M. Hicks, S. Zou, G.C. Schatz, K.G. Spears, R.P. Van Duyne, L. Gunnarsson, T. Rindzevicius, B. Kasemo, M. Käll, Nano Lett. **5**(6), 1065–1070 (2005)
49. A. Alù, N. Engheta, Phys. Rev. B **79**(23), 235412 (2009)
50. S. Linden, J. Kuhl, H. Giessen, Phys. Rev. Lett. **86**(20), 4688–4691 (2001)
51. C.L. Haynes, A.D. McFarland, L.L. Zhao, R.P. Van Duyne, G.C. Schatz, L. Gunnarsson, J. Prikulis, B. Kasemo, M. Käll, J. Phys. Chem. B **107**(30), 7337–7342 (2003)
52. L. Zhao, K.L. Kelly, G.C. Schatz, J. Phys. Chem. B **107**(30), 7343–7350 (2003)
53. V.G. Kravets, F. Schedin, A.N. Grigorenko, Phys. Rev. Lett. **101**(8), 087403 (2008)
54. V. Giannini, G. Vecchi, J. Gómez Rivas, Phys. Rev. Lett. **105**(26), 266801 (2010)
55. D. Dregely, R. Taubert, J. Dorfmüller, R. Vogelgesang, K. Kern, H. Giessen, Nat. Commun. **2**, 267 (2011)
56. N. Liu, H.C. Guo, L.W. Fu, S. Kaiser, H. Schweizer, H. Giessen, Nat. Mater. **7**(1), 31–37 (2008)
57. N. Liu, L.W. Fu, S. Kaiser, H. Schweizer, H. Giessen, Adv. Mater. **20**(20), 3859–3865 (2008)
58. C. Rockstuhl, T. Paul, F. Lederer, T. Pertsch, T. Zentgraf, T.P. Meyrath, H. Giessen, Phys. Rev. B **77**(3), 035126 (2008)
59. S.L. Zou, G.C. Schatz, Chem. Phys. Lett. **403**(1–3), 62–67 (2005)
60. B. Auguie, W.L. Barnes, Phys. Rev. Lett. **101**(14), 143902 (2008)
61. W. Zhou, T.W. Odom, Nat. Nanotechnol. **6**(7), 423–427 (2011)
62. T. Rindzevicius, Y. Alaverdyan, M. Käll, W.A. Murray, W.L. Barnes, J. Phys. Chem. C **111**(32), 11806–11810 (2007)
63. C. Helgert, C. Rockstuhl, C. Etrich, C. Menzel, E.B. Kley, A. Tünnermann, F. Lederer, T. Pertsch, Phys. Rev. B **79**(23), 233107 (2009)
64. N. Papasimakis, V.A. Fedotov, Y.H. Fu, D.P. Tsai, N.I. Zheludev, Phys. Rev. B **80**(4), 041102(R) (2009)
65. G. Dolling, C. Enkrich, M. Wegener, J.F. Zhou, C.M. Soukoulis, S. Linden, Opt. Lett. **30**(23), 3198–3200 (2005)
66. M. Decker, N. Feth, C.M. Soukoulis, S. Linden, M. Wegener, Phys. Rev. B **84**(8), 085416 (2011)
67. M. Decker, S. Burger, S. Linden, M. Wegener, Phys. Rev. B **80**(19), 193102 (2009)
68. S. Mubeen, S. Zhang, N. Kim, S. Lee, S. Krämer, H. Xu, M. Moskovits, Nano Lett. **12**(4), 2088–2094 (2012)
69. J.J. Mock, R.T. Hill, Y.-J. Tsai, A. Chilkoti, D.R. Smith, Nano Lett. **12**(4), 1757–1764 (2012)
70. Y. Xia, N.J. Halas, Mater. Res. Soc. Bull. **30**, 338–348 (2005)
71. Y. Xia, G.M. Whitesides, Annu. Rev. Mater. Sci. **28**(1), 153–184 (1998)
72. C.L. Haynes, R.P. Van Duyne, J. Phys. Chem. B **105**(24), 5599–5611 (2001)
73. A.J. Haes, J. Zhao, S. Zou, C.S. Own, L.D. Marks, G.C. Schatz, R.P. Van Duyne, J. Phys. Chem. B **109**(22), 11158–11162 (2005)
74. H. Fredriksson, Y. Alaverdyan, A. Dmitriev, C. Langhammer, D.S. Sutherland, M. Zäch, B. Kasemo, Adv. Mater. **19**, 4297–4302 (2007)
75. J. Henzie, J. Lee, M.H. Lee, W. Hasan, T.W. Odom, Annu. Rev. Phys. Chem. **60**, 147–165 (2009)

76. S. Cataldo, J. Zhao, F. Neubrech, B. Frank, C.J. Zhang, P.V. Braun, H. Giessen, ACS Nano **6**(1), 979–985 (2012)
77. A.B. Dahlin, S. Chen, M.P. Jonsson, L. Gunnarsson, M. Käll, F. Höök, Anal. Chem. **81**(16), 6572–6580 (2009)
78. L. Yang, X. Li, X. Tuo, T.T. Van Nguyen, X. Luo, M. Hong, Opt. Express **19**(S4), A657–A663 (2011)
79. P. Spinelli, M.A. Verschuuren, A. Polman, Nat. Commun. **3**, 692 (2012)
80. A.V. Kabashin, P. Evans, S. Pastkovsky, W. Hendren, G.A. Wurtz, R. Atkinson, R. Pollard, V.A. Podolskiy, A.V. Zayats, Nat. Mater. **8**(11), 867–871 (2009)
81. B. Luk'yanchuk, N.I. Zheludev, S.A. Maier, N.J. Halas, P. Nordlander, H. Giessen, C.T. Chong, Nat. Mater. **9**(9), 707–715 (2010)
82. R. Vogelgesang, A. Dmitriev, Analyst **135**(6), 1175–1181 (2010)
83. E. Cefali, S. Patane, S. Spadaro, R. Gardelli, M. Albani, M. Allegrini, *Near Field Probes: From Optical Fibers to Optical Nanoantennas* (Springer, New York, 2008)
84. A. Bek, R. Vogelgesang, K. Kern, Rev. Sci. Instrum. **77**(4), 043703 (2006)
85. A. Bek, R. Vogelgesang, K. Kern, Appl. Phys. Lett. **87**(16), 163115 (2005)
86. M. Esslinger, J. Dorfmüller, W. Khunsin, R. Vogelgesang, K. Kern, Rev. Sci. Instrum. **83**(3), 033704 (2012)
87. R. Esteban, R. Vogelgesang, K. Kern, Nanotechnology **17**(2), 475–482 (2006)
88. R. Esteban, R. Vogelgesang, K. Kern, Phys. Rev. B **75**(19), 195410 (2007)
89. R. Esteban, R. Vogelgesang, K. Kern, Ultramicroscopy **111**(9–10), 1469–1474 (2011)
90. R. Vogelgesang, R. Esteban, K. Kern, J. Microsc. **229**(2), 365–370 (2008)
91. J. Dorfmüller, D. Dregely, M. Esslinger, W. Khunsin, R. Vogelgesang, K. Kern, H. Giessen, Nano Lett. **11**(7), 2819–2824 (2011)
92. J. Dorfmüller, R. Vogelgesang, W. Khunsin, C. Rockstuhl, C. Etrich, K. Kern, Nano Lett. **10**(9), 3596–3603 (2010)
93. J. Dorfmüller, R. Vogelgesang, R.T. Weitz, C. Rockstuhl, C. Etrich, T. Pertsch, F. Lederer, K. Kern, Nano Lett. **9**(6), 2372–2377 (2009)
94. R. Esteban, R. Vogelgesang, J. Dorfmüller, A. Dmitriev, C. Rockstuhl, C. Etrich, K. Kern, Nano Lett. **8**(10), 3155–3159 (2008)
95. C. Rockstuhl, C. Etrich, C. Helgert, C. Menzel, T. Paul, S. Fahr, T. Pertsch, J. Dorfmüller, R. Esteban, W. Khunsin, R. Vogelgesang, K. Kern, A. Dmitriev, K. Bittkau, T. Beckers, R. Carius, F. Lederer, Proc. SPIE **7604**, 760401 (2010)
96. M.I. Stockman, S.V. Faleev, D.J. Bergman, Phys. Rev. Lett. **87**(16), 167401 (2001)
97. B. Lamprecht, G. Schider, R.T. Lechner, H. Ditlbacher, J.R. Krenn, A. Leitner, F.R. Aussenegg, Phys. Rev. Lett. **84**(20), 4721–4724 (2000)
98. S.M. Nie, S.R. Emery, Science **275**(5303), 1102–1106 (1997)
99. K. Kneipp, Y. Wang, H. Kneipp, L.T. Perelman, I. Itzkan, R. Dasari, M.S. Feld, Phys. Rev. Lett. **78**(9), 1667–1670 (1997)
100. I.R. Khan, D. Cunningham, S. Lazar, D. Graham, W.E. Smith, D.W. McComb, Faraday Discuss. **132**, 171–178 (2006)
101. M. Aeschlimann, M. Bauer, D. Bayer, T. Brixner, F.J. de Abajo, W. Pfeiffer, M. Rohmer, C. Spindler, F. Steeb, Nature **446**(7133), 301–304 (2007)
102. A. Rashidi, H. Mosallaei, Phys. Rev. B **82**(3), 035117 (2010)
103. G. Volpe, G. Molina-Terriza, R. Quidant, Phys. Rev. Lett. **105**(21), 216802 (2010)
104. T.S. Kao, S.D. Jenkins, J. Ruostekoski, N.I. Zheludev, Phys. Rev. Lett. **106**(8), 085501 (2011)
105. T.S. Kao, E.T.F. Rogers, J.Y. Ou, N.I. Zheludev, Nano Lett. **12**(6), 2728–2731 (2012)
106. J. Venermo, A. Sihvola, J. Electrost. **63**(2), 101–117 (2005)
107. P.B. Johnson, R.W. Christy, Phys. Rev. B **6**(12), 4370–4379 (1972)
108. P.G. Etchegoin, E.C. Le Ru, M. Meyer, J. Chem. Phys. **125**(16), 164705 (2006)
109. B.R.A. Nijboer, L. Van Hove, Phys. Rev. **85**(5), 777–783 (1952)
110. S.B. Yuste, A. Santos, J. Chem. Phys. **99**(3), 2020–2023 (1993)

Chapter 8
Structure and Properties of Photonic Amorphous Diamond

Keiichi Edagawa

Abstract Photonic band gap (PBG) formation and light propagation properties of an amorphous photonic structure named "photonic amorphous diamond (PAD)" are reviewed in this chapter. It has been demonstrated numerically and experimentally that a full three-dimensional (3D) photonic band gap is formed in the photonic amorphous diamond, in spite of complete absence of lattice periodicity. This proves that lattice periodicity is not essential to the realization of a 3D photonic band gap. The 3D photonic band gap in photonic amorphous diamond is clean with no trace of localized photonic states within it. This clean 3D photonic band gap should enable strong light confinement at an introduced defect, which has actually been demonstrated numerically. The 3D photonic band gap in photonic amorphous diamond is completely isotropic, regardless of the wavevector orientation and polarization direction, which, in principle, cannot be realized in conventional photonic crystals. In passbands, the photonic amorphous diamond exhibits diffusive light-propagation, where the scattering strength increases significantly as the frequency approaches the band edge. In frequency ranges near the band edge, the scattering strength is so high that light localization is realized. We discuss new insights given by these findings into the physical origin of photonic band gaps and issues such as light diffusion and localization in photonic materials.

8.1 Introduction

In 1987, Yablonovitch [1] and John [2] proposed the idea that a three-dimensional (3D) photonic band gap (PBG), in which electromagnetic wave propagation is forbidden in all directions, can be realized in artificial periodic dielectric structures, namely, photonic crystals. The 3D PBG was predicted to enable 3D light confinement within a wavelength-sized volume without accompanying substantial loss, which is difficult to be achieved by other media. Since such strong light confinement will enhance various light-matter interactions and also will realize very ef-

K. Edagawa (✉)
Institute of Industrial Science, The University of Tokyo, Komaba, Meguro-ku, Tokyo 153-8505, Japan
e-mail: edagawa@iis.u-tokyo.ac.jp

C. Rockstuhl, T. Scharf (eds.), *Amorphous Nanophotonics*,
Nano-Optics and Nanophotonics, DOI 10.1007/978-3-642-32475-8_8,

© Springer-Verlag Berlin Heidelberg 2013

ficient nanophotonic devices, the photonic crystals have gained broad interest and have been studied extensively [3].

Because Bragg scattering of light due to lattice periodicity was considered to be the origin of the PBG formation, it was previously believed that lattice periodicity is indispensable for the realization of PBGs. However, in 2001, a photonic amorphous structure consisting of dielectric cylinders were shown to form a 2D PBG for the TM polarization [4]. Since then, the formation of 2D TM PBGs have been reported in this type of amorphous structures [5–7], and the formation mechanism of such PBGs has been discussed in terms of evanescent coupling of Mie resonances of individual dielectric cylinders [5–11]. This mechanism is closely related with the tight-binding model [5, 8, 9] frequently used for the explanation of the electronic band-gap formation in semiconductors. Recently, the formation of a sizable "complete" 2D PBG (PBG both for TM and TE modes) has been shown to be realizable in 2D photonic amorphous structures [7].

On the other hand, it is, in principle, much more difficult to form complete 3D PBGs. With regard to photonic crystals, only a limited number of structures, including diamond and diamond-related structures, have been shown to form sizable 3D PBGs [12]. As for photonic amorphous structures, considerable suppression of photonic density of states have recently been shown to occur in the systems of randomly arranged dielectric spheres [13–15], the mechanism of which has been explained based on the Mie resonances of the spheres. However, it remains to be clarified whether complete 3D PBGs can be realized in these systems. The calculations of photonic density of states for random close packing of dielectric spheres with a supercell periodicity by Dong et al. [14], and by Liew et al. [15] have shown only a reduction of the states, i.e., a pseudogap. In contrast, the density of states calculated by Rockstuhl and Lederer [13] for a series of finite-size systems of randomly arranged dielectric spheres has shown an exponential decrease with the system size, which is claimed to indicate the formation of a complete 3D PBG in the infinite structure.

On the other hand, the formation of a complete 3D PBG has been confirmed in an amorphous network structure of dielectrics by the density-of-states calculations using a supercell structure [16]. Subsequently, the 3D PBG formation has been demonstrated by microwave transmission experiments [17]. This structure has a diamond-like local tetrahedral configuration and is therefore named "photonic amorphous diamond (PAD)". It has been shown that the 3D PBG in PAD is clean with no trace of localized photonic states within it [16–18]. This indicates that strong light confinement is realizable in the PAD as well as conventional photonic crystals, which has actually been demonstrated numerically [18]. More recently, Liew et al. have numerically shown the 3D PBG formation in a PAD-like structure [15].

Light propagation in random media has been a long-standing subject in optics [19, 20]. In random media, light undergoes multiple scattering. Then, light propagates diffusively as long as the scattering is not very strong. As the scattering intensifies, i.e., the scattering mean free path l reduces, light interference intensifies, leading to a halt of diffusive light propagation. This phenomenon is known as the Anderson localization of light [19, 21–29].

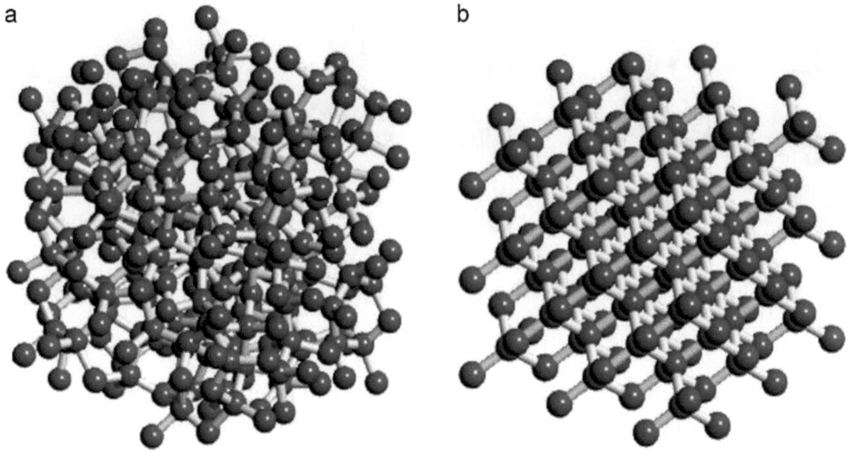

Fig. 8.1 The CRN structure constructed by Barkema and Mousseau [32] (**a**) and the crystalline diamond structure (**b**)

In the passbands, the PAD exhibits diffusive light-propagation, where the scattering strength increases significantly as the frequency approaches the band edge. In frequency ranges near the band edge, the scattering strength is high enough for light localization to be realized. Numerical studies have shown that localized states are indeed formed near the band edges. The formation of localized photonic states near the band edges in the PAD is analogous to the formation of localized electronic states in amorphous semiconductors.

In this chapter, we review PBG formation and light propagation properties of PAD. This chapter is organized as follows. In Sect. 8.2, the structure of PAD and its characteristics are explained. In Sect. 8.3, finite-difference time-domain (FDTD) calculations of photonic eigenstates in PAD are presented. In Sect. 8.4, light propagation properties of PAD are described, mainly focusing on the results of microwave transmission measurements. Finally, Sect. 8.5 is devoted to concluding remarks.

8.2 Structure

The PAD structure is based on the continuous random network (CRN) model [30–32], which has been developed to represent an atomic-structure of amorphous Si or Ge. An example of a CRN structure is shown in Fig. 8.1(a), which should be compared with the atomic-structure of crystalline Si or Ge, i.e., the crystalline diamond structure in Fig. 8.1(b). The requirement of the CRN model is simply that it should be a random network in which all the junction points are fourfold. The quality of the model can be evaluated by the amount of local strain, as measured by the degree of deviations from the regular-tetrahedral configuration; the "ideal" CRN structure should be characterized by the lowest spread in the bond-length and bond-angle distributions around those values in the crystalline diamond structure.

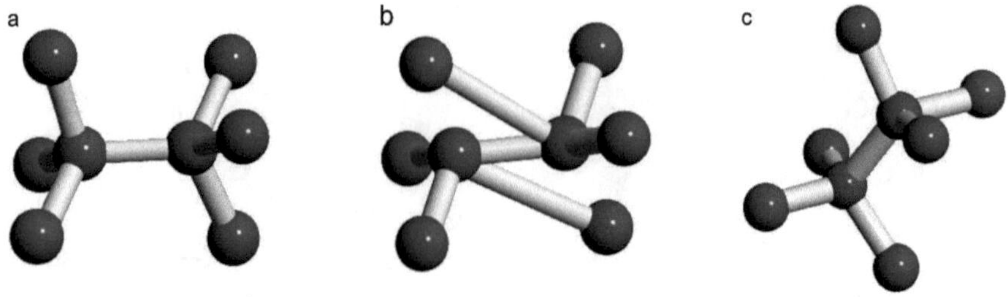

Fig. 8.2 Shematics of the elemental process to generate CRN structures in the WWW-algorithm [31]

In 1985, Wooten, Winer and Weaire [31] proposed an efficient algorithm to generate CRN structures of high-quality, which has been widely used since then. This WWW-algorithm starts from the crystalline diamond structure (Fig. 8.1(b)). First, we randomly pick up a local structure shown in Fig. 8.2(a), and transpose the bonds as in Fig. 8.2(b). This bond transposition preserves the fourfold connectivity of the network. After the bond transposition, the structure is relaxed with an interatomic potential such as the Keating potential [33] that favors the local regular-tetrahedral configuration:

$$E = \frac{3}{16}\frac{\alpha}{d^2}\sum_{\langle ij\rangle}\left(\mathbf{r}_{ij}\cdot\mathbf{r}_{ij} - d^2\right)^2 + \frac{3}{8}\frac{\beta}{d^2}\sum_{\langle ijk\rangle}\left(\mathbf{r}_{ij}\cdot\mathbf{r}_{ik} + \frac{1}{3}d^2\right)^2, \tag{8.1}$$

where α and β are the bond-stretching and bond-bending force constants, and d denotes the equilibrium bond length in the crystalline diamond structure. The relaxed configuration is schematically shown in Fig. 8.2(c). This relaxation reduces the distortion of the two tetrahedrons in the figure but increases the distortion of the tetrahedrons around them. The elemental process of Figs. 8.2(a)–8.2(c) is sequentially repeated for randomly chosen local structures of the type of Fig. 8.2(a). This results in a CRN structure but it is usually distorted rather severely. To obtain a CRN structure of high-quality, Wooten et al. [31] have proposed to introduce the Metropolis acceptance probability:

$$P = \begin{cases} 1 & (E_b > E_a), \\ \exp[(E_b - E_a)/k_B T] & (E_b < E_a), \end{cases} \tag{8.2}$$

where E_b and E_a are the total energies of the system before and after the elementary process of Figs. 8.2(a)–8.2(c), and $k_B T$ has the usual meaning. First, we adopt a high enough temperature, where almost all the trials are accepted. This leads to a sufficiently randomized CRN structure with a considerably high energy. After the randomization, we anneal the structure at a relatively low temperature, where the trials to decrease the energy are mainly accepted, leading to a CRN structure with less distortion. We may continue the annealing at an even lower temperature if necessary. By choosing an optimal thermal history, we could obtain an ideal CRN structure with a minimum distortion.

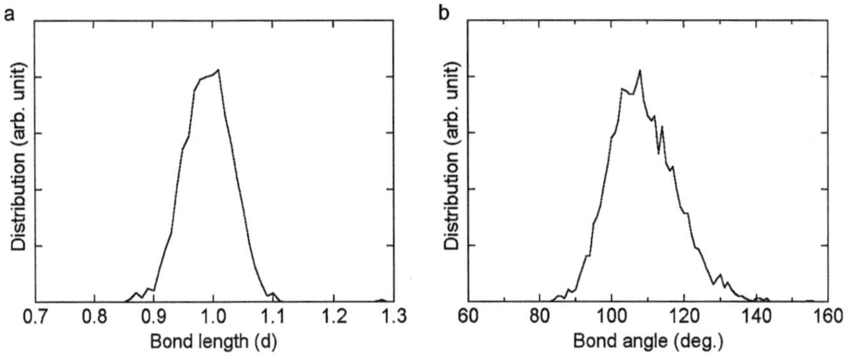

Fig. 8.3 Bond-length distribution (**a**) and bond-angle distribution in the CRN structure (**b**)

In our study, we used the CRN structure constructed and provided by Barkema and Mousseau [32], where an improved algorithm is adopted. Here, we start from a random configuration, instead of the crystalline diamond structure. This guarantees that the constructed structure is not contaminated by any memory of the initial crystalline structure. Therefore, this CRN structure cannot be regarded as a disordered diamond-crystal; the network topology is completely different from that in the diamond-crystal, and no trace of lattice periodicity should get into it. This structure consists of a periodic arrangement of a cubic supercell with the size $(11.5d)^3$. The supercell contains 1000 atoms. The CRN structure shown in Fig. 8.1(a) corresponds to a part of this structure. Indeed, we can see no trace of periodic order of the diamond-lattice whereas we notice a definite local tetrahedral order.

The presence of the local tetrahedral order can be verified by calculating the bond-length and bond-angle distributions in the CRN structure. The bond-length distribution in Fig. 8.3(a) shows a sharp peak at d with the full-width-at-half-maximum (FWHM) value of about $0.1d$. The bond-angle distribution in Fig. 8.3(b) also shows a considerably small spread around the ideal value of 109 deg; the FWHM value of the peak is around 20 deg. From these two facts, we can conclude that this CRN structure is of high-quality, i.e., it has a good local regular-tetrahedral order.

On the other hand, the absence of long-range order is evidenced by the features in the radial distribution function and the diffraction intensity function calculated in Figs. 8.4(a) and 8.4(b). Here, the radial distribution function $f(r)$ is defined as the average point density in the spherical shell $(r, r + dr)$ around an arbitrarily selected point. In Fig. 8.4(a), $f(r)$ is normalized to the average point density f_0 in the whole structure. We notice in Fig. 8.4(a) that the first peak at $r = d$ is very sharp. This peak represents the bond-length distribution already presented in Fig. 8.3(a). The second peak at $r \approx 1.7d$ is also relatively sharp but the subsequent peaks at $r \approx 2.4d$ and $3d$ are very broad; they are marginally observed. Then, no peaks can be detected in the range $r > 3.5d$. This indicates that this CRN structure has no order in the range beyond $r \approx 3.5d$. It should be noted that the range $r > 3.5d$ corresponds to $r > 1.5a$, where $a = 4d/\sqrt{3}$ denotes the lattice constant of the crystalline diamond.

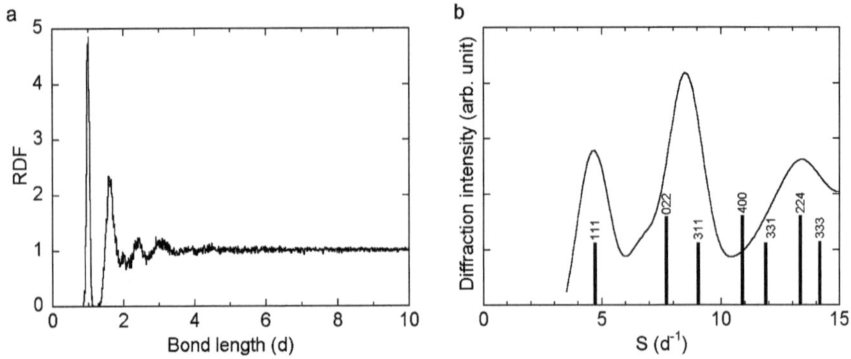

Fig. 8.4 Radial distribution function (**a**) and diffraction intensity function (**b**) for the CRN structure. In (**b**), the Bragg peak positions for the diamond crystal are also shown

This verifies the fact that this CRN structure has no trace of the diamond-lattice periodicity.

Figure 8.4(b) presents the diffraction intensity function defined as

$$I(\mathbf{S}) \equiv \left| \frac{1}{N} \sum_{j=1}^{N} \exp(i\mathbf{S} \cdot \mathbf{r}_j) \right|^2 \tag{8.3}$$

where **S** is the scattering vector, N is the number of points and $\mathbf{r}_j$ denotes the position of the jth point. Because of the isotropy of the structure, $I(\mathbf{S})$ only depends on $S = |\mathbf{S}|$. In Fig. 8.4(b), the peak positions for the crystalline diamond structure are also shown. $I(S)$ in Fig. 8.4(b) shows no sharp peaks, indicating the absence of long-range order; we can only see broad peaks that reflect the short-range order. The FWHM value of the first peak at $S = 4.7d^{-1}$ is $\Delta S \approx 2.0d^{-1}$. Then, the correlation length ξ (the range in which the structural order is kept) can be evaluated to be $\xi = 2\pi/\Delta S \approx \pi d$, which is consistent with the result of the radial distribution function in Fig. 8.4(a).

We modeled the PAD structure by connecting the tetrahedral bond in the CRN structure with dielectric rods in air background (see Fig. 1(a) in Ref. [16]). For comparison, we also constructed a structure by connecting the tetrahedral bonds in the crystalline diamond structure with dielectric rods, which we hereafter call photonic crystalline diamond (PCD) structure (see Fig. 1(b) in Ref. [16]). The PCD structure is known to be the best 3D PBG structure [12, 34]; it exhibits the largest 3D PBG among all the photonic crystals studied thus far. One might regard PAD and PCD as the photonic versions of amorphous and crystalline Si, respectively, in electronic systems.

8.3 Photonic Eigenstates

In this section, we present FDTD calculations of the frequency distribution and electromagnetic field distributions of photonic eigenstates in PAD [16, 17]. In

Sect. 8.3.1, the computational method that we adopted is explained in detail. Subsequently, the results and discussion on the frequency distribution and electromagnetic field distributions of photonic eigenstates are given in Sects. 8.3.2 and 8.3.3, respectively. The calculated frequency distributions of eigenstates has proved the formation of a 3D PBG in PAD. On the other hand, the calculated field distributions of eigenstates have indicated the features of dielectric and air bands for the two bands below and above the gap, respectively. The formations of localized photonic states have also been indicated by these calculations.

8.3.1 Computational Method

As described in Sect. 8.2, our PAD structure consists of a periodically arranged supercell. This is convenient for examining whether or not a PBG is formed. The method that we adopted is an "order-N" method originally developed by Chan et al. [35], which is suitable for the systems requiring a large supercell. Besides PAD, we have also made calculations for PCD for comparison. The cubic supercell in the PAD structure has the size $(11.5d)^3$. For PCD, we used a crystalline diamond structure with a fictitious supercell of $(5a)^3 = (11.5d)^3$. Using this fictitious supercell structure for PCD enables us to compare the results obtained under exactly the same conditions for PAD and PCD. We assumed that the refractive index of rods was $n = 3.0$, and the absorption was neglected, i.e., $\kappa = 0$ (κ: extinction coefficient). The rod radius was assumed to be $r = 0.26d$. Then, the volume fraction of the rod in the structure is 22 %. These conditions correspond to those in the experiments described in Sect. 8.4.1.

In this method, we calculate the time evolutions of the magnetic and electric fields, $\mathbf{H}(\mathbf{r}, t)$ and $\mathbf{E}(\mathbf{r}, t)$, by an FDTD method to find the steady states, i.e., eigenstates in the system. Here, the initial fields must have nonzero projections onto all the eigenstates in the frequency range of interest. The eigenstates in our structures should satisfy the boundary conditions of the Bloch theorem:

$$\mathbf{H}(\mathbf{r}+\mathbf{a}_i, t) = e^{i\mathbf{k}\cdot\mathbf{a}_i}\mathbf{H}(\mathbf{r}, t) \quad \text{and} \quad \mathbf{E}(\mathbf{r}+\mathbf{a}_i, t) = e^{i\mathbf{k}\cdot\mathbf{a}_i}\mathbf{E}(\mathbf{r}, t) \tag{8.4}$$

where $\mathbf{a}_i$ ($i = 1, 2$ and 3) represent the translational vectors of the cubic supercell, and $\mathbf{k}$ is the Bloch wave vector within the first Brillouin zone. Considering these facts and the requirement of transverse-wave for $\mathbf{H}$, we selected the initial fields [9]:

$$\mathbf{H}_\mathbf{k}(\mathbf{r}, 0) = \sum_{|\mathbf{G}|<G_{\max}} \mathbf{H}_\mathbf{G} e^{i(\mathbf{k}+\mathbf{G})\cdot\mathbf{r}+i\phi_\mathbf{G}} \quad \text{and} \quad \mathbf{E}_\mathbf{k}(\mathbf{r}, 0) = \mathbf{0} \tag{8.5}$$

where $\{\mathbf{G}\}$ are the reciprocal-lattice vectors corresponding to the direct-lattice spanned by $\mathbf{a}_i$ ($i = 1, 2$ and 3), and $\phi_\mathbf{G}$ is a random phase. $\mathbf{H}_\mathbf{G}$ is a unit vector, where the direction of $\mathbf{H}_\mathbf{G}$ is randomly chosen within the plane perpendicular to $(\mathbf{k}+\mathbf{G})$. We adopted a sufficiently large value for $G_{\max}$ to cover the frequency range of interest. The time evolutions of the fields $\mathbf{H}_\mathbf{k}(\mathbf{r}, t)$ and $\mathbf{E}_\mathbf{k}(\mathbf{r}, t)$ were calculated

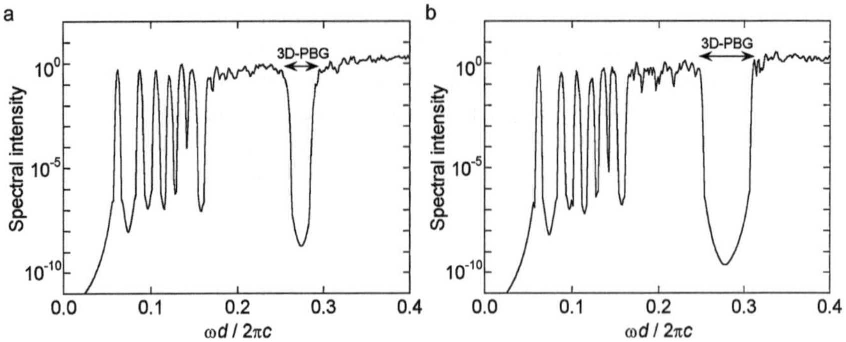

Fig. 8.5 Spectral intensities calculated for PAD (**a**) and PCD (**b**)

by an FDTD method under the boundary conditions of (8.4), where the supercell was discretized into 141^3 meshes and we typically used 2^{14} time steps with each step of about $0.05d/c$ (c: speed of light). The calculated time evolutions $\mathbf{H}_\mathbf{k}(\mathbf{r},t)$ and $\mathbf{E}_\mathbf{k}(\mathbf{r},t)$ were Fourier-transformed into frequency-domain to obtain $\tilde{\mathbf{H}}_\mathbf{k}(\mathbf{r},\omega)$ and $\tilde{\mathbf{E}}_\mathbf{k}(\mathbf{r},\omega)$, where we used a Blackman function as the window function. These functions give the field distributions of the eigenstate at $\mathbf{k}$ and at ω. In actual fact, $\tilde{\mathbf{H}}_\mathbf{k}(\mathbf{r},\omega)$ and $\tilde{\mathbf{E}}_\mathbf{k}(\mathbf{r},\omega)$ were calculated for a subset of 141^3 points due to the memory limitation. For example, for the calculations of (8.6), (8.7) and (8.8), we selected typically 42^3 evenly distributed points $\{\mathbf{r}_i\}$ in the supercell.

For the frequency distributions of the eigenstates, we calculated the function:

$$I_\mathbf{k}(\omega) = \int_{V_{\mathrm{tot}}} \left|\tilde{\mathbf{E}}_\mathbf{k}(\mathbf{r},\omega)\right|^2 d\mathbf{r}, \tag{8.6}$$

where V_{tot} is the total volume of the system. This function should consist of peaks at the eigenstate frequencies at $\mathbf{k}$ in the first Brillouin zone. In principle, we need to calculate $I_\mathbf{k}(\omega)$ for all the $\mathbf{k}$-vectors in the zone to pick up all the eigenfrequencies. However, we found that only the $I_\mathbf{k}(\omega)$ for $\mathbf{k}=\mathbf{0}$ (Γ-point sampling) is necessary to identify the gap because the supercell is large enough, as shown in Sect. 8.3.2. Also for the field distributions $\tilde{\mathbf{H}}_\mathbf{k}(\mathbf{r},\omega)$ and $\tilde{\mathbf{E}}_\mathbf{k}(\mathbf{r},\omega)$ of the eigenstates, the difference among the fields for different $\mathbf{k}$-vectors was found to be negligible, and therefore it is enough to examine the results for $\mathbf{k}=\mathbf{0}$ only. For PCD, the photonic band structure was calculated also by a plane-wave expansion method using a software "BandSOLVE" (RSoft Design Group, Inc.).

8.3.2 Frequency Distribution of Photonic Eigenstates

Figures 8.5(a) and 8.5(b) show the spectral intensities ($I_{\mathbf{k}=\mathbf{0}}(\omega)$ in (8.6)) calculated for PAD and PCD, respectively. First, the spectral intensity for PCD in Fig. 8.5(b) should be compared with the photonic band structure calculated by a plane-wave

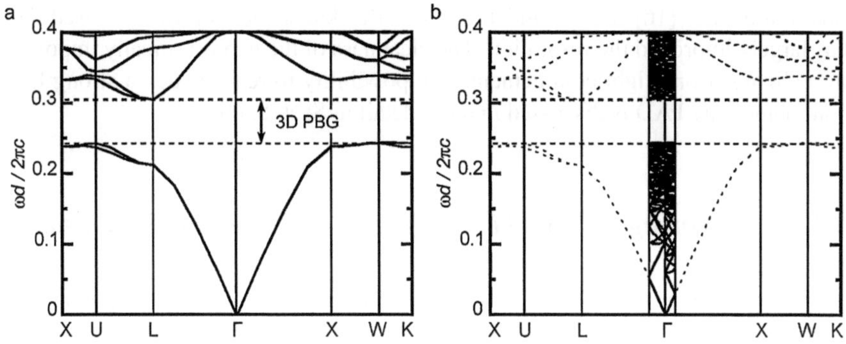

Fig. 8.6 (**a**) Photonic band structure of PCD calculated by a plane-wave expansion method; (**b**) a schematic drawing of the same band structure using the reduced zone

method in Fig. 8.6(a). As described in Sect. 8.3.1, we used a fictitious supercell of $(5a)^3$ in the calculation of the spectrum in Fig. 8.5(b). This is 4×5^3 times larger than the volume of the primitive unit cell of the fcc diamond lattice. This fact indicates that the volume of the first Brillouin zone is reduced by the factor $(4 \times 5^3)^{-1}$. Then, we can redraw the band-structure in the reduced zone, as schematically shown in Fig. 8.6(b). Here, each branch in the band-structure in the original zone is broken up into 4×5^3 pieces, which have fallen into the reduced zone. As a matter of course, this is merely a different drawing of physically the same band structure. The spectrum in Fig. 8.5(b) should consist of the peaks at the eigenfrequencies at $\mathbf{k} = \mathbf{0}$ in the reduced zone in Fig. 8.6(b). We should note here that the eigenfrequencies at $\mathbf{k} = \mathbf{0}$ in the reduced zone correspond to those at 4×5^3 **k**-points in the original zone; the larger the supercell is, the more eigenfrequencies we can pick up. In Fig. 8.5(b), we notice well-isolated peaks in the low frequency region below $\omega d/2\pi c \approx 0.17$. In contrast, above $\omega d/2\pi c \approx 0.17$, the peaks are so dense that the spectrum appears to be a continuous curve with spikes on them. This is because in a higher frequency region more branches fall into the reduced zone. Thanks to such a continuous feature in the spectrum above $\omega d/2\pi c \approx 0.17$, we can identify unambiguously the gap in the range $\omega d/2\pi c = 0.250$–0.308. The gap width to the midgap frequency ratio $(\Delta\omega/\omega_c)$ is 21 %. This frequency range agrees quite well with the gap frequency range shown in the band-structures in Figs. 8.6(a) and 8.6(b), indicating the validity of the present calculations.

The spectrum for PAD in Fig. 8.5(a) shows similar features to that for PCD in Fig. 8.5(b); we have isolated peaks in the frequency range below $\omega d/2\pi c \approx 0.17$ and a continuous curve in the range above $\omega d/2\pi c \approx 0.17$. A clear spectral gap is seen in the range $\omega d/2\pi c = 0.260$–0.292 $(\Delta\omega/\omega_c = 12\ \%)$. The position of the gap is approximately the same as that for PCD, indicating that the supercell periodicity is not relevant to the PBG formation in PAD. The gap depth is also nearly the same as that of PCD, which is 10^{-9}–10^{-10}. This fact indicates that the gap in PAD is as clean as that in PCD, with no trace of localized-state formations in the gap. It should be noted that the selection of the window function used for Fourier transformation is important to attain high resolution. The gap depth calculated by use of a rectangular function in Ref. [16] was around 10^{-3}. The Blackman function that we used here has largely improved the resolution. The cleanness of the gap is critically important in realizing strong light confinement. The possibility to realize such a strong light confinement by PAD is discussed in more detail in Sect. 8.4.3.

8.3.3 Electromagnetic Field Distributions of Photonic Eigenstates

In this subsection, we examine the characteristic features of electromagnetic field distributions of photonic eigenstates, focusing on the light localization and the picture of dielectric and air bands, by which the physical origin of the photonic band-gap formation is often explained in conventional photonic crystals.

To evaluate quantitatively the degree of localization of the photonic eigenstates, we calculated the inverse participation ratio (IPR) defined as

$$\mathrm{IPR} \equiv \frac{\int_{V_{\mathrm{tot}}} |\tilde{\mathbf{E}}(\mathbf{r})|^4 \, d\mathbf{r}}{(\int_{V_{\mathrm{tot}}} |\tilde{\mathbf{E}}(\mathbf{r})|^2 \, d\mathbf{r})^2} \cdot V_{\mathrm{tot}}, \tag{8.7}$$

where $|\tilde{\mathbf{E}}(\mathbf{r})|^2 \equiv |\tilde{\mathbf{E}}_{\mathbf{k}=\mathbf{0}}(\mathbf{r}, \omega)|^2$ is the intensity distribution of the electric field of the eigenstate (see Sect. 8.3.1), and V_{tot} is the total volume of the system. When $|\tilde{\mathbf{E}}(\mathbf{r})|^2$ is concentrated in a portion V_0 of the total volume V_{tot} ($|\tilde{\mathbf{E}}(\mathbf{r})|^2 = const. \neq 0$ for $\mathbf{r} \in V_0$ and $|\tilde{\mathbf{E}}(\mathbf{r})|^2 = 0$ for $\mathbf{r} \notin V_0$), $\mathrm{IPR} = V_{\mathrm{tot}}/V_0$. This indicates that $\mathrm{IPR} = 1$ when the electric-field intensity $|\tilde{\mathbf{E}}(\mathbf{r})|^2$ is constant throughout the total volume V_{tot} and that IPR increases when $|\tilde{\mathbf{E}}(\mathbf{r})|^2$ is concentrated in a small volume. Therefore, the IPR can be employed as a measure of the localization of the eigenstates. The calculated IPR values are plotted against frequency in Figs. 8.7(a) and 8.7(d) for PAD and PCD, respectively. In Fig. 8.7(a), peaks of IPR are apparent at the band edges. The field distributions $|\tilde{\mathbf{E}}(\mathbf{r})|^2$ at the frequencies $\omega d/2\pi c = 0.263$ and 0.287, which correspond to the two peaks in IPR for PAD, are shown in Figs. 8.8(a) and 8.8(b), respectively. These are field distributions on a two-dimensional cross section cut from the 3D structure. The regions surrounded by thick black lines correspond to the dielectric components. We see that the electric-field intensity $|\tilde{\mathbf{E}}(\mathbf{r})|^2$ is concentrated in a small area, thereby exhibiting the feature of localization. It is also apparent that a peak exists at one of the band edges for PCD in Fig. 8.7(d); however, it is very small and no feature of localization was able to be identified in the corresponding field distributions.

In textbooks [36, 37], the physical origin of photonic band gap formation in photonic crystals is often explained in the picture of dielectric and air bands. Here, let us consider a simple one-dimensional photonic crystal consisting of alternate stacking of dielectric and air layers. At the edge of the Brillouin zone with $k = \pi/a$ (a: period), the eigenstates are stationary waves with a wavelength $2a$. We have two ways to position such a wave so as to conform to the symmetry: in one way the maximum points of $|\tilde{\mathbf{E}}(\mathbf{r})|^2$ sit at the dielectric part and in the other they sit at the air part. The eigenfrequency of the former should be lower than that of the latter, leading to the

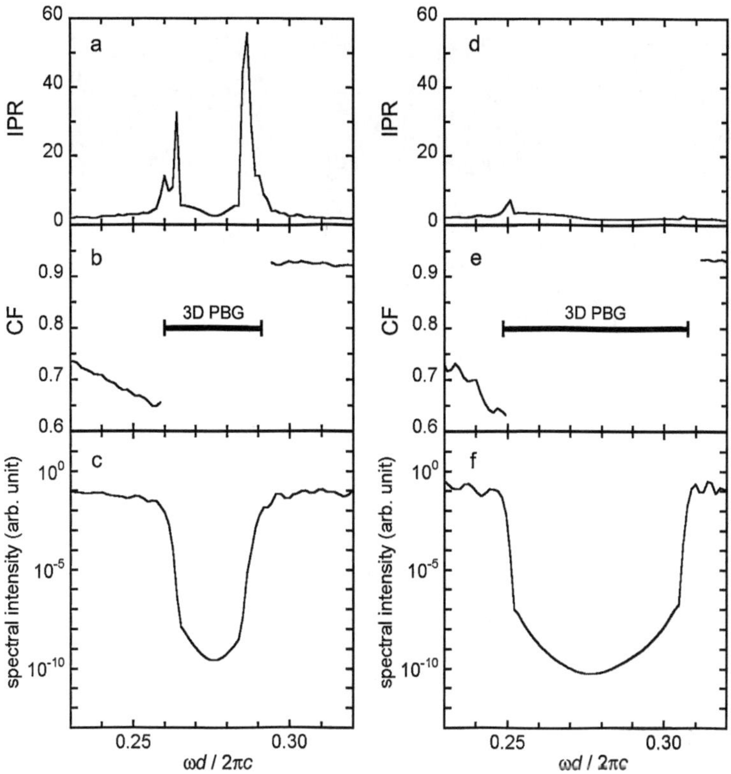

Fig. 8.7 Frequency dependences of inverse participation ratio (**a**), concentration factor (**b**) and spectral intensity (**c**) for PAD, and those (**d–f**) for PCD

formation of a gap. To examine whether this picture can also be applied to PAD, we have evaluated the concentration factor (CF) of the electric field in the air region, which is defined as

$$\mathrm{CF} \equiv \frac{\int_{V_{\mathrm{air}}} |\tilde{\mathbf{E}}(\mathbf{r})|^2 \, d\mathbf{r}}{\int_{V_{\mathrm{tot}}} |\tilde{\mathbf{E}}(\mathbf{r})|^2 \, d\mathbf{r}}, \tag{8.8}$$

for the photonic eigenstates in PAD and PCD. Here, V_{air} and V_{tot} represent the air volume and total volume, respectively. If the electric field is distributed evenly for the dielectric and air regions, CF should equal the volume fraction of the air part, that is, 78 %. The frequency dependences of CF for PAD and PCD are plotted in Figs. 8.7(b) and 8.7(e), respectively. At the lower-band top, CF is approximately 65 % for both PAD and PCD. As the frequency decreases, CF increases and approaches to the volume fraction value of 78 %. On the other hand, at the higher-band bottom, CF is approximately 93 %, much higher than the volume fraction value. Typical field distributions at frequencies near the lower-band top for PAD and PCD are shown in Figs. 8.8(c) and 8.8(d), respectively. The tendency of the

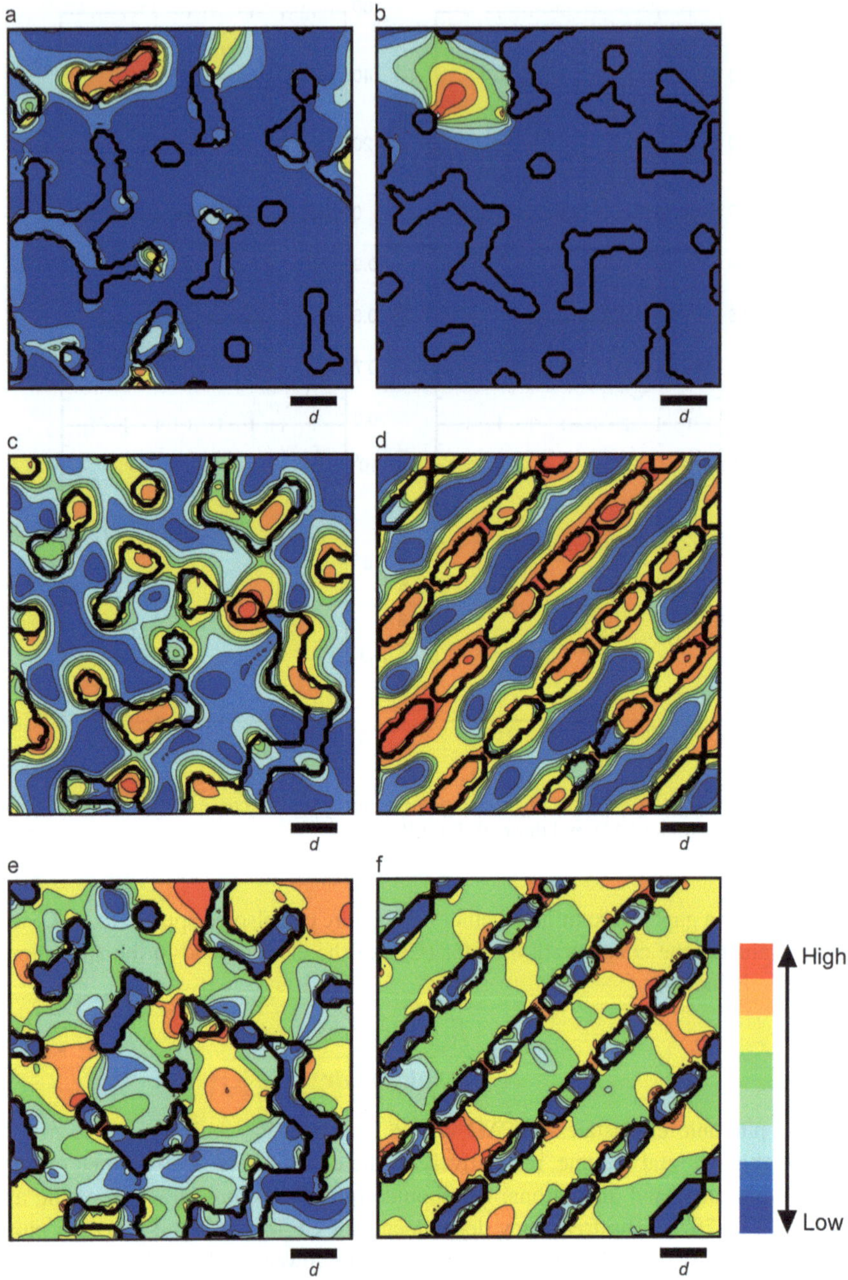

Fig. 8.8 (**a**) and (**b**): The electric field distributions at $\omega d/2\pi c = 0.263$ and 0.287 for PAD, respectively, which correspond to the two peaks of inverse participation ratio shown in Fig. 8.7(a); (**c**) and (**d**): those at 0.251 for PAD, and at 0.240 for PCD, respectively, both of which are near the lower-band top; (**e**) and (**f**): those at 0.301 for PAD and at 0.316 for PCD, respectively, near the higher-band bottom. The *regions surrounded by thick black lines* correspond to the dielectric components

electric fields concentrating on the dielectric regions is clearly seen in these field distributions. On the other hand, typical field distributions at frequencies near the higher-band bottom for PAD and PCD are presented in Figs. 8.8(e) and 8.8(f), respectively. Here, in contrast to the field distributions in Figs. 8.8(c) and 8.8(d), the electric field intensity is obviously low in the dielectric regions. These results clearly show that the picture of the dielectric and air bands can be applied also to the gap formation in PAD, despite the complete absence of lattice periodicity.

In 3D photonic crystals, the structures in which both dielectric and air regions form connected and percolating networks are known to favor the gap formation [12, 15, 34, 38]. Actually, our PAD, as well as PCD, has such a structure. The importance of the connected network structure for gap formation should be related to the vector nature of electromagnetic fields; in order for the field to be confined in dielectric or air region to realize dielectric or air band, each region must be a connected network to accommodate the continuous field lines [36]. Actually, we have confirmed an extreme fragility of bandgap against the segmentation of the dielectric network for PAD and PCD by numerical calculations. When the dielectric network has been segmented, there is no longer continuous pathway to accommodate the whole **E** field lines in the dielectric regions; they are forced to penetrate the air regions. This should raise the eigenfrequency at the lower-band top, leading to the closing of the gap.

On the contrary, we should note that some photonic crystals consisting of isolated dielectric spheres also form a complete 3D PBG. Actually, photonic crystalline diamond structures of isolated (not touching) dielectric spheres (sphere-PCDs) form 3D PBGs [39]. However, in these sphere-PCDs, the gap between the second and third bands (2–3 gap) seen for the network-PCD (see Fig. 8.6) is closed, as long as the refractive index of the spheres is not extremely high. Instead, they form a 8–9 gap. The formation mechanism of the 8–9 gap should be considerably different from that of the 2–3 gap. Recently, we have found a PBG formation in a high frequency region for a sphere-PAD, which should correspond to the 8–9 gap in the sphere-PCD.

8.4 Light Propagation Properties

In this section, we describe light propagation properties of PAD, mainly focusing on the results of microwave transmission measurements [17]. We have observed the frequency range with severely reduced transmission, which corresponds to a PBG. The PBG has been found to be isotropic, regardless of the light polarization direction. In passbands, the PAD has exhibited diffusive light propagation, where the scattering strength increases significantly as the frequency approaches the band edge. In Sect. 8.4.1, we describe the procedures and the results of microwave transmission experiments. In Sect. 8.4.2, we analyze the transmission spectra to examine a diffusive light propagation in passbands. In Sec. 8.4.3, we present FDTD calculations of light evanescence in the band gap. Here, we also demonstrate a strong light confinement in PAD.

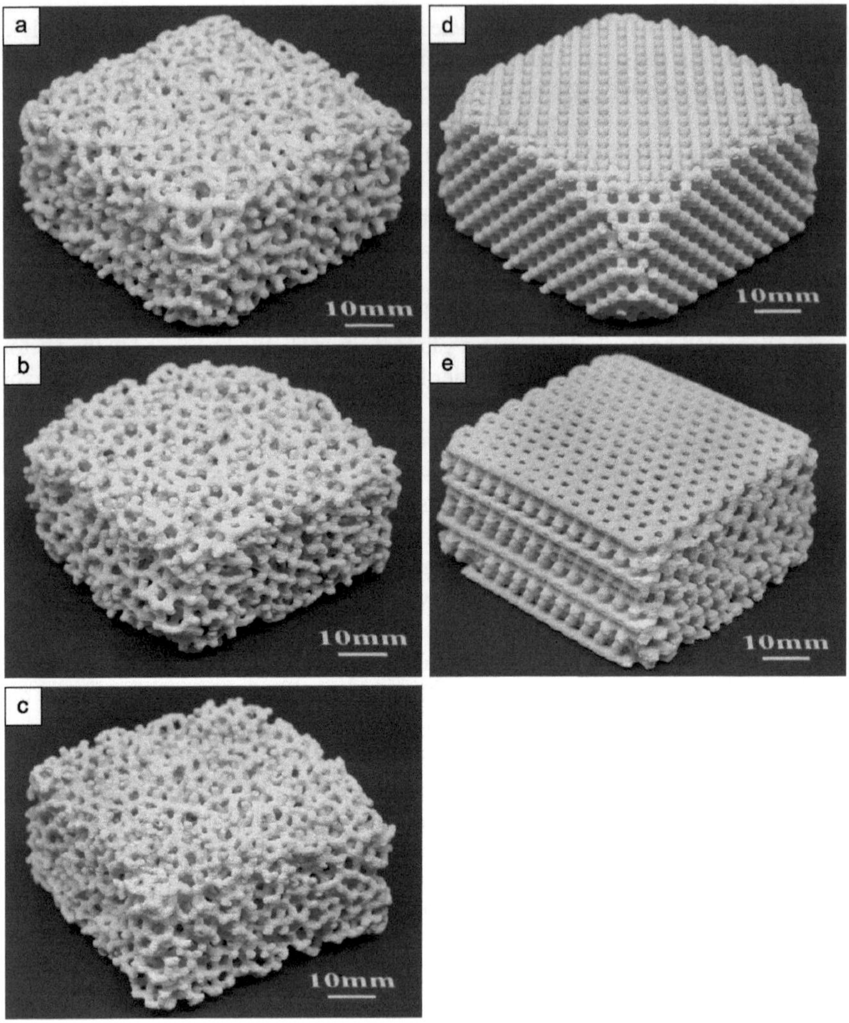

Fig. 8.9 Fabricated PAD structures (**a**–**c**), and PCD structures (**d** and **e**)

8.4.1 Microwave Transmission Experiments

PAD and PCD structures were fabricated by a selective laser sintering (SLS) method [40]. Figures 8.9(a)–8.9(c) show the PAD structures while Figs. 8.9(d) and 8.9(e) present the PCD structures. The rod length and diameter are $d \approx 3$ mm and $r \approx 0.78$ mm ($0.26d$), respectively. The size of the fabricated structures is approximately $x \times y \times z = 70 \times 70 \times 35$ mm^3 ($23.3 \times 23.3 \times 11.7d^3$). Microwave signals were incident on the x–y plane, along the z-axis. The PAD structure is defined as a periodic arrangement of a cubic supercell with the size $(11.5d)^3$ (see Sect. 8.2). The three PAD samples of Figs. 8.9(a)–8.9(c) have different orientations, cut out of the infinite periodic supercell structure. The two PCD structures have the z directions along [100] (d) and [111] (e) of the diamond crystal.

Mixtures of nylon (PA12) powders (typically 50 μm in diameter) and TiO_2 powders (75–100 μm in diameter) were used for fabrication. TiO_2 powders were added to increase the refractive index n of the rods. The rod had a porosity of approximately 40 %. The fabricated structures were soaked in water for 24 h, dried for 2 h, and cooled to 268 K. This resulted in the formation of ice in the pores of the rods, leading to a further increase in the refractive index n. Ice is suitable for increasing n without increasing the extinction coefficient κ. The final n value of the rods was determined experimentally to be 3.0. The volume fraction of the rod in the constructed structures is 22 % and the rest is air. Then, the volume-weighted average $\bar{n}$ of refractive index is approximately 1.4. The values of κ of nylon, TiO_2 and ice are 3×10^{-2}, 2×10^{-3} and 3×10^{-3}, respectively, in the GHz range. This indicates that the κ value of the rods is approximately 2×10^{-2}. Then, the $(1/e)$-absorption length L_a in the structure can be estimated to be 1.7 m for 30 GHz microwaves, which is so long that the effect of absorption is negligible.

For microwave transmission measurements, we employed a free-space method using a vector network analyzer (HP Model 8722D, Agilent Technologies) in the frequency range 18–35 GHz in the configuration schematically shown in Fig. 8.10(a). Conical horns with Teflon microwave lenses were positioned 163 mm in front of and behind the sample. Linearly polarized microwave radiation was emitted from one horn (transmitter), and the parallel- and cross-polarized components of the transmitted field were detected using the other horn (receiver) by changing the relative orientations of the mode selectors attached to the two horns, as schematically shown in Figs. 8.10(b) and 8.10(c). The transmittances T_p and T_c of the parallel- and cross-polarized components, respectively, were obtained by normalizing the measured intensities to the intensity of the parallel component measured before mounting the sample. The background level was evaluated to be 3×10^{-6}. 10–20 measurements were performed by changing the x–y positions of the sample to obtain the averaged spectra. Not only the intensity but also the phase of the transmitted waves, i.e., the complete complex transmission spectra were measured. They were used for the calculation of the response to an input pulse, the results of which are presented in Sect. 8.4.2.

Figures 8.11(a) and 8.11(b) present the spectra of T_p and T_c for PAD, and Figs. 8.11(c) and 8.11(d) present T_p and T_c for PCD, respectively. First, let us examine the results for PCD. The blue and red spectra in (c) and (d) are for the microwave incident directions along [100] and [111], respectively. These spectra exhibit large decreases reaching 10^{-4}–10^{-6} in the transmittance. The dip positions in the spectra for the two incidences are slightly different from each other, indicating an anisotropic gap formation. In Fig. 8.6(a) in Sect. 8.3.2, we have shown the photonic band structure calculated for the PCD by a plane-wave expansion method. The dip positions for the [100] and [111] incidences are expected to correspond to the bandgaps in the Γ–X and Γ–L directions, respectively. The calculated bandgap positions in Fig. 8.6(a) are indicated by blue and red bars in Figs. 8.11(c) and 8.11(d), which indeed agree quite well with the observed dip positions.

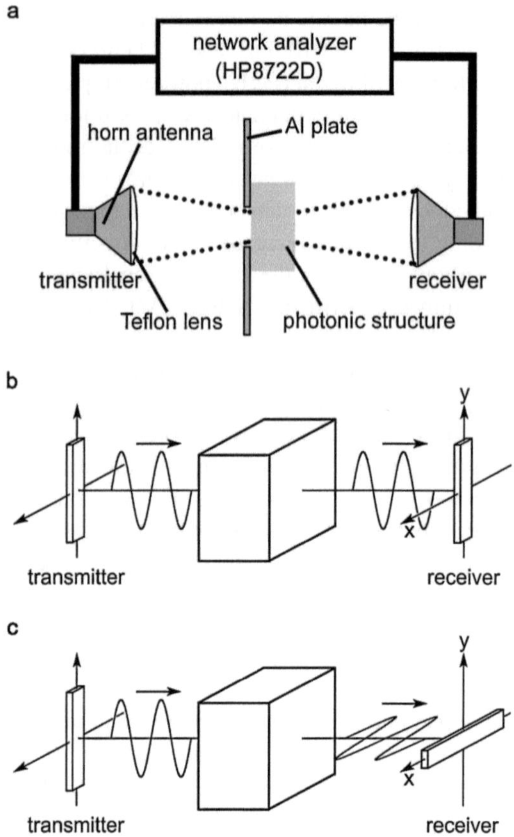

Fig. 8.10 Schematic illustrations of the experimental setting for microwave transmission measurements. (**a**) The illustration of the sample and apparatus configurations; (**b** and **c**) The illustrations of the parallel- and cross-polarized components, respectively, measured in the experiments

In the spectra of T_p and T_c for PAD in Figs. 8.11(a) and 8.11(b), respectively, the six curves are for the three samples with different orientations shown in Figs. 8.9(a)–8.9(c), and for two mutually orthogonal directions of the incident microwave polarizations for each sample. The black bars indicate the frequency range of the 3D PBG deduced by an FDTD calculation (Fig. 8.7(c)). First, in the spectra of T_c in Fig. 8.11(b), all the six spectra exhibit clear dips and the frequency ranges of the dips coincide well. This is in contrast to the case of PCD, in which the two spectra for the different incident directions show the dips in different frequency ranges. In addition, the frequency ranges of the dips in Fig. 8.11(b) agree well with that of the calculated 3D PBG. These facts indicate that the measured dips in T_c in Fig. 8.11(b) represent an isotropic 3D PBG in PAD. On the other hand, we notice a peculiar feature in the spectra of T_p in Fig. 8.11(a): the decrease in T_p is not restricted to the frequency range of the calculated 3D PBG. This feature has been successfully explained by the existence of a 3D PBG and strong light diffusion outside it, as described in detail in the following Sect. 8.4.2. Thus, from this fact, along with the result of T_c, we finally conclude the experimental substantiation of the formation of an isotropic 3D PBG in PAD.

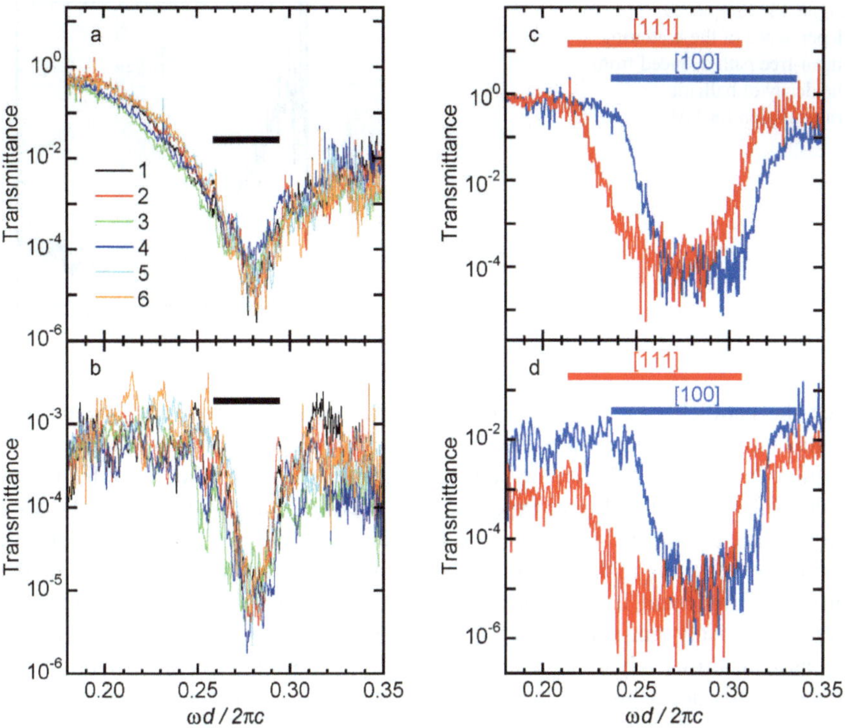

Fig. 8.11 Measured transmission spectra; (**a**) T_p for PAD, (**b**) T_c for PAD, (**c**) T_p for PCD and (**d**) T_c for PCD. The *six curves* in (**a**) and (**b**) are for three samples with different orientations, and for two mutually orthogonal directions of the incident microwave polarizations for each sample. The *blue and red spectra* in (**c**) and (**d**) are for the microwave incident directions along [100] and [111] of the diamond lattice, respectively. The *black bars* in (**a**) and (**b**) indicate the frequency range of the 3D PBG in PAD deduced by an FDTD calculation. The *blue and red bars* in (**c**) and (**d**) indicate the gap positions of PCD in the Γ–X and Γ–L directions, respectively, deduced by a photonic band-structure calculation using a plane-wave expansion method

The isotropic nature of the 3D PBG, regardless of the wavevector orientation and polarization direction, should originate in the structural isotropy, and this is a unique feature of PAD. In general, anisotropy in the bandgap is inevitable for conventional photonic crystals, as demonstrated for PCD in Figs. 8.11(c) and 8.11(d), because their structures are inevitably anisotropic. Such anisotropy restricts the orientations of waveguides and other optic devices in designing an optical circuit in photonic crystals. PAD with a completely isotropic 3D PBG should be free from such restrictions.

8.4.2 Light Diffusion

In our experimental setting in Fig. 8.10(a), the receiver has a relatively small receiving angle: approximately −10 dB attenuation at 15° deviation. This indicates

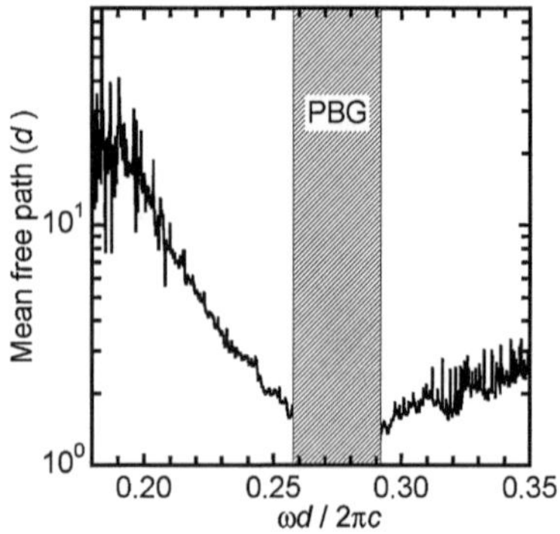

Fig. 8.12 Frequency dependence of the scattering mean-free path deduced from the decay of ballistic transmission for PAD

that the parallel-polarized component T_p reaching the receiver corresponds predominantly to the microwaves traveling directly from the transmitter to the receiver without being scattered; i.e., the ballistic component. On the other hand, the cross-polarized waves should have experienced scatterings that change the polarization direction; the component T_c corresponds to the waves that were multiply scattered, ultimately returning to the direct path between the transmitter and receiver; i.e., the diffusive component. It should be noted that this does not represent the total intensity of the diffusively transported waves; rather it corresponds to a small fraction of them because the diffusively transported waves diverge in various directions. In principle, at the low frequency limits with sufficiently large wavelengths, the wave propagation should be entirely ballistic. In the case of PAD, T_p (i.e., the ballistic component) in Fig. 8.11(a) begins decreasing at $\omega d/2\pi c \approx 0.20$, which is far from the band edge, rapidly declining to 10^{-3} at the band-edge frequency of $\omega d/2\pi c \approx 0.26$. Above the gap, the transmission remains at 10^{-3}–10^{-2}. This is in contrast to PCD, in which the wave propagation is dominantly ballistic within the entire passbands, as evidenced by large T_p values close to unity in Fig. 8.11(c).

The large reduction in the ballistic component of PAD should be due to the scattering that changes the propagation and polarization directions of the microwaves. Here, the scattering mean free path l can be estimated by the relation $T_p = \exp(-L/l)$ (L: sample thickness) [20, 28]. The estimated l is plotted in Fig. 8.12. At the lowest frequency of $\omega d/2\pi c \approx 0.18$, $l \gg d$ holds and the microwave transport is dominantly ballistic. As the frequency increases, l decreases to approach d and the diffusive transport with multiple scattering progressively becomes dominant. At the band edge of $\omega d/2\pi c \approx 0.26$, l reaches $1.7d \approx 5$ mm, and the transport is diffusive with strong scattering. Above the gap, $l \approx 5$–8 mm, indicating that the diffusive transport is dominant. The criterion for the emergence of Anderson localization is given as $2\pi l/\lambda \approx 1$ (Ioffe-Regel (IR) condition), where λ denotes the light wavelength. Here, the quantity $2\pi l/\lambda$ is estimated to be as low as three, where we use $\lambda = 10$ mm at 30 GHz in air for a rough estimation. This is only slightly larger than the threshold value of unity in the IR condition. The quantity $2\pi l/\lambda$ has been evaluated previously for strongly scattering systems close to the Anderson transition in random media comprising GaAs powders [25] ($2\pi l/\lambda = 1.5$), macroporous GaP networks [2.4 (Ref. [41]) and 3.2 (Ref. [27])], Ge

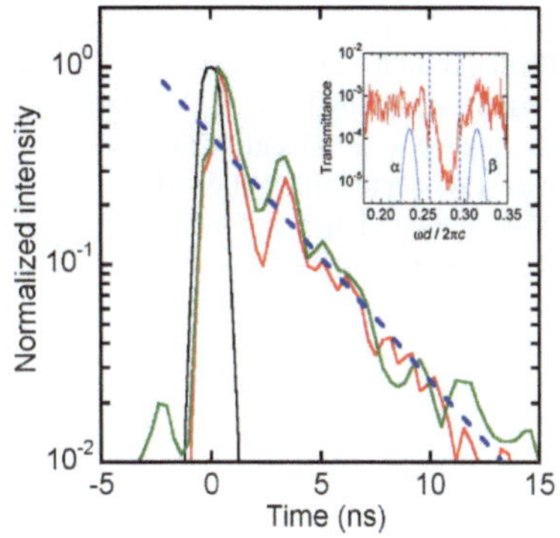

Fig. 8.13 Intensity profiles (*green and red*) of transmitted pulses responding to input Gaussian pulses given by the spectra α and β, respectively, in *inset*. The intensity profile (*black*) of the input pulse is also shown. The intensities are normalized to the peak top values. The exponential tail $I \propto \exp(-t/\tau_0)$ is indicated by a *broken line*

powders [3 (Ref. [28])], and TiO_2 powders [2.5 (Ref. [26]) and 5.6 (Ref. [42])]. The value $2\pi l/\lambda \approx 3$ evaluated for our PAD is comparable to these values.

From the measured complex transmission spectra, we have extracted the response to a pulse input to the PAD. Here, one of the measured spectra of the cross-polarized component for PAD shown in inset in Fig. 8.13 was multiplied with a Gaussian envelope of width $\Delta\omega$ centered at ω_c and then Fourier transformed to give the response to an input Gaussian pulse in the time domain. Note that we used here the complex transmission spectrum, although only its intensity is plotted in the inset. The results with $\Delta\omega d/2\pi c = 0.01$ and $\omega_c d/2\pi c = 0.235$ (α in inset), and with $\Delta\omega d/2\pi c = 0.01$ and $\omega_c d/2\pi c = 0.315$ (β) are shown in Fig. 8.13 by red and green lines, respectively. Here, the input pulse profile is shown by a black line. The transmitted pulses exhibit an exponential tail $I \propto \exp(-t/\tau_0)$ (broken line) with $\tau_0 \approx 1.5$ ns in the delay side, which is typical of diffusive transport; the tail represents the path-length distribution in the diffusive transport. Here, the diffusion constant D can be estimated to be 6.5×10^4 m^2/s using the relation [26]: $1/\tau_0 = \pi^2 D/L^2 + c/(\bar{n}L_a)$. In general, the diffusion constant can be written as $D = v_t l_t/3$ (v_t: transport velocity; l_t: transport mean free path). Using the values of the scattering mean free path l in Fig. 8.12 ($l \approx 2.7d \approx 8$ mm at $\omega_c d/2\pi c = 0.235$ and $l \approx 1.7d \approx 5$ mm at $\omega_c d/2\pi c = 0.315$) for a rough estimation, we obtain $v_t = 0.08$–$0.13c$. This is comparable to the transport velocity v_t previously evaluated for strongly-scattering random media [28, 42].

8.4.3 Light Evanescence and Light Confinement

In this subsection, we first present the results of FDTD calculations for the attenuation behavior of the light within the gap for PAD and PCD. Here, the PAD and PCD structures used for the calculations are the same as those described in Sect. 8.3.1, except for the refractive index of $n = 3.6$ (Si) used in the present calculations. Figure 8.14(a) shows a schematic illustration of the configuration used for the calculations. Here, the $(11.5d)^3$ cell was placed in a rectangular parallelepiped box with

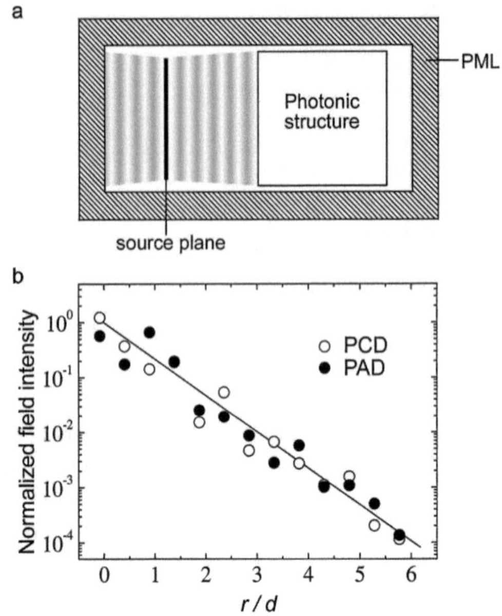

Fig. 8.14 Attenuation behavior of the evanescent wave with the midgap frequency for PAD and PCD. (**a**) A schematic illustration of the configuration used for the calculations; (**b**) The dependences of the incident electromagnetic wave intensity on the distance from the surface

absorbing boundaries of perfectly matched layers (PMLs). A plane wave of a Gaussian pulse was generated from the "source plane" located outside of the cell and time evolutions of the fields were monitored at various points in the cell, from which we evaluated the spectral intensity at the gap center frequency. Here, we averaged the intensities over several sampling points near the center of the cell for a given distance from the incident surface. Figure 8.14(b) shows the dependence of the field intensity on the distance from the incident surface. Both data show exponential decay, i.e., $I = I_0 \exp(-r/r_c)$ (r_c: the attenuation length) with nearly the same r_c ($\approx 0.6d$). This fact suggests that strong light confinement is realizable in PAD as well as conventional photonic crystals such as PCD. To demonstrate this, we have calculated the mode volume (V_m) and the quality (Q) factor of cavities introduced in PAD and have compared them with those of a cavity in PCD [18].

In general, cavities introduced in conventional photonic crystals with a 3D PBG disrupt the lattice periodicity, and form localized cavity modes within the PBG. Though the PAD lacks lattice periodicity, its structure has a well-defined local order of tetrahedral configuration, as described in Sect. 8.2. Then, we would expect that cavities disrupting such an order form localized modes as well. Here, we introduced a cavity in PAD and PCD by removing a rod from the structures. The spherical region with radius R containing the cavity at the center was cut out from the structure. The sphere was placed in a cubic cell with the edge length $2R$. The region outside the sphere and inside the cubic cell was filled with the dielectrics with the volume-weighted average $\bar{n}$ of refractive index in the structure, which is $\bar{n} \approx 1.6$. Then, FDTD calculations were performed with PML absorbing boundary conditions applied to the walls of the cubic cell, where the cavity modes were exited by a Gaussian pulse with frequency range covering the mode frequencies of interest.

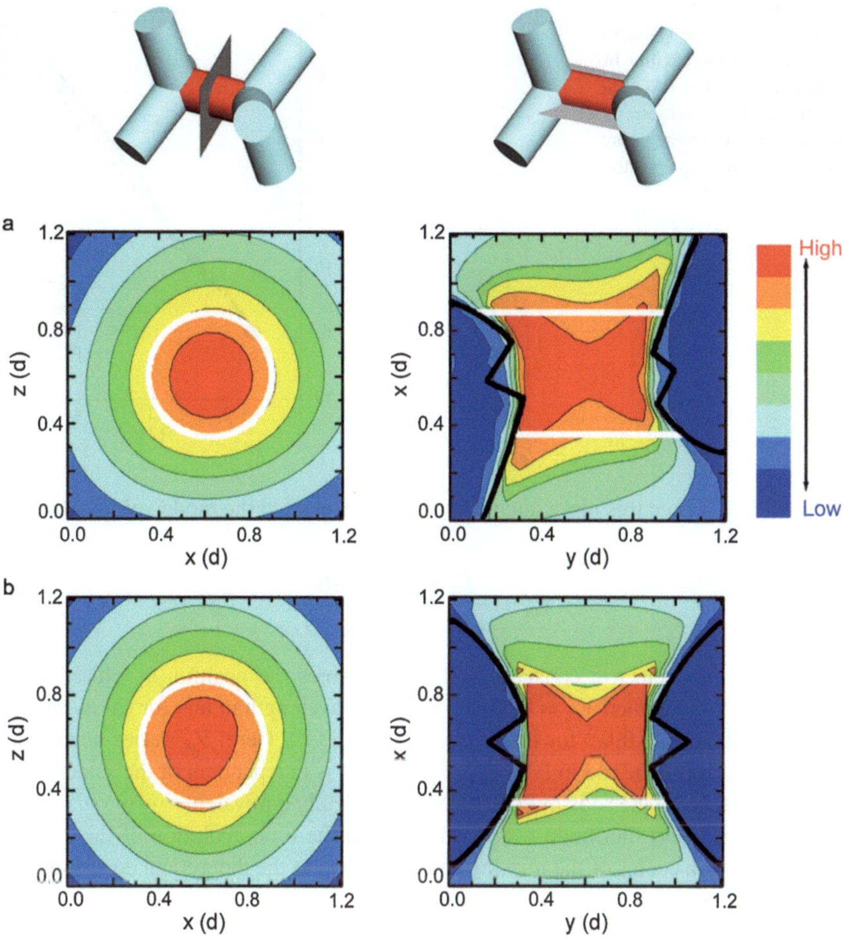

Fig. 8.15 Electric field distributions of the cavity mode formed in PAD (**a**) and PCD (**b**). The *left and right profiles* show the field distributions on the two-mutually orthogonal planes, which are illustrated above the profiles. Here, the missing rods are drawn in *red*. In the profiles, the position of the removed rod is indicated by *white circles* or *white lines*. *Black lines* indicate the boundaries of the dielectric parts

An example of the electric field distribution of the cavity mode formed in PAD is shown in Fig. 8.15(a), together with that for PCD in Fig. 8.15(b). The left and right profiles show the field distributions on the two-mutually orthogonal planes (see the schematic illustrations above the profiles), where the position of the removed rod is indicated by white circles or white lines. Black lines indicate the boundaries of dielectric parts. The direction of the electric field vector was found to be approximately parallel to the rod axis of the missing rod. The cavity modes are monopole and overall features of the field distributions of them are very similar for PAD and PCD. We can see that the cavity modes are well confined at and around the missing

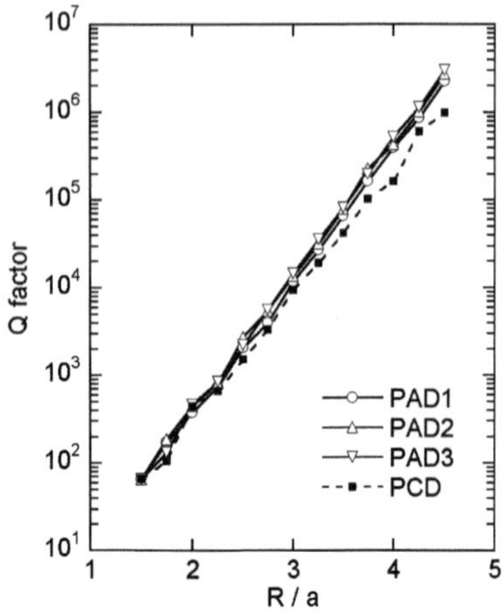

Fig. 8.16 Structure-size dependences of the Q-factors calculated for the cavities in PAD and PCD. For PAD, the results for three cavities whose mode-frequencies are close to the midgap are shown

rod position. To see this more quantitatively, we evaluated the mode volume defined as

$$V_m = \frac{\int \varepsilon(\mathbf{r})|\tilde{\mathbf{E}}(\mathbf{r})|^2\, d\mathbf{r}}{\varepsilon(\mathbf{r}_{\max})|\tilde{\mathbf{E}}(\mathbf{r}_{\max})|^2}, \tag{8.9}$$

where ε is the dielectric constant, and $\mathbf{r}_{\max}$ denotes the position at which $|\tilde{\mathbf{E}}|$ is maximum. We obtained $V_m = 0.09$–$0.11\lambda^3$ (λ: wavelength in air) for cavity modes in PAD and $V_m = 0.09\lambda^3$ for the cavity mode in PCD. Here, V_m varies depending on the selection of the rod in PAD while in PCD it is independent of the selection of the rod because all the rods in PCD are symmetrically equivalent.

The Q-factor of the mode was deduced from the time evolution of the electric field $\mathbf{E}(t)$ at the center of the missing rod according to the theoretical equation:

$$\mathbf{E}(t) = \mathbf{E}_0 \exp\big[-i\omega_0\big\{1 - (1/2Q)i\big\}t\big], \tag{8.10}$$

where $\mathbf{E}_0$ and ω_0 denote the initial field and the resonant frequency, respectively. Figure 8.16 shows the structure size dependences of the Q-factors of the cavity modes. For PAD, the results for three cavities whose mode-frequencies are close to the center of the gap are shown. All the data show exponential increase of Q with the structure size, i.e., $Q = A \cdot \exp(B \cdot R)$ without any sign of saturation. Here, the values of A and B are approximately the same for PAD and PCD, and $Q \geq 10^6$ at $R = 4.5a$. These facts lead to the conclusion that PAD can confine light as strongly as PCD.

8.5 Concluding Remarks

We have shown that the PAD forms a complete 3D PBG with no trace of localized states within it, and that strong light confinement is realizable in PAD as well as conventional photonic crystals. The 3D PBG is isotropic, regardless of the light polarization direction, which, in principle, cannot be realized in conventional photonic crystals. In passbands, the PAD exhibits diffusive light propagation, where the scattering strength increases significantly as the frequency approaches the band edge. Localized photonic states are formed in the vicinity of the band edges.

There are two typical models used for explaining the origin of the electronic band gaps in solids: "nearly-free electron model" and "tight-binding model." In the former, we start from the free electron states, and add a periodic potential as a perturbation, resulting in gap formations along the Bragg scattering planes in the **k**-space. A similar mechanism can be applied to PBG formations in photonic systems. Though this has been considered to be the main origin of PBG formations in photonic crystals, it cannot be applied to the PBG formation in the PAD because of the absence of lattice periodicity. On the other hand, in the latter, i.e., in the tight-binding model, we start from isolated localized electronic states. Then, we assume additionally couplings between adjacent states, leading to the formation of electronic bands. If we start from two localized states with different energies, we obtain two bands, and under certain conditions an electronic band gap remains between the two bands. This model requires no long-range structural order. In the photonic systems, the Mie resonances of isolated dielectric elements may act as the localized states. Actually, the origin of the 2D TM gaps in photonic amorphous structures consisting of dielectric cylinders can be explained reasonably in terms of the tight-binding-like model based on the Mie resonances [5–11]. In contrast, it appears to be difficult to apply this model to the PBG formation in the PAD that consists of a continuous network structure.

In Sect. 8.3.3, we have shown that the picture of dielectric and air bands can be applied to the PBG formation in PAD as well. This picture originally comes from the nearly-free-electron-type model, and therefore the lattice periodicity is assumed here. The results in Sect. 8.3.3 indicate that in some cases this picture holds even for amorphous systems. As discussed in Sect. 8.3.3, network structures should be favorable in this picture because they can feasibly accommodate the continuous field lines. With regard to the role of fourfold connections of the network, it is not well understood. In any case, the physical origin of the PBG formation in PAD remains to be clarified in future.

Another important subject to be tackled in future is the fabrication of PAD in an optical-wavelength size. Recently, the microfabrication techniques have been advancing drastically. In particular, the technique of direct laser writing has become widely used for this purpose [43]. This technique allows for the fabrication of almost arbitrary 3D structures including extremely complex structures like the PAD. We may also be able to utilize some self-organized structures. For example, 3D random network structures are known to form by viscoelastic phase separation in soft matters such as polymer solutions, colloidal suspensions, protein solutions, etc. [44] Some of those random networks have been shown to possess local tetrahedral-bonding configurations like the PAD [45].

The nature may have already fabricated the PAD structure. Zi et al. have found a random network structure in parrot feather barbs [46]. They have shown that the network structure consists mostly of fourfold connections, and therefore it should be close to our PAD structure. The blue color of the feather has been shown to originate in the PAD-like structure. A PAD-like network structure is also seen in kingfisher feather barbs, giving its blue color [47]. Such colorations by disordered structures are characterized by a noniridescent nature, and have recently attracted much attention; they are seen in biological systems [14, 46–52] and also in artificial systems [53–55]. The studies to clarify the coloration mechanisms are now in progress.

Acknowledgements This work was done in collaboration with S. Imagawa (IIS, Univ. Tokyo), K. Morita (IIS, Univ. Tokyo), T. Niino (IIS, Univ. Tokyo), Y. Kagawa (RCAST, Univ. Tokyo), and M. Notomi (NTT). We thank Y. Kamimura (IIS, Univ. Tokyo) for his help in manuscript preparation. We acknowledge the financial supports by a Grant-in-Aid for Scientific Research (B) (21360027) and a Grant-in-Aid for Scientific Research on Priority Areas "Nano Materials Science for Atomic Scale Modification 474" from the Ministry of Education, Culture, Sports, Science and Technology (MEXT). This work was also supported by the Mitsubishi Foundation and the Sumitomo Foundation.

References

1. E. Yablonovitch, Phys. Rev. Lett. **58**, 2059 (1987)
2. S. John, Phys. Rev. Lett. **58**, 2486 (1987)
3. M. Notomi, Rep. Prog. Phys. **73**, 096501 (2010)
4. C. Jin, X. Meng, B. Cheng, Z. Li, D. Zhang, Phys. Rev. B **63**, 195107 (2001)
5. H. Miyazaki, M. Hase, H.T. Miyazaki, Y. Kurokawa, N. Shinya, Phys. Rev. B **67**, 235109 (2003)
6. C. Rockstuhl, U. Peschel, F. Lederer, Opt. Lett. **31**, 1741 (2006)
7. M. Florescu, S. Torquato, P.J. Steinhardt, Proc. Natl. Acad. Sci. USA **106**, 20658 (2009)
8. E. Lidorikis, M.M. Sigalas, E.N. Economou, C.M. Sokoulis, Phys. Rev. Lett. **81**, 1405 (1998)
9. E. Lidorikis, M.M. Sigalas, E.N. Economou, C.M. Soukoulis, Phys. Rev. B **61**, 13458 (2000)
10. L. Shi, X. Jiang, C. Li, J. Phys. Condens. Matter **19**, 176214 (2007)
11. L. Jia, I. Bita, E.L. Thomas, Phys. Rev. Lett. **107**, 193901 (2011)
12. M. Maldovan, E.L. Thomas, Nat. Mater. **3**, 593 (2004)
13. C. Rockstuhl, F. Lederer, Phys. Rev. B **79**, 132202 (2009)
14. B.Q. Dong, X.H. Liu, T.R. Zhan, L.P. Jiang, H.W. Yin, F. Liu, J. Zi, Opt. Express **18**, 14430 (2010)
15. S.F. Liew, J.K. Yang, H. Noh, C.F. Schreck, E.R. Dufresne, C.S. O'Hern, H. Cao, Phys. Rev. A **84**, 063818 (2011)
16. K. Edagawa, S. Kanoko, M. Notomi, Phys. Rev. Lett. **100**, 013901 (2008)
17. S. Imagawa, K. Edagawa, K. Morita, T. Niino, Y. Kagawa, M. Notomi, Phys. Rev. B **82**, 115116 (2010)
18. S. Imagawa, K. Edagawa, M. Notomi, Appl. Phys. Lett. **100**, 151103 (2012)
19. P. Sheng, *Introduction to Wave Scattering, Localization, and Mesoscopic Phenomena* (Academic Press, San Diego, 1995)
20. A. Ishimaru, *Wave Propagation and Scattering in Random Media* (IEEE Press, New York, 1997)
21. P.W. Anderson, Philos. Mag., B **52**, 505 (1985)
22. K. Arya, Z.B. Su, J.L. Birman, Phys. Rev. Lett. **57**, 2725 (1986)
23. C.M. Soukoulis, E.N. Economou, G.S. Grest, M.H. Cohen, Phys. Rev. Lett. **62**, 575 (1989)
24. N. Garcia, A.Z. Genack, Phys. Rev. Lett. **66**, 1850 (1991)

25. D.S. Wiersma, P. Bartolini, A. Lagendijk, R. Righini, Nature **390**, 671 (1997)
26. M. Storzer, P. Gross, C.M. Aegerter, G. Maret, Phys. Rev. Lett. **96**, 063904 (2006)
27. F.J.P. Schuurmans, M. Megens, D. Vanmaekelbergh, A. Lagendijk, Phys. Rev. Lett. **83**, 2183 (1999)
28. J.G. Rivas, R. Sprik, A. Lagendijk, L.D. Noordam, C.W. Rella, Phys. Rev. E **63**, 046613 (2001)
29. F. Scheffold, R. Lenke, R. Tweer, G. Maret, Nature **398**, 206 (1999)
30. D.E. Polk, J. Non-Cryst. Solids **5**, 365 (1971)
31. F. Wooten, K. Winer, D. Weaire, Phys. Rev. Lett. **54**, 1392 (1985)
32. G.T. Barkema, N. Mousseau, Phys. Rev. B **62**, 4985 (2000)
33. P.N. Keating, Phys. Rev. **145**, 637 (1966)
34. C.T. Chan, S. Datta, K.M. Ho, C.M. Soukoulis, Phys. Rev. B **50**, 1988 (1994)
35. C.T. Chan, Q.L. Yu, K.M. Ho, Phys. Rev. B **51**, 16635 (1995)
36. J.D. Joannopoulos, S.G. Johnson, J.N. Winn, R.D. Meade, *Photonic Crystals—Molding the Flow of Light* (Princeton University Press, Princeton, 2008)
37. K. Inoue, K. Ohtaka, *Photonic Crystals—Physics, Fabrication and Applications* (Springer, Berlin, 2004)
38. E.N. Economou, M.M. Sigalas, Phys. Rev. B **48**, 13434 (1993)
39. A. Moroz, Phys. Rev. B **66**, 115109 (2002)
40. S. Kumar, J. Miner. Met. Mater. **55**, 43 (2003)
41. P.M. Johnson, A. Imhof, B.P.J. Bret, J.G. Rivas, A. Lagendijk, Phys. Rev. E **68**, 016604 (2003)
42. M.P. van Albada, B.A. van Tiggelen, A. Lagendijk, A. Tip, Phys. Rev. Lett. **66**, 3132 (1991)
43. G. von Freymann, A. Ledermann, M. Thiel, I. Staude, S. Essig, K. Busch, M. Wegener, Adv. Funct. Mater. **20**, 1038 (2010)
44. H. Tanaka, J. Phys. Condens. Matter **12**, R207 (2000)
45. H. Tanaka, Y. Nishikawa, Phys. Rev. Lett. **95**, 078103 (2005)
46. J. Zi, H. Yin, B. Dong, X. Liu, L. Shi, E. Yablonovitch, in *Abstracts Book*, PECS-IX (2010)
47. S. Kinoshita, S. Yoshioka, J. Miyazaki, Rep. Prog. Phys. **71**, 076401 (2008)
48. R.O. Prum, R.H. Torres, S. Williamson, J. Dyck, Nature **396**, 29 (1998)
49. E.R. Dufresne, H. Noh, V. Saranathan, S.G.J. Mochrie, H. Cao, R.O. Prum, Soft Matter **5**, 1792 (2009)
50. L.P. Biro, Mater. Sci. Eng. B **169**, 3 (2010)
51. H. Noh, S.F. Liew, V. Saranathan, S.G.J. Mochrie, R.O. Prum, E.R. Dufresne, H. Cao, Adv. Mater. **22**, 2871 (2010)
52. S.F. Liew, J. Forster, H. Noh, C.F. Schreck, V. Saranathan, X. Lu, L. Yang, R.O. Prum, C.S. O'Hern, E.R. Dufresne, H. Cao, Opt. Express **19**, 8208 (2011)
53. Y. Takeoka, M. Honda, T. Seki, M. Ishii, H. Nakamura, Appl. Mater. Interfaces **1**, 982 (2009)
54. I. Lee, D. Kim, J. Kal, H. Baek, D. Kwak, D. Go, E. Kim, C. Kang, J. Chung, Y. Jang, S. Ji, J. Joo, Y. Kang, Adv. Mater. **22**, 4973 (2010)
55. N. Kumano, T. Seki, M. Ishii, H. Nakamura, Y. Takeoka, Angew. Chem., Int. Ed. **50**, 4012 (2011)

Chapter 9
Lasing in Amorphous Nanophotonic Structures

Hui Cao and Heeso Noh

Abstract We review the recent experimental and numerical studies on lasing in photonic nanostructures with short-range order in this chapter. Despite the lack of long-range order, photonic bandgaps can be formed in such structures, and they are isotropic. Our numerical studies show that the photonic bandgaps depends not only on the spatial range of geometric order, but also on the structural topology. The photonic bandedge modes may be spatially localized, in contrast to those of photonic crystals. Lasing has been realized experimentally in semiconductor nanostructures with short-range order. The nature of lasing modes are illustrated, and the lasing characteristic can be controlled by the short-range order.

9.1 Introduction

Over the past decade lasing has been realized in various types of nanostructures. The most common one is the photonic crystal (PhC) with periodic modulation of the refractive index that result in the formation of photonic bandgap (PBG) [1]. The distributed feedback PhC laser relies on the slow group velocity at a photonic bandedge to enhance light amplification [2]. The PhC defect mode laser utilizes light localization at a structural defect to minimize the lasing threshold [3, 4]. In addition to PhCs, lasing action in photonic quasi-crystals with and without defects has been reported [5–9]. Despite lack of periodicity, the quasi-crystalline photonic structure has long-range orientational order and can possess a PBG.

The photonic amorphous structure (PAS) has neither long-range translational order nor orientational order. However, the density of optical states (DOS) may be depleted through coupled Mie resonances [10–15] or short-range structural order [16, 17]. PBGs also exist in amorphous photonic structures that consist of

H. Cao (✉)
Department of Applied Physics, Yale University, New Haven, CT 06520, USA
e-mail: hui.cao@yale.edu

H. Noh
Department of Nano and Electronic Physics, Kookmin University, Seoul, Korea
e-mail: heesonoh@kookmin.ac.kr

C. Rockstuhl, T. Scharf (eds.), *Amorphous Nanophotonics*,
Nano-Optics and Nanophotonics, DOI 10.1007/978-3-642-32475-8_9,

© Springer-Verlag Berlin Heidelberg 2013

strong Mie scatterers such as dielectric rods or spheres [10–14]. These structures are termed photonic glasses [15], in analogy to glassy silica, which has an electronic band gap spanning the entire visible frequency range. In the absence of Mie resonance, Bragg scattering of propagating waves by local domains can produce PBGs in structures with only short-range positional order. For example, complete PBGs exist in photonic amorphous diamond structures—three-dimensional (3D) continuous random networks with diamond-like tetrahedral-bonding between particles [16]. Recently hyper-uniform disordered materials with short-range geometric order and uniform local topology have been shown to posses large PBGs [17]. Unique optical features of amorphous media have also been investigated experimentally [18].

Despite these studies, little is known about the transition from PhCs to amorphous optical materials, e.g., how does the density of optical states (DOS) evolve as the structural properties of the material change from ordered to amorphous? Is there a critical size of ordered domains in polycrystalline materials below which the system becomes optically amorphous? Our recent work aimed to answer the above questions by mapping out the transition from photonic polycrystalline to amorphous structures [19]. Moreover, we have performed systematic study on lasing characteristic in polycrystalline and amorphous structures [20, 21]. It reveals what determines the lasing frequencies as well as the nature of lasing modes. These studies provide a physical insight to lasing mechanism in photonic structures lacking long-range order.

In fact, nature utilizes both crystalline and amorphous photonic structures for color generation [22–25]. Periodic structures are intrinsically anisotropic, thus the colors they produce are iridescent (i.e., change with viewing angle). In photonic polycrystals, the cumulative effect of a large number of randomly orientated crystallites makes the color non-iridescent [26]. Photonic amorphous media can also produce vivid non-iridescent colors via short-range structural order [27]. Although the refractive index contrast is usually too low to form PBGs in most biological systems, the interference of scattered light selects the color whose wavelength corresponds to the structural correlation length [28]. Therefore, short-range positional order can significantly modify photonic properties [29–31], leading to unique applications [32]. Lately we have utilized the short-range order to enhance optical confinement and improve the lasing efficiency in PAS.

9.2 Photonic Polycrystals and Amorphous Structures

9.2.1 Photonic Bandgap Effects and Enhanced Light Confinement

We start with the numerical study on PBGs and high-quality (Q) modes in two-dimensional (2D) photonic structures with short-range positional order.

Structure Generation and Characterization

To create 2D photonic polycrystals and amorphous structures, we have developed numerical simulation methods to generate N cylinders of circular shape in a square box with periodic boundaries. For the purpose of generating configurations with varying positional order, we assume that the cylinders interact elastically via the purely repulsive short-range linear spring potential

$$V(r_{ij}) = \frac{b}{2}\left(1 - \frac{r_{ij}}{d_{ij}}\right)^2 \theta\left(1 - \frac{r_{ij}}{d_{ij}}\right), \tag{9.1}$$

where r_{ij} is the center-to-center distance between cylinders i and j, b is the characteristic energy scale of the interaction, $\theta(x)$ is the Heaviside function, and $d_{ij} = (d_i + d_j)/2$ is the average diameter of cylinders i and j. To vary the degree of positional order, the cylinders are polydisperse—with a uniform distribution of diameters between d_0 and $d_0(1+p)$, where p is the polydispersity that ranges from 0 to 0.5. The average diameter $\langle d \rangle = d_0(1 + p/2)$.

Initially d_0, or the packing fraction

$$\phi = \left(\frac{1}{L}\right)^2 \sum_{i=1}^{N} \pi \left(\frac{d_i}{2}\right)^2, \tag{9.2}$$

is set to a small value $\phi_0 = 0.2$, and we place N cylinders randomly within a square of side length L. We then gradually increase the diameters of all cylinders while maintaining the relative size distribution to create a jammed packing of cylinders [33]. Each increment in diameter is followed by minimization of the total potential energy $V = \sum_{i>j} V(r_{ij})$ of the system. The energy minimization process is similar to moving each cylinder along the direction of the total force on it using overdamped dynamics. When V drops below a threshold value or the difference in energy between successive minimization steps is less than a small tolerance, the minimization process is terminated. If V is zero and gaps exist between cylinders, the system is unjammed, and it is compressed with a further increase of d_0. If $V > 0$ after the energy minimization process, a large system-spanning number of cylinders are overlapped. To eliminate overlap, the system is decompressed, i.e., d_0 is uniformly decreased for all cylinders. The energy minimization process is repeated after the decompression step to find the local potential energy minimum. If $V = 0$, the system is compressed; if not, the system is decompressed again. The increment by which the packing fraction of the cylinders is changed at each compression or decompression step is gradually reduced to zero. Eventually when all of the cylinders are just touching and the net force on each cylinder is nearly zero, the system is considered "jammed", and the process to increase the packing fraction is stopped.

For each polydispersity p, we generated at least 100 static, jammed packings of cylinders from random initial configurations. The values of ϕ are typically in the range between 0.82 and 0.85 with varying degrees of positional order. After generating jammed packings, we reduce the diameters of all cylinders to the same value D (with $\phi = 0.5$) to eliminate the size polydispersity. Thus, in the final configurations, the *structural disorder* exists only in the positions of the cylinders with order decreasing monotonically with increasing p. Figure 9.1 (a) shows the typical configurations of $N = 1024$ cylinders generated with $p = 0.1$ (left), 0.3 (middle),

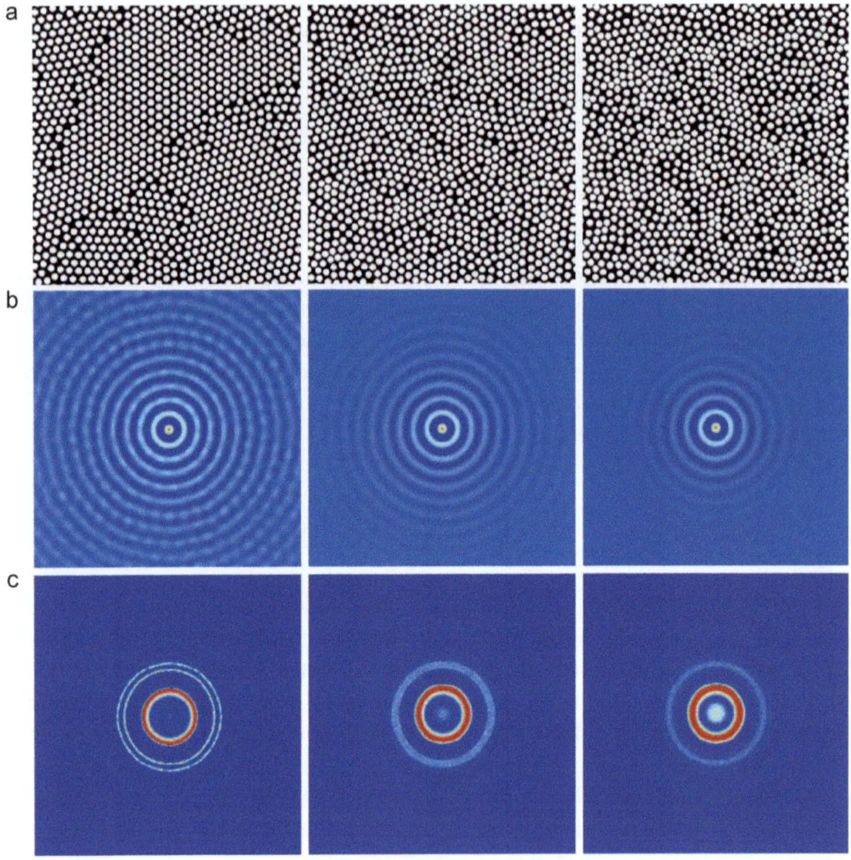

Fig. 9.1 (**a**) Typical configurations of two-dimensional arrays of air cylinders (*white*) in a dielectric host (*black*), generated using Protocol 9.1 described in Sect. 9.2.1 with polydispersity $p = 0.1$ (*left*), 0.3 (*middle*), and 0.5 (*right*). (**b**) Contour plot of the ensemble-averaged density spatial autocorrelation function $C(\Delta \boldsymbol{r})$ and (**c**) Power spectra $|f(\boldsymbol{q})|^2$ from Fourier transformed density for the same polydispersities in (**a**). The *scale bars* in (**b**) and (**c**) are $5D$ and $1/D$, where D is the diameter of cylinders in (**a**). Reprinted with permission from [19], Copyright 2008, Wiley-VCH

and 0.5 (right). For $p = 0.1$, the system contains several domains of cylinders with crystalline order, but each possesses a different orientation. With increasing p, the domains have reduced positional order and decrease in size.

To quantify the structural order, we calculate the ensemble-averaged spatial correlation function of the density, the Fourier transform of the density, the radial distribution function $g(r)$, and the local and global bond orientational order parameters. The spatial autocorrelation function of density $\rho(\mathbf{r}) = L^{-2} \sum_{i=1}^{N} \theta(\mathbf{r} - \mathbf{r}_i)$ is given by

$$C(\Delta \mathbf{r}) = \frac{\langle \rho(\mathbf{r}) \rho(\mathbf{r} + \Delta \mathbf{r}) \rangle - \langle \rho(\mathbf{r}) \rangle^2}{\langle \rho(\mathbf{r}) \rangle^2}. \tag{9.3}$$

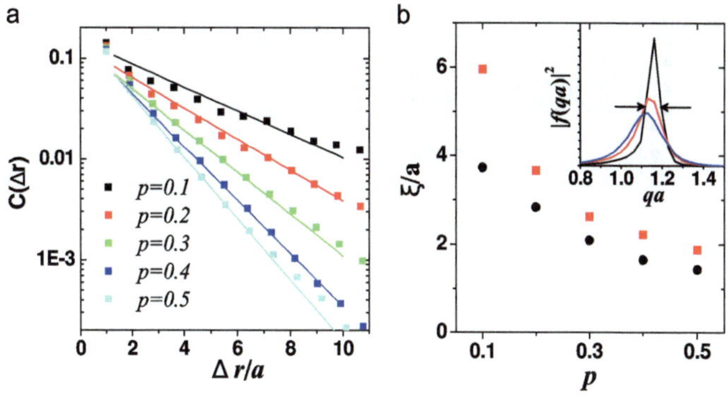

Fig. 9.2 (**a**) Logarithmic plot of the peak amplitudes for the ensemble- and angle-averaged density spatial autocorrelation function $C(\Delta\mathbf{r})$ for $p = 0.1$, 0.2, 0.3, 0.4, and 0.5. $a = L/N^{1/2}$ is the average distance between adjacent cylinders. The exponential fits of the data (*solid lines*) give the decay length ξ_r. (**b**) *Inset*: The first peaks of the angle- and ensemble-averaged Fourier transformed density $|f(qa)|^2$ for $p = 0.1$ (*solid line*), 0.2 (*dashed line*), and 0.4 (*short-dashed line*), whose width gives the average domain size ξ_q. *Main panel*: ξ_r (*circles*) and ξ_q (*squares*) versus p. Reprinted with permission from [19], Copyright 2008, Wiley-VCH

$C(\Delta\mathbf{r})$ is averaged first over the spatial coordinates of the cylinders $\mathbf{r}$ within one configuration, and then over at least 100 independent configurations. A contour plot of $C(\Delta\mathbf{r})$ is displayed in Fig. 9.1 (b) as a function of increasing p (from left to right) used to generate the configurations. For $p = 0.1$, $C(\Delta\mathbf{r})$ displays a large number of concentric rings and a modulation of the amplitude within a given ring, which indicates strong positional order. As p increases the system becomes more disordered and isotropic, since the number of visible concentric rings decreases and the amplitude within a given ring becomes more uniform. After integrating $C(\Delta\mathbf{r})$ over the polar angle, we plot in Fig. 9.2 (a) the peak amplitudes of the rings as a function of $\Delta r/a$, where $a = L/N^{1/2}$ is the average center-to-center distance between neighboring cylinders. The peak amplitudes decay more rapidly with Δr at larger p. The decay is approximately exponential, if we exclude the first peak near $\Delta r = a$. The faster decay from the first peak to the second arises from correlations induced by the just-touching jammed cylinders. The decay length ξ_r is extracted from the exponential fit $\exp[-\Delta r/\xi_r]$ of peak amplitudes after excluding the first peak. As shown in Fig. 9.2 (b), ξ_r is smaller for larger p, indicating the range of spatial order becomes shorter.

We also calculated the spatial Fourier transform of the structures, $f(\mathbf{q}) = \int d^2\mathbf{r} \exp[-i\mathbf{q}\cdot\mathbf{r}]\rho(\mathbf{r})$, where $\mathbf{q}$ is the wavevector. Figure 9.1 (c) displays the ensemble-averaged power spectra $|f(\mathbf{q})|^2$ for $p = 0.1$, 0.3 and 0.5, which consist of concentric rings. The radial width of the rings increases with p, as can be seen clearly for the first ring (with the smallest radius). The second and third rings are distinct for $p = 0.1$, which indicates the six-fold symmetry of the cylinders within each domain. For $p = 0.3$ and 0.5, these rings become wider and merge together. We integrate $|f(\mathbf{q})|^2$ over all directions of $\mathbf{q}$ to obtain the intensity as a function of the amplitude q. The inset of Fig. 9.2 (b) displays the intensity of the first ring versus

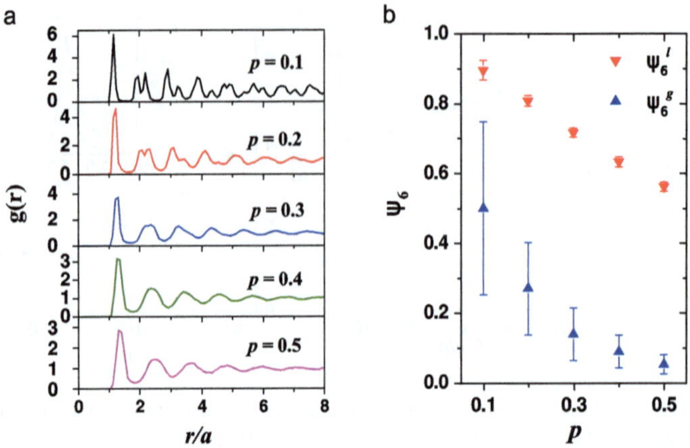

Fig. 9.3 (**a**) Radial distribution function $g(r)$ for $p = 0.1$, 0.2, 0.3, 0.4, and 0.5. (**b**) Local ψ_6^l (*downward triangles*) and global ψ_6^g (*upward triangles*) bond-orientational order parameters versus polydispersity p. Reprinted with permission from [19], Copyright 2008, Wiley-VCH

q for $p = 0.1$, 0.3 and 0.5. The center position of the peak q_0 gives the dominant spatial correlation length $s = 2\pi/q_0$. The peak becomes broader at larger p. The full width at half maximum (FWHM) of the peak Δq gives the average size of ordered domains $\xi_q = 2\pi/\Delta q$. As shown in Fig. 9.2 (b), ξ_q decreases with increasing p, similar to ξ_r.

The radial distribution function $g(r)$, plotted in Fig. 9.3 (a) for several values of p, gives the probability for a cylinder to be located a distance r from another cylinder at the origin relative to that for an ideal gas. The strong first peak, splitting of the second peak, and existence of peaks at large r for $p = 0.1$ indicate that the structure possesses crystalline order. With increasing p, the peaks are broadened, decay faster with r, and $g(r)$ resembles that for a dense liquid [34].

In addition to the translational order, we also characterized the orientational order of the configurations. The bond-orientational order parameter ψ_6 measures the hexagonal registry of nearest neighbors [35]. ψ_6 can be calculated 'locally', which does not include phase information, or 'globally', which allows phase cancellations. Equations (9.4) and (9.5) provide expressions for the global and local bond-orientational order parameters in 2D structures.

$$\psi_6^g = \frac{1}{N}\left|\sum_{i=1}^{N}\frac{1}{m_i}\sum_{j=1}^{m_i} e^{6\iota\theta_{ij}}\right| \tag{9.4}$$

$$\psi_6^l = \frac{1}{N}\sum_{i=1}^{N}\frac{1}{m_i}\left|\sum_{j=1}^{m_i} e^{6\iota\theta_{ij}}\right|, \tag{9.5}$$

where θ_{ij} is the polar angle of the bond connecting the cylinder i to its neighbor j, and m_i denotes the number of nearest neighbors of i. Two cylinders are deemed nearest neighbors if their center-to-center distance $r_{ij} < r_{\min}$, where $r_{\min}$ is the location of the minimum between the first two peaks in $g(r)$.

As shown in Fig. 9.3 (b), both ψ_6^l and ψ_6^g decrease as p increases. ψ_6^l is larger than ψ_6^g, because of the different orientations of the ordered domains. The error bars represent the standard deviations from 100 configurations. For $p = 0.1$, there is a significant fluctuation of ψ_6^g, because some configurations have only a few distinct domains while others contain many domains with different orientations. With increasing p, the number of domains N_d increases, thus the mean and standard deviation of ψ_6^g decrease. For $p = 0.5$, $\psi_6^g \approx 0$, the structures possess only local bond orientational order with $\psi_6^l \approx 0.55$ as found in dense liquids [35].

Density of Optical States

We calculate the DOS with transverse electric (TE) polarization using the order-N method [36]. The magnetic field is parallel to the axis of the air cylinders, and the electric field exists in the 2D plane. Since the cylinders are generated in a square with periodic boundary conditions, we can use it as a supercell for the DOS calculation. For the initial conditions, we choose a superposition of Bloch waves with random phases for the magnetic field and set the electric field to zero [37]. The temporal evolution of electromagnetic fields is obtained by solving the Maxwell's equations with the finite-difference time-domain (FDTD) method. We record the time series of fields at 400 positions which are randomly distributed across the structure, and Fourier transform them to the frequency domain. The sum of their Fourier intensities (i.e., spectral intensities) consists of a number of peaks that correspond to the resonant modes. Adding the contributions of many Bloch wave vectors and averaging over many configurations result in a smooth function for the DOS. We have tested our code by reproducing the DOS for two-dimensional photonic structures in the literature [37].

In Fig. 9.4, we plot the DOS as a function of the normalized frequency $\omega a/2\pi c = a/\lambda$ for the structures generated by the first protocol with $p = 0.1, 0.2, 0.3, 0.4$, and 0.5, and a triangular lattice ($p = 0$) with identical density and diameter air cylinders. The refractive index of the dielectric host in which the air cylinders are embedded is also varied with $n = 3.4$, 1.8, and 1.4 from left to right in Fig. 9.4. The values of n are taken from those of commonly used semiconductors, oxides and polymers. For $n = 3.4$ and $p = 0$, a complete depletion of the DOS from $a/\lambda = 0.235$ to 0.365 results from the full PBG between the first and second bands of the triangular lattice. With the introduction of positional disorder, defect modes are created inside the gap, and the frequency region of depleted DOS becomes shallower and narrower. The higher frequency side of the gap (air band edge) is affected more than the lower frequency side (dielectric band edge). Because the air holes are isolated and the dielectric host is connected, the dielectric bands below the gap are more robust to the disorder than the air bands above the gap. For $n = 1.8$, the PBG of the periodic structure becomes smaller, and thus the depleted region of the DOS is narrower. For the perfect crystal with $n = 1.4$, the first photonic band at the K point ($K1$) has the

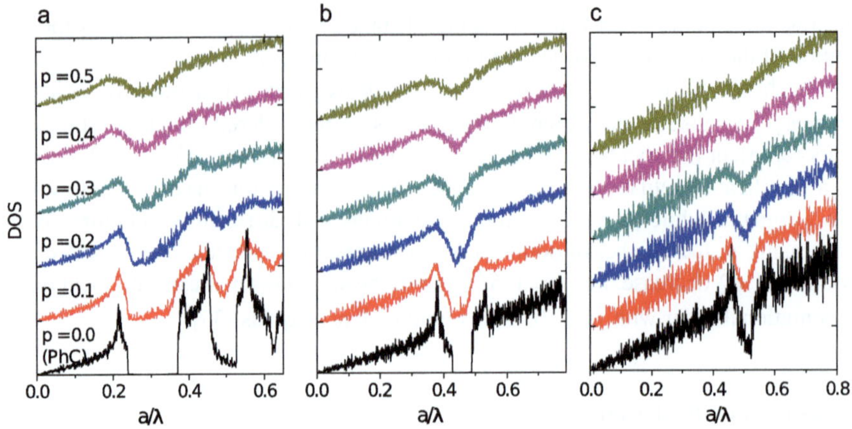

Fig. 9.4 Density of optical states (DOS) as a function of the normalized frequency $\omega a/2\pi c = a/\lambda$ for the 2D structures generated by the first protocol with $p = 0.1$, 0.2, 0.3, 0.4, and 0.5, and a triangular lattice ($p = 0$) of identical density and diameter of air cylinders. The dielectric media, in which the air cylinders are embedded, have refractive indexes (**a**) 3.4, (**b**) 1.8, and (**c**) 1.4. Reprinted with permission from [19], Copyright 2008, Wiley-VCH

same energy as the second band at the M point ($M2$), thus the full PBG disappears. As a result, the DOS displays a dip, rather than a complete depletion. As shown in Fig. 9.4, the addition of positional disorder causes the dip in the DOS to become shallower and eventually disappear at large disorder.

To quantify the strength of the DOS depletion, we introduce the normalized depth S, which is defined as the ratio of the maximal depth of DOS reduction to the DOS of a random structure at the same frequency. The density and diameter of air cylinders as well as the refractive index of the dielectric host in the random structure are identical to those of the structures under investigation. The DOS of the random structure increases almost linearly with frequency, similar to that of a homogeneous 2D dielectric medium. We investigated the dependence of S on various order parameters, e.g. the local bond orientational order ψ_6^l. S increases gradually with ψ_6^l. However, the variation depends on the refractive index contrast n, and is therefore not universal.

To obtain universal behavior for a given degree of positional order, we must account for the effect of refractive index contrast on the DOS. The refractive index contrast determines the strength of the PBG, which is reflected in the attenuation length of Bragg diffraction, or the Bragg length l_b. Roughly speaking, the Bragg length gives an order of magnitude estimate for the minimal size of a periodic structure that is necessary to form a PBG via Bragg scattering. Since periodic structures are anisotropic, l_b varies with direction. However, since the DOS is a sum of optical modes in all directions, the relevant Bragg length is an average over all directions. To obtain the value of l_b in the numerical simulations, we place a continuous dipole source of frequency ω_d in the middle of a large triangular array of air cylinders. We then calculate the electric field intensity at a distance r from the source, and inte-

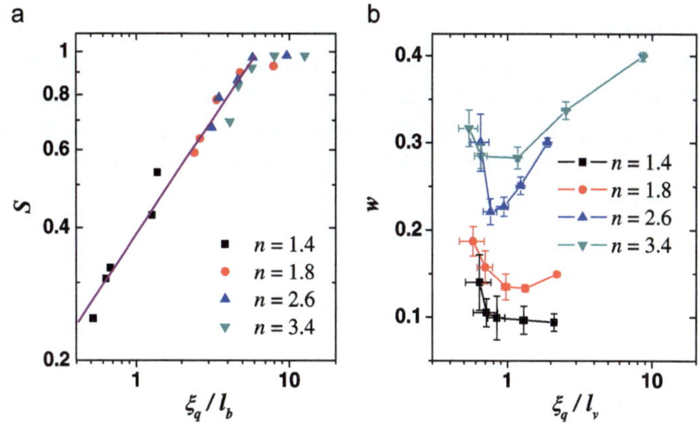

Fig. 9.5 (**a**) Normalized depletion depth of the DOS S for arrays of air cylinders in a dielectric host of refractive index n versus the ratio of the average domain size ξ_q to the angle-averaged Bragg length l_b. A linear fit (*solid gray line*) of the data on the log-log plot for $\xi_q/l_b \lesssim 5$ gives a power-law scaling exponent of 0.52. (**b**) Relative width w of the frequency region where there is a reduction in the DOS as a function of ξ_q/l_v, where l_v is the frequency- and angle-averaged Bragg length. The *error bars* are obtained from the standard deviation of ξ_q for different configurations and fitting errors in the FWHM of the DOS reduction zones. Reprinted with permission from [19], Copyright 2008, Wiley-VCH

grate it over the polar angle. The Bragg length l_b is extracted from the exponential decay of the angle-integrated field intensity I with r.

In Fig. 9.5 (a), we plot the depth S versus the average size of the ordered domains normalized by the Bragg length, ξ_q/l_b. All data points for different refractive index contrasts fall on a single curve. When ξ_q/l_b is above a threshold value (~5), S is almost unity, which implies that the depletion in the DOS is nearly complete as in a perfect crystal. When $\xi_q/l_b \lesssim 5$, S decreases rapidly. The drop can be fit by a straight line on a log-log plot, which reveals a power-law decay with an exponent ~0.52. This result can be understood qualitatively as follows. If the domain size is larger than the Bragg length, Bragg scattering in a single domain is strong enough to form a PBG. The DOS in systems with large ξ_q/l_b is nearly equal to the DOS of a perfect crystal, and these structures can be regarded as photonic polycrystals. In addition, an average over many domains of different orientations makes the directional DOS isotropic. If the domain size is smaller than the Bragg length, individual domains are too small to form PBGs. In this case, the effect of Bragg scattering is reduced due to a limited number of periodic units, and the depletion of the DOS is weakened. This is the amorphous photonic regime, where short-ranged order leads to a partial depletion of the DOS. The well-defined threshold in ξ_q/l_b demonstrates a clear transition from polycrystalline to amorphous photonic structures.

In addition to the depth of the DOS reduction, we also studied the spectral width of the reduction region. The relative width w is defined as the ratio of the full width at half minimum (FWHM) of the dip in the DOS $\delta\omega$ to the frequency ω_0 at the center of the dip. Since the Bragg length varies within the spectral region of DOS reduction, we average its value over the frequency range from $\omega_0 - \delta\omega$ to $\omega_0 + \delta\omega$. The average domain size is normalized by the average Bragg length l_v. Figure 9.5 (b) shows a plot of w versus ξ_q/l_v for several values of n. Although the curves for different n do not coincide, their trends are similar. As ξ_q/l_v increases, w first drops and then rises (except for $n = 1.4$). The turning point is at $\xi_q/l_v \sim 1$. To understand this behavior, we first examine the DOS for periodic systems. At $n = 3.4$, the DOS is enhanced at the photonic band edges due to the slow group velocity [Fig. 9.4 (a)]. When positional disorder is introduced to the structure, the DOS peak at the air band edge is quickly lowered and the higher frequency part of the PBG is filled by defect modes. In contrast, the peak at the dielectric band edge decreases more slowly, because the dielectric bands are more robust against disorder as mentioned earlier. The gap width is reduced, until the DOS peak at the dielectric band edge diminishes at a certain degree of disorder. Then the DOS below the dielectric band edge starts decreasing with further increases in disorder. The DOS reduction region becomes wider. As n decreases, the strength of DOS reduction by PBGs is weakened, and $\delta\omega/\omega_0$ is lowered. At $n = 1.4$, w no longer rises beyond $\xi_q/l_v \sim 1$; instead it tends to a plateau.

Enhanced Scattering and Mode Confinement by Short-Range Order

In nature, the refractive index contrast is typically low, nevertheless photonic amorphous structures are used to manipulate light scattering and color generation. In this section, we investigate the effects of short-range order on light scattering and mode confinement in amorphous structures with low index contrast. We consider the structures generated by the first protocol with $p = 0.5$. When we set $n = 1.4$, the DOS possesses an extremely shallow dip as shown in Fig. 9.6 (a). For $n = 1.2$, the DOS in Fig. 9.6 (e) is nearly featureless. We calculate the resonant modes in these structures using the finite element method. Instead of periodic boundary conditions, the structures have finite size and open boundaries. Each structure contains 1024 air cylinders in a dielectric medium. The open boundaries are terminated by perfectly matched layers that absorb all outgoing waves. Because of light leakage from the finite-sized structure, the resonant modes have finite lifetimes. We calculate the complex frequencies of all resonances $\omega_r + i\omega_i$. The amplitude of ω_i is inversely proportional to the lifetime. The quality factor is defined as $Q = \omega_r/2|\omega_i|$. We obtain the maximal quality factors Q_m of modes within small frequency intervals, and plot them in Fig. 9.6 (b, f). Although the dip in the DOS is barely visible at $n = 1.4$, Q_m is enhanced by a factor of three at a frequency near the center of the dip. Further, even though there is essentially no dip in the DOS for $n = 1.2$, Q_m displays a peak. Figure 9.6 (d, h) shows the spatial distributions of electric field intensities $|E(x, y)|^2$ for the modes with maximal Q_m (marked by the arrows in Fig. 9.6 (b, f)). It is evident that the mode of maximal Q_m at $n = 1.4$ is localized within the structure. For $n = 1.2$ the mode is more delocalized, but the field intensity near the boundary (marked by white dashed line) is still weaker than that in the interior. To determine the degree of localization, we calculate the mode size which is inverse of the inverse

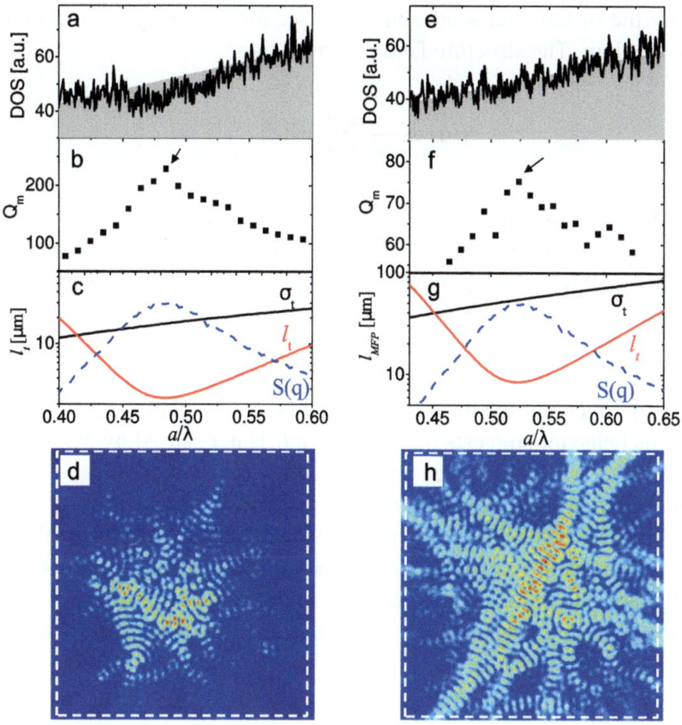

Fig. 9.6 The DOS (**a**, **e**) and maximal quality factors Q_m of resonant modes (**b**, **f**) for the amorphous photonic structures with low refractive index contrast $n = 1.4$ for (**a**–**d**) and 1.2 for (**e**–**h**). *Grey backgrounds* in (**a**, **e**) represent the DOS for 2D homogeneous media. (**d**, **h**) Spatial distribution of electric field intensities for the modes of maximal Q_m [marked by *arrows* in (**b**, **f**)]. (**c**, **g**) Transport mean free path l_t (*solid line*), total scattering cross sections of a single scatterer σ_t (*dashed line*), and the structure factor $S(q)$ at $q = 2k$ (*dotted line*), where k is the wavevector of light. Reprinted with permission from [19], Copyright 2008, Wiley-VCH

participation ratio,

$$s \equiv \frac{1}{L^2} \frac{(\int |E(x,y)|^2 \, dx \, dy)^2}{\int |E(x,y)|^4 \, dx \, dy}, \tag{9.6}$$

where a mode uniformly distributed over the sample gives $s = 1$. We find that the mode in Fig. 9.6 (d) has $s = 0.18$ and is thus highly localized, while the one in Fig. 9.6 (f) has $s = 0.44$ and is only partially localized.

To illustrate the physical mechanism that leads to mode confinement, we calculate the transport mean free path

$$\frac{1}{l_t} = \frac{\pi}{k^6} \int_0^{2k} \rho F(q) S(q) q^3 \, dq, \tag{9.7}$$

where k is the wavevector of light, ρ is the number density of air cylinders, $S(q)$ is the structure factor, $F(q)$ is the form factor, and q is the spatial frequency. $F(q)$

is given by the differential scattering cross section of a single air cylinder in the dielectric medium. The structure factor is given by

$$S(\mathbf{q}) \equiv \frac{1}{N} \sum_{i,j=1}^{N} e^{i\mathbf{q}\cdot(\mathbf{r}_i - \mathbf{r}_j)}, \tag{9.8}$$

where $\mathbf{r}_i$ denotes the center position of the ith cylinder. Since the structures are isotropic, $S(\mathbf{q})$ is invariant with the direction of $\mathbf{q}$ and is only a function of the magnitude q. In Fig. 9.6 (c, g), we show that l_t displays a significant drop at a frequency that coincides with the peak in Q_m. This indicates that the enhancement of scattering strength improves mode confinement. In Fig. 9.6 (c, g) we also plot the total scattering cross section σ_t of a single air cylinder, which increases monotonically with frequency and does not exhibit any resonant behavior within the frequency range studied. This behavior suggests that the dip in l_t is not caused by Mie resonance of individual scatterers. Instead, we contend that the short-range order enhances Bragg backscattering at certain wavelengths and shortens l_t. To prove this, we also plot $S(q)$ for the backscattering $q = 2k$ in Fig. 9.6 (c) and (g). $S(q)$ is peaked near the dip of l_t, which confirms that collective backscattering from local domains of ordered cylinders causes a dramatic decrease in l_t. Therefore, the spatial confinement of resonant modes is enhanced by short-range order through constructive interference of scattered light that occurs at specific frequencies.

9.2.2 *Lasing in Photonic Polycrystalline and Amorphous Structures*

Next we present the experimental study on lasing characteristics in photonic polycrystalline and amorphous structures.

Sample Fabrication and Lasing Experiment

The computer generated patterns of polycrystalline and amorphous arrays of circular holes were transferred to a GaAs membrane. A 190-nm-thick GaAs layer and a 1000-nm thick $Al_{0.75}Ga_{0.25}As$ layer were grown on a GaAs substrate by molecular beam epitaxy. Inside the GaAs layer there were three uncoupled layers of InAs QDs equally spaced by 25 nm GaAs barriers. 2D arrays of cylinders were written on a 300-nm-thick ZEP layer with the electron-beam lithography. The patterns were transferred to the GaAs layer by chlorine-based inductive-coupled-plasma reactive-ion-etching with the ZEP layer as a mask. The ZEP layer was subsequently removed in an oxygen plasma cleaning process. Finally the $Al_{0.75}Ga_{0.25}As$ layer was selectively removed by a dilute HF solution. Figure 9.7 (a, b) shows the top-view scanning electron microscope (SEM) images of two fabricated patterns. The lateral dimension of a pattern is 9.3 μm, and it has 1024 air holes. The radius of air holes r

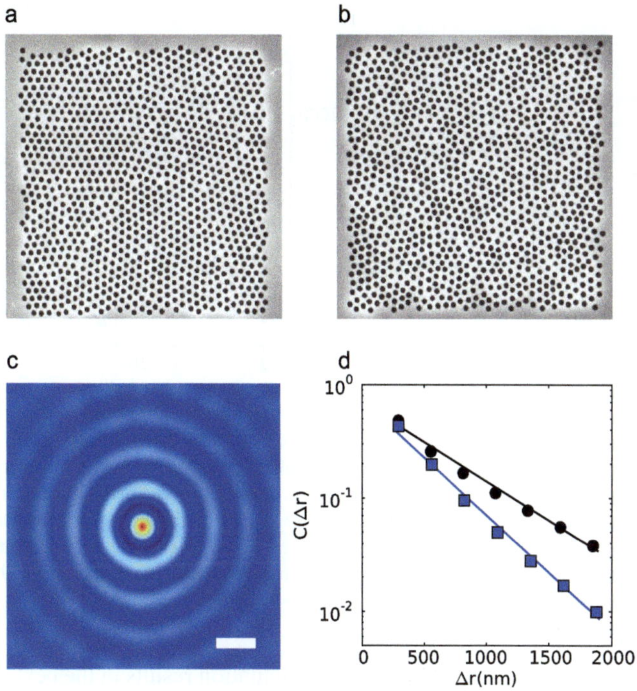

Fig. 9.7 (**a, b**) Top-view SEM images of fabricated arrays of air holes in a GaAs membrane. The *scale bar* is 2 μm. Ordered domains of different orientations can be seen in (**a**), but barely in (**b**). (**c**) 2D spatial correlation function $C(\Delta\mathbf{r})$ averaged over five configurations including the pattern in (**b**) and equivalent ones. *Scale bar* = 300 nm. (**d**) Log-linear plot of the peak amplitude of azimuthal-averaged spatial correlation function $C(\Delta r)$, *squares* for the pattern in (**a**), and *circles* for (**b**). The *straight lines* are exponential fit, giving the decay length $\xi = 2.3$a for (**a**) and 1.4a for (**b**). Reprinted with permission from [21], Copyright 2008, Wiley-VCH

is 100 nm. In the pattern of Fig. 9.7 (a), we can clearly see domains of the triangular lattice of holes, each domain has a different orientation. In Fig. 9.7 (b), the domains are so small that barely visible.

For qualitative understanding of the short-range order, the 2D spatial correlation function $C(\Delta\mathbf{r})$ for these patterns were calculated. A typical ensemble-averaged $C(\Delta\mathbf{r})$ is presented in Fig. 9.7 (c). It consists of rings whose amplitudes decrease with increasing spatial separation $\Delta r \equiv |\Delta\mathbf{r}|$. Since $C(\Delta\mathbf{r})$ is isotropic, we calculated the azimuthal-averaged $C(\Delta r)$ for the two patterns in (a, b) and plotted the peak amplitudes in Fig. 9.7 (d). The first peak away from $\Delta r = 0$ is produced by the nearest cylinder, thus its position corresponds to the average spacing a of the nearest neighbors. The two patterns in Fig. 9.7 (a, b) have $a = 290$ nm. The faster damping of $C(\Delta r)$ for the pattern in Fig. 9.7 (b) reflects the spatial correlation is shorter-ranged. They both fall on straight lines in a log-linear plot, indicating exponential decays. The decay length $\xi = 2.3a$, and $1.4a$ for the patterns in Fig. 9.7 (a, b). Ac-

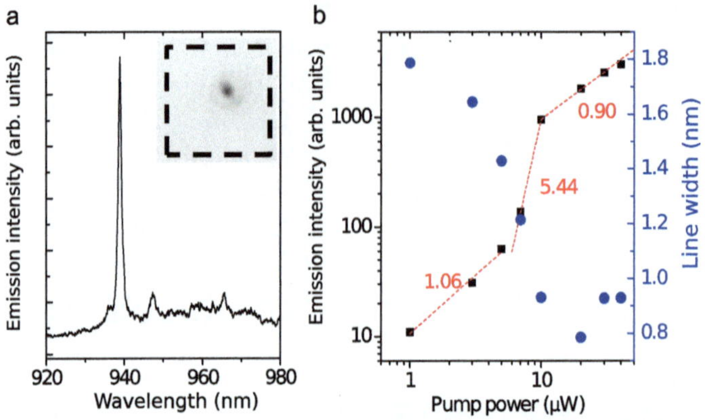

Fig. 9.8 (**a**) Measured spectrum of emission from a pattern of $\xi = 4.4a$ and $a = 290$ nm at the incident pump power $P = 30$ μW. The *inset* is an optical image of the lasing mode corresponding to the sharp peak in the emission spectrum. *Black dashed square* marks the pattern boundary. (**b**) Intensity I (*black square*) and width $\Delta\lambda$ (*blue circle*) of the lasing peak in (**a**) versus the incident pump power P in a logarithmic plot. The *red dotted lines* represent linear fit of $\log I$ vs. $\log P$ in three regimes, and the numbers next to them are those slopes. Reprinted with permission from [21], Copyright 2008, Wiley-VCH

cording to our previous study [19] and the calculation results in the next section, the pattern in Fig. 9.7 (a) is polycrystalline and (b) amorphous.

In the lasing experiments, the samples were cooled to 10 K in a continuous-flow liquid Helium cryostat, and optically pumped by a mode-locked Ti:Sapphire laser (pulse width ~200 fs, center wavelength ~790 nm, and pulse repetition rate ~76 MHz). A long working distance objective lens (numerical aperture =0.4) focused the pump light to a pattern at normal incidence. The diameter of pump spot on the sample surface was about 2 μm. The emission from the sample was collected by the same objective lens. The emission spectrum was measured by a high resolution spectrometer with a liquid-nitrogen-cooled coupled-charged-device (CCD) array detector. Simultaneously the spatial distribution of emission intensity across the sample surface was projected onto a thermoelectric-cooled CCD camera.

Figure 9.8 (a) is part of a time-integrated spectrum of emission from a polycrystalline pattern ($\xi/a = 4.4$, $a = 290$ nm). It features a sharp peak on top of a broad QD emission band. Figure 9.8 (b) plots the intensity I and linewidth $\Delta\lambda$ of this peak as a function of the incident pump power P. The variation of $\log I$ with $\log P$ exhibits a S-shape with two kinks. The slopes in the three regimes separated by the two kinks were obtained from curve fitting and written on the graph. In the first and last regimes, the slopes of $\log I$ over $\log P$ are very close to unity, indicating a linear growth of I with P. The second regime has a slope of 5.44, meaning I scales as $P^{5.44}$. The first regime corresponds to spontaneous emission of QDs to a resonant mode, thus the intensity increases linearly with the pump power. When the pump is high enough, the emission peak grows superlinearly as a result of light amplification by stimulated emission. This is the second regime. In the third regime, lasing occurs in this mode, and the gain saturation reduces the slope to one. Also

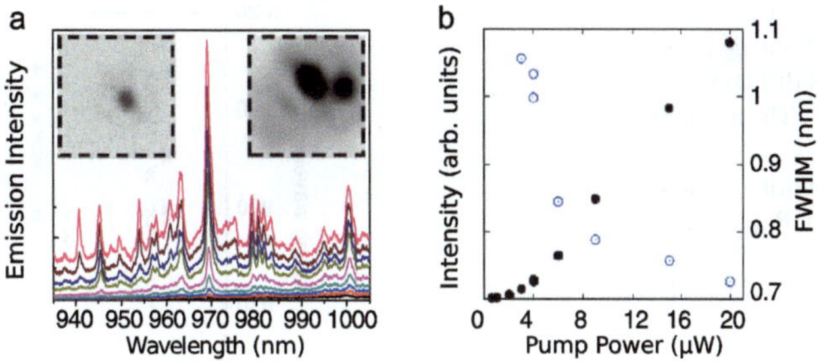

Fig. 9.9 (**a**) Evolution of emission spectra of a pattern with $\xi/a = 1.4$ and $a = 270$ nm as the pump power increases (*from bottom to top*). *Inset* is an optical image of the lasing mode at $\lambda = 970$ nm. (**b**) Intensity and width of the emission peak at $\lambda = 970$ nm as a function of the pump power. Reprinted with permission from [21], Copyright 2008, Wiley-VCH

seen in Fig. 9.8 (b), the spectral width $\Delta\lambda$ of the peak first drops quickly with increasing P, then levels off and increases slightly at higher P. The rapid decrease of $\Delta\lambda$ is expected at the onset of lasing oscillation. The gradual increase above the lasing threshold results from the hot carrier effect. Due to the short pulse pumping, the carrier density keeps changing in time. It causes a temporal change of the refractive index, and consequently a continuous red-shift of lasing frequency [38, 39]. In our time-integrated measurement of lasing spectrum, the transient frequency shift results in a broadening of the lasing line. Such broadening increases with the hot carrier density and becomes dominant at high pumping level. The inset of Fig. 9.8 (a) is an optical image of the lasing mode, revealing its strong localization inside the pattern whose boundary is marked by the black dashed line.

We also realized lasing in photonic amorphous structures. As seen in Fig. 9.9 (a), there are many peaks of comparable height in the emission spectra of a pattern with $\xi = 1.4a$. Figure 9.9 (b) is a plot of the intensity and width of an emission peak at $\lambda = 970$ nm versus the pump power. The rapid increase of peak intensity and dramatic reduction of the peak width illustrates the onset of lasing action. A tunable interference filter was placed in front of the CCD camera to select this lasing mode for imaging. Inset of Fig. 9.9 (a) are the optical images of two lasing modes. The left one is spatially localized inside the pattern whose boundary was drawn by the dashed line. The double-peaked intensity distribution of the mode on the right is similar to that of a 2D necklace state in [40], suggesting it is a hybrid of two localized states. Further study is needed to confirm it is a necklace state [40, 41].

As we moved the pump beam spot across a pattern, new lasing peaks replaced the existing ones, and they have distinct frequencies. This phenomena indicate that resonant modes are localized in different positions of the samples, and brought to lasing when overlapped with the pump spot. We repeated the lasing experiment on several patterns with different arrangement of air holes but same ξ/a, and found the lasing peaks varied from pattern to pattern. We measured the patterns of different ξ/a and recorded the lasing wavelengths. Figure 9.10 plots the wavelengths of lasing peaks for five values of ξ/a. For a fixed ξ/a, there is a wide spread of lasing

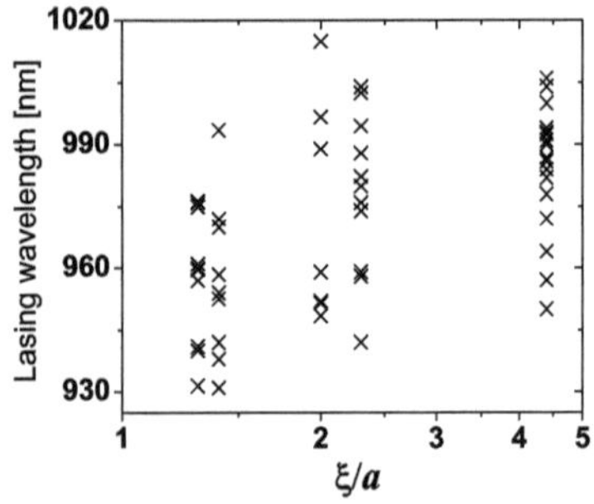

Fig. 9.10 Measured wavelengths of lasing peaks for patterns of different domain size ξ. The average spacing of nearest-neighboring air holes is $a = 290$ nm. Reprinted with permission from [21], Copyright 2008, Wiley-VCH

wavelengths due to the broad QD gain spectra. Nevertheless, it is evident that the lasing peaks shift to shorter wavelength as the average domain size ξ decreases.

Numerical Simulation of Lasing Modes

To explore the nature of lasing modes in photonic polycrystalline and amorphous structures, we performed numerical simulation. The perforated GaAs membrane is approximated as a 2D array of infinitely long air cylinders embedded in a uniform dielectric host with an effective index of refraction n_w. The value of n_w was obtained in the following steps. First we calculated the photonic band structure of a triangle lattice of air holes in a free-standing GaAs membrane of thickness 190 nm using the plane wave expansion method [42]. The density and size of air holes were identical to those of the fabricated samples. Next we calculated the photonic band structure of the approximate 2D system with n_w as a parameter. By adjusting the value of n_w, we matched the center frequency of the fundamental PBG obtained in the above two cases. In our calculation, we considered only the transverse-electric (TE) polarization (electric field perpendicular to the air cylinder axis), because experimentally the laser emission is TE polarized due to stronger gain of the InAs QDs for the TE polarized light.

Using the finite-difference frequency-domain (FDFD) method, we calculated the resonant modes with long lifetime in the passive structures. Due to the finite size of a pattern, light may escape through the open boundary. The outgoing wave is absorbed by the perfectly matched layer that surrounds the pattern. The resonant modes have finite lifetime, and their frequencies are complex numbers $\omega_r + i\omega_i$. The magnitude of ω_i is inversely proportional to the mode lifetime. The quality factor is defined as $Q = \omega_r/2|\omega_i|$. We calculated the complex frequencies of TE modes, and found the highest quality factor Q_m within small frequency bins. Figure 9.11 (a) plots Q_m versus the normalized frequency a/λ for $\xi/a = 2.3$ and 1.4. Q_m drops quickly as ξ/a decreases. Nevertheless, it reaches the maximum at the same frequency $a/\lambda = 0.3$ for different ξ/a. This result can be explained by the DOS shown in Fig. 9.11 (b). For $\xi/a = 2.3$, there is a significant depletion of DOS, almost comparable to that of a PhC. The system can be regarded as a photonic polycrystal, as individual domains are large enough to form the PBG via Bragg scattering. Defect modes are formed

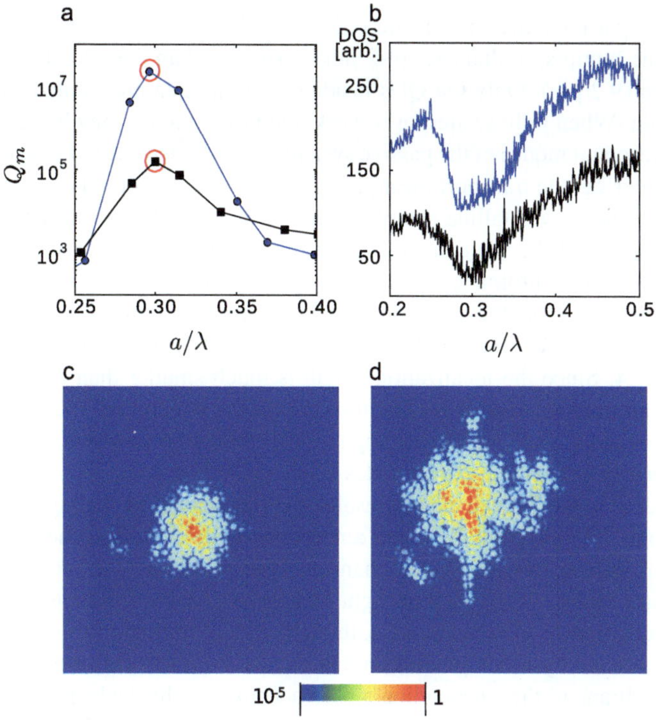

Fig. 9.11 (**a**) FDFD calculation of approximate 2D structures ($n_w = 2.73$) giving the maximal quality factor Q_m versus the normalized frequency a/λ for $\xi/a = 2.3$ (*blue circle*) and 1.4 (*black square*). (**b**) Calculated DOS for $\xi/a = 2.3$ (*top*) and 1.4 (*bottom*). The curves are shift vertically for clarification. Calculated spatial intensity distribution of the highest-Q_m mode for $\xi/a = 2.3$ (**c**) and 1.4 (**d**). They are marked with *circles* in (**a**). Reprinted with permission from [21], Copyright 2008, Wiley-VCH

inside the PBG. The closer their frequencies to the center of PBG, the higher their quality factors. Therefore, the highest-Q resonances in the photonic polycrystal are defect modes. They are strongly localized in space, as can be seen from a typical mode profile in Fig. 9.11 (c). The mode size, computed from the inverse participation ratio of the field distribution, is about 0.6 μm. When ξ/a is reduced to 1.4, Bragg scattering from each domain is not enough to produce a PBG. Consequently, the DOS has a dip instead of a gap. In this photonic amorphous structure, the peak of Q_m coincides with the dip of DOS. The reduction in DOS results from the short-range order, which enhances optical confinement and produces the maximum of Q_m [19, 20]. It leads to the maximum of Q_m. The spatial profile of a typical high-Q resonance is shown in Fig. 9.11 (d). The mode size is about 1.0 μm, much smaller than the lateral size of the structure (9.3 μm). This means the mode is still localized spatially, but the degree of localization is less than the defect mode in a polycrystal [Fig. 9.11 (c)]. The maximal Q_m drops quickly as the average domain size ξ decreases.

In our experiment the optical gain is distributed nonuniformly across the sample, because the pump spot diameter (~2 μm) is smaller than the lateral dimension of the pattern (9.2 μm). Only the QDs inside the pump area are excited and provide optical gain. When light scattering is weak, the lasing modes may be very different from the resonant modes of the passive systems [43–45]. However, our samples have strong scattering, the transport mean free path is estimated [46] at $a/\lambda = 0.30$ to be 0.36 μm in the polycrstalline sample ($\xi/a = 2.3$) and 0.53 μm in the amorphous structure ($\xi/a = 1.4$). We extracted the localization length ξ_L in these structures by placing a monochromatic source at the center and calculating the steady-state field distribution. From the decay of the field intensity away from the source, we get $\xi_L = 0.38$ μm at $a/\lambda = 0.30$ in the sample of $\xi/a = 2.3$, and 0.46 μm in the sample of $\xi/a = 1.4$. Since the localization length is much smaller than the system size, the resonant modes are strongly confined within the structures. The typical size of high-Q modes is smaller than or comparable to the pump spot size, thus the lasing modes correspond to the high-Q modes inside the pump area.

According to the 2D calculation results, the frequencies of the highest-Q modes are the same for polycrystalline and amorphous structures, thus the lasing modes should not shift in frequency as ξ changes. This prediction contradicts the experimental data in Fig. 9.10, because light leakage in the third dimension is ignored in the calculation of 2D structures. Although light is confined in the free-standing GaAs membrane by index guiding, it can escape from the top or bottom surfaces of the membrane to the surrounding air. To account for this leakage, we performed the three-dimensional (3D) finite-difference time-domain (FDTD) calculation. The structural parameters used in the calculation are identical to those of the fabricated samples.

We calculated the high-Q resonances of TE polarization in the absence of gain or absorption. The quality factors Q_t of all modes within the frequency range of interest were found and plotted versus the normalized frequency a/λ in Fig. 9.12 for $\xi/a = 2.3$ and 1.4. The values of Q_t are orders of magnitude lower than those in Fig. 9.11 (c) for the same value of ξ/a. This result illustrates that the vertical leakage of light is much larger than the lateral leakage for the high-Q resonances. Such strong vertical leakage results from tight confinement in the lateral dimension. Namely, spatial localization of a mode in the xy-plane (parallel to the membrane) results in a broad distribution of in-plane wavevector $\mathbf{k}_{||}$ (projection of $\mathbf{k}$ vector to the xy-plane). The $\mathbf{k}_{||}$ components within the light cone ($|\mathbf{k}_{||}| \leq \omega/c$) can escape from the membrane in the $\pm z$ directions (normal to the membrane). The vertical leakage rate is characterized by the out-of-plane energy loss per optical cycle Q_v^{-1}, and the lateral by Q_h^{-1}. The total loss is described by $Q_t^{-1} = Q_h^{-1} + Q_v^{-1}$.

For the amorphous structure of $\xi/a = 1.4$, the mode of maximal Q_t at $a/\lambda \simeq 0.32$ has $Q_h = 4.88 \times 10^4$, and $Q_v = 1.62 \times 10^3$. Thus, the vertical leakage rate is an order of magnitude larger than the lateral one. In a polycrystalline structure, the tighter in-plane confinement of defect modes within the PBG makes the vertical leakage even stronger. Consequently, the modes at the center of PBG no longer have the highest Q_t, even though their Q_h is maximal. The highest Q_t modes are away from the PBG center, as shown in Fig. 9.12 for $\xi/a = 2.3$. These modes are less

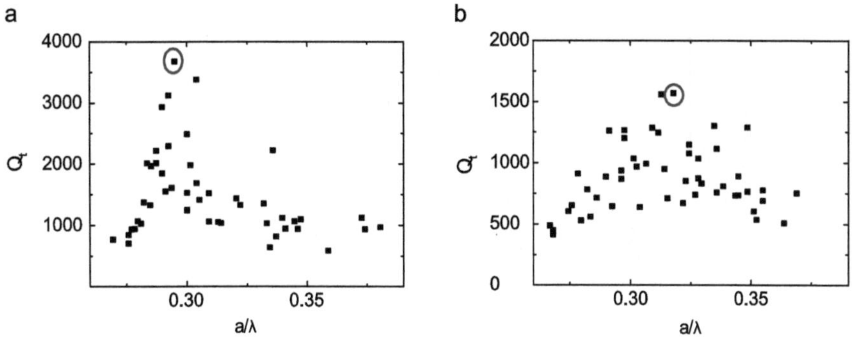

Fig. 9.12 3D FDTD calculation of resonances in the perforated GaAs membrane giving their quality factor Q_t as a function of the normalized frequency a/λ for $\xi/a = 2.3$ (**a**) and 1.4 (**b**). Reprinted with permission from [21], Copyright 2008, Wiley-VCH

confined in-plane, meaning their Q_h is lower. The larger spread in the xy-plane narrows the $\mathbf{k}_{||}$ distribution. The $\mathbf{k}_{||}$ components within the light cone is reduced, so is the out-of-plane leakage. The maximal-Q_t mode at $a/\lambda = 0.3$ has $Q_h = 7.55 \times 10^4$, and $Q_v = 3.87 \times 10^3$. Although its Q_h is lower than that of the defect modes at $a/\lambda = 0.32$ (PBG center), the Q_v is higher, so is the Q_t. Intuitively, the mode at the higher frequency side of the PBG center, i.e., at $a/\lambda = 0.34$, should have comparable Q_t to the maximal-Q_t mode at the lower frequency side ($a/\lambda = 0.30$), as their spectral distance to the PBG center is the same. However, the Q_t is lower at $a/\lambda = 0.34$, as seen in Fig. 9.12. This is because the air holes are isolated and the dielectric medium is connected in the membrane. The dielectric band edge at the lower frequency side of PBG is more robust to disorder, as evident in the DOS shown in Fig. 9.11 (b). Consequently, light confinement is stronger in the lower frequency part of the PBG, giving higher Q_t at $a/\lambda = 0.3$.

The 3D numerical simulation reveals that the maximal-Q_t modes shift to higher frequency as ξ decreases. This prediction agrees with the experimental observation that the wavelengths of lasing modes decreases from polycrystalline to amorphous patterns.

Control of Lasing with Short-Range Order

To characterize the effect of short-range structural order on lasing, we can either measure the threshold of individual lasing modes at different wavelength, or compare the intensity of different lasing modes at the same pumping level. In principle, the spectral variation of gain coefficient and mode competition for gain would influence the lasing threshold and emission intensity. However, the gain spectrum of InAs QDs is very broad, and the gain coefficient has little variation over a wide spectral range. More importantly, the gain spectrum is dominated by inhomogeneous broadening, which significantly weakens mode competition. Hence, the interaction of lasing modes at different wavelength or spatial location is negligible. To increase

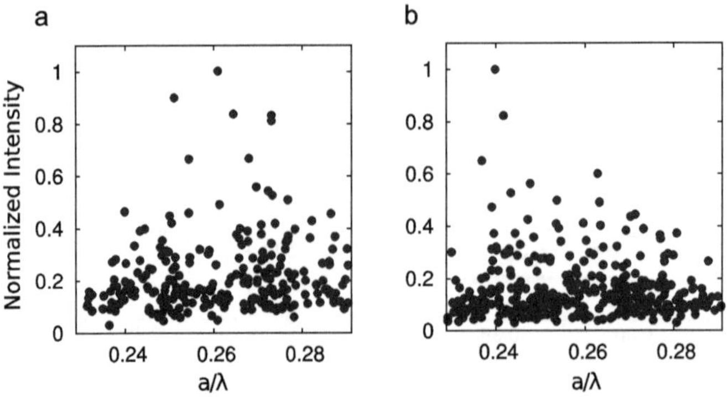

Fig. 9.13 Measured intensities of lasing peaks from three PAS of $a = 235$ nm, 255 nm and 275 nm. The air filling fraction $f = 0.33$ in (**a**) and 0.28 in (**b**). Laser emission intensity reaches the maximum at the normalized frequency $a/\lambda = 0.26$ in (**a**) and 0.24 in (**b**). Reprinted with permission from [20], Copyright 2008, Wiley-VCH

the range of normalized frequency $\omega a/2\pi c = a/\lambda$, we probe lasing in multiple arrays of distinct a. The ratio of r over a is kept constant, so that the filling fraction f of air does not change. Figure 9.13 (a) plots the intensities of numerous lasing modes collected from three arrays of $a = 235$ nm, 255 nm and 275 nm. The incident pump power is fixed at 16 μW. The pump spot is scanned across the interior of each array to probe lasing modes at different locations. The size of pump spot is kept at 1.2 μm. The range of a/λ, covered by the three patterns, is from 0.23 to 0.29. As a/λ increases, the lasing mode intensity first increases and then decreases. It reaches the maximum at $a/\lambda = 0.26$. Thus lasing becomes the strongest, or equivalently the lasing threshold is the lowest at $a/\lambda = 0.26$. Next we change the air filling fraction f from 0.33 to 0.28 by varying r/a, and measure three samples of $a = 235$ nm, 255 nm and 275 nm. As shown in Fig. 9.13 (b), the maximum intensity of lasing modes is shifted to $a/\lambda = 0.24$. These results confirm that there exists an optimal frequency for lasing in the PAS, and its value can be tuned by the structural parameters.

To interpret the experimental data, we calculate the quality (Q) factor of resonant modes in the samples. The higher the Q factor, the longer the photon lifetime in the PAS, the stronger the amplification of light. We use the finite-difference frequency-domain (FDFD) method in the numerical simulation. Since the 3D FDFD calculation is computationally expensive, we calculate the resonant modes in the approximate 2D structures. More specifically, we first compute the effective refractive index n_w of the waveguided mode in the GaAs layer free standing in air. We consider only the transverse-electric (TE) polarization (electric field parallel to the GaAs layer), because the laser emission is TE polarized. Then the value of n_w as a function of λ is assigned to the refractive index of the dielectric host in which the air cylinders are embedded. The air cylinders are assumed to be infinitely long in the 2D FDFD calculation. Figure 9.14 (a) plots Q vs. a/λ for three PAS of $f = 0.33$. The values of a and r are equal to those of the fabricated samples. The Q factor is peaked at

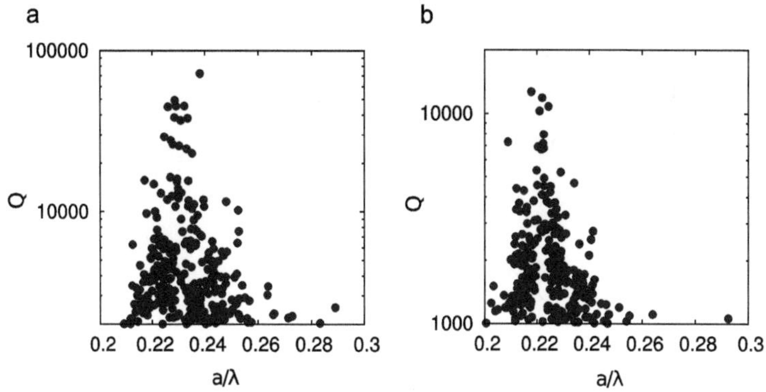

Fig. 9.14 Calculated Q factor of resonant modes in the PAS whose structural parameters are identical to those in Fig. 9.13 (a) and (b). As the air filling fraction f changes from 0.33 (**a**) to 0.28 (**b**), the maximum of Q factor shifts to lower normalized frequency a/λ. Reprinted with permission from [20], Copyright 2008, Wiley-VCH

$a/\lambda = 0.24$. When the air filling fraction f is changed to 0.28, the maximum of Q factor shifts to a lower frequency (Fig. 9.14 (b)). These trends agree qualitatively to those of laser emission intensity measured experimentally (Fig. 9.13). The maxima occur at slightly different frequencies, due to 2D approximation in the numerical simulation and uncertainty in determining the refractive index of GaAs at low temperature. Note that the 2D simulation only takes into account light leakage through the edges of an array. Experimentally, light can also escape through the top or bottom surface of the GaAs membrane to air. The vertical leakage is included in our finite-difference time-domain (FDTD) simulation of a free-standing GaAs membrane. For the PAS, the total Q factor (including both lateral and vertical leakage) displays a similar trend to that in Fig. 9.13. However, the actual Q value is notably lower than that in Fig. 9.14 as a result of the vertical leakage. Hence, the variation of Q with a/λ is determined by the lateral leakage. The existence of Q maximum indicates light confinement is the strongest at specific wavelength λ that scales with the characteristic length scale a of the structure. Thus the Q enhancement originates from the short-range order. Since stronger optical confinement increases the dwell time of light in the structure, light experiences more amplification and the lasing threshold is reduced. In other words, the maximum of the Q factor leads to a minimum of the lasing threshold, or equivalently, a maximum of laser emission intensity at a fixed pump power above the threshold.

To confirm the origin of Q enhancement, we estimate the transport mean free path l_t, which is a measure of the scattering strength. For the approximate 2D structure,

$$l_t(\lambda) = \left(\frac{\pi}{k^6} \int_0^{2k} \rho F(q) S(q) q^3 \, dq \right)^{-1}, \qquad (9.9)$$

where q is the spatial frequency of the structure, $F(q)$ is the form factor, $S(q)$ is the structure factor, ρ is the density of scatterers, and k is the wave vector [47].

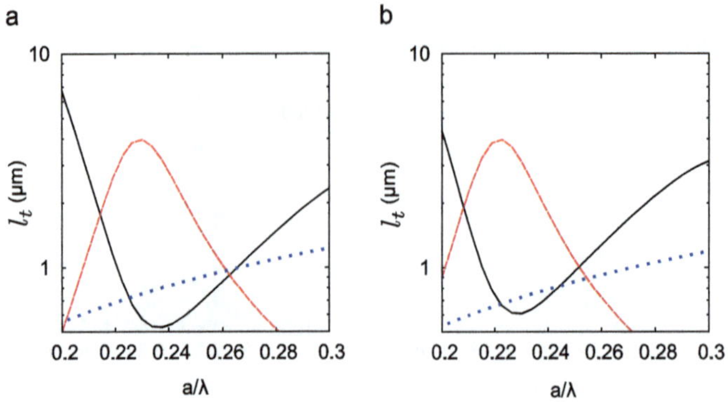

Fig. 9.15 (**a**) Estimated transport mean free path l_t (*black solid curve*) as a function of the normalized frequency a/λ in the PAS of air filling fraction $f = 0.33$ (**a**) and 0.28 (**b**). *Red dashed curve* is the structure factor $S(q = 2k)$ of PAS, and the *blue dotted curve* the total scattering cross section σ_t of a single air cylinder embedded in a dielectric host. σ_t does not exhibit any Mie resonance. The dip in l_t almost coincides with the peak in $S(q)$, confirming the enhanced light scattering is caused by short-range order

$k = 2\pi n_e/\lambda$, where λ is the wavelength in vacuum, and n_e is the effective index of refraction of the 2D structure. If the structure is completely random, $S(q)$ becomes 1. For the PAS, we compute the structure factor from the center positions of air holes. Since the structure is isotropic, S depends only on the magnitude of $\mathbf{q}$. $F(q)$ is obtained from the differential scattering cross section of a single air cylinder (infinitely long) embedded in a dielectric host of refractive index n_w. n_e is estimated by the Maxwell-Garnett formula with the air filling fraction f.

Figure 9.15 (a) plots l_t (black solid curve) as a function of a/λ for the PAS of $f = 0.33$. It has a significant dip at $a/\lambda = 0.24$, where light scattering becomes the strongest. The minimum of l_t almost coincides with the maximum of Q in Fig. 9.14 (a), indicating the optimal light confinement is caused by the strongest scattering. To find the origin of enhanced light scattering, we plot the total scattering cross section σ_t of a single air cylinder (blue dotted curve) in Fig. 9.15 (a). It increases monotonically with a/λ, and does not exhibit any resonant behavior within the frequency range of study. Hence, the dip in l_t is not caused by any Mie resonance of individual scatterers. In the same figure we also plot $S(q)$ for the backscattering $q = 2k$. Its value is peaked near the dip of l_t, confirming the short-range order enhances backscattering and shortens l_t [29]. Similar results are obtained for the PAS of $f = 0.28$ [Fig. 9.15 (b)]. The dip of l_t moves to lower frequency, consistent with the shift of Q maximum [Fig. 9.14 (b)] and the strongest laser emission intensity [Fig. 9.13 (b)]. Note that in the estimation of l_t with (9.1), the near-field coupling of adjacent scatterers is neglected. Within the frequency range of interest, there is no scattering resonance of individual air cylinders, thus the coupling of neighboring cylinders via evanescent wave is weak.

9.3 Photonic Network Structures and Lasers

The photonic amorphous structures have two types of topology: (i) an aggregate of dielectric spheres/cylinders, (ii) a continuous network of dielectric material. The photonic bandgaps can be formed in (i) via evanescent coupling of Mie resonances of individual scatterers [10–14, 18]. However, two-dimensional and three-dimensional (3D) realizations of (ii) can possess larger PBGs [16, 17, 48] For example, a photonic amorphous diamond structure formed by a 3D network of silicon has a 18 % PBG [16, 48]. Since the PAS is isotropic, the PBG is identical in all directions. The photonic bandedge (BE) modes can be strongly localized without introducing any defect in a PAS [17, 18, 48]. It is dramatically different from a periodic structure where the PBG is anisotropic and the BE modes are spatially extended.

Most studies on amorphous network structures have focused on passive systems that have no gain or nonlinearity. What happens if we introduce optical gain to the 2D amorphous network structures? Can we achieve lasing? What are the lasing modes? These questions will be addressed in the section.

9.3.1 2D Photonic Network Laser

To generate a 2D trivalent network structure (each junction having three bonds), we first created a jammed packing of polydisperse cylinders in a computer simulation [19]. The center positions of cylinders are marked by black solid circles in Fig. 9.16 (a). Then we performed a Delaunay tessellation that provides triangular partitioning (blue thin lines) [17]. Associated with each triangle is a centroidal point. We connected the centroids of neighboring triangles with line segments. The resulting structure is a trivalent network shown in red thick lines. The spatial Fourier spectra of the structure [inset of Fig. 9.16 (b)] exhibits a circular ring pattern, indicating the structure is isotropic and there exits a dominant spatial frequency that corresponds to the radius of the ring. We also calculated the 2D spatial correlation function $C(\Delta r)$, which is plotted in Fig. 9.16 (b). The characteristic length scale of the structure a is obtained from the first peak of the correlation function. The rapid decay of the amplitude of $C(\Delta r)$ with Δr reveals the structural correlation is short-ranged.

As to be detailed later, we fabricated 2D trivalent network structures in a GaAs membrane that was 190 nm thick and free-standing in air. To obtain the effective index of refraction n_e of the GaAs segments for the 2D simulation, we computed the fundamental PBG in a triangle lattice of air holes in a free-standing GaAs membrane (using the plane wave expansion method), then adjusted n_e of an approximate 2D structure to get a similar PBG. The value of n_e depends on the filling fraction of air in the GaAs membrane f, which is chosen to be 0.53 to have the maximal PBG. The 2D network structure was assigned the same $f = 0.53$ and $n_e = 2.68$. We calculated the 2D density of optical states (DOS) using the order N method in

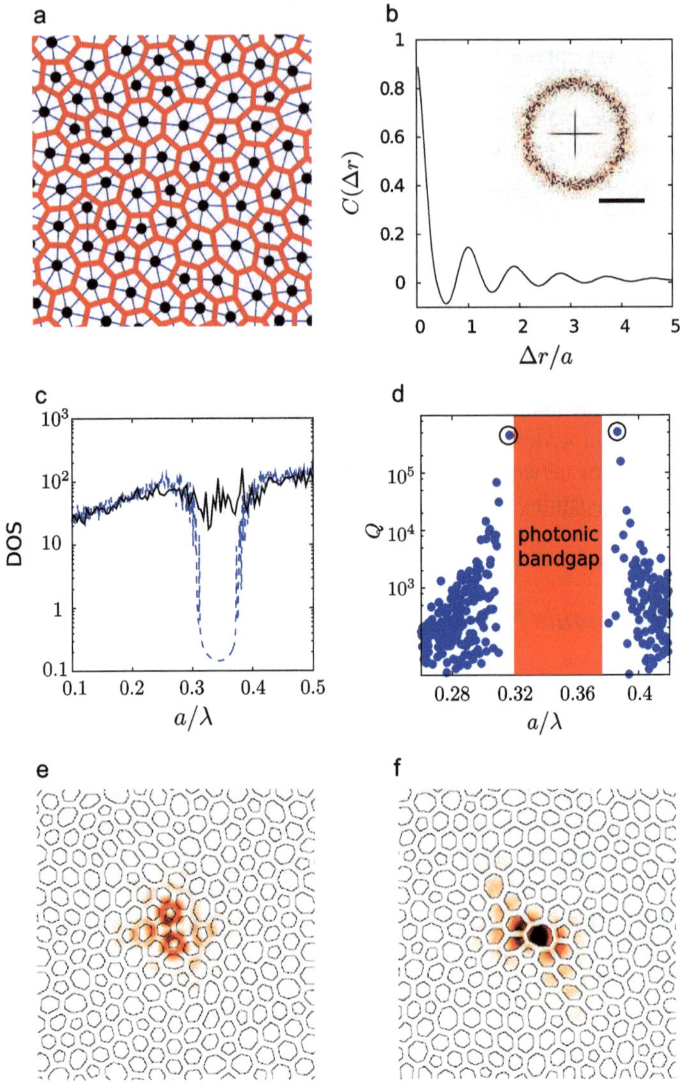

Fig. 9.16 (**a**) *Black solid circles* are the center positions of polydisperse cylinders produced by jammed packing in a computer simulation. *Blue thin lines* represent the Delaunay tessellation, that leads to the formation of a trivalent network structure shown in *red thick lines*. (**b**) Spatial Fourier spectra (*inset*) and spatial correlation function $C(\Delta r)$ (*main panel*) of the trivalent network structure. (*scale bar*: $2\pi/a$) (**c**) Calculated density of optical states for a trivalent network structure (*blue dashed line*) and an amorphous array of air cylinders (*black solid line*) with the same air filling fraction $f = 0.53$ and dielectric refractive index $n_e = 2.68$. (**d**) Calculated Q factor of the resonant modes in the trivalent network structure as a function of the normalized frequency a/λ. (**e**) and (**f**) are calculated intensity distributions of two modes [*circled* in (**d**)] at the low and high frequency sides of the bandgap. Reprinted with permission from [49], Copyright 2008, Wiley-VCH

a finite-difference time-domain (FDTD) simulation [36]. The boundary condition is periodic, and the supercell contains 2048 vertices. Only TE modes are considered, as the lasing modes in the GaAs membrane are TE polarized due to stronger amplification by QDs. Figure 9.16 (c) displays a significant depletion of DOS in the 2D trivalent network. For comparison, we calculated the DOS for an amorphous array of monodisperse air cylinders in GaAs. This structure was derived from the jammed packing of polydisperse cylinders (from which the network structure was created) by reducing the radii of all cylinders to a constant value. The air filling fraction f is identical to that in the network structure. As seen in Fig. 9.16 (c), the DOS has only a shallow dip in the amorphous array of air cylinders. Hence, the PBG effect is greatly enhanced in the trivalent network structure. This is attributed to the uniform topology of each junction.

Next we calculated the quality factor ($Q \equiv \omega\tau$, ω is the frequency, τ is the lifetime) of the resonant modes in the 2D structure with the finite-element method (using the commercial software Comsol) in 2D. The network has 2048 vertices and open boundary. As shown in Fig. 9.16 (d), there are no guided modes with TE polarization in the frequency range $a/\lambda = 0.32$ and 0.38, as a result of the complete PBG for TE guided modes. The highest-Q modes are located at the BEs. They are tightly confined within the structure, as seen in Fig. 9.16 (e, f). The spatial localization of BE modes reduces light leakage through the boundary of the structure, resulting in high Q factor. This is distinct from the periodic structure whose BE modes are extended. Figure 9.16 (e)/(f) also reveals that the intensity of the BE mode at the low/high frequency side of the PBG is mostly concentrated in the dielectric/air. Thus it can be labeled as the dielectric/air BE mode.

The computer generated patterns of 2D trivalent network were transfered to a GaAs membrane containing InAs quantum dots (QDs). Figure 9.17 (a) is a top-view scanning electron microscope (SEM) image of one pattern with $a = 315$ nm. The lateral dimension of each pattern is 9.7 μm × 9.7 μm. A series of patterns with different a were fabricated. The lasing experimental setup is the same as that in the previous Sect. 9.2.2. We realized lasing in a network structure of $a = 315$ nm with optical excitation of InAs QDs. The emission spectrum consists of a few narrow peaks on top of a broad background [top curve in Fig. 9.18 (a)]. The background originates from the broad-band amplified spontaneous emission, while the narrow peaks correspond to the resonant modes. Figure 9.17 (b) plots the intensity of one emission peak at wavelength $\lambda = 1000$ nm versus the incident pump power P. It displays a threshold behavior. When P exceeds 9.8 μW, the emission intensity increases much more rapidly with P. The full width at half maximum (FWHM) of the peak also decreases dramatically with increasing P and reaches the value of 0.28 nm at $P = 16$ μW. Such behaviors indicate the onset of lasing action. The optical image of the lasing mode, shown in Fig. 9.19 (a), reveals the mode is located in the interior of the pattern. Its lateral size was approximately 2.4 μm, significantly smaller than the pattern size (9.7 μm). With a further increase of pump power, we observed lasing in multiple modes.

We repeated the lasing experiment on different network configurations with the same a. They were generated from different jammed packings of polydisperse cylin-

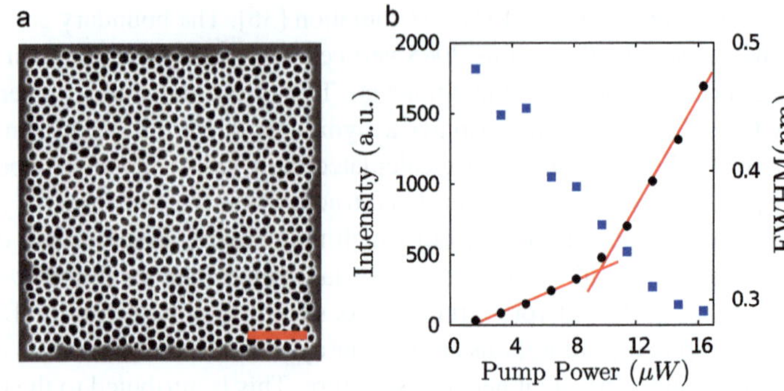

Fig. 9.17 (**a**) A plane-view scanning electron microscope image of the fabricated trivalent network structure in a GaAs membrane with $a = 315$ nm. The *scale bar* is 2 μm. (**b**) Measured intensity (*black circle*) and spectral width (*blue square*) of one emission peak at $\lambda = 1000$ nm as a function of the incident pump power P. Reprinted with permission from [49], Copyright 2008, Wiley-VCH

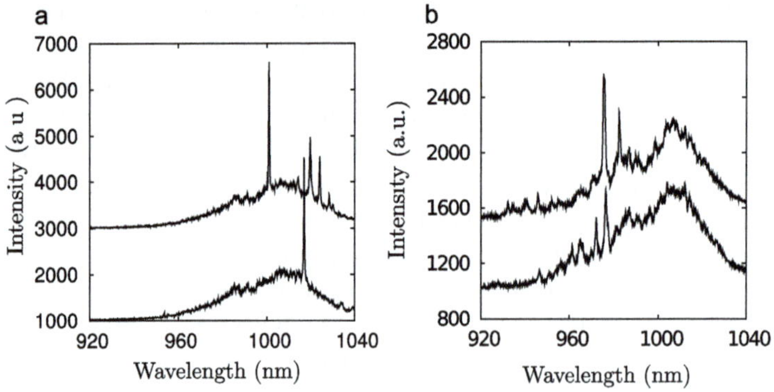

Fig. 9.18 Measured emission spectra for the trivalent network structures of $a = 315$ nm in (**a**), and 275 nm in (**b**). The two spectra in each panel are taken from different configurations. With decreasing a, the lasing modes blue shift. Reprinted with permission from [49], Copyright 2008, Wiley-VCH

ders. Their spatial Fourier spectra and spatial correlation functions are nearly identical, indicating these configurations are statistically equivalent. Lasing was realized in these patterns within the same spectral range, although the frequencies of individual lasing modes varied from pattern to pattern [Fig. 9.18 (a)].

We performed 3D FDTD simulation of the real structures that were extracted from the digitized SEM images. The results illustrated that the lasing modes are located near the dielectric BE of the PBG. The air BE is located at much shorter wavelength $\lambda \simeq 720$ nm. The air BE modes could not lase as they are beyond the gain spectrum of InAs QDs. Although they can be tuned into the gain spectrum by increasing a, the dielectric BE modes are preferred for lasing as they are concentrated in the GaAs and experience more gain from the InAs QDs. We calculated a dielectric BE mode in the 3D FDTD simulation. As shown in Fig. 9.19 (b), the

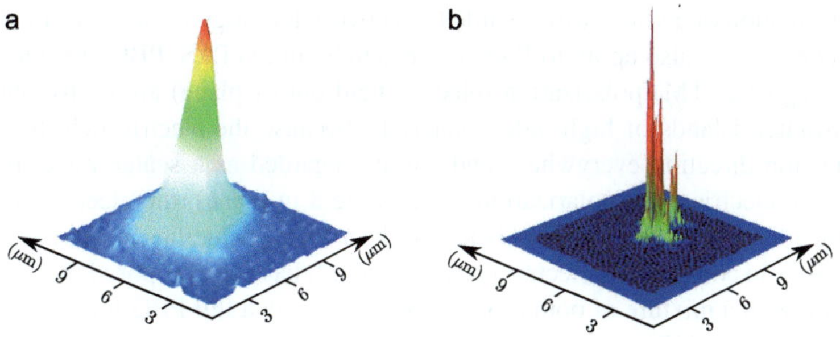

Fig. 9.19 (**a**) Optical image of the lasing mode in Fig. 9.17 (**b**). (**b**) Spatial intensity distribution of the dielectric BE mode calculated in a 3D FDTD simulation of the real structure

mode is spatially localized and has a size is similar to the measured one Fig. 9.19 (a). However, the fine features, e.g., closely-spaced intensity maxima, were smeared out in the optical image due to a finite resolution of our imaging system. The Q factor of this mode is about 6000, which is limited by the out-of-plane leakage of light.

Finally we probed lasing in trivalent network structures of different a. Figure 9.18 (b) displays the emission spectra taken from two configurations of $a =$ 275 nm. Lasing peaks shift to shorter wavelength. This move is consistent with the blue shift of the dielectric BE as a decreases. However, the 3D FDTD calculation predicted that the dielectric BE shift to $\lambda \sim 900$ nm, but the lasing peaks were seen around 975 nm. Although the dielectric BE modes at 900 nm have higher Q, they are far from the peak of gain spectrum and experience much lower gain than the modes at 975 nm. Thus the latter have lower lasing threshold and dominated the lasing spectra.

9.3.2 3D Photonic Network Structures

Although many studies have demonstrated that PBGs can be formed in 2D and 3D PAS with short-range order [10, 14, 16–18, 48, 50], the exact physical mechanism or condition for the PBG formation in PAS is not well understood. Our recent numerical study aims to improve the fundamental understanding of PBG formation in PAS, which would allow researchers to design photonic amorphous materials with optimized and tunable PBGs.

In addition to geometric order, structural topology plays an important role in forming a PBG. For the composite dielectric materials consisting of two components with different refractive indices, there are two cases regarding the topology of the high-index component. (i) Cermet topology: the high-index material consists of isolated inclusions, each of which is completely surrounded by the low-index material. (ii) Network topology: the high-index material is connected and forms a continuous network running through the whole composite. Previous studies of periodic structures have indicated that the cermet topology is more favorable for the

PBG formation of a scalar wave, while the network topology for a vector field [51]. Such conclusions also apply to PAS. For example, in 2D PAS, PBGs for the transverse magnetic (TM) polarization (electric field out of plane) are easily obtained with isolated islands of high-index materials, because the electric field has same polarization direction everywhere and can be regarded as a scalar wave. For the transverse electric (TE) polarization (electric field in plane), the electric field has varying polarization direction and behaves like a vector field, thus it is easier to produce PBGs in connected dielectric networks [52]. It has been proposed that a hybrid structure with a mixture of both topologies can possess a full PBG for both TE and TM polarizations [17].

It is much more difficult to form complete PBGs in 3D structures. Substantial reductions in the density of optical states (DOS) have been demonstrated in PAS composed of randomly packed dielectric spheres of uniform size [14], as a result of evanescent coupling of the Mie resonances of individual spheres. Dielectric network structures, for example, the photonic amorphous diamond (PAD), exhibit much stronger depletion of the DOS [16, 48]. It was conjectured that the tetrahedral bonding configuration in the PAD plays an important role in the formation of isotropic PBG. However, the PAD is constructed from a "continuous-random-network" (CRN) originally developed for modeling of amorphous Si or Ge [53], thus it is difficult to separate the relative contributions of tetrahedral bonding and local geometric order to the PBG formation. Identifying the key parameters that determine when a PBG will form in PAS is important not only for developing novel photonic glasses [54], but also for understanding color generation in nature [24]. Both cermet and network topologies have been found in color-producing PAS of many animal species [27, 28]. It is also conjectured that pseudo PBGs may be formed and responsible for non-iridescent coloration of many PAS [55].

This section presents a detailed numerical study of the DOS and PBGs in 3D PAS. We vary the topology, short-range geometric order, refractive index contrast, and filling fraction to maximize the depletion of DOS and the strength of PBG in the absence of long-range structural order. This study allows us to identify the essential elements for the formation of PBGs in PAS.

Network Generation and Characterization

We first study dielectric composites with the cermet topology—high-index dielectric spheres embedded within a low-index host material (air). We employ a two-stage numerical protocol to generate 'just-touching', jammed sphere packings in a cubic simulation cell with varying positional order [33, 56]. First, liquid states of monodisperse spheres are cooled at fixed packing fraction $\phi = 0.60$ from an initial high temperature T_0 to zero temperature at different rates. In the second step, each zero-temperature configuration is compressed in steps of $\Delta\phi = 10^{-3}$ followed by minimization of the total energy until a static packing with infinitesimal particle overlaps is obtained. By varying the cooling rate, we are able to create static packings with a range of positional order and packing fractions from random close

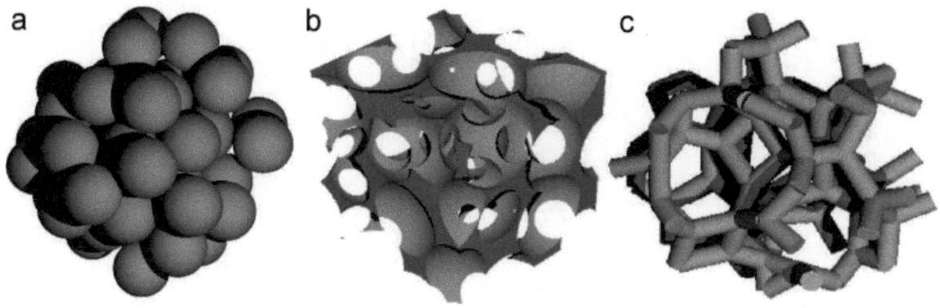

Fig. 9.20 Three examples of photonic amorphous structures: (**a**) jammed packing of dielectric spheres at $\phi = 0.64$, (**b**) inverse structure of (**a**) with air fraction $\gamma = 0.8$, and (**c**) tetrahedral network of dielectric rods with $\gamma = 0.8$ obtained from the Delaunay tessellation of (**a**). Reprinted with permission from [58], Copyright 2008, Wiley-VCH

packing at $\phi = 0.64$ to the face centered cubic (FCC) structure at $\phi = 0.74$. In general, the slowly cooled samples can be compressed to higher packing fractions. Figure 9.20 (a) shows a cluster of 50 spheres from the interior of a jammed sphere packing containing 1000 spheres at $\phi = 0.64$. For comparison, we generate completely disordered configurations by placing spheres randomly in the cubic box with no overlaps at $\phi = 0.35$.

We also generate structures with network topologies, where the high-index dielectric material forms the continuous network, using two methods. For the first method, we invert the cermet structure of jammed dielectric spheres in air. The inverse structure consists of low-index (air) spherical inclusions in a continuous high-index dielectric network. By adjusting the radius R of the spheres (but fixing their positions), we can vary the air fraction γ in the inverse structure. An inverse structure with $\gamma = 0.8$ is shown in Fig. 9.20 (b). At this γ, adjacent air spheres begin to overlap and the dielectric material exhibits an irregular topology.

The second method, which is based on an algorithm described in Refs. [17, 57], produces more uniform network topologies than those from the first method. In this method, a 3D Delaunay tessellation is performed on the sphere centers from the cermet structures. Each tetrahedron of the tessellation has four facets shared with four neighbors. We then calculate the center of mass of each tetrahedron, and connect the centers of mass of nearest neighbors by a dielectric rod. This creates a tetrahedrally connected dielectric network, where each junction (vertex) has four dielectric bonds. All dielectric rods have same radius W, but different lengths d. By changing W, we can vary the air fraction γ. A tetrahedral network with $\gamma = 0.8$ is shown in Fig. 9.20 (c).

We now calculate the density autocorrelation function and spatial Fourier spectra of the cermet and network structures described above. Since the dielectric spheres embedded in air and the corresponding inverted structure possess identical geometrical properties, we focus only on the air spheres and tetrahedral network structures below.

As shown in the inset to Fig. 9.21 (a), the 3D spatial Fourier transform of the tetrahedral network structures displays concentric spherical shells without discrete Bragg peaks, which reflects structural isotropy and a lack of long-range order. The

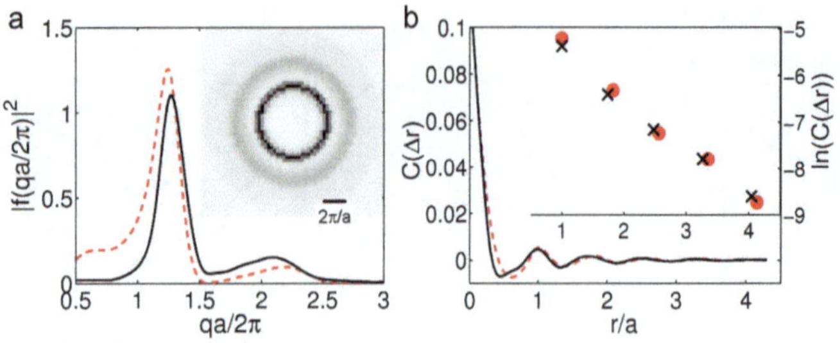

Fig. 9.21 Structural characterization of photonic amorphous structures. (**a**) Angle-averaged power spectra of the spatially Fourier transformed density for jammed sphere packings (*dashed line*) and tetrahedral networks (*solid line*) versus $qa/2\pi$, where q is the spatial frequency and a is the mean spacing between spheres. The *inset* shows a cross-section of the 3D power spectrum for the tetrahedral network. (**b**) Angle-averaged density autocorrelation for the sphere packing and network structures. The *inset* shows the amplitudes of the oscillatory peaks of $C(\Delta r)$ for sphere packings (*circles*) and tetrahedral networks (*crosses*). Reprinted with permission from [58], Copyright 2008, Wiley-VCH

radii of the shells provides the characteristic spatial modulation frequencies of the structures. Similar results are obtained for the tetrahedral networks generated from the jammed sphere packings. The angle-averaged power spectra for both sphere and network structures are plotted in Fig. 9.21 (a). The main peak represents the dominant spatial frequency, and its width is inversely proportional to the average size of ordered domains [19]. The sphere and network structures have similar peak widths, and thus comparable domain sizes.

We also calculated the real-space density autocorrelation function $C(\Delta r)$ averaged over all angles for the sphere and network structures [19]. As shown in Fig. 9.21 (b), both structures display highly damped oscillations of $C(\Delta r)$. The first peak away from $\Delta r = 0$ is located at the average spacing a between nearest neighbors. We find that the amplitudes of the oscillatory peaks decay exponentially [inset to Fig. 9.21 (b)] with a decay length (excluding the first peak) $\xi_r \approx 0.9a$ for the sphere packings and $1.1a$ for the tetrahedral networks. Hence, there are weak spatial correlations and short-range order in these PAS.

DOS of PAS with Cermet and Network Topologies

In this section, we compare the DOS for jammed dielectric spheres in air, the inverse structure, and the tetrahedral networks of dielectric rods using the order-N method [36]. We choose a cubic supercell with size $8.7a$ containing 1000 spheres and refractive indices $n = 3.6$ and 1 for the high- and low-index materials, respectively. We find that the optimal air fraction that yields the largest reduction of the DOS is $\gamma = 0.75$ for the dielectric sphere packings and 0.80 for both the inverse structure and tetrahedral network. The DOS was ensemble-averaged over five distinct configurations at the optimal γ for each topology, and then normalized by the

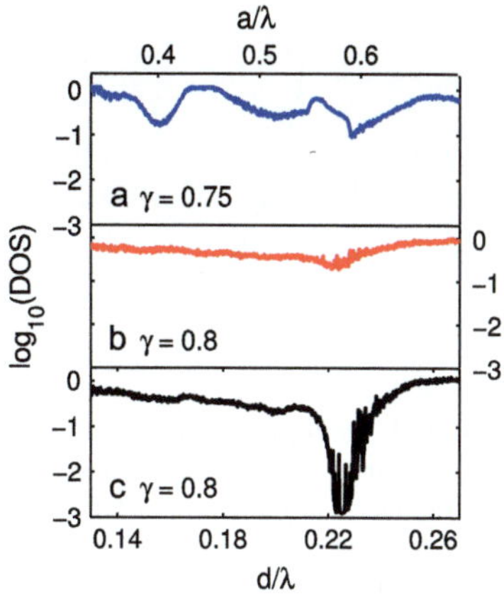

Fig. 9.22 DOS for (**a**) jammed dielectric spheres in air with $\gamma = 0.75$, (**b**) inverted structures with $\gamma = 0.8$, and (**c**) tetrahedral networks with $\gamma = 0.8$. The wavelength λ is normalized by the mean spacing between spheres a (average bond length d) on the *top* (*bottom*) *scale*. Reprinted with permission from [58], Copyright 2008, Wiley-VCH

DOS of a "homogeneous" medium with the same γ. The latter structure is generated by placing cubic dielectric voxels (with lateral dimension $0.043a$, which is much smaller than the wavelength of light λ) randomly in the supercell.

As shown in Fig. 9.22, the maximal DOS reduction occurs in the tetrahedral network structure, which is two orders of magnitude larger than that for the dielectric spheres and inverse structures. For the tetrahedral networks, the PBG is formed at normalized frequency $d/\lambda \approx 0.22$, where d is the average length of dielectric rods and $d/a = 0.39$. The width of the PBG normalized by the gap center frequency is $\sim$5.5 %. The modest reduction in the DOS at $a/\lambda \approx 0.41$ for the dielectric spheres stems from Mie resonances of individual spheres [14]. The uniformity of the dielectric spheres allows the coupling of their Mie resonances, which of the lowest order for isolated dielectric spheres in air occurs at $a/\lambda \approx 0.41$. In contrast, the air sphere structures have only a small reduction of the DOS in the frequency range where the tetrahedral networks show a pronounced PBG, despite the fact that both structures have dielectric network topology and similar degree of spatial correlation. It is clear that the dramatic difference in the DOS cannot be explained by the small differences in spatial correlations.

Our studies of jammed dielectric sphere packings show that uniformity in the size of dielectric spheres leads to strong coupling of Mie resonances that result in a depletion of the DOS. In the inverse structure of air spheres, the basic scattering unit is the dielectric filling between air spheres. For the tetrahedral network structure, the basic scattering unit is centered at each junction where four dielectric rods meet. Note that in the network topology, the adjacent scattering units are connected, in contrast with the cermet topology. To compare the uniformity of local scattering units in dielectric networks, we calculate the average refractive index near the center of each unit. For the tetrahedral network structure, we calculate the mean refractive index $\bar{n}$ within a sphere of radius r whose center coincides with the center of each junction. We then compute the average $\langle \bar{n}(r) \rangle$ and its variance $V(r)$ over all junctions. For the air spheres, the dielectric junction center is set at the center of refractive index distribution within each tetrahedron obtained from the 3D Delaunay

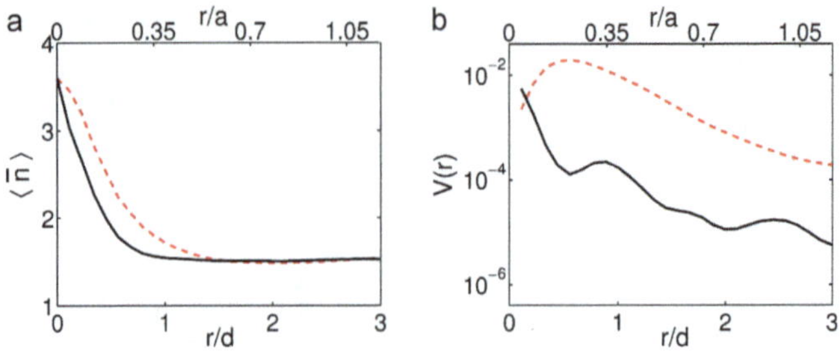

Fig. 9.23 Uniformity of the local scattering environment for the dielectric networks of tetrahedral bonding (*solid line*) and air spheres (*dashed line*). (**a**) Mean index of refraction $\langle \bar{n}(r) \rangle$ and (**b**) variance $V(r)$ within a distance r from the dielectric junction center. r is normalized by the mean spacing of spheres a (average bond length d) on the *top* (*bottom*) *scales*. Reprinted with permission from [58], Copyright 2008, Wiley-VCH

tessellation of the sphere centers. Similarly, we calculate the mean refractive index $\bar{n}$ around each junction center, $\langle \bar{n}(r) \rangle$, and $V(r)$ averaged over all junctions.

In Fig. 9.23 (a) we show that on average the tetrahedral network and air spheres structure have similar distributions of the mean refractive index $\langle \bar{n}(r) \rangle$ around each dielectric junction. In addition, the average refractive index for both networks approaches the same value at large r since the air fraction γ is the same for both structures. However, the variance $V(r)$ of $\bar{n}$ for the two network structures shows marked differences for all r as shown in Fig. 9.23 (b). The tetrahedral network possesses much smaller fluctuations in $\bar{n}$ from one junction to another. Thus, the scattering units are much more uniform for the tetrahedral network than those in the air spheres. The uniformity of local refractive index distribution ensures similar scattering characteristic of individual dielectric junctions and facilitates their coupling which leads to a dramatic depletion of the DOS.

The formation of a PBG in the tetrahedral network structure also depends on the air fraction γ and the refractive index of the dielectric material n. In Fig. 9.24 (a), we show the variation of the PBG for different values of γ while keeping n at 3.6. Reducing the air fraction below 0.8 leads to a decrease in the PBG. A reduction in γ increases the average refractive index of the structure, thus reducing the ratio of the index difference $(n - 1)$ to the average refractive index. It leads to a decrease of the overall scattering strength, and a weakening of the PBG. In contrast, if γ is increased to above 0.8, there is an insufficient amount of high-index material to scatter light. Thus, there exists an optimal air fraction γ at which the scattering strength is maximal and the PBG is the largest. The optimal value of γ varies with the refractive index contrast. As shown in Fig. 9.24 (b), as n decreases, the maximal DOS reduction shifts to smaller γ value. In addition, the DOS dip becomes shallower, reflecting the PBG effect is weaker at lower refractive index contrast. While the depth of DOS reduction changes slightly when n varies from 3.6 to 3.2, it drops by nearly two orders of magnitude with a further reduction of n from 3.2 to 2.8. This threshold behavior indicates there is a cut-off value of n for the PBG formation in the tetrahedral network structure.

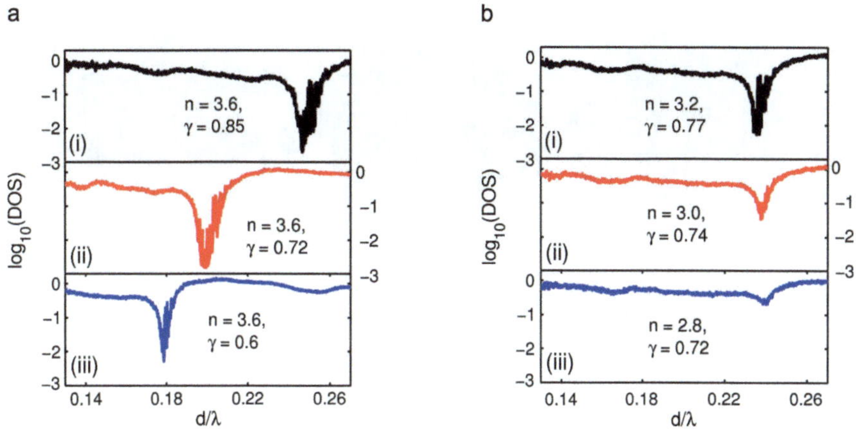

Fig. 9.24 DOS of tetrahedral networks for different values of the air fraction γ and refractive index n. (**a**) $n = 3.6$, (*i*) $\gamma = 0.85$, (*ii*) $\gamma = 0.72$, and (*iii*) $\gamma = 0.6$. (**b**) (*i*) $n = 3.2$, $\gamma = 0.77$, (*ii*) $n = 3.0$, $\gamma = 0.74$ and (*iii*) $n = 2.8$, $\gamma = 0.72$. Reprinted with permission from [58], Copyright 2008, Wiley-VCH

Effects of Short-Range Order

In addition to the factors studied above, short-range positional order and tetrahedral bond order play important roles in the formation of PBGs in PAS. In this section, we focus on the dielectric network of tetrahedral bonding, which yields the largest PBGs, and vary the amount of positional and tetrahedral bond order. In particular, we tune the positional order of the original sphere packings from which the tetrahedral networks are formed. The degree of positional order increases with the volume fraction of spheres ϕ, which varies from 0.35 to 0.69. We label the tetrahedral networks (Fig. 9.25 (a)–(c)) generated from the sphere packings at $\phi = 0.35$, 0.64, and 0.69 as A, B, and C. 2D cross-sections of the 3D spatial Fourier spectra for these structures are presented in Fig. 9.25 (d)–(f). The power spectra of networks A and B consist of concentric shells, but the shell width is notably larger for A. Thus both A and B are isotropic structures, but B possesses more positional order than A. In contrast to A and B, network C features discrete diffraction peaks in the Fourier spectrum, and the structure is no longer isotropic.

In Fig. 9.26, we compare the DOS of the tetrahedral networks A, B, and C, with the refractive index of the dielectric rods set to $n = 3.6$. By adjusting the dielectric rod radius W, we find that the optimal air fraction for all three structures is $\gamma = 0.8$. As expected, network A, with the least positional order, possesses the smallest depletion in the DOS. However, network C with the strongest degree of positional order has a smaller DOS depletion than network B. This result contrasts with recent findings for 2D PAS with air cylinders embedded in dielectric materials that show increasing positional order leads to stronger DOS depletion [19]. To understand these results, we must also compare the uniformity of the local refractive index distribution and the structural topology of the three network structures at

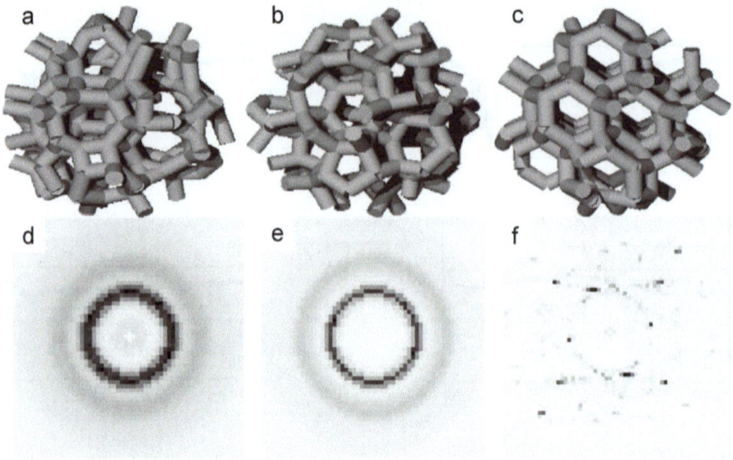

Fig. 9.25 Tetrahedral dielectric networks generated from sphere packings with packing fraction (**a**) $\phi = 0.35$, (**b**) 0.64, and (**c**) 0.69. 2D cross-sections of the 3D spatial Fourier spectra of the corresponding tetrahedral networks are shown in (**d**), (**e**), and (**f**). Reprinted with permission from [58], Copyright 2008, Wiley-VCH

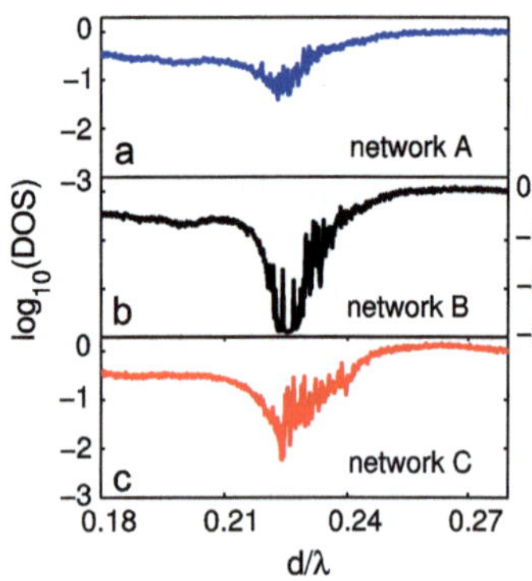

Fig. 9.26 The DOS for three tetrahedral dielectric networks (**a**) A, (**b**) B, and (**c**) C with positional order increasing from A to C. Reprinted with permission from [58], Copyright 2008, Wiley-VCH

fixed radius W of the dielectric rods. We find that networks B and C have comparable fluctuations in $\bar{n}$ over all the junctions. Thus, local uniformity does not explain the difference in the depletion of the DOS for networks B and C.

To investigate the effects of local topology on the depletion of the DOS, we compute the tetrahedral order parameter [17, 48]

$$\zeta = 1 - \frac{3}{8} \sum_{j=1}^{3} \sum_{k=j+1}^{4} \left(\cos\psi_{jk} + \frac{1}{3} \right)^2, \tag{9.10}$$

where ψ_{jk} is the angle between two dielectric rods joined at a junction in the tetrahedral network [59]. For a periodic diamond network, $\psi_{jk} = 109.5°$, $\cos(\psi_{jk}) = -1/3$ for all j and k, and thus $\zeta = 1$ at each junction. If the dielectric rods are ran-

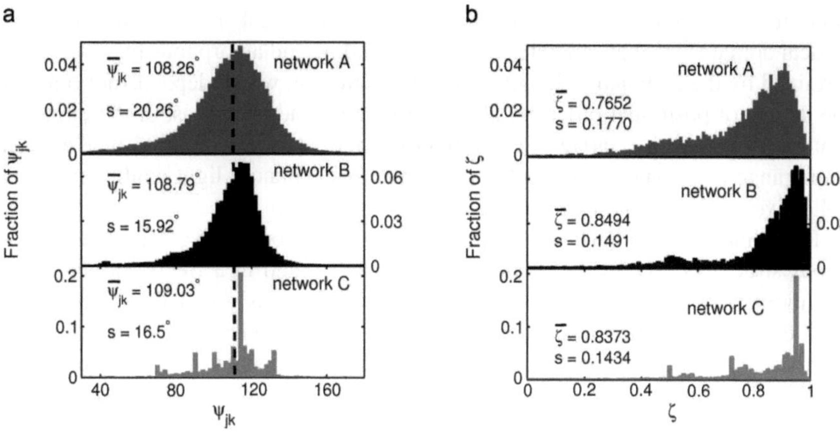

Fig. 9.27 Characterization of the local topology for networks A, B and C. (**a**) The distribution of angles ψ_{jk} between dielectric rods j and k at each tetrahedral junction. The *vertical dashed line* indicates the angle for the periodic diamond structure, $\psi_{jk} = 109.5°$. (**b**) Distribution of the tetrahedral order parameters ζ at each junction. The average $\bar{\psi}_{jk}$ and $\bar{\zeta}$ and standard deviations s_{ψ} and s_{ζ} are also provided. Reprinted with permission from [58], Copyright 2008, Wiley-VCH

domly orientated, $\langle\zeta\rangle = 0$. In Fig. 9.27, we plot the distributions of ψ_{jk} and ζ for the A, B, and C networks, and provide the mean values ($\bar{\psi}_{jk}$ or $\bar{\zeta}$), and standard deviations s_{ψ} and s_{ζ}.

Network A possesses the widest distributions for both ψ_{jk} and ζ, which indicates that the local topology varies significantly from one junction to another and the bond angles within each junction are not uniform. The distributions of ψ_{jk} and ζ are narrower for network B, and are peaked at $\psi_{jk} = 114°$ and $\zeta = 0.95$, which indicates that most of the junctions have a similar topology to that in a diamond lattice. In contrast, network C displays multi-modal distributions for ψ_{jk} and ζ. For example, the ζ distribution possesses peaks at $\zeta = 0.95, 0.72$, and 0.5. The first peak reveals that there are many junctions with strong tetrahedral order, while the second and third peaks reflect the existence of many "defect" junctions with low ζ. Such defect junctions are likely located at domain boundaries, and introduce irregularity in the local configuration of scattering units. Figures 9.26 and 9.27 show that photonic amorphous networks with strong tetrahedral order and few defect junctions have broad PBGs.

9.4 Conclusion

In our numerical study on the density of optical states (DOS) in 2D photonic structures with short-range positional order, we observe a transition from polycrystalline to amorphous photonic systems. In polycrystals, photonic band gaps (PBGs) are formed within individual domains, which leads to a depletion of the DOS similar to that in periodic structures. In amorphous photonic media, the domain sizes are too small to form PBGs, thus the depletion of the DOS is weakened significantly. The critical domain size that separates the polycrystalline and amorphous regimes is determined by the attenuation length of Bragg scattering, which depends not only on the degree of positional order but also the refractive index contrast of the photonic material. Even with relatively low refractive index contrast, we find that modest short-range positional order in photonic structures enhances light confinement via collective scattering and interference.

Experimentally we have demonstrated lasing in photonic polycrystals and amorphous structures. 2D arrays of air holes were fabricated in a free-standing GaAs membrane, and the average size of ordered domains was gradually varied. InAs QDs embedded in the GaAs membrane provide gain under optical pumping. In a photonic polycrystal, defect modes at the center frequency of PBG are tightly confined in-plane, causing strong light leakage out of the plane. The lasing modes shift away from the PBG center to reduce the out-of-plane leakage. In a photonic amorphous structure, the depletion of DOS is significantly weakened, and the lasing modes have less in-plane confinement. Nevertheless, the short-range structural order improves optical confinement and enhances the Q factor at certain frequency. Consequently, lasing becomes the most efficient, i.e., the laser emission becomes the strongest, at such frequency. The optimal lasing frequency can be tuned by the structure factor. The photonic polycrystal laser and amorphous laser are in between the photonic crystal laser and random laser. Our study demonstrates that lasing can be manipulated by varying the short-range order of the nanostructures.

In addition, we compare the DOS in 3D photonic amorphous structures with cermet and network topologies. We find that interconnected networks of high-index material with uniform dielectric junctions and tetrahedral bonding give rise to large isotropic PBGs. Further, reduced fluctuations in the refractive index around each junction and strong tetrahedral order for the angles between the dielectric rods that form the junctions enhance isotropic PBGs. High refractive index contrast and a low fraction of high-index material are also important to PBG formation. We have thus identified several parameters that can be tuned to create broad isotropic PBGs in photonic amorphous structures in the absence of long-range structural order.

We also fabricate 2D trivalent network structures with short-range order in a free-standing GaAs membrane. Such structures display wide isotropic photonic bandgaps. We have realized lasing in the dielectric bandedge modes with optical pumping. The bandedge modes are spatially localized, different from the extended bandedge modes in photonic crystals. By varying the characteristic length scale of the network structure, we can tune the lasing frequency within the gain spectrum of InAs quantum dots.

The future work includes fabrication of 3D photonic network structures, which is more difficult than fabricating 2D structures. Since such structures have only short-range order, they may be fabricated by self-assembly, which is much easier than the fabrication of 3D periodic structures. In fact, 3D photonic network structures have already been produced routinely in nature, and our preliminary studies suggest they are formed by phase separation, e.g., spinodal decomposition [27]. We can mimic nature to make photonic amorphous structures in mass quantity at room temperature [60]. Furthermore we can outperform the natural structures by using inorganic materials with large refractive index contrast to enhance the interaction with light, or incorporating active materials with gain or nonlinearity to achieve new functionalities.

Acknowledgements We thank our co-workers who have contributed to the studies presented in this chapter. Dr. Jin-Kyu Yang and Seng-Fatt Liew performed experimental and numerical studies on PBG and lasing in photonic nanostructures with short-range order. Prof. Corey O'Hern and Carl Schreck developed the numerical methods to generate the amorphous nanostructures. Prof. Glenn Solomon grew the semiconductor samples by molecular beam epitaxy, Dr. Michael Rooks worked with us in fabricating the nanostructures with the electron-beam lithography, and Dr. Mikhail Guy assisted us in the computer simulation. We acknowledge Professors Eric Dufresne, Richard Prum, A. Douglas Stone, Simon Mochrie, and Dr. Vinodkumar Saranathan for many stimulating discussions and the collaborative studies on coloration of photonic amorphous structures in nature. Our research program has been sponsored by the National Science Foundation, and the Materials Research and Engineering Center at Yale University.

References

1. S. Noda, T. Baba (eds.), *Roadmap on Photonic Crystals* (Kluwer Academic, Dordrecht, 2003)
2. M. Meier, A. Mekis, A. Dodabalapur, A. Timko, R.E. Slusher, J.D. Joannopoulos, O. Nalamasu, Laser action from two-dimensional distributed feedback in photonic crystals. Appl. Phys. Lett. **74**(1), 7–9 (1999)
3. O. Painter, R.K. Lee, A. Scherer, A. Yariv, J.D. O'Brien, P.D. Dapkus, I. Kim, Two-dimensional photonic band-gap defect mode laser. Science **284**(5421), 1819–1821 (1999)
4. H.-G. Park, S.-H. Kim, S.-H. Kwon, Y.-G. Ju, J.-K. Yang, J.-H. Baek, S.-B. Kim, Y.-H. Lee, Electrically driven single-cell photonic crystal laser. Science **305**(5689), 1444–1447 (2004)
5. M. Notomi, H. Suzuki, T. Tamamura, K. Edagawa, Lasing action due to the two-dimensional quasiperiodicity of photonic quasicrystals with a penrose lattice. Phys. Rev. Lett. **92**(12), 123906 (2004)
6. K. Nozaki, T. Baba, Quasiperiodic photonic crystal microcavity lasers. Appl. Phys. Lett. **84**(24), 4875–4877 (2004)
7. S.-K. Kim, J.-H. Lee, S.-H. Kim, I.-K. Hwang, Y.-H. Lee, S.-B. Kim, Photonic quasicrystal single-cell cavity mode. Appl. Phys. Lett. **86**(3), 031101 (2005)
8. K. Nozaki, T. Baba, Lasing characteristics of 12-fold symmetric quasi-periodic photonic crystal slab nanolasers. Jpn. J. Appl. Phys. **45**(8A), 6087–6090 (2006)
9. X. Xu, H. Chen, D. Zhang, Enhancement of stimulated emission in 12-fold symmetric quasicrystals. Appl. Phys. B **89**, 29–34 (2007)
10. C. Jin, X. Meng, B. Cheng, Z. Li, D. Zhang, Photonic gap in amorphous photonic materials. Phys. Rev. B **63**(19), 195107 (2001)
11. J. Ballato, J. Dimaio, A. James, E. Gulliver, Photonic band engineering through tailored microstructural order. Appl. Phys. Lett. **75**(11), 1497–1499 (1999)
12. C. Rockstuhl, U. Peschel, F. Lederer, Correlation between single-cylinder properties and bandgap formation in photonic structures. Opt. Lett. **31**(11), 1741–1743 (2006)
13. Y. Wang, S. Jian, Band-gap engineering of amorphous photonic materials. Phys. Lett. A **352**(6), 550–553 (2006)
14. C. Rockstuhl, F. Lederer, Suppression of the local density of states in a medium made of randomly arranged dielectric spheres. Phys. Rev. B **79**(13), 132202 (2009)
15. P.D. García, R. Sapienza, Á. Blanco, C. López, Photonic glass: a novel random material for light. Adv. Mater. **19**(18), 2597–2602 (2007)
16. K. Edagawa, S. Kanoko, M. Notomi, Photonic amorphous diamond structure with a 3D photonic band gap. Phys. Rev. Lett. **100**(1), 013901 (2008)

17. M. Florescu, S. Torquato, P.J. Steinhardt, Designer disordered materials with large, complete photonic band gaps. Proc. Natl. Acad. Sci. **106**(49), 20658–20663 (2009)
18. M. Rechtsman, A. Szameit, F. Dreisow, M. Heinrich, R. Keil, S. Nolte, M. Segev, Amorphous photonic lattices: band gaps, effective mass, and suppressed transport. Phys. Rev. Lett. **106**, 193904 (2011)
19. J.-K. Yang, C. Schreck, H. Noh, S.-F. Liew, M.I. Guy, C.S. O'Hern, H. Cao, Photonic-band-gap effects in two-dimensional polycrystalline and amorphous structures. Phys. Rev. A **82**(5), 053838 (2010)
20. H. Noh, J.-K. Yang, S.F. Liew, M.J. Rooks, G.S. Solomon, H. Cao, Control of lasing in biomimetic structures with short-range order. Phys. Rev. Lett. **106**, 183901 (2011)
21. J.-K. Yang, H. Noh, S.F. Liew, M.J. Rooks, G.S. Solomon, H. Cao, Lasing modes in polycrystalline and amorphous photonic structures. Phys. Rev. A **84**, 033820 (2011)
22. P. Vukusic, R.J. Sambles, Photonic structures in biology. Nature **424**(6950), 852–855 (2003)
23. S. Kinoshita, S. Yoshioka, Structural colors in nature: the role of regularity and irregularity in the structure. ChemPhysChem **6**(8), 1442–1459 (2005)
24. R.O. Prum, Anatomy, physics, and evolution of structural colors, in *Bird Coloration*, vol. 1 (Harvard University Press, Cambridge, 2006), pp. 295–353
25. P. Simonis, J.P. Vigneron, Structural color produced by a three-dimensional photonic polycrystal in the scales of a longhorn beetle: Pseudomyagrus waterhousei (coleoptera: Cerambicidae). Phys. Rev. E **83**(1), 011908 (2011)
26. K. Kertész, Z. Bálint, Z. Vértesy, G.I. Márk, V. Lousse, J.P. Vigneron, M. Rassart, L.P. Biró, Gleaming and dull surface textures from photonic-crystal-type nanostructures in the butterfly Cyanophrys remus. Phys. Rev. E **74**, 021922 (2006)
27. E.R. Dufresne, H. Noh, V. Saranathan, S.G.J. Mochrie, H. Cao, R.O. Prum, Self-assembly of amorphous biophotonic nanostructures by phase separation. Soft Matter **5**(9), 1792 (2009)
28. H. Noh, S.F. Liew, V. Saranathan, S.G.J. Mochrie, R.O. Prum, E.R. Dufresne, H. Cao, How noniridescent colors are generated by quasi-ordered structures of bird feathers. Adv. Mater. **22**(26–27), 2871–2880 (2010)
29. L.F. Rojas-Ochoa, J.M. Mendez-Alcaraz, J.J. Sáenz, P. Schurtenberger, F. Scheffold, Photonic properties of strongly correlated colloidal liquids. Phys. Rev. Lett. **93**(7), 073903 (2004)
30. A. Derode, V. Mamou, A. Tourin, Influence of correlations between scatterers on the attenuation of the coherent wave in a random medium. Phys. Rev. E **74**, 036606 (2006)
31. M. Reufer, L.F. Rojas-Ochoa, S. Eiden, J.J. Sáenz, F. Scheffold, Transport of light in amorphous photonic materials. Appl. Phys. Lett. **91**(17), 171904 (2007)
32. Y. Takeoka, M. Honda, T. Seki, M. Ishii, H. Nakamura, Structural colored liquid membrane without angle dependence. ACS Appl. Mater. Interfaces **1**(5), 982–986 (2009)
33. G.-J. Gao, J. Blawzdziewicz, C.S. O'Hern, Frequency distribution of mechanically stable disk packings. Phys. Rev. E **74**, 061304 (2006)
34. J.Q. Broughton, G.H. Gilmer, J.D. Weeks, Molecular-dynamics study of melting in two dimensions. inverse-twelfth-power interaction. Phys. Rev. B **25**, 4651–4669 (1982)
35. P.J. Steinhardt, D.R. Nelson, M. Ronchetti, Bond-orientational order in liquids and glasses. Phys. Rev. B **28**, 784–805 (1983)
36. C.T. Chan, Q.L. Yu, K.M. Ho, Order-N spectral method for electromagnetic waves. Phys. Rev. B **51**, 16635–16642 (1995)
37. E. Lidorikis, M.M. Sigalas, E.N. Economou, C.M. Soukoulis, Gap deformation and classical wave localization in disordered two-dimensional photonic-band-gap materials. Phys. Rev. B **61**, 13458–13464 (2000)
38. G. Pompe, T. Rappen, M. Wehner, F. Knop, M. Wegener, Transient response of a short-cavity semiconductor laser. Phys. Status Solidi (b) **188**(1), 175–180 (1995)
39. F. Jahnke, S.W. Koch, Many-body theory for semiconductor microcavity lasers. Phys. Rev. A **52**, 1712–1727 (1995)
40. C. Vanneste, P. Sebbah, Complexity of two-dimensional quasimodes at the transition from weak scattering to Anderson localization. Phys. Rev. A **79**, 041802 (2009)

41. J.B. Pendry, Quasi-extended electron states in strongly disordered systems. J. Phys. C **20**(5), 733 (1987)
42. S.G. Johnson, J.D. Joannopoulos, Block-iterative frequency-domain methods for maxwell's equations in a planewave basis. Opt. Express **8**(3), 173–190 (2001)
43. X. Wu, J. Andreasen, H. Cao, A. Yamilov, Effect of local pumping on random laser modes in one dimension. J. Opt. Soc. Am. B **24**(10), A26–A33 (2007)
44. J. Andreasen, H. Cao, Creation of new lasing modes with spatially nonuniform gain. Opt. Lett. **34**(22), 3586–3588 (2009)
45. J. Andreasen, C. Vanneste, L. Ge, H. Cao, Effects of spatially nonuniform gain on lasing modes in weakly scattering random systems. Phys. Rev. A **81**, 043818 (2010)
46. S.F. Liew, J. Forster, H. Noh, C.F. Schreck, V. Saranathan, X. Lu, L. Yang, R.O. Prum, C.S. O'Hern, E.R. Dufresne, H. Cao, Short-range order and near-field effects on optical scattering and structural coloration. Opt. Express **19**(9), 8208–8217 (2011)
47. S. Fraden, G. Maret, Multiple light scattering from concentrated, interacting suspensions. Phys. Rev. Lett. **65**(4), 512–515 (1990)
48. S. Imagawa, K. Edagawa, K. Morita, T. Niino, Y. Kagawa, M. Notomi, Photonic band-gap formation, light diffusion, and localization in photonic amorphous diamond structures. Phys. Rev. B **82**(11), 115116 (2010)
49. H. Noh, J.-K. Yang, S.F. Liew, M.J. Rooks, G.S. Solomon, H. Cao, Photonic network laser. Opt. Lett. **36**(18), 3560–3562 (2011)
50. H. Miyazaki, M. Hase, H.T. Miyazaki, Y. Kurokawa, N. Shinya, Photonic material for designing arbitrarily shaped waveguides in two dimensions. Phys. Rev. B **67**, 235109 (2003)
51. E.N. Economou, M.M. Sigalas, Classical wave propagation in periodic structures: cermet versus network topology. Phys. Rev. B **48**, 13434–13438 (1993)
52. R.D. Meade, A.M. Rappe, K.D. Brommer, J.D. Joannopoulos, Nature of the photonic band gap: some insights from a field analysis. J. Opt. Soc. Am. B **10**(2), 328–332 (1993)
53. G.T. Barkema, N. Mousseau, High-quality continuous random networks. Phys. Rev. B **62**, 4985–4990 (2000)
54. P.D. García, R. Sapienza, C. López, Photonic glasses: a step beyond white paint. Adv. Mater. **22**(1), 12–19 (2010)
55. B.Q. Dong, X.H. Liu, T.R. Zhan, L.P. Jiang, H.W. Yin, F. Liu, J. Zi, Structural coloration and photonic pseudogap in natural random close-packing photonic structures. Opt. Express **18**(14), 14430–14438 (2010)
56. C.S. O'Hern, L.E. Silbert, A.J. Liu, S.R. Nagel, Jamming at zero temperature and zero applied stress: the epitome of disorder. Phys. Rev. E **68**, 011306 (2003)
57. O. Sigmund, K. Hougaard, Geometric properties of optimal photonic crystals. Phys. Rev. Lett. **100**, 153904 (2008)
58. S.F. Liew, J.-K. Yang, H. Noh, C.F. Schreck, E.R. Dufresne, C.S. O'Hern, H. Cao, Photonic band gaps in three-dimensional network structures with short-range order. Phys. Rev. A **84**, 063818 (2011)
59. P.-L. Chau, A.J. Hardwick, A new order parameter for tetrahedral configurations. Mol. Phys. **93**(3), 511–518 (1998)
60. J.D. Forster, H. Noh, S.F. Liew, V. Saranathan, C.F. Schreck, L. Yang, J.-G. Park, R.O. Prum, S.G.J. Mochrie, C.S. O'Hern, H. Cao, E.R. Dufresne, Biomimetic isotropic nanostructures for structural coloration. Adv. Mater. **22**(26–27), 2939–2944 (2010)

Chapter 10
Amorphous Nanophotonics in Nature

Stephen Luke and Peter Vukusic

Abstract Visual appearance generates stimuli associated with many biological functions, including interspecies and intra species communication. A range of biological structural colour mechanisms has been identified. These mechanisms include highly periodic microstructures associated with bright and saturated colours, and amorphous structures which produce broadband colours and generally diffuse reflectances. In this chapter several highly functional amorphous structures found in biological systems are detailed, and their optical characteristics are described.

10.1 Introduction

Colour production in nature can be broadly separated into two distinct categories, based on two levels of animal morphology [1]. The majority of bright appearances found in nature are derived through chemical pigmentation. These are achieved through selective absorption (or scattering) of particular wavelengths of light. An alternative strategy, one which often results in ultra-bright appearances, is the use of sub-micrometer structures. The interaction of light with these intricate microstructures, constructed from material which itself shows little or no intrinsic optical absorption, can produce vivid colours and often stunning visual effects.

Although this chapter will focus primarily on amorphous structures and broadband appearances, this section will introduce the reader to the field of natural photonics as a whole and will include detail of periodic structures. Presentation of details of all structural colour identified to date would be an enormous task and require considerable space. As such a very brief overview of many of the structural colour mechanisms will be given, using examples to highlight the optical effects associated with the structural types. The majority of colour mechanisms that have been identified employ a periodic structure, in either 1, 2 or all 3 dimensions, and result in bright, narrow-band reflectance. Whilst not strictly the remit of this chapter,

S. Luke (✉) · P. Vukusic
School of Physics, University of Exeter, Exeter, UK
e-mail: S.M.Luke@ex.ac.uk

P. Vukusic
e-mail: P.Vukusic@exeter.ac.uk

C. Rockstuhl, T. Scharf (eds.), *Amorphous Nanophotonics*,
Nano-Optics and Nanophotonics, DOI 10.1007/978-3-642-32475-8_10,
© Springer-Verlag Berlin Heidelberg 2013

an introduction to natural photonics would not be complete without describing the microstructure and optical mechanisms which are responsible for the majority of biological structural colour.

The saturated colours and often brilliant visual effects found in nature have inspired significant scientific study. The earliest record of such study is the work of Robert Hooke, reported in his 1665 publication 'Micrographia' [2]. Hooke correctly concluded that the bright colouration of duck and peacock feathers is a structural effect that arises as the result of the interaction of light with alternating layers of two different materials. Newton also undertook studies of avian colour appearance mechanisms and reached a similar conclusion to that of Hooke, this was described in his work 'Opticks' [3], published in 1704. Significant progress in the field was achieved by Lord Rayleigh who used Maxwell's equations of electromagnetism to develop a theory of reflection from stratified surfaces [4]. A subsequent review of the field, published in 1919, concluded that many bright colours found in birds, beetles and beetles were 'structural colours', resulting from thin film interference, rather than pigmentary colours [5].

Since these early works the invention of the electron microscope has revealed a broad range of structural colour mechanisms associated with many intricate structural designs. Whether the purpose is high reflectivity for intra-specific signalling, cryptic colouration or one of several other possible purposes, such optical systems have evolved to perform a function and have been under continuous evolution, in response to selection pressures for, in some cases, hundreds of millions of years. It is therefore unsurprising to find a diverse array of structural colouration mechanisms in the natural world which are seemingly optimised to perform specific biological functions.

Reviewing the literature concerning natural photonic structure will identify a broad range of optical effects in a broad range of animals and insects. From this literature it is apparent that a large proportion of natural photonics arise from either optical interference associated with multilayer structures or with diffraction associated with surface or bulk periodic structures [1]. Multilayer systems, or 1D photonic crystals comprising alternating layers of high and low refractive index materials, are prolific in nature. While some systems exist in which the multilayering exists in broad flat layers, it is more often incorporated within a more specialist design that is best suited to a specific function, physiology or environment [1]. The *Morpho* genus of butterflies is an often quoted example of a specialised 1D photonic structure. The brilliant blue wings (see Fig. 10.1(a)), quoted as being visible from a quarter of a mile away, arise from the interference of light reflected from the discrete multilayer system contained on *Morpho* wing scales (Fig. 10.1(b)). Absorbing pigments and imperfect periodicity amongst the discrete stacks of layers enhances the saturation of the blue colour and the spread the reflection into a wider angle range [6].

Periodicity in 2D is also a geometry that leads to colour production in nature. An interesting example of structural colour effected by a 2D periodic structure is that exhibited by the marine polychaete worm *Aphrodita* (sea mouse). The sea mouse has a body which is covered in short hair-like structures known as setae which display exceptional iridescence (see Fig. 10.1(c)). The core of each setae comprises a bundle of thin wall chitin tubes held together in a 2D hexagonal packed array. Coherent

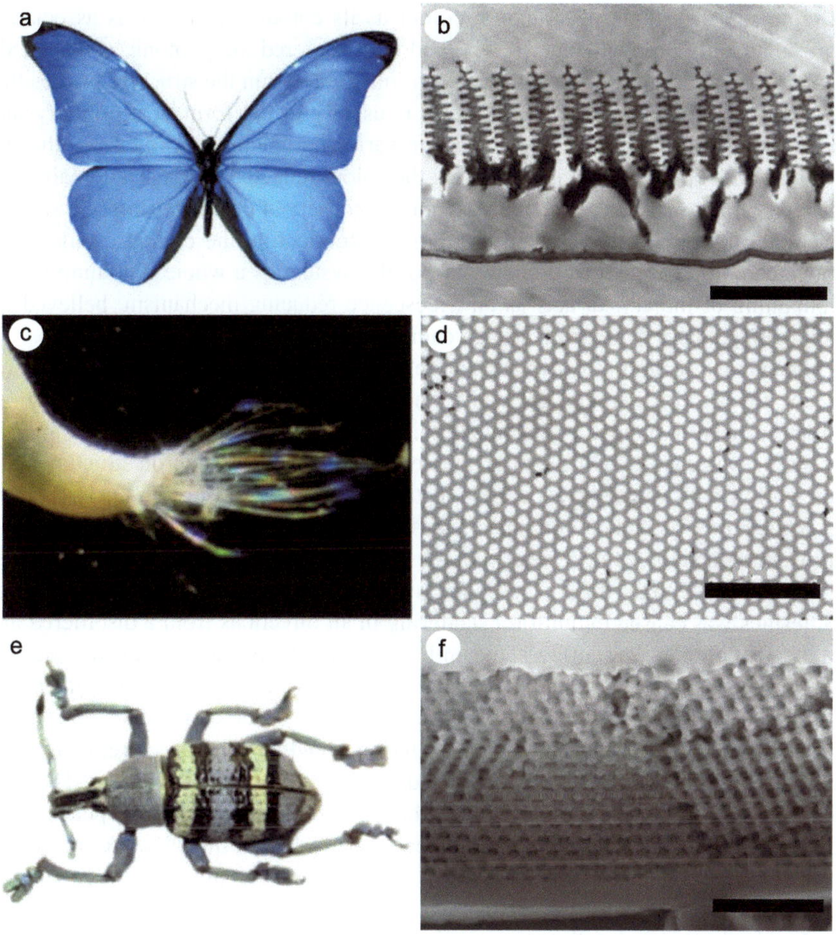

Fig. 10.1 (**a**) Optical image showing the blue iridescent colour of a typical *Morpho* species. (**b**) TEM image of a transverse section through a typical *Morpho* wing scale. Multilayer interference from the 1D periodicity results in the bright blue reflection from the wings. (**c**) Optical image of the 'sea mouse' *Aphrodita*, note the iridescent hair like structures, called setae. (**d**) This iridescence is effected by diffraction from a 2D photonic structure within in setae. (**e**) The weevil *Eupholus magnificus* displays bright blue and green stripes, both colours are structural in origin. (**f**) The green colour results from Bragg scattering of light from a highly ordered 3D photonic crystal within the scales covering the weevils body. *Scale bars*: All 2 μm. (**d**) Reprinted with permission, ©2010, American Physical Society. (**e**) and (**f**) Reprinted with permission, ©2011, Optical Society of America

scattering from this periodic structure results in the coloured reflection. In *Aphrodita* the diameter of the cylindrical tubes varies from 100 nm up to 360 nm, resulting in the large range of reflected colours [1, 7].

Remarkable 3D periodicity has been identified in certain insect species, including both Lepidoptera and weevils. The weevil *Eupholus magnificus* (Fig. 10.1(e)) is a remarkable example of structural colour as its colour appearance is associated with both fully ordered (Fig. 10.1(f)) and quasi-ordered 3D photonic crystals [8]. Coherent scattering of light from planes of material within the structure creates the yellow–green reflection. The 3D crystal is usually in the form of a poly-crystal, smaller domains of the lattice structure in various orientations formed together to form a solid. For any given illumination, the colour reflected from each domain will depend on its orientation, the macroscopically observed colour is simply a spatial average of the reflection from each domain. Reorienting the crystal changes the reflected colour from each domain but not the system as a whole. Put simply, the domaining of the crystal acts as an iridescence reducing mechanism, believed to help with predator avoidance through camouflage [1, 8].

Broadband scattering systems in nature have been identified and studied but there is not a wealth of literature devoted to the subject. Mason [9] dedicated a portion of his early work on insects to white species. Without the aid of the electron microscope he was unable to inspect the detailed nature of the sub-micron structures but his work has proven to be accurate nonetheless. His work inspected many species of white butterfly with colour appearances ranging from chalky white to highly lustrous, near-metallic whites. He concluded that the differences between the observed effects were due to the differences in microstructure. The chalky matt white of pierid butterflies was caused by uniform scattering in all directions from a disordered arrangement of reflecting surfaces, the uniform nature of the scatter results in a lack of 'high lights' (gloss) [9]. Electron microscopy has since revealed this to be the case. Lustrous whiteness, such as that exhibited by species of the moth genus *Euproctis*, is created by a combination of optical scattering from fine scale structures contained within its relatively transparent scales, and stacking of the scales in layers. The difference between matt white and pearly white, Mason concluded, was dependent on the amount of fine scattering structure. Scales that exhibit minimal scattering structure and scales that overlap in a layered manner often exhibit minimal diffuse scatter and therefore display lustrous, near-metallic whiteness [9].

In this chapter recent discoveries of broadband scattering structures found in nature will be discussed. As for the narrow band examples detailed above the focus will be on insects. Broadband scattering microstructure found in both butterflies and beetles will be discussed, including details of the key structural parameters and the mechanisms by which they function.

10.2 Amorphous Nanostructures in Insects

10.2.1 Lepidopteran Colouration

Some of the brightest colours seen in nature are those displayed by butterflies and moths (order Lepidoptera). Such colouration has inspired a significant amount of scientific interest leading to lepidopteran colour appearance strategies becoming widely studied e.g. [6, 10, 11]. The extensive study of lepidopteran systems has led to them being amongst the best understood and has led to the identification of a number of diverse structural types.

The name of the order Lepidoptera is derived from the Ancient Greek words for 'scale' and 'wing'. Almost without exception, all members of the order exhibit scales on their wings. It is the scale covering which is largely responsible for wing colouration, either through pigmentation or through microstructure on the surface of, or contained within, the wing scales [10, 11]. Each scale is a flattened projection of cuticle from an epidermal cell within the surface of the wing. Although a large range of scale structures have been identified, most scales conform to a basic pattern, even the most specialised scales share features with basic unspecialised scales. The scales feature a stalk (*petiole*), which fits into a socket on the wing surface, and a main body of the scale (*blade*). The *blade* has two surfaces, the upper and lower *laminae*. The lower *lamina*, which usually faces the wing surface and is not exposed, is usually relatively flat and featureless. The upper *lamina* usually features prominent ridges running from root to tip [10]. These ridges are often seen to be connected by a perpendicular series of cross-ribs, creating *windows* to the empty scale interior. Pillar-like *trabeculae* appear throughout the scale, with the apparent role of separating the upper and lower *laminae* [10].

Pieris Rapae, the Small Cabbage White Butterfly

The small cabbage white butterfly (*Pieris rapae*) is a member of the Pieridae family and is common across much of the world (see Fig. 10.2(a)). Its chalky white colour is structural in origin, resulting from a disordered structure within its wing scales (see Fig. 10.2(b)). *P. rapae* wing scales conform to the basic scale structure previously described, with longitudinal ridges and perpendicular cross-ribs. The specialised part of the scale structure, seen in many pierid species, comprises a dense array of ellipsoidal beads of pterin, contained within the empty void between the upper and lower scale *lamina* (see Fig. 10.2(c)). The array appears to 'hang' from both the ridges and cross-ribs [12, 13]. A gender difference has been noted with males pierids seen to exhibit a much denser bead-array within their scales [14].

Wing colour of the pierid family has been the subject of numerous studies and significant literature exists. Identification of the scale microstructure was not possible prior to the invention of the electron microscope. Studies, therefore, focused on the macroscopically observable properties of the wing. This work revealed the marked gender difference, identified as significantly reduced UV reflectance for males [15, 16]. The strong UV absorption identified is associated with the pterin class of pigments. Pterins are a class of pigments with differing absorption properties. Erythopterin absorbs wavelengths up to approximately 500 nm resulting in the bright yellow colour of many pierid species [17]. Leucopterin absorbs exclusively in the UV and is the likely pigment present in the scales of *P. rapae*.

Early scanning electron microscope (SEM) studies allowed microscopic analysis of the wing scales and revealed the extent of the presence of the ellipsoidal beads suspended from the scale ridges and cross-ribs [18]. Further research later

Fig. 10.2 (**a**) Optical image showing the chalky white wings of the small cabbage white *P. rapae*. (**b**) As with the majority of lepidoptera, the wing colouration is effected by a covering of scales. (**c**) *P. rapae* wing scales possess ridges and orthogonal cross-ribs, common to most butterfly species scales. The dense array of ellipsoidal beads is a common feature amongst pierid species. The beads effect both strong UV absorption and enhance visible wavelength light scatter. *Scale bar*: (**c**) 1 μm

concluded that the UV absorbing pigment was is isolated within these dense bead-arrays [19–21].

More recent work by Stavenga et al. has suggested that these beads also play an important role in enhancing broadband light scatter from the wing scales [22]. Morehouse et al. presented evidence for this when they detected a correlation between pterin bead-array density and absolute reflectance of the wings of the pired butterfly *Pontia protodice* [19]. To establish a similar trend for the species *P. rapae* it was necessary to alter the bead-array density of the wing scales of that species. Using a method outlined by Rutowski et al. [21] the pterin beads were dissolved using an immersion technique involving isopropyl alcohol and dilute ammonium hydroxide (NH_4OH). SEM imaging of the wing scales post immersion confirmed the reduction in the bead-array density (see Fig. 10.3).

Initially beads suspended in the scale windows are removed (Fig. 10.3(b)); it is likely that these beads are removed first due to the fragile supporting links which hold them within the windows being dissolved. For longer immersion times (Fig. 10.3 (c) and (d)) the beads appear to dissolve, thereby reducing in volume and eventually number until no beads remain. Bead-array density was calculated for all wing samples after immersion in the NH_4OH solutions. Bead-array density was found to be negatively correlated with immersion time in NH_4OH.

Removal of the bead-array had two principle effects on wing reflectance (Fig. 10.4 (a)). At short wavelengths (400–450 nm) reflectance increased from ca. 8 % for untreated wings to ca. 40 % for wings from which all the beads had been

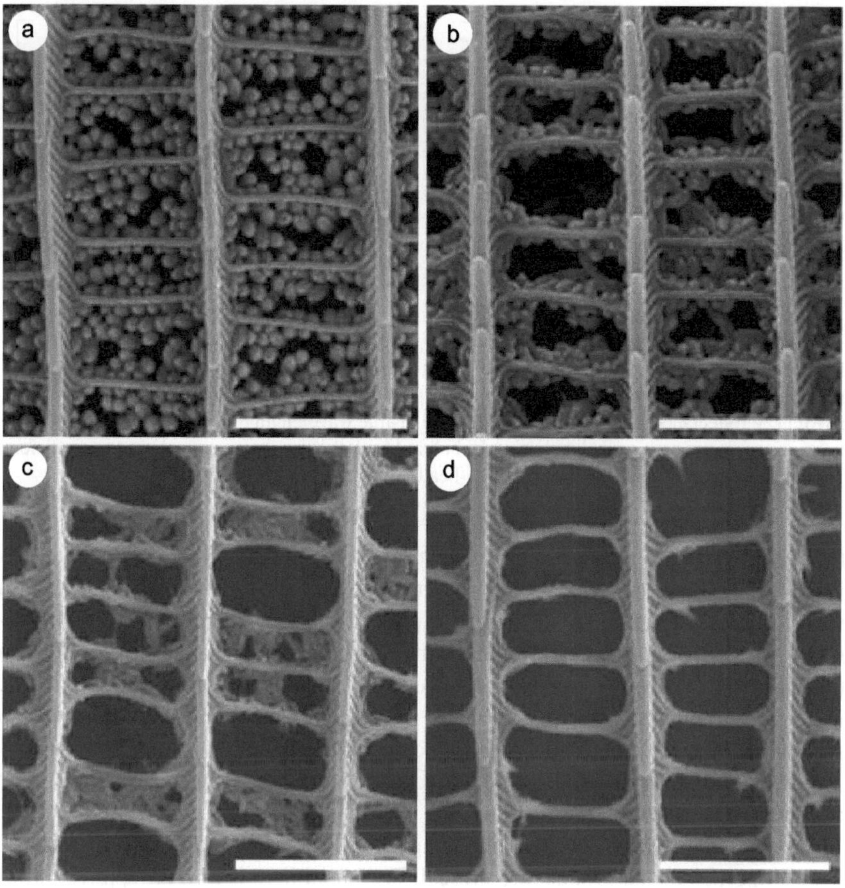

Fig. 10.3 Immersion of *P. rapae* wings in dilute ammonium hydroxide preferentially dissolves the beads over the wing superstructure. Longer immersion times result in greater bead removal. SEM images depicting the extent of the pterin bead array in male *P. rapae wings*: (**a**) Untreated wing; (**b**) after immersion for a short period of time (15–30 seconds); (**c**) after immersion for an intermediate period (60–90 seconds); (**d**) complete bead removal after immersion for approximately 120–150 seconds. *Scale bars*: 2 μm

removed. This is entirely expected due to the removal of the absorption of the pterin pigment associated with the beads. Conversely, reflectance at longer wavelengths was reduced from ca. 65 % for untreated wings to ca. 50 % for treated wings. This is symptomatic of the decrease in optical scatter due to the removal of the pterin bead scattering centres.

At short wavelengths where UV absorption by the pterin pigment is significant, there is a strong negative correlation between total wing reflectance and bead-array density. At a wavelength of ca. 430 nm there appears to be a critical point where the reduction in optical scatter associated with pterin bead removal is balanced by the reduction in optical absorption by the pterin. At approximately this wavelength,

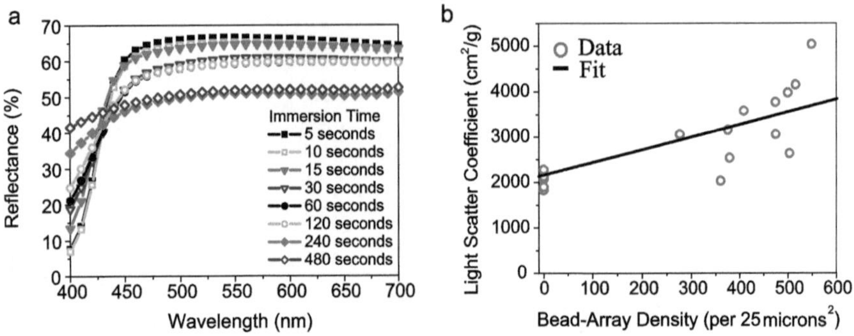

Fig. 10.4 Optical effects of removing the beads from the wing scales of male *P. rapae* butterflies. (**a**) The dual role of the beads is apparent from the reflectance spectra, as the beads are removed UV reflectance significantly increases and visible wavelength reflectance decreases. (**b**) Reflectance spectra were recalculated using Kubelka–Munk theory into an optical scatter coefficient. This quantitatively reveals the extent of non-bead optical scatter

the NH_4OH solution has very little effect on the wing reflectance, despite the major modification of the wing scale structures. At longer wavelengths ($\lambda > 430$ nm), where optical scattering rather than absorption dominates, there is a positive correlation between bead-array density and wing reflectance. This data is consistent with the previous work on *P. rapae* conducted by Morehouse [19] and Stavenga [22], it strongly implies that for *P. rapae*, the wing scale pterin bead-array density is a significant factor in explaining wing reflectance and thus its appearance.

Calculation, using Kubelka–Munk theory [23], of the light scatter coefficient (S) of the wing samples enables their optical properties to be understood more clearly. This technique is regularly used in the paper manufacturing industry where the optical properties of the product are critical. Presented here is a novel application of the theory to a natural scattering system. Figure 10.4(b) shows the relationship between the light scatter coefficient and pterin bead-array density for male *P. rapae* wing scales. Light scatter is observed to increase with increasing bead-array density; this again confirms that the beads are a major contributor to wing reflectance through enhanced optical scatter. This analysis, however, also shows quantitatively that non-bead intra-scale structures contribute to the optical scatter observed [24]. The remaining light scatter results from the bi-grating-like superstrate, or the planar substrate of the wing scale, or a combination of both.

Finite element modelling further indicates the optical scattering role of the scale superstructure. Models consisting of 2D cross-sections through the wing structure were constructed and the simulated scattered power from the system calculated. In the absence of better data, a refractive index of 1.56, similar to that of generic cuticle was used [6]. The pterin absorption profile was also incorporated. The primary focus on the modelling was the presence of the planar scale substrate, an approximately 100 nm thick layer of chitin forming the lower surface of the wing scales. The effect of the presence of the scale substrate is clearly demonstrated in the simulations. Scattered power density in the dorsal hemisphere is significantly increased when the scale substrate is included in the model. A mean enhancement of approximately 2.54 is observed when the scale substrate is included compared to when it was omitted (over a wavelength range of 300–800 nm). The enhancement factor appears to follow the reflectance spectrum of the substrate indicating that the planar substrate acts as a mirror, reflecting light scattered by the beads and enhancing the scale reflectance. The key result of this modelling is the indication that while the pterin beads enhance optical scatter from the scales, they are not solely responsible for it, in line with the optical data presented by previous research and within this section. The remaining scale structure, specifically the scale substrate, is also a vital component in the scattering system.

Significant research has been conducted on the mating habits of pierid butterflies. Obara [25] and Rutowski [26] suggest that UV absorption by the wings is crucial in distinguishing between males and females. This is particularly the case for *P. r. crucivora* where there is a distinct sexual dimorphism. This dimorphism is not so distinct for British *P. rapae* species, the wing scales of both male and female of which are adorned with beads and subsequently exhibit low UV reflectance [14]. Only at longer, visible wavelengths does the magnitude of the reflectance significantly differ between males and females. Obara and Majerus conducted experiments on British *P. r. rapae* and observed that the lack of distinct sexual dimorphism occasionally led to males approaching other males mistakenly in courtship [27].

Rutowski et al. and Kemp et al. suggest that the UV reflectance generated by wing scales in the butterfly *Colias eurytheme* contains information about the quality of the lamellae-based nanostructure that generates the signal [21, 28], giving an honest indication of the condition of the male butterfly. The contrast in reflectance of short wavelengths ($\lambda < 430$ nm) to that of the longer wavelength region from male *P. rapae* wings is directly correlated to the wing scale pterin bead array-density. This may give a similar indication of the fitness of the individual; this hypothesis would need to be confirmed by behavioural studies.

Morpho Cypris

The brilliant blue colour of the wings of *Morpho* butterfly species is one of the most famous examples of so-called structural colour. In the introduction to this chapter we saw that the scale ridges exhibit discrete multilayering in a ‘Christmas tree’ like microstructure. Multilayer interference of light reflected from the periodic structure results in the vivid blue colouration. A theoretical model has been proposed which explains the extraordinary reflective properties of *Morpho* by introducing the cooperation of the regularity and irregularity of the structure [29].

Among the species of the *Morpho* genus, a few, e.g. *M. cypris* and *M. rhetenor helena*, have a white stripe pattern on their brilliantly coloured blue wings. As whiteness is generally associated with disordered structures and it is found here amongst the vivid blue from a highly ordered structure it begs the question, what structural difference is responsible for the optical difference between the blue and white regions of the wing?

Spatial analysis of both the dorsal (inner surface on closed wings) and the ventral (outer surface) indicate that the white stripe pattern is almost identical on both sides of the wing. Removal of the scales reveals that the pattern is repeated on the wing membrane. SEM analysis reveals that the dorsal scales show little structural variation irrespective of blue or white colouration, all dorsal scales exhibit the tree-like structure with little or no variation in ridge number, size or spacing. The ventral scales exhibit the unspecialised lepidopteran scale structure and again little variation is seen between brown and white scales [30].

Optical measurements of the blue and white dorsal scales reveal in both cases a strong reflectance peak at a wavelength of 470 nm. In spite of this spectral spike, the wing stripe still appears white. The crucial difference between the blue and white scales is revealed in the longer visible wavelength reflectance. Strong absorption in the blue scales results in *ca.* 5 % reflectance for wavelengths longer than 550 nm [30]. For white scales this minimum is *ca.* 30 %. The blue light is still strongly reflected from the white scales (as they have the tree-like structure) but the special reflection properties of the *Morpho* wing negate the effect. The blue light is reflected into a very narrow angular range, due to the anisotropy in the ridge lamella structure. The higher reflectance of blue wavelengths is only applicable in a limited viewing angle, in the rest of the viewing hemisphere diffusely reflected light of all wavelengths is visible, resulting in the white colouration. The dorsal scales also contribute to the reflectance, removal of these scales sees a 10 % drop in reflectance for all visible wavelengths [30].

10.2.2 Disordered Nanostructure in Coleoptera

Coleoptera form the largest and one of the most visually diverse animal orders on earth. Their colour appearances range from extremely matt black through highly saturated visible colours to broadband specularly reflecting surfaces [31, 32], many also have detailed colour patterning [33]. A broad range of diverse structural colour systems has been identified in many different beetles; these include two dimensional surface diffraction gratings such as those found in a number of beetles in the *Serica* genus [34, 35] and three dimensional photonic crystals in the scales of many weevils [36, 37]. However, by far the most common structural colour system in beetles is based on multilayers [33]. Different macroscopic appearances, including iridescent colours and broadband reflection, have been attributed to variations in multilayer design [38–40].

In contrast to the structures described in these highly periodic systems, which create saturated colours and sometimes metallic appearances, the production of whiteness requires an absorption-free microstructure that scatters all optical wavelengths equally. Intense optical scatter from a structure containing many disordered reflecting surfaces is require; it is this disorder that is responsible for the uniformity with which all wavelengths are scattered.

Cyphochilus is a scarab beetle reported to display brilliant whiteness from a surface layer of scales with a thickness of only 5 μm [41]. This remarkable optical performance arises from the interaction of light with a random filamentary microstructure contained within the scales. The *Cyphochilus* system exhibits considerably higher scattering efficiency than a typical synthetic counterpart, delivering comparable optical performance for a significantly smaller layer thickness [41].

The remarkable optical properties of *Cyphochilus* suggest optimisation of the scale design for optimal light scatter. The filament system previously identified can be considered as a collection of discrete scattering centres. In such a system, the phenomenon known as optical crowding reduces scattering efficiency if the scatterers are spaced too closely together [42, 43], the brilliant whiteness of *Cyphochilus* indicates that the scale design is optimised to limit optical crowding.

In this section the optimisation of the light scattering nanostructure within *Cyphochilus* scales and those of two further white beetle species, *Lepidiota stigma* and *Calothyrza margaritifera*, is detailed. *Cyphochilus* (Fig. 10.5(a)) is a member of the Scarabaeidae family, often known as scarab beetles. *Lepidiota stigma* (Fig. 10.5(c)) is a chafer beetle also of the Scarabaeidae family. *Calothyrza margaritifera* (Fig. 10.5(e)) is a member of the Cerambycidae family, also known as longhorn beetles or longicorns. All three species are native to south-east Asia.

In each case an amorphous nanostructure, that forms the interior of the scales, has been identified as the source of the specimens' white appearance. Examining the scales of the three species by scanning electron microscope reveals the scales' internal structure. In each case an outer envelope of cuticle surrounds the disordered structure within. Both *Cyphochilus* and *L. stigma* exhibit a dense filamentary network within their scales (Fig. 10.5 (b) and (d)). Typically these filaments have diameters in the range 250–350 nm with a spacing of the order of 700 nm. Image analysis indicates a scale void filling fraction of approximately 50–70 %. The white elytral patches of *C. margaritifera* result from a thick covering of hair-like scales. The internal structure of these is unlike that previously described; rather than filamentary, this microstructure comprises disordered spheres aggregated together into clumps and filaments (Fig. 10.5(f)). Imaging of the structure indicates a sphere diameter of 300 nm and that the accumulated microstructure fills approximately 55 % of the scales internal void.

The amorphous nature of these microstructures results in broadband optical scatter, i.e. all wavelengths of light are reflected with approximately equal efficiency and as a result all three species exhibit a bright white colour appearance. In fact these species all exhibit remarkably efficient reflectance. A sheet of typical office paper has a thickness of 100 microns; this is approximately ten times the thickness of any of these beetle scales. Despite this, the paper only reflects between 5 % and 10 % more light when compared to the beetle scales. Such remarkable optical scatter from so little material indicates an exceptionally efficient and likely optimised scattering system. Optimisation of the scale microstructure design could manifest itself in one of three key parameters; the scale void filling fraction, the scattering centre size and the scattering centre spacing. Where optimisation of these key parameters is evident, strong optical scatter and bright white reflectance occur.

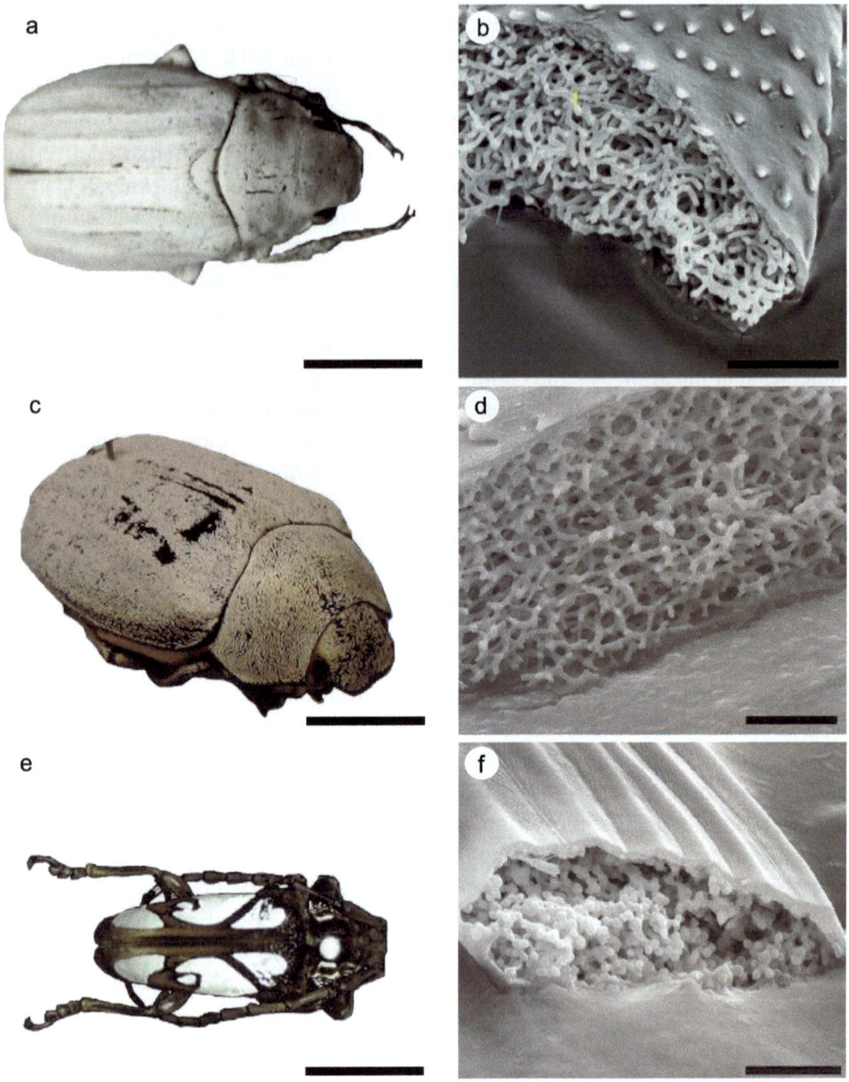

Fig. 10.5 (**a**) The scarab beetle *Cyphochilus* derives its exceptional whiteness from a covering of scales. (**b**) SEM of a single *Cyphochilus* scale revealing its highly disordered microstructure. (**c**) and (**d**) Another scarab beetle *Lepidiota stigma* exhibits both similar whiteness and a similar disordered scale microstructure. (**e**) The gloss white eltyral patches of the longhorn beetle *Calothyrza margaritifera* are also the result of a scale covering. (**f**) With the scales of *C. margaritifera* is a disordered structure unlike that of the scarab species previously described. *Scale bars*: (**a**), (**c**) and (**e**): 1 cm. (**b**), (**d**) and (**f**): 4 μm

The amount of material contained within the scale void appears to be the key parameter that influences the scattering efficiency of the scale. Finite element modelling has been used to simulate the scale structures of the three species and to

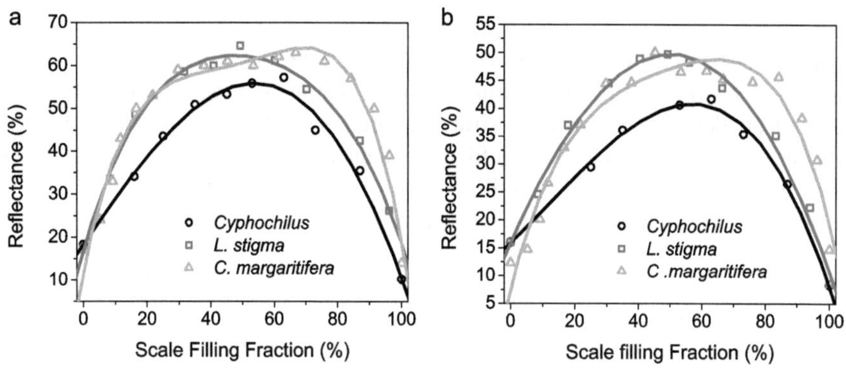

Fig. 10.6 Finite element modelling results which indicate that the scales of each of the three beetles are approximately optimised in terms of scale void filling fraction. In each case the peak in the curve approximately coincides with the scale void filling fraction of each species measured from TEM images. (**a**) TE polarised illumination, (**b**) TM polarised illumination. In each case the *solid line* represents a best fit to the modelled data points

probe the influence of scale void filling fraction on optical scatter. Computational limitations and the highly complex nature of the structures limited modelling to two dimensions. The models were generated by image manipulation of TEM images of transverse slices through the structure. Varying the size of the air pores within these original models varied the filling fraction [44].

Transmission and reflectance were simulated for both TE and TM incident light at normal incidence. The 2D model geometry is set out in the XY plane with the Z-plane defined as into the page. In this geometry, TE incident radiation is defined as the electric vector perpendicular to the XY plane. TM radiation is therefore defined as the electric vector parallel to the XY plane. Whilst reducing the dimensionality of the model was a computational requirement, it does not appear to change the conclusions that can be drawn from the modelled data. In reducing the dimensionality we have removed a component of disorder from the model. The remaining 2D model, however, remains fundamentally disordered. This method may result in a slight reduction in the intensity of the optical scatter modelled, but otherwise accurately models the optical scattering phenomena in question and does not modify the conclusions that can be drawn from the models. The modelled reflectance profiles obtained from the process described above are displayed in Fig. 10.6 for both incident TE radiation (Fig. 10.6(a)) and incident TM radiation (Fig. 10.6(b)).

While the absolute values of the modelled reflectance are not the same for both polarisations, the modelled reflectance vs. scale void filling fraction profiles are very similar. For each species there is a good agreement between the simulated filling fractions at which peak reflectance occurs and the actual measured filling fractions. For TE incident radiation, the modelled filling fraction of *Cyphochilus* scales which results in peak reflectance is 63 ± 6 %. This is very close to the 68 ± 7 % filling fraction of actual *Cyphochilus* scales, measured over a large number of TEM sections. This indicates that approximately 63 % is the optimum filling fraction for the filamentary system within *Cyphochilus* scales. Analogous modelling was carried out for the scale microstructures of *L. stigma* and *C. margaritifera* (also plotted in Fig. 10.6 (a) and (b)). These results indicate that the optimum filling fraction is not independent of the other microstructure parameters; the optimum filling fraction appears to change when a different filament diameter is considered. *L. stigma* scales also appear to exhibit an optimal filling fraction, which at 48 ± 3 % is significantly different to the optimised filling fraction of *Cyphochilus* scales. *L. stigma* scales have filaments with a diameter of approximately 350 nm, compared to a diameter of 250 nm for *Cyphochilus* filaments. It appears that this difference in filament design is responsible for the difference in optimal filling fraction.

The different microstructure of *C. margaritifera* appears to result in a different reflectance vs. filling fraction profile. For both polarisations it is apparent that for this structure, a small variation in reflectance is observed for a wide range of scale filling fractions. This is noticeable for scale filling fractions in the range 35 to 75 %, where very little change in modelled reflectance is observed. This suggests that while a filling fraction in the range 35–75 % is essential for optimal light scatter, it is far less critical for the particle based scattering system of *C. margaritifera* than it is for the filamentary based systems of *Cyphochilus* and *L. stigma*.

Within the microstructures identified, the individual filaments and spheres can be considered as individual scattering units. The size and spacing of these units influences the optical scatter from the system as a whole. The effect of individual scattering element diameter on overall reflectance was investigated using a Mie theory model of a homogeneous sphere with a varying diameter. Three wavelengths representing key colours of the visible spectral range were used in the modelling process. For each sphere diameter, the radiation which was scattered into the reflection hemisphere was integrated and is depicted for each of the three wavelengths in Fig. 10.7(a).

The results of this modelling imply that for wavelengths covering the entire visible spectrum there is an optimum scattering centre diameter which leads to maximised optical scatter. The data suggest that for visible wavelengths, this optimum diameter is in the range 200–300 nm. SEM image analysis carried out previously has revealed that the filaments within the scales of *Cyphochilus* and the spheres within the scales of *C. margaritifera* fall within this size range, implying that their scattering structures are optimally sized to ensure maximal optical scatter from their scales. Interestingly, the filaments within the scales of *L. stigma* have a diameter which does not fall into the optimum range identified here, this is assumed to be the reason for *L. stigmas* apparent poor optical efficiency when compare to *Cyphochilus* and *C. margaritifera*.

The effect of scattering centre spacing on the optical properties of a sample was investigated using synthetic partical-based scattering systems. Blends of titanium dioxide (TiO_2) and solid latex spheres were coated onto base paper in different blend ratios. The fabricated system could be considered as a set of monosized spheres in which the blend ratio between TiO_2 and latex determined the average spacing between TiO_2 particles. Each fabricated system was immersed in matching fluid, blended to optically match the refractive index of the latex spheres, leaving the TiO_2

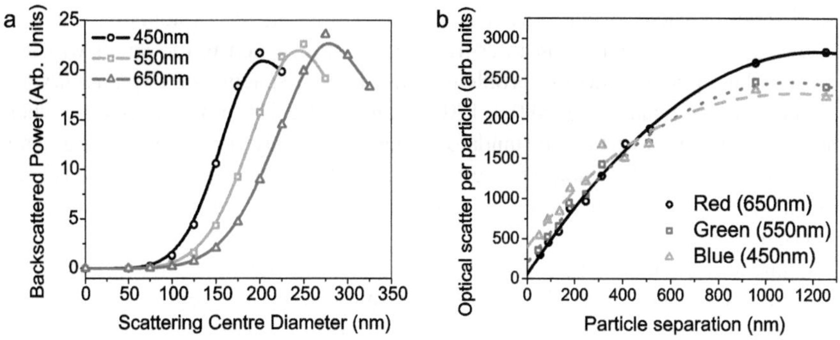

Fig. 10.7 (**a**) Mie theory back scattering results which indicate an optimum centre size for maximum scattering efficiency of visible wavelengths to be approximately 200–300 nm. The beetle scale microstructures have scattering elements (fibrils or particles) that fall within this size range, suggesting optimisation for maximal optical scatter. (**b**) Similar optimisation is apparent with the scattering centre spacing, in the beetle species this is typically 600–700 nm. Optical scatter from synthetic systems indicate this is within an approximately optimal range of 500–700 nm. In both cases, the *lines* represent best fits to the data points

particles in an optically homogeneous medium. Reflectance spectra were taken from each system and Kubelka–Munk theory [23] was used to calculate the light scatter contribution of the individual TiO_2 particles as a function of their spacing.

The optical scatter data collected during this experiment is presented in Fig. 10.7(b). The data reveal that the contribution to optical scatter from each individual scattering particle falls as the particles are moved closer together. The optimum surface-to-surface spacing appears to be approximately 1200 nm, but the most dramatic reduction in light scatter contribution occurs below a surface-to-surface separation of approximately 500 nm. This reduction in scattering efficiency is due to the phenomenon of optical crowding [42, 43]. As the particles are moved closer together the incident light is no longer scattered by the individual particles, instead it is effectively scattered from a larger single particle with subsequent lower scattering efficiency. The data show that the wavelength of incident light has very little effect on this phenomenon.

While avoiding optical crowding is vital for strong optical scatter from a particle-based system, a reciprocal problem must also be addressed. To ensure high optical scatter per scattering centre, the scatterers must be relatively sparsely distributed. In a finite sample, however, this leads to a low spatial density of scattering centre leading to low overall light scatter. Therefore any attempt to overcome the loss in scattering efficiency due to optical crowding must be balanced against the reciprocal loss in total scatter associated with fewer scattering centres per unit volume. The optimum compromise between the two effects appears to occur at an inter-particle spacing of approximately 500–600 nm. At this spacing the scattering particle spatial density is relatively high and dominates over the negative effects of optical crowding.

The filamentary structures identified within the scales of *Cyphochilus* and *L. stigma* differ from these systems as they are based on a connected system, unlike the discrete particle systems fabricated here. However, the systems can be considered analogous because an inter-filament spacing which is too small will lead to a similar optical crowding problem. The filament spacing of *Cyphochilus* and *L. stigma* were measured at 580 ± 120 nm and 700 ± 180 nm respectively, where the error margin quoted is the standard deviation of many ($n = 50$) measurements. This indicates that the filaments are approximately optimally spaced to maximise optical scatter. Optical crowding is largely avoided, while scattering centre density is kept high.

In any amorphous medium designed for efficient broadband optical scatter, there is a fine balance between achieving the maximum scattering-centre density, while maintaining sufficient scattering-centre spacing. If the scattering centres are placed too close together, optical crowding will occur. Balanced against this is the need to maintain sufficient scattering centre number density. If the scattering centres are spaced too far apart, their number density will fall and this will also result in a loss of light scatter. There must clearly be a compromise between these two effects in order to create the most efficient scattering system. Experimental data and theoretical modelling indicate that the scattering microstructures within the scales of *Cyphochilus*, *L. stigma* and *C. margaritifera* have evolved into systems with structural variations that achieve an optimal compromise in this way. Further to this, Mie theory modelling and experimental scattering data indicate that scattering centres with a diameter of 200–300 nm and mutual distance of approximately 500 nm reach an optimal size and spacing. Our SEM and TEM analyses indicate that the microstructures within the scales of all three species are typically within these size and spacing ranges, further implying structural optimisation.

Further compromises in design may also hamper the optical function of beetle scale. Specifically here, for these white beetles, the ultra-thin microstructures in their scales must scatter light efficiently for bright whiteness but must concurrently be relatively light so as not to add unnecessary weight. Additionally, the material itself must be robust enough to limit abrasive damage during the beetle's lifetime.

The bright white colour of these beetles is understood to serve the same primary functions ascribed to most colouration strategies used by animals; namely to communicate with conspecifics or to avoid predation [45]. While bright whiteness may seem counter-intuitive as a cryptic colour in *Cyphochilus*, their natural habitat has been linked to areas rich in white fungi [41]. Their white appearance may therefore aid in predator avoidance. Colour appearance often has a multifunctional role and colours may evolve in response to two or more, often contradictory, selection pressures [46]. While acting as a cryptic colour, the bright whiteness may also aid conspecific recognition. Colour appearance is known to serve other non-visual functions such as thermoregulation [33]. Certainly, it has been shown that any form of dorsal colouration has consequences for thermoregulation in specific species of diurnal beetles [33]. Hadley et al. [46] found that white morphs of the beetle *Neocicindela perhispida* were able to forage for longer without overheating when transferred from their natural white sand habitat to a black sand terrain, compared to the native black morphs of the same species, which escaped the heat by burrowing into the sand.

10.3 Non-insect Examples

10.3.1 White Hairs on the Edelweiss Plant

Leotopodium nivale alpinium (edelweiss) is an herbaceous plant found in the European mountains at altitudes up to 3400 m. In this harsh environment the plant is subject to very dry and windy conditions. The whole plant, including stems, leaves, bracts and flowers are found to be abundantly covered in felt-like hair. This hair is thought to protect the plant through limiting water evaporation. Recent work suggests that the white hairs also play a secondary protection role; they limit the plants' exposure to the high flux of harmful UV radiation in the rarefied high-altitude atmosphere [47].

Investigation of the white covering reveals a 'criss-cross' of hairs, some millimetres in length and with a diameter of approximately 10 μm. SEM imaging of the hair reveals longitudinal filaments running the length of the hairs, spaced in such a manner that they form a diffraction grating-like structure on the surface. The filaments have a size of 180 nm, appropriate for manipulation of UV radiation [47].

Optical measurements of the hair both *in vivo* and when removed from the plant reveal exceptionally broadband reflectance, in line with their white colour. When reflection and transmission are compared similar trends are observed, namely very little UV reflection or transmission. Nearly 100 % of the UVA radiation (300–400 nm) is absorbed by the white hair covering [47].

In this case the term 'absorbed' is used to describe radiation which loses energy while transiting through the filament walls and also radiation which first turns into guided modes in a Fano-resonance process and then disappears with the lifetime of the guided mode. Theoretical models, which approximate the hair surface as a flat diffraction grating, indicate that the filaments allow the incident radiation to excite guided modes which allows the electromagnetic energy to propagate over a much larger distance. It is unclear what the actual absorption mechanism is, whether it takes place within the filament wall itself or whether the core of the hair contains an absorbing compound. One suggested compound is water which has excellent UV absorbing properties [47].

Photonic structures are not common in plants but this example highlights the potential gain from employing photonic strategies. The fleece covering of the plant protects the organism from not only both cold and drought, but harmful UV radiation as well. A photonic structure has evolved which uses diffraction effects to preferentially dissipate UV radiation by allowing the effective energy exchange from the incident wave to guided modes.

10.3.2 Disordered Multilayer Systems

Camouflage, as a means of survival, is highly developed in fish and provides another interesting example of a 'chaotic' reflector system. The most common fish colour appearance is broadband silver; this provides camouflage against the silvery hue of the water surface when viewed from below. In fish, colour appearance is usually attributed to reflection from multilayer stacks (1D photonic crystals) [48]. Multilayering is typically associated with bright saturated (narrowband) colours, but with subtle alterations to the periodic system can bring out a broadband response. Three different perturbations to an ideal quarter-wave plate stack result in broadband reflectance; the use of multiple filters with different spacing, a chaotic multilayer with a random distribution of layer thicknesses and finally a 'chirped' multilayer with an increasing or decreasing layer thickness [31, 33]. The first method is common and is thought to explain the silver colour of many fish species. Of more interest to this chapter is the 'chaotic' system where a random distribution of later thicknesses results in an ultra-broadband response.

Two species of the family *Trichiurodae*, the 'hairtail' and the 'ribbonfish', exhibit such a chaotic reflector. Identified by McKenzie et al. [49], these species were found to reflect approximately 65 % of light from 400 nm right up to well over 1000 nm, an incredibly broad range over which to exhibit such a constant reflectance. Computational modelling using the multiple filter method (with 3 filters with responses centres on 500, 800 and 1100 nm) was found to recreate the observed colour, but not the actual reflectance spectrum. Multilayer systems which consist of alternating layers of high and low refractive index layers and randomly varying optical thickness have been studied and give rise to a phenomenon known as *optical localisation.* The reflected waves generated at the interfaces interfere in such a way that the electromagnetic wave is confined to a region known as the *localisation length.* If the thickness of the system is greater than the localisation length then the propagation of light through the system is prohibited. In the case of non-absorbing material, the incident radiation can then only be reflected [49]. Computational models using high refractive index layers of 1.83 (Guanine) with thicknesses in the range 55–165 nm and low index layers of 1.33 (Cytoplasm) with thicknesses in the range 75–225 nm, and a total thickness of 200 layers (26 μm) were generated. Ten multilayer stacks were generated and the results averaged, giving a good qualitative match to optical measurements. TEM imaging of cross-sections through the skin of the ribbonfish confirms the chaotic reflector system [49].

The random multilayer stack is distinctly different from a quarter wave stack in that the quarter-wave stack is notable for its exceptional efficiency with very few layers but only over a very narrow band. This is undesirable in a system which is trying to achieve a broadband response with a limited material choice. In fish there is little constraint on the thickness of the system so the efficiency of a quarter wave stack brings no advantage, a disordered system is a much more likely result given the beneficial broadband response it brings [49].

10.4 Conclusions

In response to a range evolutionary selection pressures many species have developed complex microstructure based colour appearance mechanisms. In the majority of systems, alternating layers of high and low refractive index result in a surface with a high reflectivity and a bright saturated colouration. Higher degrees of order (2D and 3D photonic crystals) typically results in a similar optical narrow-band coloured reflectance. In order to achieve a broadband reflectance, usually resulting in a white colour appearance, an amorphous (disordered) structure is typically required. Such amorphous structures are not as common in nature as their more ordered counterparts, but some such systems have been identified. The Pierid family of butterflies ('whites and sulphurs') contains many examples of such colour appearance. In this case the disorder is in the form of a dense array of nanoscopic beads which significantly enhance optical scatter of all visible wavelengths, resulting in a broadband, and thus white, response. Amorphous structures have also been identified in several species of beetle. In these cases ultra-thin scales exhibit exceptional whiteness and brightness using a minimal thickness of scattering structure. Such efficiency suggests an optimised scattering structure. Investigation of the structure has confirmed the high degree of optimisation.

With such efficient systems identified it is of little surprise to learn of possible bio-mimetic applications of such microstructure. The filamentary structure of *Cyphochilus* in particular has been of interest. With only 5 μm of material this structure out performs many synthetic systems containing far more material, mimicking its structure and its optimisation could potentially lead to optical advances applicable to many technologies. With so many species from which to choose, including those of plants, fish and insects, there is still a wealth of natural resources that will inspire emerging and future optical technology.

References

1. P. Vukusic, *Optical Interference Coatings*, Natural Coatings (Springer, Berlin, 2003)
2. R. Hooke, Micrographia: or some physiological descriptions of minute bodies made by magnifying glasses, in *Project Gutenberg* (1665)
3. I. Newton, Opticks: Or a Treatise of the Reflexions, Refractions, Inflexions and Colours (Royal Society, London, 1704)
4. L. Rayleigh, On the reflection of light from a regularly stratified medium. Proc. R. Soc. Lond., a Contain. Pap. Math. Phys. Character **93**(655), 565–577 (1917)
5. L. Rayleigh, On the optical character of some brilliant animal colours. Philos. Mag. **37**(217) (1919). doi:10.1080/14786440108635867
6. P. Vukusic, J. Sambles, C. Lawrence, R. Wootton, Quantified interference and diffraction in single Morpho butterfly scales. Proc. R. Soc. Lond. B, Biol. Sci. **266**(1427), 1403 (1999)
7. T. Trzeciak, P. Vukusic, Photonic crystal fiber in the polychaete worm *pherusa sp.* Phys. Rev. E **80**(6), 061908 (2009)
8. C. Pouya, D. Stavenga, P. Vukusic, Discovery of ordered and quasi-ordered photonic crystal structures in the scales of the beetle *eupholus magnificus*. Opt. Express **19**(12), 11355–11364 (2011)
9. C. Mason, Structural colors in insects. I. J. Phys. Chem. **30**, 383–395 (1926)
10. H. Ghiradella, Structure of *iridescent lepidopteran* scales: variations on several themes. Ann. Entomol. Soc. Am. **77**(6), 637–645 (1984)
11. H. Ghiradella, Structure of butterfly scales: patterning in an insect *cuticle*. Microsc. Res. Tech. **27**(5), 429–438 (1994)

12. H. Ghiradella, D. Aneshansley, T. Eisner, R. Silberglied, H. Hinton, Ultraviolet reflection of a male butterfly: interference color caused by thin-layer elaboration of wing scales. Science **178**(4066), 1214 (1972)
13. H. Ghiradella, Hairs, bristles, and scales. Microsc. Anat. Invertebr. **11**, 257–287 (1998)
14. M. Giraldo, D. Stavenga, Sexual dichroism and pigment localization in the wing scales of *Pieris rapae* butterflies. Proc. R. Soc. Lond. B, Biol. Sci. **274**(1606), 97 (2007)
15. F. Lutz, 'Invisible' colors of flowers and butterflies. Nat. Hist. **33**, 565–576 (1933)
16. K. Makino, K. Satoh, M. Koike, N. Ueno, *Sex in Pieris Rapae L. and the Pteridin Content of Their Wings* (1952)
17. B. Wijnen, H. Leertouwer, D. Stavenga, Colors and pterin pigmentation of pierid butterfly wings. J. Insect Physiol. **53**(12), 1206–1217 (2007)
18. J. Kolyer, A. Reimschuessel, Scanning electron microscopy on wing scales of *Colias eurytheme*. J. Res. Lepid. **8**, 1–15 (1970)
19. N. Morehouse, P. Vukusic, R. Rutowski, Pterin pigment granules are responsible for both broadband light scattering and wavelength selective absorption in the wing scales of pierid butterflies. Proc. R. Soc. Lond. B, Biol. Sci. **274**(1608), 359 (2007)
20. N. Yagi, Note of electron microscope research on pterin pigmentation in *pierid* butterflies. Annot. Zool. Jpn. **27**, 113–114 (1954)
21. R. Rutowski, J. Macedonia, N. Morehouse, L. Taylor-Taft, Pterin pigments amplify iridescent ultraviolet signal in males of the orange sulphur butterfly, *Colias eurytheme*. Proc. R. Soc. B **272**(1578), 2329 (2005)
22. D. Stavenga, S. Stowe, K. Siebke, J. Zeil, K. Arikawa, Butterfly wing colours: scale beads make white pierid wings brighter. Proc. R. Soc. Lond. B, Biol. Sci. **271**(1548), 1577 (2004)
23. P. Kubelka, F. Munk, Ein beitrag zur optik der farbanstriche. Z. Tech. Phys. **12**, 593–601 (1931)
24. S. Luke, P. Vukusic, B. Hallam, Measuring and modelling optical scattering and the colour quality of white pierid butterfly scales. Opt. Express **17**(17), 14729–14743 (2009)
25. Y. Obara, Studies on the mating behavior of the white cabbage butterfly, *Pieris rapae crucivora Boisduval*. J. Comp. Physiol. A: Neuroethol. Sens. Neural Behav. Physiol. **69**(1), 99–116 (1970)
26. R. Rutowski, The use of visual cues in sexual and species discrimination by males of the small sulphur butterfly *Eurema lisa* (lepidoptera, pieridae). J. Comp. Physiol. A: Neuroethol. Sens. Neural Behav. Physiol. **115**(1), 61–74 (1977)
27. Y. Obara, M. Majerus, Initial mate recognition in the British cabbage butterfly, *Pieris rapae rapae*. Zool. Sci. **17**(6), 725–730 (2000)
28. D. Kemp, P. Vukusic, R. Rutowski, Stress-mediated covariance between nano-structural architecture and ultraviolet butterfly coloration. Ecology **20**, 282–289 (2006)
29. S. Kinoshita, S. Yoshioka, K. Kawagoe, Mechanisms of structural colour in the *Morpho* butterfly: cooperation of regularity and irregularity in an iridescent scale. Proc. R. Soc. Lond. B, Biol. Sci. **269**(1499), 1417 (2002)
30. S. Yoshioka, S. Kinoshita, Structural or pigmentary? Origin of the distinctive white stripe on the blue wing of a *Morpho* butterfly. Proc. R. Soc. Lond. B, Biol. Sci. **273**(1583), 129 (2006)
31. A. Parker, D. Mckenzie, M. Large, Multilayer reflectors in animals using green and gold beetles as contrasting examples. J. Exp. Biol. **201**(9), 1307 (1998)
32. J. Vigneron, J. Colomer, N. Vigneron, V. Lousse, Natural layer-by-layer photonic structure in the squamae of *Hoplia coerulea* (*Coleoptera*). Phys. Rev. E **72**(6), 61904 (2005)
33. A. Seago, P. Brady, J. Vigneron, T. Schultz, Gold bugs and beyond: a review of iridescence and structural colour mechanisms in beetles (Coleoptera). J. R. Soc. Interface **6**(suppl 2) (2009). doi:10.1098/rsif.2008.0354.focus
34. T. Anderson, A. Richards Jr, An electron microscope study of some structural colors of insects. J. Appl. Phys. **13**, 748 (1942)
35. C. Mason, Structural colors in insects. II. J. Phys. Chem. **31**(3), 321–354 (1927)
36. A. Parker, V. Welch, D. Driver, N. Martini, Structural colour: opal analogue discovered in a weevil. Nature **426**(6968), 786–787 (2003)

37. V. Welch, J. Vigneron, Beyond butterflies—the diversity of biological photonic crystals. Opt. Quantum Electron. **39**(4), 295–303 (2007)
38. P. Vukusic, R. Kelly, I. Hooper, A biological sub-micron thickness optical broadband reflector characterized using both light and microwaves. J. R. Soc. Interface **6**(Suppl 2), S193 (2009)
39. M. Srinivasarao, Nano-optics in the biological world: beetles, butterflies, birds, and moths. Chem. Rev. **99**(7), 1935–1962 (1999)
40. T. Hariyama, M. Hironaka, H. Horiguchi, D.G. Stavenga, The leaf beetle, the jewel beetle and the damselfly; insects with a multilayers show case, in *Structural Colors in Biological Systems: Principles and Applications* (Osaka University Press, Osaka, 2005)
41. P. Vukusic, B. Hallam, J. Noyes, Brilliant whiteness in ultrathin beetle scales. Science **315**(5810), 348 (2007)
42. F. Steig, Ending the 'crowding/spacing theory' debate. J. Coat. Technol. **59**, 96–97 (1987)
43. J. Braun, Crowding and spacing of titanium dioxide pigments. J. Coat. Technol. **60**(758), 67–71 (1988)
44. S. Luke, B. Hallam, P. Vukusic, Structural optimization for broadband scattering in several ultra-thin white beetle scales. Appl. Opt. **49**(22), 4246–4254 (2010)
45. S. Doucet, M. Meadows, Iridescence: a functional perspective. J. R. Soc. Interface **6**(Suppl 2), S115 (2009)
46. N. Hadley, A. Savill, T. Schultz, Coloration and its thermal consequences in the New Zealand tiger beetle *Neocicindela perhispida*. J. Therm. Biol. **17**(1), 55–61 (1992)
47. J. Vigneron, M. Rassart, Z. Vértesy, K. Kertész, M. Sarrazin, L. Biró, D. Ertz, V. Lousse, Optical structure and function of the white filamentary hair covering the edelweiss bracts. Phys. Rev. E **71**(1), 011906 (2005)
48. E. Denton, M. Land, Mechanism of reflexion in silvery layers of fish and cephalopods. Proc. R. Soc. Lond. B, Biol. Sci. **178**(1050), 43–61 (1971)
49. D. McKenzie, Y. Yin, W. McFall, Silvery fish skin as an example of a chaotic reflector. Proc. R. Soc. Lond. Ser. A, Math. Phys. Sci. **451**(1943), 579 (1995)

Chapter 11
Random Light Scattering

Franz-Josef Haug

Abstract This chapter presents optical properties of surface textures without any long range order in their geometry. Such textures are usually called random, as opposed to periodic ones like gratings. The random nature is extremely beneficial for applications where the optical response extends over a wide frequency range; suitable operation over an extended spectral domain is thus ensured by the absence of a preferential period, which would otherwise yield undesired selectivity of light scattering into a certain angle for a fixed frequency. To introduce the reader to this burgeoning field in the context of amorphous nanophotonics, this chapter starts by briefly introducing volume and surface scattering. After introducing a few random surface textures of technological importance, Sect. 11.3 discusses the statistical description of random surfaces in terms of root mean square roughness and autocorrelation length. Light scattering is discussed in terms of scalar scattering theory in Sect. 11.4. In Sect. 11.5 a Fourier theory is presented which avoids some of the limitations of scalar theory. Finally, Sect. 11.6 illustrates the application of random surfaces and the description of their scattering properties in thin film solar cells.

11.1 Introduction

Random light scattering is important for a variety of applications like paper and projection screens, to name just two every day examples. In order to facilitate reading without irritating glare, paper must completely diffuse the incident light. The same holds for projection screens which should reflect the incident light uniformly into all directions, independently of the wavelength. The situation of uniform scattering from an illuminated area into all directions was first described by Lambert and carries his name ever since. For Lambertian scattering, the quantity called bidirectional reflection distribution function (*BRDF*) is thus a constant, independent of angle and wavelength. Scattering properties of surfaces are normally measured by mounting a

F.-J. Haug (✉)
École Polytechnique Fédérale de Lausanne (EPFL), Institute of Microengineering (IMT), Photovoltaics and Thin Film Electronics Laboratory, Rue A.-L. Breguet 2, 2000 Neuchâtel, Switzerland
e-mail: franz-josef.haug@epfl.ch

C. Rockstuhl, T. Scharf (eds.), *Amorphous Nanophotonics*, Nano-Optics and Nanophotonics, DOI 10.1007/978-3-642-32475-8_11,
© Springer-Verlag Berlin Heidelberg 2013

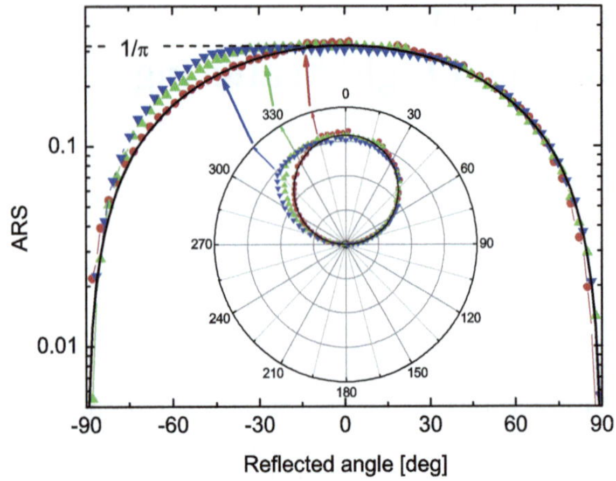

Fig. 11.1 Angle resolved scattering of white paper, measured with a laser of 543 nm. The *inset* shows the same data as polar plot. *Circles, up and down triangles* denote incident angles of 15, 30 and 45 degrees, the *line* illustrates the theoretical curve of Lambertian scattering

small detector on a moving arm with the scattering surface sitting in the center of rotation. This measurement yields a quantity called angle resolved scattering (*ARS*) and since it measures zero intensity whenever the detector lies in the horizontal plane, the two quantities are related by the cosine of the observation angle with respect to the surface normal:

$$ARS = BRDF \cdot \cos\theta \tag{11.1}$$

Normalization is achieved via the total integrated scattering (*TIS*):

$$TIS = \int_{\theta=0}^{\pi/2} \int_{\varphi=0}^{2\pi} (BRDF \cdot \cos\theta) \sin\theta \, d\theta \, d\varphi \tag{11.2}$$

For a white Lambertian scattering with $TIS = 1$, evaluation of the integral requires $BRDF_L = 1/\pi$ and $ARS_L = (\cos\theta)/\pi$. Figure 11.1 shows that white paper is a good approximation for Lambertian scattering, even though the arrows illustrate that high angles of incidence tend to scatter preferentially into the angle that would be expected if the surface were a mirror. Paper is a volume scatterer where the refractive index contrast between cellulose fibers and air yields complete randomization after a series of scattering events. Similarly, the index contrast between water particles and air yields random light scattering in clouds, other examples are paints or correction fluid where the index contrast is achieved between pigment particles and an organic binder.

Mie theory predicts that scattering becomes independent of wavelength when the scattering particles are much larger than the wavelength [1]. Then, the depth of the scattering volume should extend over several scattering particles in order to reach fully random scattering. If complete reflection is desired, e.g. in white reflectors like spectralon, considerable volumes may be required because a multitude of scattering events is required to statistically cancel out all contributions of forward scattering. Moreover, it must be ensured that no absorption occurs in the relevant range of wavelengths.

Experience tells us that random scattering events are not always independent of the wavelength. For example, the sky appears blue because of Rayleigh scattering

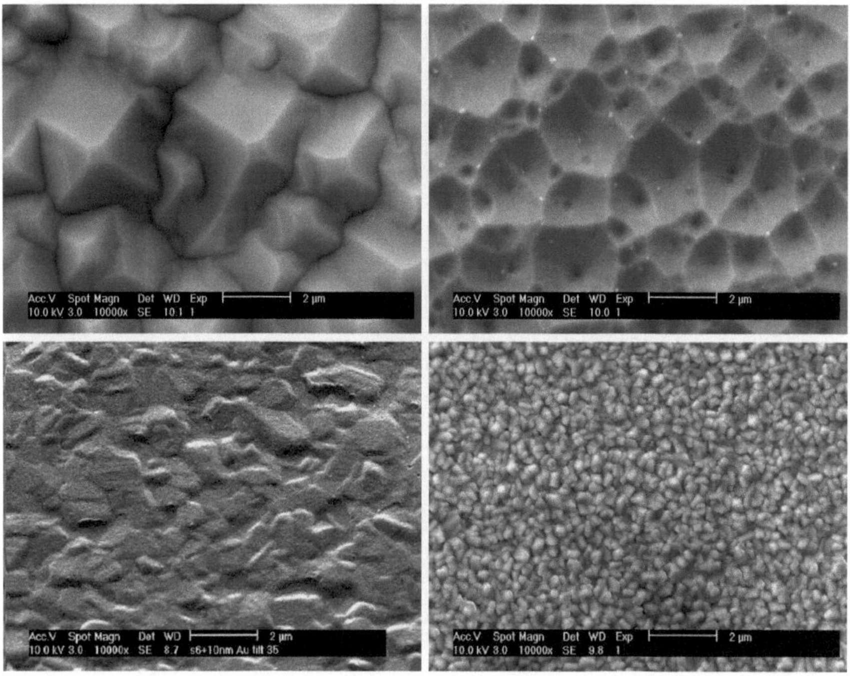

Fig. 11.2 Surface morphologies of randomly oriented pyramids etched into a Si-wafer (*upper left*) [2, 3], etch pits in a ZnO film (*upper right*) [4, 5], facets of a silver film deposited at high temperature (*lower left*) [6, 7], and triangular pyramids of SnO_2 (*lower right*) [8, 9]

which is a limiting case of Mie theory where the wavelength is longer than the circumference of the scattering particle. In this case, scattering follows a $1/\lambda^4$ law, i.e. short wavelengths are scattered more efficiently than long wavelengths.

11.2 Random Surface Textures

After these introductory remarks on volume scattering and the wavelength dependence of the scattering process, the remainder of this chapter will be devoted to surface scattering. In Fig. 11.2, some typical random surface structures are presented which are often used for light scattering in solar cells (see last section of this chapter).

There are two main routes to fabricate random surface textures, etching and faceting during growth of thin films. In most cases, etching starts with the formation of etch pits in the vicinity of imperfections like grain boundaries. If the etched material possesses a defined orientation with respect to the surface, etching is likely to proceed along a preferred orientation, resulting in etch pits with more or less defined side angles. Likewise, during the deposition of thin films, preferential growth of orientations with minimum surface energy can yield surface facets, particularly

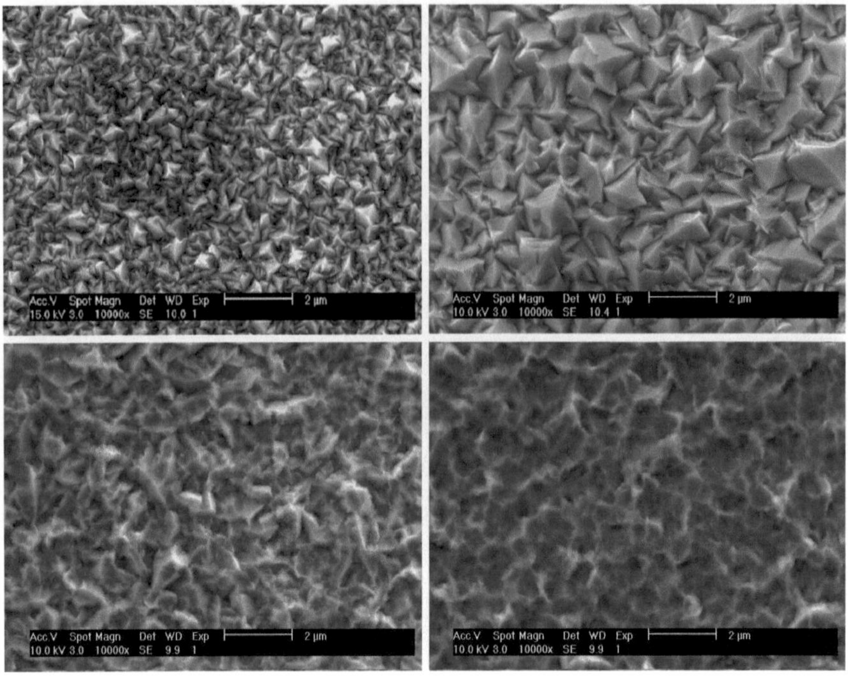

Fig. 11.3 Surface morphologies of ZnO grown by LP-CVD with 2 and 5 μm thickness (*upper left and right*, respectively). The *lower panels* show how plasma treatment can be used to change the surface of the 5 μm thick film [13]

at high substrate temperature [10]. The variability of surface textures is illustrated in Fig. 11.2.

The influence of the deposited thickness is illustrated in Fig. 11.3. For boron-doped ZnO grown by low pressure chemical vapor deposition (LP-CVD), the shape of the facets stays essentially the same since they represent a preferential growth direction, but their sizes increases linearly with the film thickness as smaller pyramids get buried [11, 12]. The shape of the pyramids can be modified by a plasma treatment [13].

11.3 Statistical Properties of Random Surfaces

In this section, a few key aspects used in the theoretical description of random surfaces will be summarized [14, 15]. To do so, the 2 and 5 μm thick ZnO films of Fig. 11.3 are used as example. From their AFM data shown in Fig. 11.4, the root mean square roughness σ_{rms} is evaluated to 68 and 180 nm, respectively.

Lateral information can be obtained by spatial Fourier transform. The power spectral density (*PSD*), sometimes also called surface factor $g(k_x, k_y)$, contains the absolute square of the Fourier components as function of the Fourier wave vectors

Fig. 11.4 AFM images of the $Z2$ and $Z5$ surfaces shown in Fig. 11.3. Image size $10 \times 10\ \mu m^2$

k_x and k_y that they represent. For discrete AFM images which are normally squares of side length S, the wave vectors are defined as $k = m \cdot 2\pi/S$. The integer m is the Fourier order which counts how many full waves fit into the image size. Applying another Fourier transform to the *PSD* does not reproduce the surface information because phase information has been discarded. Instead, the autocorrelation function $\mathrm{Cor}(x, y)$ is obtained. For truly random surfaces, the *PSD* as well as the autocorrelation function are rotationally invariant and it is sufficient to regard only one coordinate. Figure 11.6 illustrates cuts along one of the axes through the autocorrelation functions of the two surfaces shown in Fig. 11.4. Superimposed are Gaussian characteristics with the correlation length C and the surface roughness σ:

$$\mathrm{Cor}(x, y) = \sigma^2 \cdot \exp\{-(x^2 + y^2)/C^2\} = \sigma^2 \cdot \exp\{-(r/C)^2\}. \tag{11.3}$$

For the as-grown Z2 and Z5 surface textures shown in the upper panels of Fig. 11.4, correlation lengths are 208 and 465 nm, respectively. Assuming a Gaussian autocorrelation function, the application of the Fourier transform yields also a Gaussian characteristic for the power spectral density:

$$PSD(k) = \sqrt{\pi} L\sigma^2 \cdot \exp\{-(\pi k C)^2\}. \tag{11.4}$$

While being a convenient approximation, (11.3) is not always justified on real surfaces. Elson reported that the autocorrelation function often contains a contribution of an exponential relationship [14]. This is also the case in the shown examples. Figure 11.5 shows that the Gaussian overestimates the experimental data for small displacements while it decays too fast for large displacement. The deviations become particularly clear in Fig. 11.6 where the experimental PSD and the Gaussian approximation are compared. The logarithmic presentation shows that the Gaussian approximation decays too fast. It applies to some extent in the shaded region whose significance will be explained in the next section.

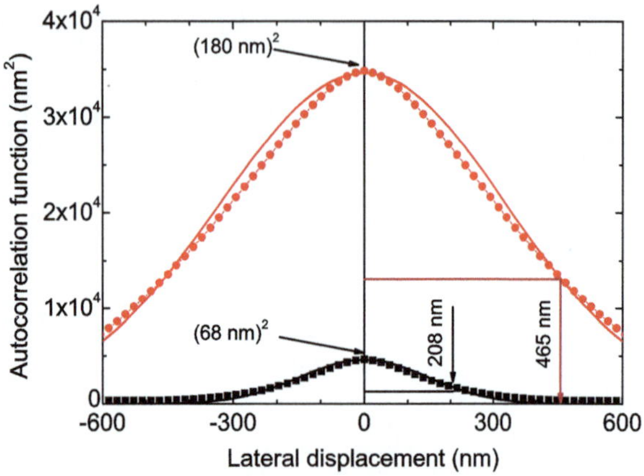

Fig. 11.5 Autocorrelation function of the $Z2$ and $Z5$ surfaces shown in Fig. 11.4. *Squares and circles* denote measured data of $Z2$ and $Z5$, respectively, *lines* illustrate Gaussian approximations

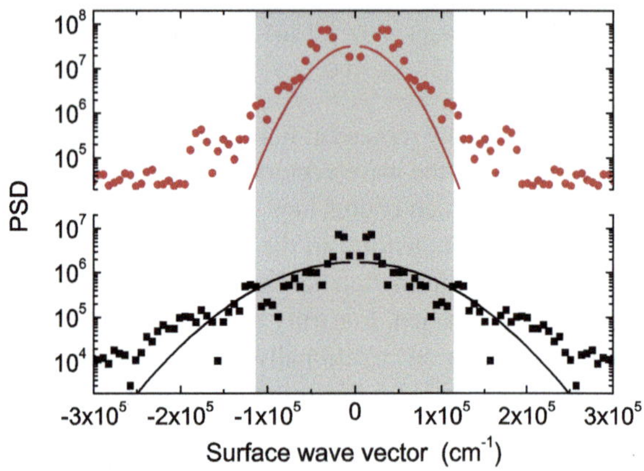

Fig. 11.6 Power spectral density of the surfaces shown in Fig. 11.4. *Squares and circles* denote measured data of $Z2$ and $Z5$, respectively. The *lines* illustrate again Gaussian approximations

11.4 Scalar Scattering Theory

Historically, surface scattering has been thoroughly investigated for remote sensing applications based on radar waves where ocean waves, forests or deserts as well as building surfaces are successfully treated as random scattering centers [16–18]. The subject has been reviewed in several monographs [19–21]. In this context, the reflected radiation is normally composed from two parts. The first is specular reflection, i.e. reflected under an angle equal to the incident angle, similar to the case of a flat surface. The second part consists of radiation that is scattered diffusely into the whole angular domain. The description is greatly simplified if the height information of the surface features with respect to the mean level of the surface follows a Gaussian distribution with standard deviation σ. For a probing wavelength λ, the intensity of the specularly reflected beam R_{spec} is then expressed in terms of the total reflectivity R_{tot} as follows [22]:

$$R_{spec} = R_{tot} \cdot \exp\left\{ -\left(\frac{4\pi \cdot n}{\lambda} \sigma \cdot \cos\theta_1 \right)^2 \right\}. \tag{11.5}$$

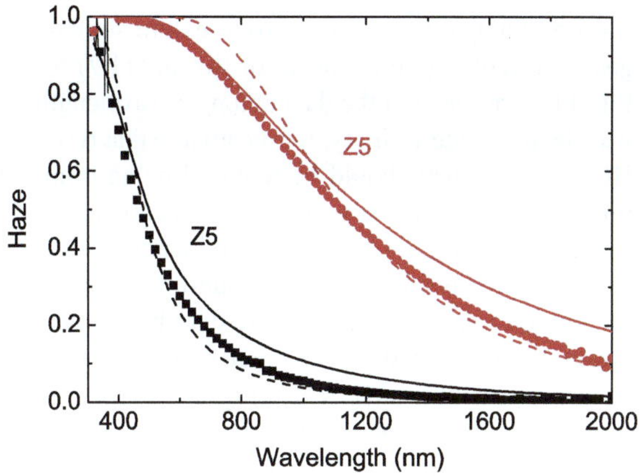

Fig. 11.7 Haze in transmission from textured ZnO into air. *Squares and circles* represent Z2 and Z5, respectively. *Full and dashed lines* are according to (11.7) with exponents of 2 (scalar theory) and 3 (empiric modification), respectively

Different from the original theory dealing with radar waves in air, (11.5) includes the refractive index n to account for situations where scattering occurs in a medium. Finally, the angle θ_1 is measured between the incident beam and the surface normal. Since polarization effects were neglected in the derivation of (11.5), the treatment is often referred to as scalar scattering theory.

Based on (11.5), the reflection haze H_R, i.e. the ratio between diffusely scattered and totally reflected light, can be defined as follows:

$$H_R = \frac{R_{tot} - R_{spec}}{R_{tot}} = 1 - \exp\left\{-\left(\frac{4\pi \cdot n}{\lambda}\sigma \cdot \cos\theta_1\right)^2\right\} \approx \left(\frac{4\pi \cdot n}{\lambda} \cdot \cos\theta_1\right)^2. \quad (11.6)$$

The right hand side of (11.6) illustrates that surface scattering is proportional to $1/\lambda^2$ rather than the $1/\lambda^4$ dependence of Rayleigh scattering [22]. Other than remote sensing, the theory of random surface scattering was also very successfully applied in photovoltaics [23–25]. Upon entering into such structures, light is scattered not only in reflection, but also in transmission. Analogously to (11.6), the haze H_T for transmission through an interface between two media with refractive indices n_1 and n_2 is given by [22]:

$$H_T = \frac{T_{tot} - T_{spec}}{T_{tot}} = 1 - \exp\left\{-\left(\frac{2\pi\,|n_1 \cdot \cos\theta_1 - n_2 \cdot \cos\theta_2|}{\lambda}\sigma\right)^2\right\}. \quad (11.7)$$

Here, θ_1 and θ_2 denote the angles of the incident and the specularly transmitted beams, respectively. Since both angles are measured with respect to the surface normal, they are related by Snell's law.

Figure 11.7 shows experimental haze data of the ZnO surfaces shown in Fig. 11.3. The figure includes theoretical results according to (11.7), using the materials' refractive index dispersion and their root mean square surface roughness as measured by AFM. The scalar theory appears to apply only in the range of short wavelengths, but overestimates the haze in the near IR region. This observation led to the proposition of empirical changes. For example, for the transmission haze of textured SnO_2:F which develops a texture similar to the ZnO surface, it was suggested to replace the exponent of two in (11.7) by three [23]. The dashed lines in Fig. 11.7 appear to fit the data at longer wavelengths, but they overestimate at short wavelengths. Interestingly, for the texture that results when ZnO is etched in diluted HCl, the exponent should be replaced by an even higher value while at the same time the roughness σ is smaller than its corresponding values of surface profilometry [26]. However, none of these modifications can be explained conclusively.

The quoted results of scalar scattering can be used to describe the haze, at least on a global scale. Apart from the distribution between specular and diffuse light, however, the theory does not predict the angular properties of scattering. This deficiency can be illustrated by looking at relief gratings which are the periodic version of surface corrugations. The diffraction efficiency, i.e. the distribution of intensity between the grating orders, depends sensitively on shape and depth of the grating while the angular properties of diffraction are governed by the period. In order to address this deficiency in the scalar theory, lateral information like the autocorrelation length must be used as well [27, 28].

In the spatial Fourier analysis, the random surface is decomposed into a sum of sinusoidal gratings and the amplitude of these individual gratings is represented by their respective Fourier components. Since the superposition principle holds, an incident light beam gets diffracted at each of these grating periods. Even monochromatic light will get diffracted into a variety of angles. Looking at the power spectral density in Fig. 11.6, it can further be assumed that large periods contribute more than very small periods. This allows one to define a (crude) estimate for the *ARS* by the following procedure:

1. Fix the wavelength and the angle of the incident light (e.g. 543 nm and perpendicular).
2. For each period represented in the *PSD*, use the grating equation to calculate the angle into which light gets scattered.
3. Weigh all of these contributions by the value of the *PSD* and multiply with the cosine of the diffraction angle to get the ARS. Note that this procedure cuts off all components with period smaller than the wavelength (for 543 nm, everything outside the shaded area in Fig. 11.6).
4. Since #3 disregards some Fourier components, normalization is needed. Integrate the scattered light intensity according to (11.2), leaving out $\theta = 0$ which represents the specular beam.
5. Finally, normalize the result to the haze value at this wavelength.

The *ARS* obtained by this procedure for the *Z*2 and *Z*5 textures is illustrated by the full lines in Fig. 11.8. Since the *ARS* is rotationally invariant for the case of perpendicular incidence, only positive angles are shown. Overall, the agreement with the experimental result is acceptable. Scattering into small angles appears to be overestimated in both cases while at high angles, the procedure works better for the *Z*5 texture. The *ARS* based on the Gaussian approximation to the *PSD* works well for the *Z*2, but not at all for *Z*5 (dashed lines). This observation is not surprising since Fig. 11.6 already showed that the Gaussian approximation underestimates the

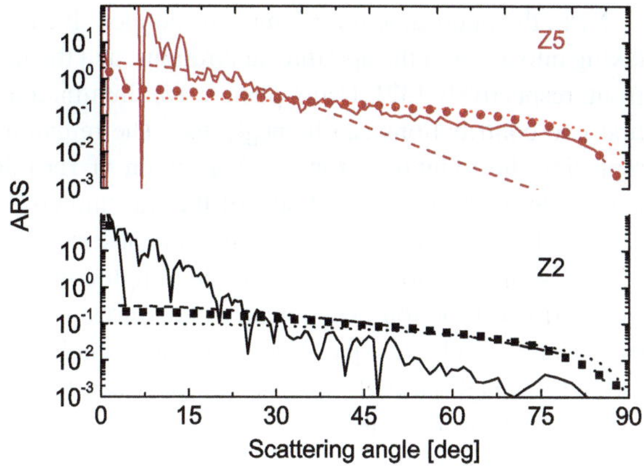

Fig. 11.8 Angle resolved scattering of the $Z2$ and $Z5$ surfaces upon scattering in air at a wavelength of 543 nm. Owing to rotational symmetry, only the radial variation is shown. Symbols refer to measured data, *lines* denote theoretical ARS data based on AFM data (*full lines*), a Gaussian approximation to the PSD (*dashed*), and a Lambertian cos-law (*dotted*)

wings of the PSD, particularly so for the $Z5$ texture. Using a sum of Gaussian and exponential should yield an improvement in this case. For illustration, the graph contains also the cosine characteristic of uniform scattering into all angles, i.e. like Lambertian but with specular part (dotted line). This fits rather well to both surfaces and over the whole range of angles except for the specular contribution.

In conclusion, it appears that the scalar theory describes the overall behavior of the haze. It seems well suited to short wavelengths, but it overestimates the haze at long wavelengths. Unfortunately, it cannot be decided a priori in which region the formula applies reliably. Likewise for the angle resolved scattering, the approach with the *PSD* based on AFM data appears to be valid globally, but rather unpredictable in detail. If predictive power is requested of a simulation tool, a more reliable approach is mandatory.

11.5 Fourier Scattering Theory

This section presents an approach based on Fourier optics whose principles are discussed in the books Goodman and Maradudin [20, 29]. Similar to the previous section, the individual steps of the procedure are illustrated using AFM data of the $Z2$ and $Z5$ surfaces [30]. The following considerations are based on an application of the Rayleigh-Sommerfeld diffraction integral [29]; after some modifications discussed by Harvey, it yields a relation for the radiance L that is received by an observer from the illuminated scattering element (or pupil) U_0 [31].

$$L(\alpha, \beta) = \frac{\lambda^2}{A_s} \left| F\{U_0(\hat{x}, \hat{y})\} \right|^2. \tag{11.8}$$

Here, α and β denote the direction cosines towards the position of the observer, λ and A_s are the wavelength and the surface of the scattering surface, respectively, and $F\{\ldots\}$ denotes the Fourier transform. Finally, $\hat{x}$ and $\hat{y}$ denote the in-plane coordinates after scaling with the wavelength.

Next, the pupil function U_0 must be defined. It can be separated into three terms taking into account the aperture, aberrations, and the modulation of the actual wavefront, respectively [32]. Using plane wave illumination without imaging optics, the first two contributions can be neglected. The remaining term can be described by replacing the roughness zone with a screen of zero depth that changes either the amplitude or the phase (or both) of the transmitted light plus as additional phase changes from the scattering element to the observer [29].

The simplest form of screen is the amplitude screen. It is conveniently used to describe diffraction phenomena of apertures, using areas which are either fully transparent or fully opaque. More generally, amplitude screens also describe transmission through partially transparent materials like photographic film. It appears inappropriate to use the height information of the ZnO film as amplitude screen since it is transparent for visible light, but this type of definition can nevertheless be found in the literature. According to (11.8), this would predict that the radiance is proportional to the power spectral density. Not surprisingly, this result was found inadequate in the previous section and the underlying reason becomes clear now. Moreover, the background subtraction of AFM routines can be set to yield a mean amplitude of zero, resulting in a value of zero for the zero frequency term. Since the height reference of surface data is arbitrary, the zero frequency term of the numeric PSD becomes meaningless [33, 34].

Alternatively, the surface texture can be described as an element that changes only the phase during the transit of the roughness zone $\zeta(x, y)$ as illustrated in Fig. 11.9. The phase change after the transit through the roughness zone depends on the path lengths which are traveled within the two different materials. For a hemispherical geometry, the following pupil function U_0 applies [32]:

$$U_0(\hat{x}, \hat{y}) = \frac{1}{(1+\varepsilon)^2} \cdot e^{i \cdot const} \cdot e^{ik\zeta(\hat{x},\hat{y})\cdot(n_1-n_2)}. \tag{11.9}$$

In (11.9), $k = 2\pi/\lambda$ denotes the wave vector in vacuum. The constant term in the phase change can be neglected because it does not contribute in a Fourier transform and drops out subsequently during the calculation of the absolute value. The term ε in (11.9) is the ratio between the distance from the origin to the points $(\hat{x}, \hat{y})$ of the integration (typically the diameter of the incident laser beam) and the distance to the detector sweeping out a hemispherical surface [32]. For typical measurement geometries ε can be neglected, making (11.9) computable with the fast Fourier transform (FFT) algorithm. The definition of (11.9) is easily generalized towards the case of reflection by replacing the difference of the refractive indices in the exponent by twice the index of the medium in which incident and reflected beam propagate [35, 36].

The definition in terms of phase screens differs from the amplitude screen in a very important aspect since it defines the contribution of zero frequency term unambiguously. In agreement with the experiment, the thickness of the sample below the actual roughness zone is not relevant to light scattering and drops out of the calculation. The zero frequency term therefore describes the intensity of the specular beam. Based on this reasoning, Harvey suggested a normalization procedure which states

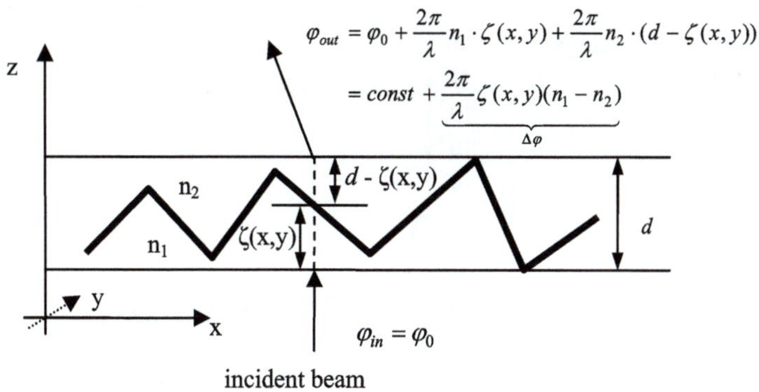

Fig. 11.9 Illustration of the phase change upon propagation through the surface roughness zone denoted by the *horizontal bars*

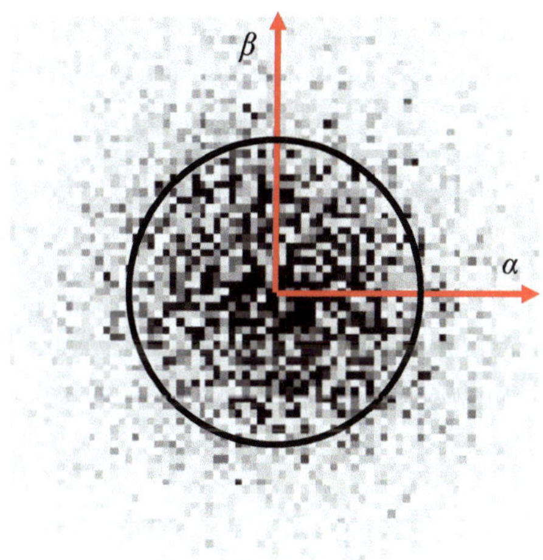

Fig. 11.10 Illustration of the normalization procedure. The greyscale of the pixel array represents the intensities of the Fourier components for *Z*2 surface and a wavelength of 543 nm (excepting the zero frequency term). The unit circle ($\gamma = \sqrt{\alpha^2 + \beta^2} = 1$) discriminates evanescent modes (*outside*) from propagating modes (*inside*). For the shown conditions, the radius of the unit circle extends over 18 points

that the total observable light intensity in air is carried only in the waves with real propagation constant; evanescent modes with imaginary propagation constant are not considered [37]. Evanescent modes are easily discriminated from the calculation because their direction cosines are located outside the unit circle ($\alpha^2 + \beta^2 > 1$), figuratively speaking they are "diffracted over the horizon". This argument is similar to point #3 of the procedure outlined in the previous section, except that the specular beam intensity is now an integral part of the radiance. The procedure is illustrated in Fig. 11.10 for the *Z*2 texture and a wavelength of 543 nm. It shows a greyscale plot of the absolute square of the Fourier components in the (α, β) plane in the vicinity of the origin.

Haze and *ARS* take on a different importance compared to scalar theory. In the previous section, the distribution between specular and diffused light was described by (11.5) and (11.6) and the angular properties of the *ARS* were calculated as secondary quantity. Using the Fourier approach of (11.8) in combination with the phase

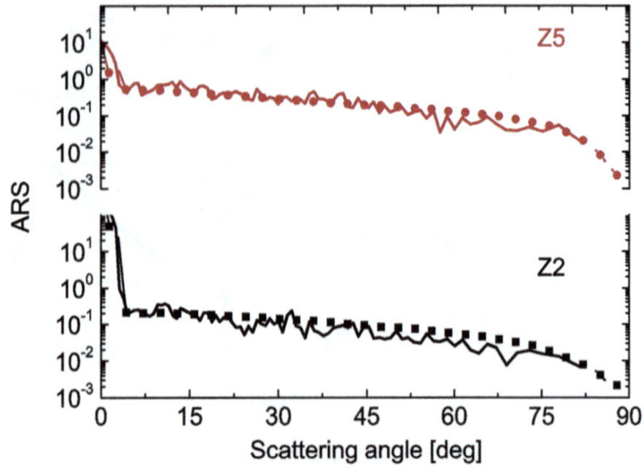

Fig. 11.11 Comparison of experimental *ARS* with results of the Fourier approach

screen of (11.9), the radiance $L(\alpha, \beta)$ takes over the role of the primary quantity. It can be related to the angle resolved scattering by accounting for the projection factor $\cos\theta$ between the radiating surface and the observer [31].

$$ARS(\varphi, \theta) = \cos\theta \cdot L(\alpha, \beta). \tag{11.10}$$

Here, $ARS(\varphi, \theta)$ is the 2D version of the angle resolved scattering. The angles φ and θ in (11.10) are related to the direction cosines α and β by $\tan\varphi = \beta/\alpha$ and $\cos\theta = (1 - \alpha^2 - \beta^2)^{1/2}$.

Using random textures and perpendicular incidence, the 2D-$ARS(\varphi, \theta)$ is again rotationally invariant and can be averaged over the azimuthal angle φ to yield the familiar 1D-$ARS(\theta)$:

$$ARS(\theta) = \frac{1}{2\pi} \int ARS(\varphi, \theta)\, d\varphi. \tag{11.11}$$

The resulting *ARS* of the *Z*2 and *Z*5 surfaces are shown in Fig. 11.11. The modeled characteristics match almost perfectly over the full angular range, including even the intensity of the specular part.

Based on the radiance $L(\alpha, \beta)$, the haze is obtained as follows.

$$H_T = \frac{T_{tot} - T_{spec}}{T_{tot}} = \frac{\sum_{\alpha^2+\beta^2<1} L(\alpha, \beta) - L(0,0)}{\sum_{\alpha^2+\beta^2<1} L(\alpha, \beta)}. \tag{11.12}$$

Figure 11.12 shows that the Fourier approach successfully describes the haze in the whole studied wavelength range. There is a small underestimation, but there can be no question that the overall correspondence is much better than the scalar results shown in Fig. 11.7. Moreover, the Fourier approach relies only on measurable quantities like the AFM profile and the refractive index dispersion, but there are no adjustable parameters.

The presented Fourier model was found to perform well in comparison with other models and calculation routines [38]. Nevertheless, there are a few limitations. The first concern addresses the replacement of the roughness zone with a screen of zero thickness. Essentially, this procedure neglects diffraction within the peak to valley

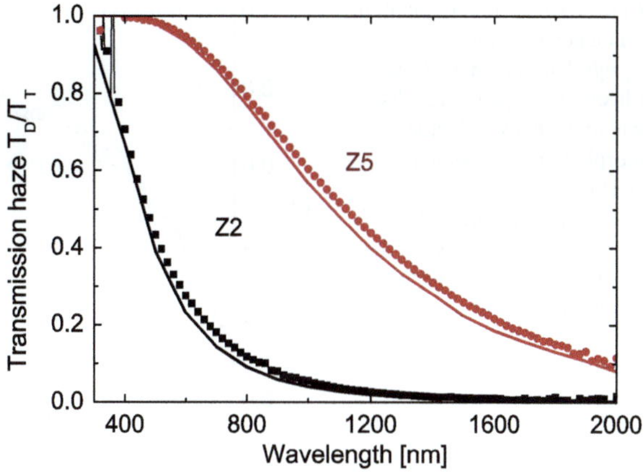

Fig. 11.12 Comparison of the experimental haze (*symbols*, cf. Fig. 11.7) with results of the Fourier approach (*lines*)

depth of the roughness zone as illustrated by the dashed light path in Fig. 11.9. General aspects of this approximation are discussed by Goodman [29], for effects from the surface textures used the current context, see Ref. [39].

A second concern addresses the validity of describing random scattering by a Fourier approach that applies to periodic phenomena by definition. It turns out that this an issue for the resolution of small scattering angles. Imagine that the Fourier expansion treats the random surface like a superposition of sinusoidal gratings. With increasing order of the expansion, one, two, three, etc. full wavelengths are fitted into the dimension of the AFM image. The smallest possible diffraction angle beside the specular beam is then related to first order diffraction at the largest possible grating period. For the shown situation with an image size of 10 μm and a laser wavelength of 543 nm, 3.1° is the smallest angle that is resolved. Accordingly, the second order beam diffracts into 6.2°, etc., resulting in a rather coarse resolution. Towards higher angles, the characteristics in Fig. 11.11 become denser by averaging over the rotational symmetry according to (11.11). Note that these limitations apply to particular conditions of the shown AFM images, smaller angles are obviously resolved if the surface profile is recorded on a larger scale.

11.6 Application to Light Scattering in Thin Film Silicon Solar Cells

As conclusion of this chapter, it is demonstrated how light scattering is used to aid solar cell development. In devices based on thin film silicon, the route towards higher efficiencies follows a trade off between two competing mechanisms. On the one hand, poor charge transport as well as throughput considerations in production suggest to keep the device as thin as possible. Acceptable thicknesses for amorphous and microcrystalline cells are 300 nm and 2 μm, respectively, less would be preferred. On the other hand, the indirect band gaps in these materials require thicknesses much in excess of the given values for complete absorption, particularly for

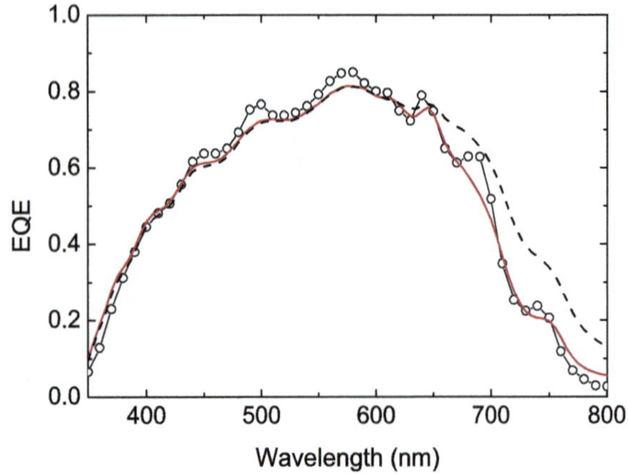

Fig. 11.13 External quantum efficiency of a solar cell (*symbols*). The *dashed line* illustrates modeling results where the back reflector morphology is assumed conformal to the front electrode while the *full line* is based on measured AFM data for front and back interface (after [36])

light with long wavelengths. The issue can be overcome by absorption enhancement using light scattering at interface textures [40] where textures can be either random or periodic [41, 42]. While photovoltaic manufacturing concentrated almost exclusively on random textures [43, 44], periodic textures are expected to outperform random structures [45]. Recent experiments demonstrated that they are at least on equal footing [46]. Nevertheless, a choice of relatively few natural textures like those shown in Figs. 11.2 and 11.3 is still used in virtually all types of thin film silicon solar cells due to their relative ease of fabrication [41, 47, 48]. Engineered structures like gratings have been applied only rarely, but this might change since low cost embossing processes are becoming increasingly common on large areas [49, 50].

Given the importance of random structures, a correct description of their impact on solar cell performance is mandatory. Rigorous solution of Maxwell's equations can be used for this task [35, 51], but continuously high demands on computing power prohibit their use as routine tool for screening. Instead, simplified ray optic models are used where the intensity of specular and diffuse contributions is followed throughout the multiple layers that constitute a solar cell [23, 25].

A recurring issue is the correct description of the interfaces within the device. First, their scattering properties must be modeled reliably since they are generally not accessible to measurement [38]. The use of empiric parameters must therefore be avoided as much as possible. Second, the scattering events at all textured interfaces must be described adequately. The widely assumed situation of conformal coverage is usually not correct since growing films can develop their own surface texture (cf. Figs. 11.2 and 11.3), or they can smooth a given texture to minimize surface area. A correct description can be obtained by interrupting the growth process regularly for measuring the actual surface topography.

Figure 11.13 shows experimental and simulated data of the external quantum efficiency (EQE) [36]. This characteristic is a key quantity for solar cells since it corresponds to the probability that an electron is created from an incident photon and optical enhancement by light scattering is important for the weakly absorbed light at long wavelengths. In the shown case, a solar cell with an amorphous absorber layer of 700 nm thickness is grown on an etched ZnO substrate (cf. upper right panel of Fig. 11.2). After growth of the silicon films, the initial rms surface roughness of 35 nm is reduced to 15 nm. The optical enhancement is clearly overestimated if this morphological change is neglected in the simulation whereas a simulation on the basis of the correct interface morphologies gives excellent correspondence with the measurement.

11.7 Conclusions

This chapter discussed light scattering at random surface textures, i.e. surfaces with little correlation on small scale and no correlation on long scales. Scalar scattering theory has been developed to describe this class of surface topography for remote sensing applications with radar waves. Scalar theory relies on essentially two parameters to characterize a surface, its rms roughness σ and an autocorrelation function which is usually approximated by a Gaussian with correlation length C. Shortcomings of this simplification are discussed using AFM data of real surface morphologies. In the second part, a quantitative Fourier theory is presented. Starting from the same surface data, it yields a more adequate description of light scattering, and it was shown how the improved model is successfully used to describe light scattering and absorption in a solar cell.

Acknowledgements This work was funded by the European Union, the Swiss National Science Foundation and the Swiss Federal Office for Energy. Dr. M. Python is thankfully acknowledged for recording the images in Figs. 11.2 and 11.3, Dr. K. Jäger of the Technical University of Delft provided the data for Fig. 11.13.

References

1. G. Mie, Beiträge zur Optik trüber Medien, speziell kolloidaler Metallösungen. Ann. Phys. **330**(3), 377–445 (1908)
2. K.E. Bean, Anisotropic etching of silicon. IEEE Trans. Electron Devices **25**(10), 1185–1193 (1978)
3. P. Campbell, M. Green, Light trapping properties of pyramidally textured surfaces. J. Appl. Phys. **62**(1), 243–249 (1987)
4. O. Kluth, B. Rech, L. Houben, S. Wieder, G. Schöpe, C. Beneking, H. Wagner, A. Löffl, H.W. Schock, Texture etched ZnO:Al coated glass substrates for silicon based thin film solar cells. Thin Solid Films **351**(1–2), 247–253 (1999)
5. J. Müller, B. Rech, J. Springer, M. Vanecek, TCO and light trapping in silicon thin film solar cells. Sol. Energy **77**(6), 917–930 (2004)
6. M. Hirasaka, K. Suzuki, K. Nakatani, M. Asano, M. Yano, H. Okaniwa, Design of textured Al electrode for a hydrogenated amorphous silicon solar cell. Sol. Energy Mater. **20**(1–2), 99–110 (1990)
7. A. Banerjee, S. Guha, Study of back reflectors for amorphous-silicon alloy solar-cell application. J. Appl. Phys. **69**(2), 1030–1035 (1991)

8. H. Sakai, T. Yoshida, T. Hama, Y. Ichikawa, Effects of surface morphology of transparent electrode on the open-circuit voltage in a-Si:H solar cells. Jpn. J. Appl. Phys. **29** (part 1), 630–635 (1990)
9. K. Sato, Y. Gotoh, Y. Hayashi, K. Adachi, H. Nishimura, Improvement of textured SnO_2:F films for a-Si solar cells. Reports of the Research Lab, Asahi Glass Co **40**(2), 233–241 (1990)
10. J.A. Thornton, Influence of apparatus geometry and deposition conditions on the structure and topography of thick sputtered coatings. J. Vac. Sci. Technol. **11**(4), 666–670 (1974)
11. S. Faÿ, U. Kroll, C. Bucher, E. Vallat-Sauvain, A. Shah, Low pressure chemical vapour deposition of ZnO layers for thin-film solar cells: temperature-induced morphological changes. Sol. Energy Mater. Sol. Cells **86**(3), 385–397 (2005)
12. S. Faÿ, J. Steinhauser, N. Oliveira, E. Vallat-Sauvain, C. Ballif, Opto-electronic properties of rough LP-CVD ZnO:B for use as TCO in thin-film silicon solar cells. Thin Solid Films **515**(24), 8558–8561 (2007)
13. J. Bailat, D. Dominé, R. Schlüchter, J. Steinhauser, S. Faÿ, F. Freitas, C. Bucher, L. Feitknecht, X. Niquille, R. Tscharner, A. Shah, C. Ballif, High efficiency pin microcrystalline and micromorph thin film silicon solar cells deposited on LPCVD ZnO coated glass substrates, in *Proc. 4th World PVSEC*, 2006, Hawaii (2006), pp. 1533–1536
14. J. Elson, J. Bennett, Relation between the angular dependence of scattering and the statistical properties of optical surfaces. J. Opt. Soc. Am. **69**(1), 31–47 (1979)
15. J. Elson, J. Bennett, Calculation of the power spectral density from surface profile data. Appl. Opt. **34**(1), 201–208 (1995)
16. H.E. Bennett, J.O. Porteus, Relation between surface roughness and specular reflectance at normal incidence. J. Opt. Soc. Am. **51**(2), 123–129 (1961)
17. H. Davies, The reflection of electromagnetic waves from a rough surface. Proc. Inst. Electr. Eng. **101**, 209 (1954)
18. S. Rice, Reflection of electromagnetic waves from slightly rough surfaces. Commun. Pure Appl. Math. **4**(2–4), 351–378 (1951)
19. P. Beckmann, A. Spizzichino, *The Scattering of Electromagnetic Waves from Rough Surfaces* (Artech House, Norwood, 1987)
20. A. Maradudin, *Light Scattering and Nanoscale Surface Roughness* (Springer, Berlin, 2007)
21. E.R. Méndez, D. Macías, Inverse problems in optical scattering, in *Light Scattering and Nanoscale Surface Roughness*, ed. by A.A. Maradudin (Springer, Berlin, 2007), pp. 435–464
22. C.K. Carniglia, Scalar scattering theory for multilayer optical coatings. Opt. Eng. **18**(2), 104–115 (1979)
23. M. Zeman, R. Van Swaaij, J.W. Metselaar, R.E.I. Schropp, Optical modeling of a-Si:H solar cells with rough interfaces: effect of back contact and interface roughness. J. Appl. Phys. **88**, 6436 (2000)
24. J. Krc, F. Smole, M. Topic, Analysis of light scattering in a-Si:H solar cells by a one-dimensional semi-coherent optical model. Prog. Photovolt. **11**(1), 15–26 (2003)
25. J. Krc, M. Zeman, O. Kluth, F. Smole, M. Topic, Effect of surface roughness of ZnO:Al films on light scattering in hydrogenated amorphous silicon solar cells. Thin Solid Films **426**(1–2), 296–304 (2003)
26. H. Stiebig, T. Brammer, T. Repmann, O. Kluth, N. Senoussaoui, A. Lambertz, H. Wagner, Light scattering in microcrystalline silicon thin film solar cells, in *Proc. 16th EU-PVSEC*, 2000, Glasgow (2000), pp. 549–552
27. G. Brown, V. Celli, M. Haller, A. Maradudin, A. Marvin, Resonant light scattering from a randomly rough surface. Phys. Rev. B **31**(8), 4993–5005 (1985)
28. V. Celli, A.A. Maradudin, A.M. Marvin, A.R. McGurn, Some aspects of light-scattering from a randomly rough metal surface. J. Opt. Soc. Am. A **2**(12), 2225–2239 (1985)
29. J. Goodman, *Introduction to Fourier Optics* (Roberts & Company Publishers, Greenwood Village, 2005)
30. D. Dominé, F.J. Haug, C. Battaglia, C. Ballif, Modeling of light scattering from micro- and nanotextured surfaces. J. Appl. Phys. **107**, 044504 (2010)

31. J. Harvey, C. Vernold, A. Krywonos, P. Thompson, Diffracted radiance: a fundamental quantity in nonparaxial scalar diffraction theory. Appl. Opt. **38**, 6469–6481 (1999)
32. J. Harvey, Fourier treatment of near-field scalar diffraction theory. Am. J. Phys. **47**, 974 (1979)
33. F.J. Haug, C. Battaglia, D. Domine, C. Ballif, Light scattering at nano-textured surfaces in thin film silicon solar cells, in *Proc. 35 IEEE PVSC*, 2010, Hawaii, IEEE (2010), pp. 754–759
34. K. Bittkau, M. Schulte, M. Klein, T. Beckers, R. Carius, Modeling of light scattering properties from surface profile in thin-film solar cells by Fourier transform techniques. Thin Solid Films **519**(19), 6538–6543 (2011)
35. C. Rockstuhl, S. Fahr, F. Lederer, F.J. Haug, T. Soderstrom, S. Nicolay, M. Despeisse, C. Ballif, Light absorption in textured thin film silicon solar cells: a simple scalar scattering approach versus rigorous simulation. Appl. Phys. Lett. **98**(5), 051102 (2011)
36. K. Jäger, M. Fischer, R. van Swaaij, M. Zeman, A scattering model for nano-textured interfaces and its application in opto-electrical simulations of thin-film silicon solar cells. J. Appl. Phys. **111**, 083108 (2012)
37. J.E. Harvey, A. Krywonos, A global view of diffraction: revisited, in *Proc. SPIE AM100-26*, 2004, Denver (2004)
38. M. Schulte, K. Bittkau, K. Jäger, M. Ermes, M. Zeman, B.E. Pieters, Angular resolved scattering by a nano-textured ZnO/silicon interface. Appl. Phys. Lett. **99**, 111107 (2011)
39. D. Dominé, The role of front electrodes and intermediate reflectors in the optoelectronic properties of high-efficiency micromorph solar cells. PhD thesis, University of Neuchatel (2009)
40. E. Yablonovitch, G.D. Cody, Intensity enhancement in textured optical sheets for solar cells. IEEE Trans. Electron Devices **29**(2), 300–305 (1982)
41. H.W. Deckman, C.R. Wronski, H. Witzke, E. Yablonovitch, Optically enhanced amorphous silicon solar cells. Appl. Phys. Lett. **42**(11), 968–970 (1983)
42. P. Sheng, A.N. Bloch, R.S. Stepleman, Wavelength-selective absorption enhancement in thin-film solar cells. Appl. Phys. Lett. **43**, 579 (1983)
43. G. Yue, L. Sivec, B. Yan, J. Yang, S. Guha, High efficiency hydrogeneated nanocrystalline silicon based solar cells deposited by optimized Ag/ZnO back reflectors, in *Proc. 25th European PVSEC*, 2010, Valencia (2010)
44. J. Bailat, L. Fesquet, J.B. Orhan, Y. Djerdidane, B. Wolf, P. Madlinger, J. Steinhauser, S. Benagli, D. Borrello, L. Castens, G. Monteduro, M. Marmelo, B. Dehbozorghi, E. Vallat-Sauvain, X. Multone, D. Romang, J.F. Boucher, J. Meier, U. Kroll, M. Despeisse, G. Bugnon, C. Ballif, S. Marjanovic, G. Kohnke, N. Borelli, K. Koch, J. Liu, R. Modavis, D. Thelen, S. Vallon, A. Zakharian, D. Weidmann, Recent developments of high-efficiency micromorph tandem solar cells in Kai-M PE-CVD reactors, in *Proc. 5th World PVSEC*, 2010, Valencia (2010), p. 2720
45. J.M. Gee, Optically enhanced absorption in thin silicon layers using photonic crystals, in *Proc. 29th IEEE PVSC*, 2002, New Orleans (2002), pp. 150–153
46. C. Battaglia, C.M. Hsu, K. Söderström, J. Escarré, F.J. Haug, M. Charrière, M. Boccard, M. Despeisse, D. Alexander, M. Cantoni, Y. Cui, C. Ballif, Light trapping in solar cells: can periodic beat random? ACS Nano **6**(3), 185 (2012)
47. H. Keppner, J. Meier, P. Torres, D. Fischer, A. Shah, Microcrystalline silicon and micromorph tandem solar cells. Appl. Phys. A, Mater. Sci. Process. **69**(2), 169–177 (1999)
48. M. Green, P. Basore, N. Chang, D. Clugston, R. Egan, R. Evans, D. Hogg, S. Jarnason, M. Keevers, P. Lasswell, Crystalline silicon on glass (CSG) thin-film solar cell modules. Sol. Energy **77**(6), 857–863 (2004)
49. J. Bailat, V. Terrazzoni-Daudrix, J. Guillet, F. Freitas, X. Niquille, A. Shah, C. Ballif, T. Scharf, R. Morf, A. Hansen, D. Fischer, Y. Ziegler, A. Closset, Recent development of solar cells on low-cost plastic substrates, in *Proc. 20th European PVSEC*, 2005, Barcelona (2005)
50. A. Bessonov, Y. Cho, S.J. Jung, E.A. Park, E.S. Hwang, J.W. Lee, M. Shin, S. Lee, Nanoimprint patterning for tunable light trapping in large-area silicon solar cells. Sol. Energy Mater. Solar Cells **95**, 2886 (2011)
51. C. Haase, H. Stiebig, Thin-film silicon solar cells with efficient periodic light trapping texture. Appl. Phys. Lett. **91**(6), 061116 (2007)

Chapter 12
Active Plasmonics in Self-organized Soft Materials

Roberto Caputo, Luciano De Sio, Ugo Cataldi, and Cesare Umeton

Abstract In this chapter we show several possibilities for obtaining active plasmonics in self-organized amorphous materials. The starting point is a brief theoretical description of the physical mechanisms that allow such a tunability. Afterwards, it follows an overview of the up-to-date results obtained in this direction. Several proof-of-concept prototypes have been successfully fabricated and are reported as well.

12.1 Introduction

Nanophotonic devices have the capability to concentrate light into the nanoscale range and hold high potential for many applications as integrated optics [1], plasmonic circuits [2], biosensing [3] and quantum information processing [4]. One promising way to localize the optical radiation into a nanometer-sized volume has been obtained by using the unique properties of plasmonic nanomaterials [5]. These amorphous materials are an effective bridge between bulk materials and atomic or molecular structures and exhibit a very intense color, which is absent in the bulk material as well as in individual atoms. The physics behind this behavior can be explained by considering the collective oscillation of the free conduction electrons that can be induced by an external electromagnetic field interacting with very small metal particles: the so called Localized Surface Plasmon Resonances (LSPR). In fact, the plasmonic coupling, existing between metal nanoparticles (NPs) and light, enables a series of interesting optical phenomena, such as Surface-Enhanced Raman Scattering (SERS) [6], Resonance Light Scattering (RLS) [7] and Surface Plasmon Resonance (SPR) [8]. The dynamic control of plasmonic resonances is a hot-topic and several possibilities to obtain this result have been already demonstrated. In this chapter, we illustrate how an "active" control of the plasmonic resonance spectral position is possible when the material showing plasmonic properties is acted on by external stimuli (electric field, optical field, temperature and mechanical strain).

R. Caputo (✉) · L. De Sio · U. Cataldi · C. Umeton
LiCryL-INFM/CNR & Physics Department, University of Calabria, 87036 Rende, CS, Italy
e-mail: roberto.caputo@fis.unical.it

C. Rockstuhl, T. Scharf (eds.), *Amorphous Nanophotonics*,
Nano-Optics and Nanophotonics, DOI 10.1007/978-3-642-32475-8_12,

© Springer-Verlag Berlin Heidelberg 2013

In the following we will start with a brief introduction to the concept of localized plasmonic resonance, giving a particular mention on how actively modify the plasmonic spectral position. This is possible by acting on the medium surrounding the sub-entities, responsible for the plasmonic response, but also modifying their inter-distance. The chapter will then analyze the functioning of several selected amorphous nanomaterials that show an active plasmonic response.

12.2 Tunability of Localized Plasmonic Resonances

A convenient way to dynamically modify the plasmon resonance frequency of a homogeneous (surface or bulk) distribution of mono-dispersed metal NPs is to vary the dielectric permittivity of the medium surrounding the NPs. Indeed, the optical properties of spherical particle dispersions can be predicted by the Mie theory [9] through the expression of the extinction cross section.

$$\sigma_{ext}(\omega) = 9\frac{\omega}{c}\varepsilon_m^{3/2} V_0 \frac{\varepsilon_2(\omega)}{[\varepsilon_1(\omega) + 2\varepsilon_m]^2 + \varepsilon_2(\omega)^2} \qquad (12.1)$$

Where $V_0 = (4\pi/3)R^3$ is the nanoparticle volume, R being its radius, ω is the angular frequency of the exciting radiation, ε_m is the dielectric function of the medium surrounding the metallic nanoparticles, and ε_1 and ε_2 are the real and imaginary parts of the dielectric function of the metallic nanoparticles respectively. Based on this theory, for small and isolated metal particles, the spectral position of the plasmonic absorption peak depends on the refractive index of the surrounding medium, according to the condition that minimizes the denominator of (12.1):

$$\varepsilon_1(\omega) = -2\varepsilon_m \qquad (12.2)$$

A modification of the dielectric behavior of the host material corresponds, therefore, to a tuning action of the Plasmon resonance frequency. The outstanding properties of Liquid Crystals (LCs) make them an ideal candidate for this role; indeed, these materials represent an excellent example of reconfigurable medium where the refractive index can be finely controlled by means of external stimuli.

Another way to modify the plasmonic resonance frequency of a metal NPs array is to change size, shape and inter-particle distance. Kinnan et al. have recently demonstrated this possibility by performing a very systematic study on Ag NPs layers [10]. For these experiments, two-dimensional Ag NPs layers have been fabricated by using different sizes of nano-particles. For each size, the particle density has been chemically controlled in such a way that it is possible to fix the inter-particle distance. This distance has a quite important role. Indeed, if we assume that single particles put at close distance undergo a dipolar interaction, we can argue that a change of the inter-particle distance corresponds to a modification of both the plasmonic coupling and the resonance frequency. Experiments of Kinnan et al. confirmed this hypothesis: by fixing the particle size and increasing their density, the extinction coefficient of the plasmonic peak increases. At the same time, the spectral position of this peak changes and, depending on the particle size, a blue or a red shift is observed.

12.3 Electrically and Temperature Controlled Active Plasmonics

When combined with noble metal NPs, LCs can have a two-fold role: they can be exploited as a "host-fluid" for dissolving colloidal noble nanoparticles [11] and, at the same time, if used as the medium surrounding the resonant entities, they can dynamically tune the spectral position of the plasmon resonance position. In 2005, Kossyrev et al. have provided one of the first demonstrations of this possibility when they showed that, by layering a regular array of gold nanoparticles with nematic LC (E7) it is possible to obtain a tunable plasmonic system. Indeed, by applying a low amplitude electric field (10 V/μm) to the sample, they obtained a plasmonic red shift of about 8 nm [12]. However, a drawback in using LCs is represented by the fact that, in general, they require chemical and/or mechanical treatments to be aligned; in addition, the order degree of the LC material is affected by the inclusion of nanoparticles and therefore their concentration in the host LC must be limited to a few percents [13, 14]. A convenient alternative to avoid this issue has been proposed by Smalyiukh et al. who obtained bulk self-alignment of dispersed gold nanorods imposed by the intrinsic cylindrical micelle selfassembly in nematic and hexagonal liquid crystalline phases [15]. Due to their elongated shape, nematic molecules mix quite efficiently with gold nanorods; this allows a high filling fraction of the gold component in the mixture and hence the fabrication of a system with noticeable plasmonic properties. These properties are also tunable and very pronounced effects can be observed when the system undergoes a temperature variation. In order to avoid the difficulty of mixing metal NPs and LCs, Khoo et al. have instead fabricated a system where a "carpet" of gold nanodisks lays on an Indium-Tin-Oxide (ITO) coated glass substrate [16]. Another conductive glass has been put at controlled distance on top of the previous one in such a way that a cell is fabricated. This cell has been then filled in with a dual-frequency liquid crystal (DFLC), which can change the sign of its dielectric anisotropy either from positive to negative, or from negative to positive, depending on the frequency of the electric field that is applied to the sample. This feature allows the switching of the dual-frequency liquid crystal director in both directions, parallel and perpendicular to the applied field, by using low and high frequencies. To these drastic changes of the refractive index both blue- and red-shifts in the plasmonic resonances correspond. Experiments performed on this system have demonstrated a blue-shift of about 4 nm for field frequency below an observed cross-over frequency of 14 KHz and a red-shift of about 21 nm for frequencies ranging between 15–21 KHz. Recently, we have reported on the realization of a periodic soft-composite structure, with a wide range of photonic applications, which might represent a solution to the previously illustrated drawbacks [17]. By a microfluidic etching of a composite micro/nano grating, made of polymer slices alternated to films of well oriented LC (POLICRYPS) [18], we realize a polymeric platform that is able to align and micro-confine a large variety of LC materials, without the need of any chemical and/or mechanical treatment. The same platform can be utilized for efficiently aligning Cholesteric Liquid Crystals (CLC) doped with gold nanoparticles, as well. A homogeneous mixture of metallic nanoparticles (NPs) and CLC has been prepared by mixing the Harima gold nanopaste NPG-J (from

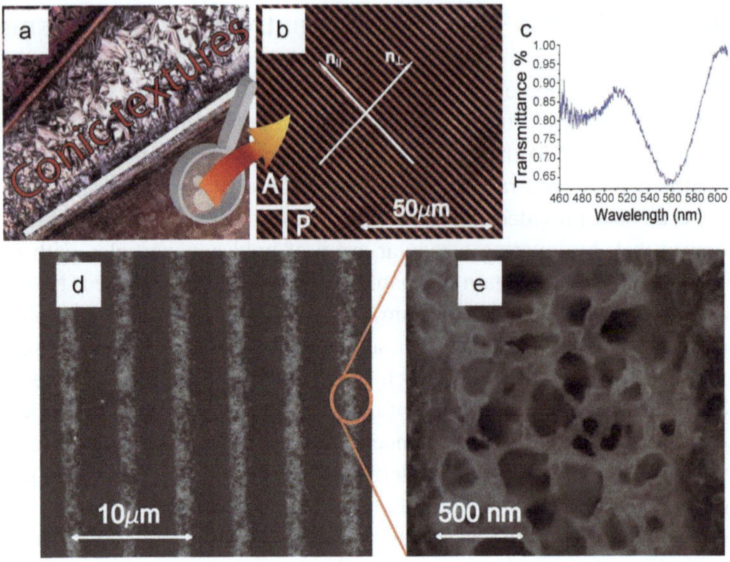

Fig. 12.1 Polarized optical microscope view of the polymeric template filled with CLC and gold NPs mixture at the edge of the grating area (**a**). The high magnification of the CLC and gold NPs area aligned in ULH geometry is shown in (**b**) while its typical reflection notch is reported in (**c**). EBSD view (**d**) and high magnification (**e**) of the polymeric template filled with CLC and gold NPs. [28]—Reproduced with permission of the Royal Society of Chemistry

Harima Chemicals, Inc. [19], generally used for other aims like ink-jet printing and laser sintering [20]), and the BL095 CLC by Merck [21] (helix pitch $p \approx 400$ nm). Harima nanopaste NPG-J contains 55 wt% of gold nanoparticles with a mean size of 7 nm and a distribution ranging in 5–12 nm. The plasmonic absorption peak in solvent (naphthen) is instead at $\lambda \approx 525$ nm. The Harima NPs have been mixed to the CLC in high concentration (about 20 wt%) and then infiltrated, by capillarity, in the polymeric template. The best optical performances were exhibited by a grating of $L = 10$ μm thickness and $\Lambda = 6$ μm periodicity. The sample was kept at a fixed high temperature (≈90 °C) during the whole filling process, thus keeping the CLC in the isotropic phase; then, by slowly (0.5 deg/min) cooling down the sample to room temperature, a self-organization process occurred, which was able to orient the axes of CLC helices to be almost parallel to the polymeric slices, in a uniform lying helix configuration [17]. Figure 12.1a shows a Polarized Optical Microscope micrograph of the sample at the edge of the photo-sculptured grating area; on the left, the photo puts into evidence the existence of a standard focal conic texture, induced by a random distribution of the helical axes.

On the right, we can see the uniform lying helix geometry induced by the polymeric structure (Fig. 12.1b), as demonstrated by the exhibition of a selective reflection, typical of a short pitch CLC (Fig. 12.1c). In order to study the effects produced by the presence of gold NPs inside the microstructure, we first tried to figure out their distribution within the host CLC by performing a Scanning Electron Microscope (SEM) and an Electron Back-Scattering Diffraction (EBSD) characterization of the sample. In particular, EBSD analysis is valuable to distinguish gold from other materials, since the yield of backscattered electrons increases with the specimen atomic number (Z); in our case, the presence of gold ($Z = 79$) produces a high contrast with the polymer microstructure (a thiol based system with $Z \approx 18$). As shown in the EBSD view of Fig. 12.1d, the bright stripes confirm the presence of gold along the microchannels only, while in the high magnification (Fig. 12.1e), we can notice, that CLC branches are literally "wrapped" by densely packed gold particles. Despite the quite high concentration of gold NPs, the high magnification in Fig. 12.1e indicates that the CLC is uniformly aligned, with helices confined between polymeric slices. This is a clear indication that the CLC exhibits good "host-fluidity" properties for the gold NPs; at the same time, its high degree of order indicates that the CLC phase is weakly perturbed by the presence of gold NPs. The obtained structure represents, therefore, a successful combination of a "top-down" approach (polymeric template fabrication) with a "bottom-up" one, represented by the self-organization of gold NPs in CLC helices; in this sense, the microscale meets the nanoscale. Considering the size of involved NPs ($\approx$7 nm), the SEM resolution is not able to show their morphological details.

In order to check the influence of gold NPs on the optical properties of the whole structure, we have investigated its spectral response by shining the sample with linearly polarized white light (wavelengths in the range 350–1000 nm) at normal incidence; obtained results are reported in Fig. 12.2. It is evident that p-polarized light is strongly diffracted and its transmission is almost suppressed (diffraction pattern of Fig. 12.2a (curve 3); on the other hand, s-polarized light is highly transmitted in almost the whole analyzed range (diffraction pattern of Fig. 12.2a, curve 1). This behavior has already been observed in the past with a similar structure [22] and can be easily explained by supposing that CLC helices lay, in average, along the channels of the template. It is reasonable to assume, therefore, that two different effective refractive indices exist: $n_\perp$ (whose value can be estimated to be $\approx$1.64), quite different from the polymeric refractive index n_p ($\approx$1.54), which is experienced by light whose electric field is perpendicular to the channels (p-wave), and $n_\parallel$ whose value can be estimated to be $\approx 1.56 \approx n_p$, for light whose electric field is parallel to them (s-wave). Above rough estimation of $n_\perp$ and $n_\parallel$ have been made by calculating two different averages of differently weighted ordinary and extraordinary refractive indices (n_o and n_e) of the CLC.

Due to the alternation of $n_\perp$ and n_p indices, the grating structure is therefore experienced only by the p-wave. Incidentally, the noticeable difference between $n_\perp$ and $n_\parallel$ values is confirmed by the pronounced birefringence shown by the sample (Fig. 12.1b, picture taken between crossed polarizers). In the curve 1 of Fig. 12.2a, a pronounced absorption peak at $\lambda = 532$ nm, with an extinction coefficient of 0.2, can be observed; this peak cannot be attributed to any diffractive mechanism because the structure appears optically homogeneous ($n_\parallel \approx n_p$) to the incoming s-wave. Both the spectral position and the narrow width of the measured peak suggest, instead, that it is due to the presence of gold NPs dispersed in the CLC host and, more precisely, to their localized plasmon resonance. This hypothesis is supported by similar results reported in literature [12], and observed in different systems

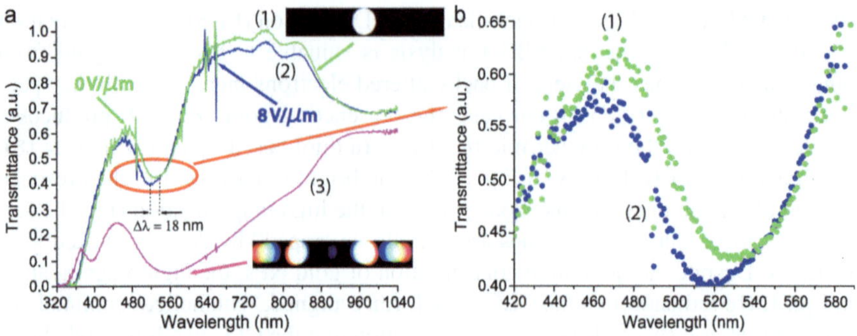

Fig. 12.2 Spectral response of the sample for two values of the external electric field (**a**) and its higher magnification detail (**b**). [28]—Reproduced with permission of the Royal Society of Chemistry

where gold NPs are also involved [23]. Even if physically remarkable, this result is not outstanding in itself; it is quite interesting, on the contrary, the circumstance that the plasmonic response of such a composite nanomaterial is also tunable. This has been demonstrated by applying an external electric field (8 V/μm, frequency 1 kHz, square wave) across the cell, perpendicularly to the helix axes. This field induces an in-plane tilt of the optical axis of the CLC (aligned in ULH texture), with a consequent variation of the value of $n_{\parallel}$, experienced by s-waves. This index variation is directly related to the tuning of the plasmon resonance frequency according to (12.2). Indeed, when applying the external electric field, the impinging probe light experiences a decreased value of the refractive index of the CLC (from the initial $n_{\parallel}$ to about n_o; typically from 1.56 to 1.5).

Following (12.2), the resonance condition is fulfilled for higher (negative) values of $\varepsilon_1(\omega)$. It is well known that, in the visible range, the real part of the electric permittivity of gold nanoparticles increases with frequency [24]; therefore, fulfillment of (12.2) takes place for higher values of ω. This yields a blue shift of the plasmonic absorption peak. This hypothesis is confirmed by results shown in Fig. 12.2b (higher magnification of the highlighted region of Fig. 12.2a): the absorption peak is blue shifted from $\lambda = 532$ nm (curve 1, Fig. 12.2b) to $\lambda = 514$ nm (curve 2, Fig. 12.2b). An additional demonstration of the tunability of the plasmonic response of our composite structure has been obtained by varying the temperature of the sample; results are reported in Fig. 12.3. By means of a miniature oven (CaLCTec S.r.l.), it was possible to vary the sample temperature from 25 °C up to 75 °C; a red-shift of the plasmonic absorption peak was observed in the range from 532 nm to 582 nm. Since the pitch p of chiral liquid crystals elongates with temperature [25]; assuming that the CLC helices are wrapped by gold NPs, the consequence of this elongation is an increase of the inter-distance between neighboring metal NPs. It has been already demonstrated that the plasmon response of NP arrays depends on particle size and density [26, 27]. In particular, results show that, for densely packed metal NPs, the absorption peak is quite broad and is centered in the blue-green range, while this peak slightly reduces its width and shifts to the red in case of well separated NPs. In

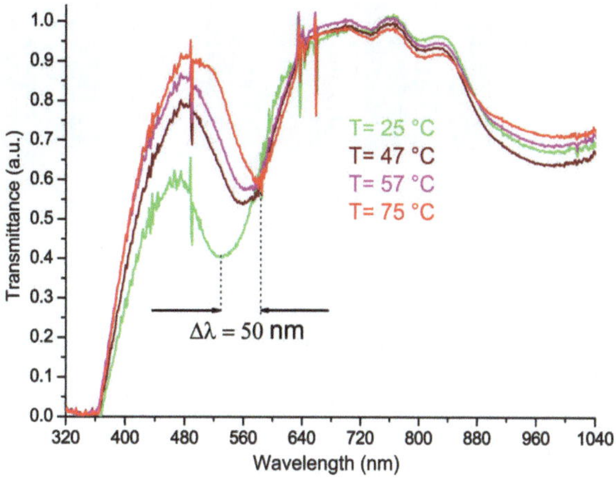

Fig. 12.3 Spectral response of the sample versus its temperature. [28]—Reproduced with permission of the Royal Society of Chemistry

fact, as observed in [10], when NP aggregates are considered whose size is comparable with the wavelength of the incoming radiation, different areas of the aggregate surface experience different phases of the incident radiation; thus, higher multipolar modes (quadrupolar, octupolar and even hexadecapolar) have to be taken into account. Given that the excitation frequency of these higher modes is larger than that of the dipolar one, a broadening of the plasmon peak occurs which has its center in the blue region of the electromagnetic spectrum.

On the contrary, single, small-sized, NPs, when hit by the incoming radiation, experience the same phase of the electromagnetic wave on their whole area; in this condition, gold NPs behave as simple dipoles: the plasmonic peak width is quite narrow and, depending on the particle size, the peak is centered in the green–red part of the electromagnetic spectrum. Based on above considerations, the temperature dependent shift of the plasmonic resonance observed in our sample can be therefore explained in terms of a temperature induced passage from densely packed (wrapping the CLC branches) to mono-dispersed gold NPs. We can conclude that a soft-matter periodic structure containing a composite mixture of CLC and gold NPs can be exploited as an "active plasmonic" system. The CLC material acts as a "host fluid" whose refractive index can be varied in a broad range, by utilizing both external electric fields on temperature variations. This has a strong influence on the position of the plasmonic absorption peak of NPs that exhibits, consequently, a broad tunability. In our opinion, this might represent the first step towards the realization of a new generation of nanostructured materials with tunable optical properties. More details of this work can be found in [28].

12.4 Optically Controlled Active Plasmonics

In the previous paragraph, we have shown that a reconfigurable medium can have a strong influence on the LSPR properties of a system of metal NPs. If LCs are

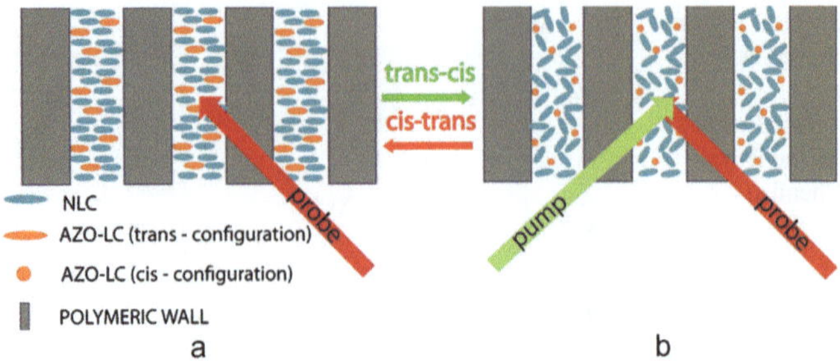

Fig. 12.4 Typical behavior of an optically controllable trans-cis system. Mechanism. (**a**) If the structure is probed with *red laser light*, the azo-dye molecules remain in the trans-configuration and the LC material remains in the nematic phase; (**b**) if an optical pumping with *green light* takes place, the molecules undergo a trans-cis conformational change and the LC makes a transition to the isotropic phase

considered, electric field and temperature can efficiently modify the orientation of molecules and tune the effective refractive index of the medium. However, there are other methods to tune the refractive index of LC materials. In particular, when an LC system is doped with small amounts of azo-dye molecules, the refractive index tuning can be achieved by using light. In Fig. 12.4, it is shown the typical mechanism that takes place in these systems. When irradiated with light of a specific wavelength, azo-dye molecules undergo a photo-chemical trans-cis conformational transformation: they modify their shape and induce disorder in the host LC material (Fig. 12.4b).

This mechanism is usually called all-optical switching and it is very convenient because it offers the possibility to commute the properties of an optical system without using any electrode [29–31]. Such a possibility can prove very appealing also with plasmonic systems. Liu et al. have provided an interesting example of this concept where they use a photo-switchable grating made of azo-dye-doped Holographic Polymer-Dispersed Liquid Crystals (HPDLCs) combined with a gold nanodisk array [32]. By putting the two elements in close proximity, they show some interactions because the grating becomes the surrounding medium of the nanodisk array. These interactions yield on one hand an LSPR-enhanced diffraction of the grating and, on the other hand, a diffraction-mediated LSPR of the gold nanodisk array. The first remarkable result is that the coupled system (grating + nanodisk array) exhibits a shifted LSPR frequency, because of the presence of the grating. Moreover, due to the absorption of the gold nanodisks, the diffraction efficiency outside the LSPR range is much lower than that of the bare grating, whereas it is higher within the LSPR range. Where the tunability of the system is considered, experiments show that pumping with green light ($\lambda = 514$ nm) has no effect on the LSPR of the bare gold nanodisk array; in contrast, a significant decrease of the extinction coefficient can be observed under different pumping intensities for the azo-based HPDLC grating. A change in the band 600–800 nm is due to the decreased diffraction from the grating caused by the Nematic to Isotropic (N-I) phase transition of the LCs. In this case, indeed, light propagating through the grating experiences an effective index of

LC droplets, which is an average of the ordinary and extraordinary indices, and only slightly higher than that of the LC material in the isotropic phase; for a 10-μm-thick grating, the index change due to the N-I phase transition gave a ~30 % reduction of the diffraction efficiency when measured by a p-polarized He–Ne laser (in this condition the index modulation between polymer and LC is almost negligible and the diffraction grating very weak).

Finally, it is worth noting that the realized variations are fully reversible because only the trans-cis isomerization transition is responsible of the isothermal switching processes of the LC state: by removing the photo-pump action, the azo-dye molecules relax in the original trans configuration and the LC order is restored. Switching times are not extremely fast but they can be eventually improved by realizing the optically switchable grating with a POLICRYPS technique [17, 33]. It has been indeed demonstrated that this technique provides more performing diffraction gratings than the HPDLC one.

Starting from a similar concept, a light-driven plasmonic absorber has been recently demonstrated [34]. There is a significant amount of research focused on developing plasmonic absorbers with wide working frequency ranges or multiple resonances [35–37]. These devices usually require an accurate design to guarantee optimal performances and it can result quite difficult to reconfigure them or tune the absorption bands without redesigning and repeating the entire fabrication process. In this regard, it would be extremely convenient to have an actively tunable plasmonic absorber that can be adapted to a specific application. As an example, such a device could be used to optimize the efficiency of solar cells, which usually exhibit narrow absorption peaks in the near IR and visible range. The tunable plasmonic absorber designed by Zhao et al. [34] comprises a photosensitive nematic liquid crystal (PNLC) layer covering an asymmetric gold nanodisk array. The photosensitivity of the PNLC derives again from the presence of azo-dye dopants that, upon light irradiation, can induce a change of the refractive index of the liquid crystal. Figure 12.5a shows a schematic of the active plasmonic absorber: there are two different sizes of gold nanodisks, which exhibit two distinct localized surface plasmon resonances (LSPR) [38]; their frequency values are determined by the size of the metal nanodisks, as well as by the inter-spacing between them.

Indeed, the specific electromagnetic energy impinging on the structure induces dipole oscillations in the nano-units present on the substrate; as such, the energy is converted into a displacement current and confined until it completely dissipates through the structure. Thus, an absorption phenomenon takes place at a specific frequency: The use of two different sizes of gold nanodisks (Fig. 12.5b) results in two different absorption bands. Figure 12.5c shows a SEM image of the fabricated plasmonic absorber: it is a three-layer structure specially designed for the top gold pattern to achieve the dual-band near perfect absorption and a 200 nm gold bottom layer to block any incident/transmitted light. The top layer consists of two nanodisk arrays arranged in a two-dimensional (2D) pattern with the same period d, as shown in Fig. 12.5b. A thin layer of SiO_2 (30 nm) is used as the middle spacing layer and its thickness has been chosen by considering two criteria: (1) the layer has to be thin enough to ensure the plasmonic coupling between the top and bottom metal layers;

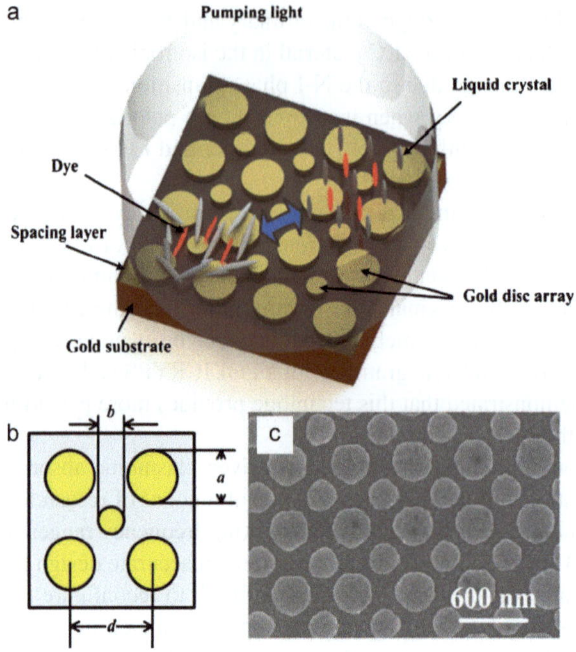

Fig. 12.5 (**a**) Schematic of light-driven tunable plasmonic absorber. Light-sensitive liquid crystals on top of the absorber can be modulated with light. (**b**) The top layer consists of two sets of alternatively arranged, two-dimensional nanodisks with the same period and different diameters, followed by a spacing layer of SiO_2 and a bottom gold layer. (**c**) SEM images of fabricated near perfect plasmonic absorber working at dual frequencies. [34]—Reproduced with permission of the American Institute of Physics

(2) a dielectric with relatively high dielectric constant is convenient in a capacitor configuration between the top and bottom metal layers, since a large capacity allows a very good confinement of light energy inside the three-layer structure. In order to check the functionality of the three-layer structure, several numerical simulations have been performed. The absorption bands strongly depend on the geometrical features of the nanodisk arrays and even slight changes can determine a noticeable frequency shift. In order to characterize this dependence, two designs have been considered with different combinations of large nanodisk and small nanodisk arrays ($a = 300$ nm, $b = 200$ nm; and $a = 280$ nm, $b = 180$ nm). The experimental check of the fabricated structures has been performed by using a Fourier Transform infrared (FTIR) spectrometer and an imaging microscope (Bruker FTIR, HYPERION 3000). Both simulated and measured curves are plotted in Figs. 12.6a,b. Experimental results match almost perfectly in terms of position of the absorption peaks, with a slightly reduced absorption when compared to simulations. Differences are likely caused by imperfections in the fabrication and different boundary conditions in simulations and real cases.

In order to incorporate a tuning mechanism, this system has been layered with a photosensitive LC material. As an active medium it has been chosen the liquid

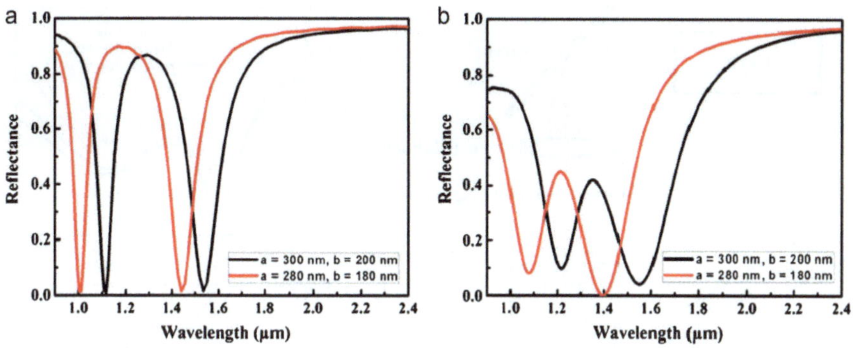

Fig. 12.6 (**a**) Two designs of the near-perfect plasmonic absorber (i) $a = 300$ nm, $b = 200$ nm, and (ii) $a = 280$ nm, $b = 180$ nm. FDTD simulations show that near perfect absorption can be achieved at selected wavelength ranges. (**b**) Experiment results of the two designs. Over 90 % absorbing efficiencies can be achieved. [34]—Reproduced with permission of the American Institute of Physics

crystal 5CB (pentyl-cyano-biphenyl) doped with ≈4 % of methyl-red dye. These materials are of real interest for possible applications, because all-optical modulation of the liquid crystal birefringence can be implemented at speeds several orders of magnitude higher than with electro- or acousto-optical approaches [39–41]. The presence of the nanodisc array perturbs the azo-dye doped LC material that assumes an isotropic phase.

This has been demonstrated by probing the reflection property of the structure with s- and p-polarized light. No appreciable variation has been observed, indicating that the liquid crystal molecules are distributed in the structure with their axis randomly distributed on the gold surfaces. The experimental setup for optical tuning of the absorption dip is shown in Fig. 12.7a, with an infrared microscope connecting to FTIR spectrometer to collect the reflection spectrum. An argon laser ($\lambda = 514$ nm) is used for the photo-pumping with its power tunable from 0 to 248 mW. The laser impinges obliquely on the sample so that its reflections are not collected by the microscope objective. When illuminated by the pumping light, several mechanisms can occur in the methyl-red doped nematic liquid crystal [42]. The most likely one is an optically induced alignment of the methyl-red molecules (and therefore the liquid crystals molecules) towards the normal to the gold surface, i.e., the probe "sees" a lower refractive index value closer to the ordinary index value n_o. Figure 12.7b shows the numerical simulation of a sample with different nanodisk sizes ($a = 300$ nm and $b = 200$ nm). It is evident that, with the liquid crystal overlayer, the two absorption dips are shifted by a considerable amount (from 1.1 μm (in air) to 1.35 μm, and from 1.52 μm (in air) to 1.7 μm, respectively). Because of the large index mismatch between the liquid crystal ($n_{\text{iso}} \approx 1.61$) and air ($n = 1.0$), the minima of the reflection bands are higher. When the refractive index of the liquid crystal is changed by an amount $\Delta n = -0.06$ (from 1.61 to 1.56), the simulation shows that a shift of about 30 nm should take place. The corresponding experimental results are plotted in Fig. 12.7c: we can clearly observe the change in the absorption dips

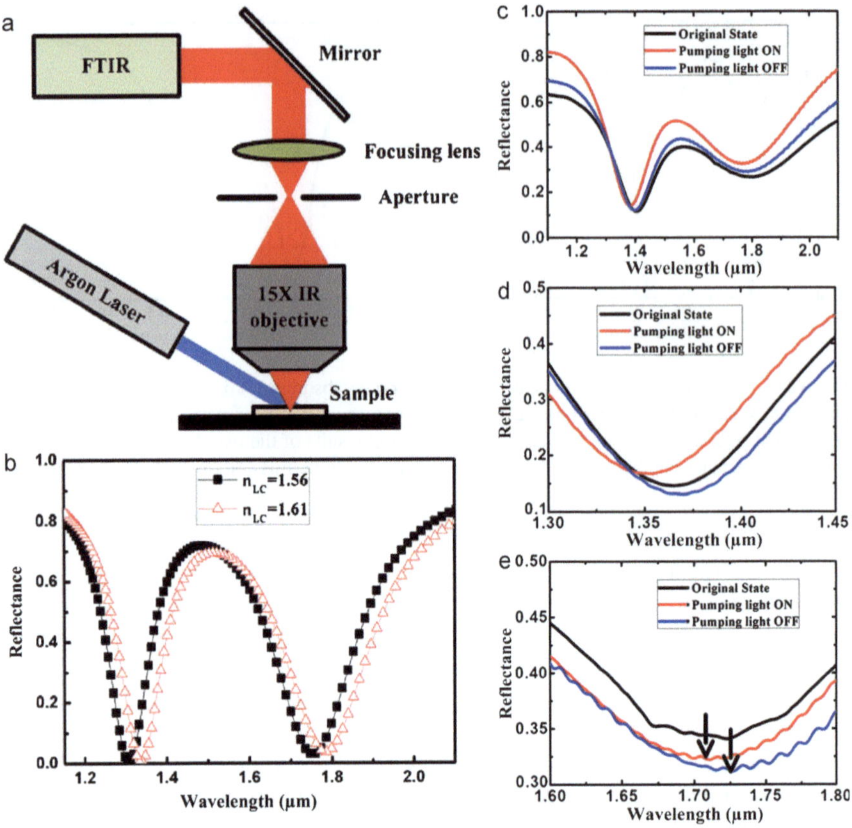

Fig. 12.7 (**a**) Experimental setup for characterizing the optically tunable plasmonic absorber. (**b**) FDTD-simulated results of the performance of the liquid crystal-plasmonic absorber; about 30 nm shift can be observed. (**c**) Measured absorption band shifts with the pumping light turning ON and OFF. (**d**) Zoom-in view of the left absorption dip. The shift is around 25 nm. (**e**) Zoom-in view of the second absorption dip. The shift is around 20 nm. [34]—Reproduced with permission of the American Institute of Physics

before and after the pumping light is applied; furthermore, in the experiment with low cw laser power illumination the tuning is reversible and repeatable and the tuning speed is ∼1 s. We also observe that the performance of the plasmonic absorber is somehow compromised after introducing the liquid crystal and dye mixture, with one absorption dip giving ∼70 % absorption compared to over 90 % before the introduction of liquid crystal. Figures 12.7d and 12.7e show a magnification of the two resonances from Fig. 12.7c; the shifts are around 20–25 nm, which is in good agreement with simulation results (shift of 30 nm). By optimizing the geometry, or the dye-concentration, and utilizing other mechanisms or liquid crystals to produce larger birefringence change, we could expect larger and faster tuning for the absorption dips. We can conclude by summarizing the main features of the two examples of photo-tunable plasmonic devices: In the first case, the combination of a photo-sensitive HPDLC grating and a gold nanodisk array gives rise to interesting interactions; results of experiments show that the diffraction efficiency of the grating is enhanced by the presence of the resonant entities. On the other side, a significant decrease of the extinction coefficient of the LSPR peak is observed because variations in the diffractive properties of the grating take place upon light irradiation.

The second example concerns a near-perfect dual-band plasmonic absorber; the two absorption bands of the device strongly depend on the geometrical features of two arrays of gold nanodisk, with different sizes, that are used in the main layer of the device. By further optimizing the device, large room for improvements can be attained in terms of tuning range, absorption efficiency and tuning response time.

12.5 Strain Controlled Active Plasmonics

There is a substantial interest in realizing nano-particulated substrates where shape, size and density of NPs can be finely controlled. The main reason is due to the dipole-dipole interaction that takes place between isolated metal particles. As already said, the interparticle separation influences the plasmon resonance frequency; roughly speaking, a change of plasmon frequency corresponds to a change of the color of the considered material. A wide scenario of applications is then foreseen for these systems as optical sensors and detectors. However, a detail that cannot be neglected is related to how pronounced this effect can be: will this color change be visible by naked eye or just detectable by means of a spectrophotometer? In order to answer this question, several groups have investigated the plasmonic shift that can derive from the application of mechanical strains to flexible substrates containing metal NPs.

A significant example has been provided by Correa-Duarte et al. who have successfully deposited gold NPs onto either planar or honeycomb structured poly(dimethylsiloxane) (PDMS) elastomeric substrates [43]. The involved process is sketched in Fig. 12.8. Planar or honeycomb structured elastomer films have been obtained by depositing a PDMS liquid solution (Sylgard 184 from Dow Corning GmbH) onto a silicon substrate and on a monolayer of polystyrene spheres (65 nm) closely packed on a hexagonal structure. Both samples were dried up in oven and then peeled off from the substrates. The planar PDMS film has been coated with 25 nm silica-coated gold NPs (Au@SiO_2, with a gold core of 15 nm) by following a layer-by-layer self-assembly technique [44]. After the peeling off, the honeycomb film contained some Poly-styrene spheres that have been dissolved by immersing it in Chloroform for a couple of hours. Afterwards, Au@SiO_2 NPs have also been deposited on this film by following the layer-by-layer technique.

In order to check an eventual dynamic plasmonic response, obtained samples (both planar and honeycomb structured) have undergone mechanical strain experiments. Samples were clamped and mounted within a uniaxial frame that allowed both stretching and compression experiments. Obtained results, reported in Fig. 12.9a, evidence a gradual reduction of the absorbance of the sample when it is

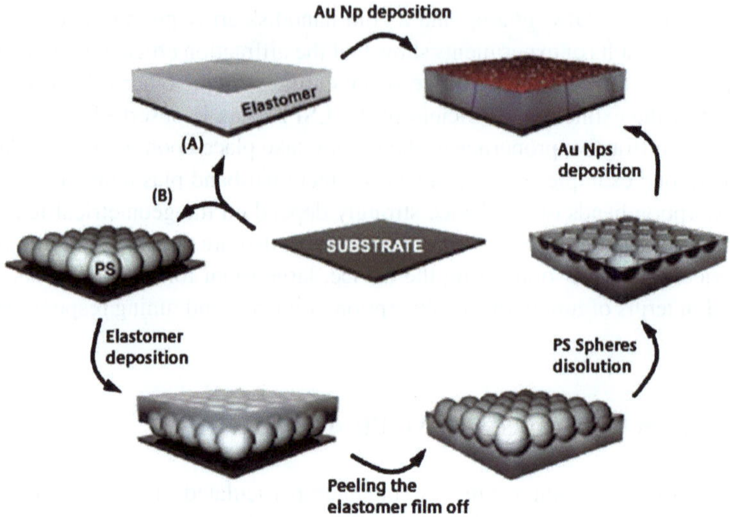

Fig. 12.8 Schematic illustration of the synthetic process followed for preparation of functionalized PDMS substrates. gold NPs are deposited onto planar (**a**) or honeycomb-like structured (**b**) PDMS substrates. The sketch is courtesy of Prof. Correa-Duarte. [43]—Reproduced with permission of Springer

subject to a stretching action; a reverse behavior is observed upon compressing the sample. For each curve, the maximum absorbance value has been plotted versus the percentage of applied strain, exhibiting a linear dependence (inset of Fig. 12.9a). This dependence can represent a quick and easy way to monitor the strain of the film and, hence, to use the system as an optical sensor. Keeping in mind that these experiments have been recorded using a standard UV-Vis spectrophotometer with a quite large beam diameter (0.5 mm), the sensitivity of the method can be considered quite high. Nevertheless, the obtained curves do not show any shift of the plasmon frequency. In fact, if we suppose that a dipolar interaction between neighboring gold NPs exists, upon stretching the PDMS film, we should expect a gradual blue-shift of the plasmon frequency because of a homogeneous increase of the interparticle separation.

On the contrary, if this homogeneous separation of NPs does not take place and, due to the stretching, there is only the formation of random cracks of the gold NPs layer (Fig. 12.9b), the absence of this shift can be justified. In this case, indeed, the stretching action would just produce the formation of big agglomerates of NPs, separated between each other, with consequent no influence on the plasmonic response of the system. Atomic force microscopy (AFM) images of the samples, before and after the stretching, have confirmed the presence of localized cracks in the sample surface and, hence, verified the exactness of this last hypothesis (images not reported). The honeycomb structured PDMS film also shows interesting features when considered as a strain sensor. Indeed, the sensing functionalities are two-fold because of the diffractive and plasmonic nature of the device. Due to the symmetry of the structure, a laser probe impinging on the structure produces a hexagonal

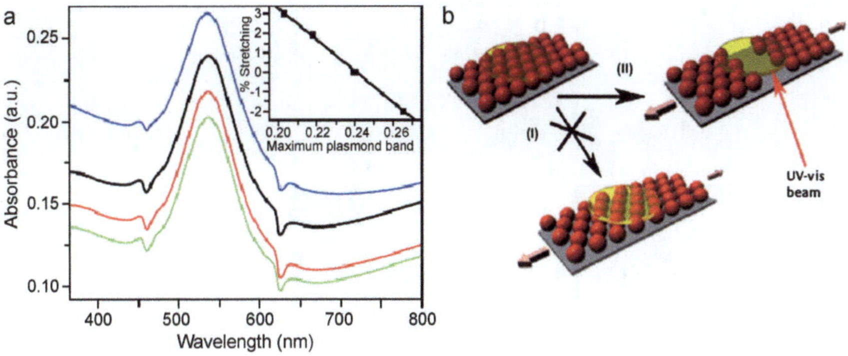

Fig. 12.9 (**a**) UV-Vis spectra of a planar PDMS films coated with Au@SiO_2 as it is (*black curve*), subject to 2 % stretching (*last curve*), 3 % stretching (*green curve*) and 2 % compression (*first curve*). (**b**) Scheme of the stretched film in presence of cracks of the NPs layer or with uniform separation of NPs. [43]—Reproduced with permission of Springer

diffractive pattern of the transmitted light. By applying a mechanical strain to the sample, a deformation of the holes takes place that has a direct consequence on the diffracted pattern. The honeycomb micro-structure offers also a higher surface area for NPs deposition; for this reason, the authors deposited 10 monolayers of Au@SiO_2 NPs and checked the UV-Vis spectra of the sample under mechanical strain.

Obtained results reveal a behavior very similar to the one obtained by using the NPs coated planar film, and the gold layers do not influence the diffractive behavior of the structure. These results suggest that both diffraction and surface Plasmon resonance can be simultaneously exploited for optical sensing.

Experiments have been also performed by including gold NPs in the bulk of flexible PDMS materials [45]. In order to avoid the cracking issues experienced by Correa-Duarte et al., which can derive from depositing a coating of NPs on the PDMS surface, the particles have been dissolved within the liquid PDMS solution. This task was not trivial: when dealing with NPs, the hardest job is to prevent their clustering. For this reason, gold NPs (from Harima Chemicals) have been, first, diluted in chloroform, stirred for about one hour and then their good dispersion has been verified by SEM. The gold NPs diluted in chloroform have been, then, mixed with PDMS. In doing this, there are, at least, three ways to proceed: particles can be directly combined with the base pre-polymer, with the curing agent or with the two components previously mixed together. At first, the very low viscosity of the curing agent was supposed that should improve its miscibility with gold NPs. Unfortunately, this hypothesis was not correct: after a short time (about 1 hour), NPs phase-separated from the curing agent. The combination of NPs with the base pre-polymer gave better results: the obtained mixture was quite stable and homogeneously colored. Probably, the high viscosity of this material prevents the diffusion of NPs and their aggregation in clusters. The further addition of the curing agent to the base pre-polymer doped with NPs did not influence the stability of the mixture that was then ready to be baked. The samples used for experiments have been prepared by

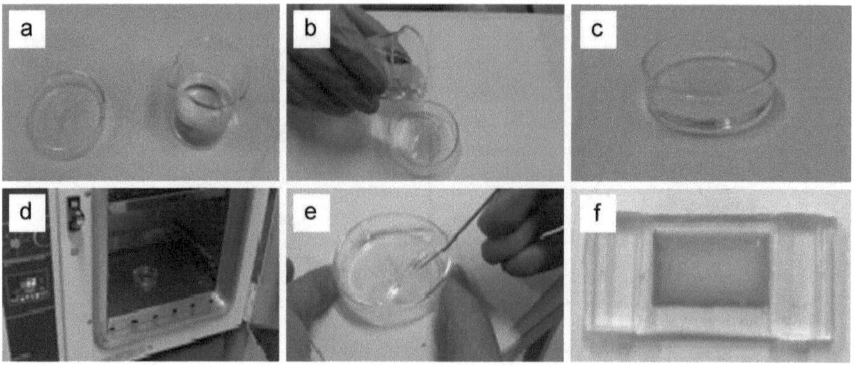

Fig. 12.10 Step-by-step procedure used for realizing PDMS samples with bulk distribution of gold NPs. (**a**) PDMS preparation, (**b**) air degassing, (**c**) master covering, (**d**) baking process, (**e**) peeling off, (**f**) PDMS template filled with gold nanoparticles. [45]—Reproduced with permission of Taylor & Francis group

following the steps depicted in Fig. 12.10: instead of preparing a homogeneous substrate made of PDMS doped with NPs, the presence of this material was limited just to a thin volume.

In order to obtain such a result, a glass substrate (1×2 cm^2, 1.1 mm thick) was glued to the bottom of a petri dish (Fig. 12.10a) and then was poured a pure PDMS mixture (curing agent/base pre-polymer in 1/10 weight ratio) in it (Figs. 12.10b,c). Afterwards, the petri dish was baked in the oven (previously warmed up at 100 °C, Fig. 12.10d) for about one hour. After the baking step, the hardened PDMS substrate was just peeled off from the petri dish (Fig. 12.10e). The obtained sample contains a hollow volume corresponding to the one occupied by the glass substrate glued in the Petri dish; this volume was filled in with a second mixture made of PDMS (same ratio between the two components) and Harima particles at 1 wt%, prepared as already explained. It is worth noting that the commercial product from Harima (NPG-J) used for these experiments contains an effective 56 wt% concentration of gold nanoparticles. As such, the real concentration of gold NPs in the samples corresponds to 0.56 wt%. The PDMS sample, which included now gold NPs, was additionally baked at 200 °C for about one hour. At the end of the process, a PDMS layer, $\sim$2 mm of thickness, was obtained containing a thin layer (about 100 μm thick) of PDMS doped with gold NPs (Fig. 12.10f). The choice of limiting the presence of NPs to a small volume of the sample is motivated by two reasons. First of all, by fabricating a thick layer of PDMS homogeneously doped with NPs, even with low gold concentrations, the extinction coefficient of the sample would be way too high, making difficult a reliable study of the plasmonic response of the system. Moreover, in future studies, it could be convenient to replicate a micro/nano structure (e.g. diffraction grating) on the surface of the pure PDMS sample; by covering afterwards this structure with NPs, it could be possible to study the influence of NPs in the optical and/or electro-optical behavior of such a structure. Such experimental condition can be realized by using the procedure indicated above for fabricating

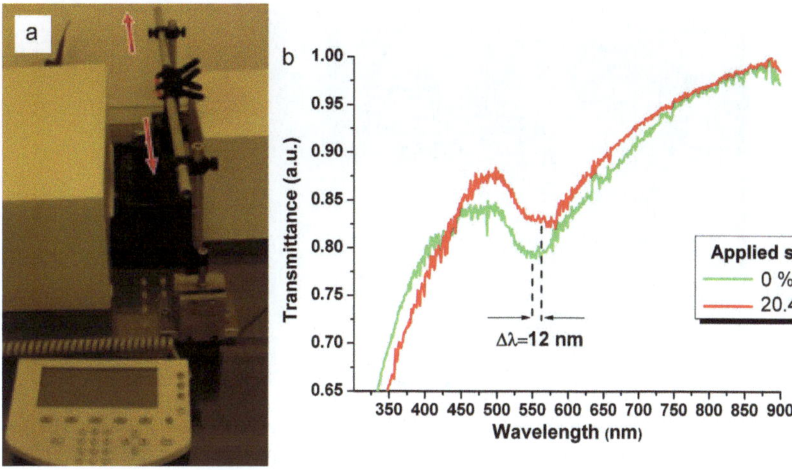

Fig. 12.11 (**a**) Experimental set-up used for the stretching characterization of PDMS samples containing gold NPs; (**b**) Behavior of the plasmon resonance wavelength as a function of the strain applied to the sample; the *dashed lines* indicate the points of minimum of the surface plasmon resonances. [45]—Reproduced with permission of Taylor & Francis group

the sample and by simply substituting the thin glass substrate in the Petri dish with another one where the micro/nano structure is present.

The prepared sample has been experimentally characterized by studying its spectroscopic response for different stretching conditions of the PDMS substrate. In order to apply a mechanical strain to the sample, this has been clamped with tweezers that can be moved apart by two micrometer mobile actuators. The set-up has been mounted between the light source and the detector of an Agilent 8453E spectrophotometer (Fig. 12.11a); for each elongation of the sample, the corresponding transmission spectrum has been acquired. Obtained normalized spectra are reported in Fig. 12.11b: from the graph, a similarity emerges with the work of Correa-Duarte; to an increase of the applied strain corresponds a reduction of the absorbance. However this is not the only effect since a red shift of the plasmon resonance wavelength of about $\Delta\lambda = 12$ nm has been observed (from $\lambda = 551$ nm, no applied strain, to $\lambda = 563$ nm, 20 % applied strain). This result confirms that, by using a PDMS matrix doped with gold NPs, it is possible to obtain a system showing tunable plasmonic properties. In order to interpret the measured red shift, results obtained by Kinnan and Chumanov can provide useful hints. Indeed, these authors have also demonstrated that, by changing the inter-distance between NPs, the electromagnetic coupling in their system is correspondingly modified [10]. In particular, spectroscopic investigations performed on substrates coated with silver NPs with different surface densities have shown a blue- or red-shift of the plasmon resonance wavelength, depending on the particle size: for small particles (59 nm or less) a red shift and a broadening of the plasmon peak takes place; in contrast, for NPs larger than 129 nm, there is a blue shift and a narrowing of the peak. In order to check which condition better fits their results, authors of Ref. [45] performed a SEM character-

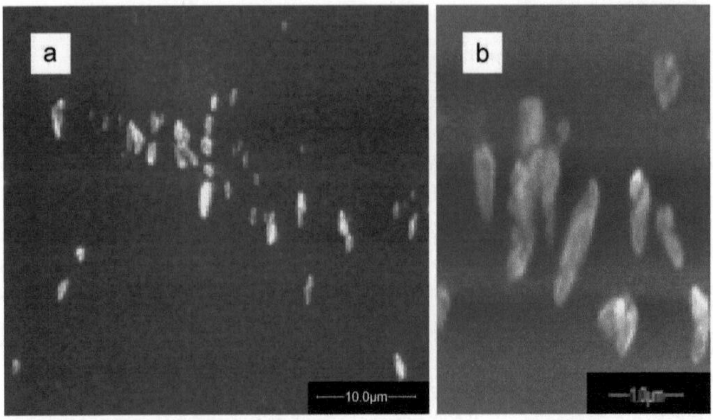

Fig. 12.12 Electron microscopy images of (**a**) NPs aggregates typical ensemble vision, (**b**) particular of the shape of NPs aggregates. [45]—Reproduced with permission of Taylor & Francis group

ization of their samples. Figure 12.12 shows two micrographs depicting the same area but with different magnifications: size and distribution of particles are quite at variance. Thus it is possible to assume that the red shift of the plasmon resonance wavelength reported in [45] is the superposition of several contributes given by the differently sized NPs aggregates.

We can conclude that the possibility of realizing active plasmonic systems tuned by mechanical actions has been demonstrated in two systems in which gold NPs are confined to the surface or dispersed in the bulk respectively. From an application oriented point of view, we can envisage several possibilities for the obtained result: when NPs are confined in the bulk, it could be convenient to reproduce the system in thin layers (100 μm) in order to realize very precise pressure control sensors. A more sophisticated application would instead result in advanced systems for solar exploitation, due to the ability of Au nanoparticles to harvest the electromagnetic radiation [46]. Finally, by better controlling size and distribution of Au nanoparticles, it should be even possible to achieve a larger tuning range of the shift of the plasmon resonance wavelength.

12.6 Conclusions

In the field of plasmonic devices the possibility of obtaining active functionalities is very appealing under several aspects. Main reason is that these devices usually require an accurate design to guarantee optimal performances and, in order to obtain a slight tuning of their properties it is sometimes necessary to complete redesign and repeat the entire fabrication process. Theory suggests, at least, two ways to dynamically tune the spectral position of the plasmon resonance frequency. One possibility is to modify the refractive index of the medium surrounding the resonant system. Depending on the involved materials, the tunability can be achieved by applying an external electric field, by changing the temperature of the device or by illuminating it with a specific wavelength. Another tuning mechanism is based on the dipolar interaction that gives rise to the resonance behavior. If metal sub-entities are moved apart, a modification of their interaction takes place and, hence, a tuning of their resonance frequency is obtained; thus, a mechanical action exerted on a flexible system containing metal nano-particles can induce a plasmonic tuning.

In conclusion, in this chapter, we have shown several examples of plasmonic systems that gain their peculiar functionalities from a very large number of amorphously arranged particles. Being the common denominator of quite different systems, amorphous materials can hold promises for the realization of innovative devices in optoelectronics, optofluidics, biosensors and metamaterials.

Acknowledgements Our sincere thanks go to Dr. Toralf Scharf and Prof. Carsten Rockstuhl for the fruitful collaboration in the framework of the NANOGOLD project. A special thank to Prof. Iam-Choon Khoo and Prof. Miguel Correa-Duarte for their helpfulness in writing this chapter. Finally, we acknowledge that part of the research leading to reported results has received funding from the European Union's Seven Framework Programme (FP7/2007-2013) under grant agreement No. 228455.

References

1. V.M. Shalaev, S. Kawata, *Nanophotonics with Surface Plasmons*. Advances in Nano-Optics and Nano-Photonics (Elsevier, Amsterdam, 2007)
2. T.W. Ebbesen, C. Genet, S.I. Bozhevolnyi, Phys. Today **61**, 44 (2008)
3. J.N. Anker, W.P. Hall, O. Lyandres, N.C. Shah, J. Zhao, R.P. Van Duyne, Nat. Mater. **7**, 442 (2008)
4. L. Childress, J.M. Taylor, A.S. Sorensen, M.D. Lukin, Phys. Rev. A **72**, 052330 (2005)
5. J. Gersten, A. Nitzan, J. Chem. Phys. **73**, 3023 (1980)
6. S. Nie, S.R. Emory, Science **275**, 1102 (1997)
7. M. Xiang, X. Xu, F. Liu, N. Li, K. Li, J. Phys. Chem. B **113**, 2734 (2009)
8. T. Okamoto, T. Yamaguchi, J. Phys. Chem. B **107**, 10321 (2003)
9. G. Mie, Ann. Phys. **25**, 377 (1908)
10. M.K. Kinnan, G. Chumanov, J. Phys. Chem. C **114**, 7496 (2010)
11. R. Pratibha, W. Park, I. Smalyukh, J. Appl. Phys. **107**, 063511 (2010)
12. P.A. Kossyrev, A. Yin, S.G. Cloutier, D.A. Cardimona, D. Huang, P.M. Alsing, J.M. Xu, Nano Lett. **5**, 1978 (2005)
13. T. Hegmann, H. Qi, V.M. Marx, J. Inorg. Organomet. Polym. Mater. **17**, 483 (2007)
14. M. Mitov, C. Portet, C. Bourgerette, E. Snoeck, M. Verelst, Nat. Mater. **1**, 229 (2002)
15. Q. Liu, Y. Cui, D. Gardner, X. Li, S. He, I.I. Smalyukh, Nano Lett. **10**, 1347 (2010)
16. Y.J. Liu, Q. Hao, J.S.T. Smalley, J. Liou, I.C. Khoo, T.J. Huang, Appl. Phys. Lett. **97**, 1 (2010)
17. L. De Sio, S. Ferjani, G. Strangi, C. Umeton, R. Bartolino, Soft Matter **7**, 3739 (2011)
18. R. Caputo, A.V. Sukhov, A. Veltri, C. Umeton, J. Opt. Soc. Am. B **21**, 1939 (2004)
19. http://www.harima.co.jp
20. K. Maekawa, K. Yamasaki, T. Niizeki, M. Mita, Y. Matsuba, N. Terada, H. Saito, Mater. Sci. Forum **638–642**, 2085 (2010)
21. http://www.merck.de
22. G. Strangi, V. Barna, R. Caputo, A. de Luca, C. Versace, N. Scaramuzza, C. Umeton, R. Bartolino, Phys. Rev. Lett. **94**, 063903 (2005)
23. S. Link, M.A. El-Sayed, J. Phys. Chem. B **103**, 4212 (1999)

24. P.B. Johnson, R.W. Christy, Phys. Rev. B **6**, 4370 (1972)
25. P.G. de Gennes, J. Prost, *The Physics of Liquid Crystals*, 2nd edn. (Clarendon Press, Oxford, 1993)
26. F. Tam, G.P. Goodrich, B.R. Johnson, N.J. Halas, Nano Lett. **7**, 496 (2007)
27. U. Kreibig, M. Vollmer, *Optical Properties of Metal Clusters* (Springer, Berlin, 1996)
28. L. De Sio, R. Caputo, U. Cataldi, C. Umeton, J. Mater. Chem. **21**(47), 18967 (2011)
29. L. De Sio, S. Serak, N. Tabiryan, C. Umeton, J. Mater. Chem. **21**, 6811 (2011)
30. K.M. Lee, H. Koerner, R.A. Vaia, T.J. Bunning, T.J. White, Macromolecules **43**(19), 8185 (2010)
31. T. Ikeda, O. Tsutsumi, Science **268**, 187 (1995)
32. Y.J. Liu, Y.B. Zheng, J. Liou, I.-K. Chiang, I.C. Khoo, T.J. Huang, J. Phys. Chem. **115**, 7717 (2011)
33. R. Caputo, L. De Sio, A.V. Sukhov, A. Veltri, C. Umeton, Opt. Lett. **29**, 1261 (2004)
34. Y. Zhao, Q. Hao, Y. Ma, M. Lu, B. Zhang, M. Lapsley, I.C. Khoo, T.J. Huang, Appl. Phys. Lett. **100**, 053119 (2012)
35. B. Zhang, Y. Zhao, Q. Hao, B. Kiraly, I. Khoo, S. Chen, T.J. Huang, Opt. Express **19**, 15221 (2011)
36. C.G. Hu, Z.Y. Zhao, X.N. Chen, X.G. Luo, Opt. Express **17**, 11039 (2009)
37. Z.H. Jiang, S. Yun, F. Toor, D.H. Werner, T.S. Mayer, ACS Nano **5**, 4641 (2011)
38. K.A. Willets, R.O. Van Duyne, Annu. Rev. Phys. Chem. **58**, 267 (2007)
39. I.C. Khoo, J. Liou, M.V. Stinger, Mol. Cryst. Liq. Cryst. **527**, 109 (2010)
40. I.C. Khoo, J.H. Park, J.D. Liou, J. Opt. Soc. Am. B **25**, 1931 (2008) and references therein
41. U.A. Hrozhyk, S.V. Serak, N.V. Tabiryan, L. Hoke, D.M. Steeves, B.R. Kimball, Opt. Express **18**, 8697 (2010)
42. I.C. Khoo, Phys. Rep. **471**, 221 (2009)
43. M.A. Correa-Duarte, V. Salgueiriño-Maceira, A. Rinaldi, K. Sieradzki, M. Giersig, L.M. Liz-Marzan, Gold Bull. **40**(1), 6 (2007)
44. M.A. Correa-Duarte, M. Giersig, N.A. Kotov, L.M. Liz-Marzan, Langmuir **14**, 6430 (1998)
45. U. Cataldi, P. Cerminara, L. De Sio, R. Caputo, C.P. Umeton, Mol. Cryst. Liq. Cryst. **558**, 22 (2012)
46. H.A. Atwater, A. Polman, Nat. Mater. **9**, 205 (2010)

Chapter 13
Plasmonic Nanoparticle-Based Metamaterials: From Electric to Magnetic Response

José Dintinger and Toralf Scharf

Abstract The self-assembly of nanoparticles into hierarchical architectures is currently attracting a lot of interest due to their potential applications in a wide range of fields like nanophotonics, nanoelectronics or catalysis. In the present chapter, we discuss the potential of metal nanospheres for the bottom-up fabrication of optical metamaterials. Controlling the spatial arrangement of the nanoparticles in these composites offers a promising route to engineer unique optical responses originating from their collective plasmonic resonance. Here we explore experimentally how different types of NP arrangements can give rise to distinct macroscopic effective properties, including both electric and magnetic optical responses. For each of the structures investigated, we propose a brief overview of the current state-of-the-art of the appropriate bottom-up fabrication methods and analyze their optical properties in details. First, the optical constants of "bulk" amorphous nanoparticle metamaterials are investigated by ellipsometry, demonstrating that controlling the nanoparticle filling fraction provides an efficient route to tune the metamaterial permittivity. As an example of a potential application, the realization of a hybrid plasmonic Bragg mirror is discussed. Finally, we focused on the fabrication and characterization of dense spherical nanoclusters that can sustain a magnetic response at optical frequencies. In doing so, we demonstrate the possibility to engineer the permeability of nanocluster-based metamaterials, thereby opening interesting perspectives for the realization of isotropic negative index materials operating in the visible.

13.1 Introduction

Although the interest in metallic nanoparticles (NPs) has grown tremendously in the scientific community over the past 30 years [1–3], their first use, to stain glass or decorate pottery, can be traced back to thousands of years ago [4, 5]. The most famous examples are the Roman Lycurgus cup [6] or the stained-glass windows of medieval cathedrals [7, 8] where luminous red and yellow colors are produced by

J. Dintinger · T. Scharf (✉)
Optics & Photonics Technology Laboratory, Ecole Polytechnique Fédérale de Lausanne (EPFL), Neuchâtel, 2000, Switzerland
e-mail: toralf.scharf@epfl.ch

C. Rockstuhl, T. Scharf (eds.), *Amorphous Nanophotonics*,
Nano-Optics and Nanophotonics, DOI 10.1007/978-3-642-32475-8_13,
© Springer-Verlag Berlin Heidelberg 2013

the gold, silver or copper NPs that these glassworks contain. The origin of these fascinating colors was, however, not well understood until the 19th century with the seminal works of Michael Faraday. In a famous lecture marking the birth of nanoscience in 1857 [9, 10], Faraday demonstrated that the vivid appearance of metal colloids was due to the "very minute size" of their constituents, which results in properties very distinct from those of the corresponding bulk metal. The underlying physical mechanism was clarified some years later with the advent of Maxwell's electromagnetic theory of light and several important works on the phenomenon of scattering and absorption of light by small objects by Rayleigh [11], Lorenz [12] and Mie [13] among others (see in [14–17]).

A clear picture of the phenomenon can be drawn within the Drude model for metals which considers the conduction electrons of metals as free from the attraction of the heavier positive ions forming the metal lattice, screened by the valence electrons. When the particles dimensions are much smaller than visible wavelengths, the optical field can polarize homogeneously the NP and its spatial phase can be considered constant over the NP volume (quasistatic or Rayleigh approximation). Under the influence of the time-varying electric field of an optical wave, the free electrons are driven collectively into oscillation with respect to the static ionic background. However, unlike their bulk counterpart, the free electrons in a metal NP are subjected to a restoring force induced by polarization charges created at the opposite surfaces of the NP, following the electronic displacement (Fig. 13.1a). This attractive force imparts a resonant behavior to the electronic cloud which acts, within this small size limit, as an electric dipole oscillator characterized by a resonant frequency. By analogy with the collective electronic oscillations in gaseous plasmas, these resonances were later termed localized surface plasmon resonances (LSPR). When the incident light wave's frequency is resonant, the amplitude of the oscillation is strongly enhanced as the optical energy is efficiently transferred to the free electrons. At the same time, the electronic motion is damped by two loss mechanisms categorized as radiative and non-radiative. Non-radiative damping, on the one hand, refers to the dissipation of motional energy in the form of heat within the metal NP and its environment, due principally to collisions between the vibrating electrons and the lattice ions (i.e. Joule heating). Radiative damping, on the other hand, corresponds to the reemission of light (in all directions) by the oscillating electrons (i.e. elastic scattering). These two resonant mechanisms, absorption and scattering, result in the different colors of stained glasses or colloidal dispersions when observed in transmission or diffuse reflectance. However, for small NPs like the ones shown in Fig. 13.1b, scattering is negligible and absorption strongly dominates.

The resonant wavelength λ_{LSP} also marks the transition between in- and out-of-phase oscillations of the electronic cloud with respect to the exciting wave. As a result, the optical light wave is not only attenuated but its phase is also affected and varies strongly around the resonance, a phenomenon often referred as *anomalous dispersion* (since the permittivity unusually decreases with the frequency). In addition to the characteristic LSPR extinction signatures observed in the far field, i.e. at a great distance from the NPs, plasmonic resonances also manifest themselves by a strong build-up of the optical near-field close to the NP surface. The NP therefore

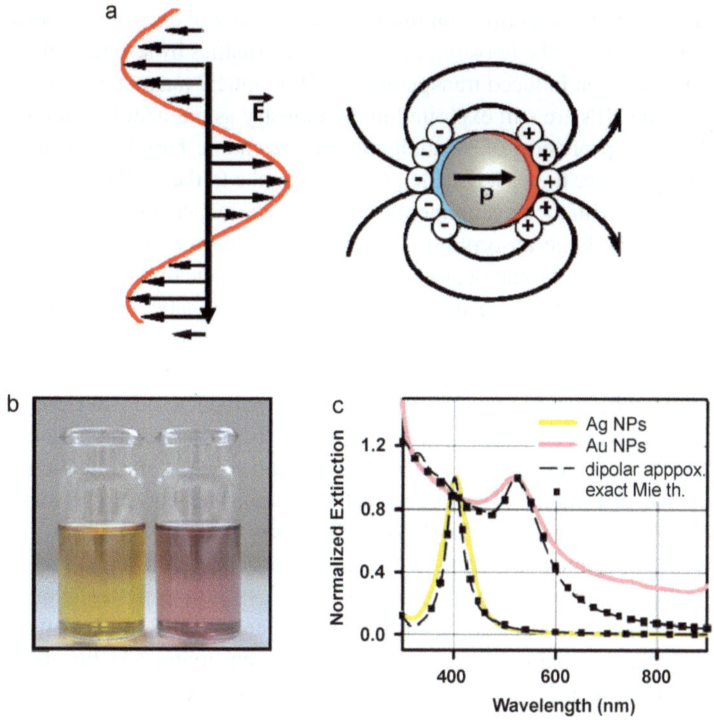

Fig. 13.1 Localized surface plasmon resonances (LSPR) of metal NPs: (**a**) Schematic representation of the excitation of the localized plasmon oscillations by an external electric field, within the quasistatic approximation. (**b**) Dispersions of silver (*left*) and gold (*right*) NPs with diameter on the order of 10 nm. (**c**) TEM of the corresponding silver NPs. (**d**) Extinction spectra of the two dispersions

acts as an optical antenna [18, 19] which confines light in nanoscopic volumes (typically on the order of the NP volume [1]), leading to a large increase of the near-field intensities in comparison with the incident light. The resonant frequency, and hence the color of the NPs, is determined by the strength of the restoring force which in turn depends on the surface charge densities and their spacing. Therefore, λ_{LSP} depends essentially on the metal and host polarizability (i.e. permittivity), and on the size and shape of the NP. Figure 13.1d shows the typical LSPR absorption band of gold (red) and silver colloids (yellow) in a liquid solution at wavelengths close to 520 nm and 400 nm respectively. Beside the nature of the metal, another important factor that affects the plasmonic resonance is the coupling among NPs. When the distance between neighboring NPs is on the order of their size, their respective optical near-fields start to interact and their electromagnetic coupling gives rise to new coupled plasmonic resonances, much like how electron atomic orbitals hybridize to form molecular orbitals [20]. Controlling the NP organization thereby offers an additional degree of freedom to tune the NP plasmonic resonances and create new properties. Recently, the collective properties of metal NP assemblies has attracted a lot of attention in the scientific community and a variety of exciting properties have been observed such as the boosting of near-field intensities by several order of magnitude [21], plasmon induced transparency [22] or optical magnetic responses [23].

While metal NPs are still exploited in the industry as pigment for paints or coatings [24, 25], a profusion of groundbreaking plasmonic based applications have been developed over the past decade [1–3, 26] thanks to these advances of modern nanotechnology with the development of new fabrication, characterization and simulation tools. The large absorption and scattering cross sections of metal NPs are attractive for various biomedical applications. When functionalized with specific targeting ligands, they can be used as powerful contrast labeling agents for dark-field bio-imaging, or as therapeutic agents for the photothermal treatment of cancer [27, 28]. Another popular application in the biomedical field is biosensing [29–31] which exploits the spectral sensitivity of the LSP frequency to local changes in the NP environment (i.e., permittivity changes) and to the NP aggregation state, to monitor the kinetics of binding events or to detect the presence of specific biomolecules such as DNA or proteins. The large local field enhancement is also exploited to improve the rate of various photophysical processes such as photoabsorption, fluorescence [32] or Raman scattering [33]. These effects have allowed to improve considerably the detection limit of various analytical spectroscopy techniques, down to the single molecule level in the case of surface enhanced Raman scattering [34] (SERS). They are also investigated as a potential means to improve the efficiency of light-emitting devices [35] or photovoltaic cells [36] (more details on this last topic are given in another chapter of this book). Other plasmonic applications include subwavelength waveguiding [37], optical trapping [38], nanoscopy [39] or lithography [40] and the list is still growing rapidly. In addition, we only addressed here applications based on the optical properties of metal NPs but they are also widely used for their catalytic or antimicrobial properties.

The focus of this chapter is on the use of metal NPs for the realization of a new class of artificial composites with unusual optical properties: the so-called metamaterials [41–44] (MMs). A general definition of an MM, is given by Cai and Shalaev [43] as *an artificially structured material which attains its properties from the unit structure rather than the constituent materials. An MM has an inhomogeneity scale that is much smaller than the wavelength of interest, and its electromagnetic response is expressed in terms of homogenized material parameters*. These parameters are usually the permittivity, $\varepsilon(\lambda)$, and permeability, $\mu(\lambda)$ which describe respectively the macroscopic polarization and magnetization of a material. Similarly to a standard material where $\varepsilon(\lambda)$ and $\mu(\lambda)$ describe the effect of its constituent atoms as the average result of their electric and magnetic polarizability, $\alpha_E(\lambda)$ and $\alpha_M(\lambda)$, the effective permittivity $\varepsilon_{\mathrm{eff}}(\lambda)$ and effective permeability $\mu_{\mathrm{eff}}(\lambda)$ of an MM stem from the response of the MM *unit structure*, the so called meta-atoms or meta-molecules. As stated in the second part of the definition, the size and spacing of the meta-atom must be small enough to prevent any diffraction effects, so that the electromagnetic wave doesn't sense the microscopic inhomogeneities of the MM and perceives an effectively homogeneous material. For visible wavelengths, this sets an upper limit for the size of the meta-atoms with a maximum on the order of 100 nm. This is still much larger than the sub-nanometric size of real atoms and it is therefore possible to control the geometry of the meta-atoms by relying on advanced nanofabrication techniques. The recent interest for MMs is largely motivated by the prospect of realizing materials with tailored properties on demand through a rational design of the meta-atom geometry. This enables the engineering of distinct optical properties that are precluded in the constituent materials, allowing potentially to extend the optical properties of conventional materials. In this respect, one of the current challenge of MM research is the realization of a bulk isotropic negative index material at optical frequencies [44, 45]. Besides its fundamental interest, this prospect also promises several exciting applications such as a superlens to image below the diffraction limit [46], or an invisibility cloak that bends light around an object [47]. This requires the engineering of both the electric and magnetic response of the meta-atoms. To achieve such properties, it is necessary to use resonant meta-atoms in order to strengthen the usually weak light matter interaction. This is especially true for the magnetic response since naturally occurring materials do not show any magnetic activity at optical frequencies (as indicated by their permeability which is set to the free space value: $\mu = 1$). Although the field of MMs is still young, huge advances have been made since the first demonstration of a negative index in the microwave regime, about a decade ago [48]. Owing to their strong plasmonic resonances in the visible, metallic nanostructures emerged as essential ingredients for the realization of optical MMs. Several structures based on metallic meta-atoms and inspired from the microwave designs, have successfully demonstrated similar effects close to the visible range [49, 50]. These mainstream MM designs usually consist in planar arrays of sub-wavelength metallic inclusions with complex geometries such as split rings [51], nanorod pairs [52] or fishnet structure [53]. The optical magnetic response of such meta-atoms results from the excitation of a coupled plasmon mode with anti-parallel currents that give rise to a magnetic moment. Despite these accomplishments, the current state-of-the-art MMs still suffer from several drawbacks inherent to their design and fabrication process. Due to their complex geometry, their fabrication usually requires advanced top-down nanofabrication methods such as electron- or ion-beam lithography which offer a great flexibility in the design of the meta-atoms and their large scale arrangement. Nevertheless, in addition to being expensive and time-consuming, these techniques, generally, only allow the realization of planar arrangements of meta-atoms. It is therefore difficult to fabricate a truly three dimensional bulk MM where light can effectively propagate. Additionally, this results in anisotropic geometries suffering moreover from strong spatial dispersion due to the rather large size of the meta-atoms and their strictly periodic arrangement. In consequence, the MM properties are angle and polarization dependent in a highly non-trivial manner. This prohibits their application for many envisioned MM devices that have been proposed while assuming to have a material at hand with an isotropic electric and magnetic response. Finally, the operation of current MMs in the visible remains limited due to the technical difficulties to further scale down the size of the meta-atoms [44]. Hence, one of the major challenges of the current MM research is to devise new strategies allowing the realization of an isotropic three dimensional MM at optical frequencies. To overcome the limitations of standard top-down planar MMs, bottom-up approaches based on the self organization of metal colloids have been recently suggested as a promising alternative [44, 45]. The huge progress of colloidal nanochemistry now allows the preparation of metal NPs with a wide range of shapes and their organization into complex ordered assemblies over various length scales [54, 55]. Nevertheless, the realization of complex architectures like the ones obtained from top down methods (nanorod pairs, split rings) still remain highly challenging. Fortunately, several theoretical works have shown that the most common type of metal colloids, i.e. nanospheres, could be used to build isotropic negative index MMs [56–60]. Depending on their organization in space, different properties can emerge from the plasmonic resonance of a nanosphere (Fig. 13.2a): both the effective permittivity $\varepsilon_{\text{eff}}(\lambda)$ and the effective permeability $\mu_{\text{eff}}(\lambda)$ (Fig. 13.2c) of the NP-MM can be engineered. For instance, metal NPs act as electric meta-atoms allowing to engineer $\varepsilon_{\text{eff}}(\lambda)$ (Fig. 13.2b) while spherical clusters of NPs act as magnetic meta-atoms (Fig. 13.2c) influencing $\mu_{\text{eff}}(\lambda)$ via the near-field interactions between NPs which give rise to a magnetic response. Although the realization of such architectures requires to control the NP organization over several length scales, a periodic ordering of the meta-atoms is not mandatory to observe the properties. Much to the contrary, amorphous arrangements are highly desirable as they provide a simple solution to the aforementioned problem of spatial dispersion and anisotropy by bearing intrinsically an isotropic response, independent of the light propagation direction. Several works have indeed shown that the MM properties are largely determined by the resonance of the individual meta-atoms, rather than by their positional arrangement and hence these properties survive the disorder [61–63] (a more detailed discussion of this problem is presented in another chapter of this book). If the meta-atoms bear an isotropic response, as in the case for spherical NPs or spherical clusters of NPs (Fig. 13.2), the strength of the MM resonance remains unaffected by the lights polarization state. Also, amorphous structures are evidently easier to fabricate, due to less severe geometrical constraints. In conclusion, the self assembly of plasmonic NPs represent a promising approach to solve some of the limitations of the current top-down MMs since it should allow the realization of isotropic MMs operating in the visible range and fabricated by up-scalable and cost effective bottom-up methods.

The aim of the work presented in this chapter is to evaluate the potential of colloidal metallic NP clusters for the bottom-up fabrication of optical MMs. To this end, we investigate how different types of NP arrangements can give rise to distinct macroscopic effective properties. The chapter is divided in three sections devoted to the bottom-up fabrication and optical characterization of different type of NP organization, like the ones depicted in Fig. 13.2. Section 13.2 is dedicated to bulk NP composites (Fig. 13.2b) and the determination of their effective optical constants by spectroscopic ellipsometry. These parameters are essential for the determination of the optical properties of more complex NP arrangements like the NP-polymer multilayers discussed in Sect. 13.3. Section 13.4 focuses on dense spherical clusters of NPs (Fig. 13.2c) and their potential magnetic response. A fabrication method to form NP cluster meta-atoms containing a large number of NPs is presented and the optical response of the fabricated clusters is carefully analyzed, confirming the occurrence of a dipolar magnetic resonance for the largest clusters.

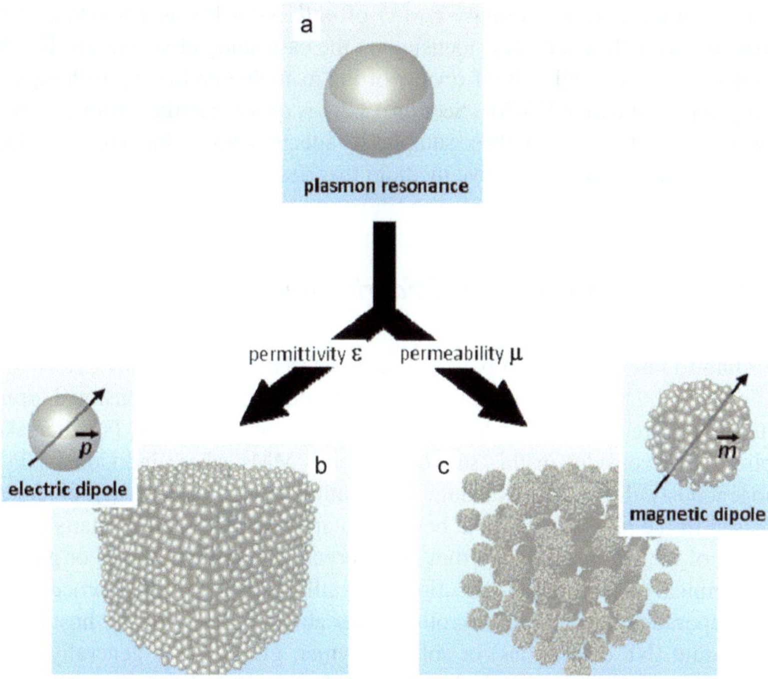

Fig. 13.2 Self-assembled amorphous plasmonic NP metamaterials: the LSPR resonance of a spherical NP (**a**) allows to engineer the permittivity of bulk NP composites (**b**) where NPs are homogeneously distributed in a host material, and the permeability when the NP forms spherical clusters (**c**)

13.2 Bulk NP Composites: From Dilute Systems to Closed Packed Assemblies

As explained above, metal nanostructures form ideal MM building blocks due to their LSP resonances. A bulky assembly of NPs, with no specific local arrangement such as the one depicted in Fig. 13.2b, represents a good example of an amorphous *electric* MM. Its effective permittivity $\varepsilon_{\text{eff}}(\lambda)$ can be tuned to attain extreme values, owing to the strong anomalous dispersion that occurs in the spectral vicinity of the LSP resonance [64]. The effective permittivity $\varepsilon_{\text{eff}}(\lambda)$ of such a bulk NP composite is a crucial parameter for evaluating the response of more elaborate metastructures such as the ones investigated in the next sections. A rigorous description that accounts for all the NP interactions in such complex structures requires demanding simulation tools but a qualitative evaluation of the emerging effective properties can be obtained by treating these complex assemblies as made out of an effective medium characterized by $\varepsilon_{\text{eff}}(\lambda)$. The effective permittivity (permeability) is the spatial average of microscopic fields and polarization (magnetization), and hence strongly depends on the volume fraction f occupied by the meta-atoms or -molecules. Consequently, we devote this first section to the experimental determination of the effective permittivity $\varepsilon_{\mathrm{eff}}(\lambda)$ of *bulk* NP MMs as a function of their NP volume filling fraction, f_{NP}, focusing on the case study of spherical silver NPs. Section 13.2.1 starts with a brief review of the available bottom-up techniques for the preparation of bulk NP MMs. Section 13.2.2 is devoted to the retrieval of the effective permittivity $\varepsilon_{\mathrm{eff}}(\lambda)$ of these samples by spectroscopic ellipsometry including a brief introduction to the effective medium theories.

13.2.1 Bulk NP Metamaterial Fabrication

According to our definition, a bulk NP metamaterial consists of any extended (dimensions much larger than the wavelengths of interest) three dimensional arrangements of NPs, regardless of whether it is ordered or amorphous. For the sake of simplicity, we consider in this section only NP MMs where individual NPs are homogeneously distributed. Random aggregation into ill-defined agglomerates is addressed in the next section. The bottom-up approaches generally starts with the synthesis of the NP meta-atoms from metal precursors in liquid, solid or gas phase using chemical (generally wet reduction of metallic salts) or physical processes (gas phase evaporation, electrolysis, pyrolisis, laser ablation, ...). The NP host medium can be liquid (NP dispersions) or solid (polymer, glass) and is generally a transparent dielectric with a passive role in the MM properties. In some cases however they can provide additional functionalities such as optical gain when doping the host matrix with dyes [65], or an active tunability under external stimuli (temperature, light, electric field, ...) like when using for instance liquid crystals [66]. The principal interest of colloidal self organization methods is to control the NP organization over various length scales, from the nano- to the macroscale. In general this involves coating the NP surface with specifically designed ligands to balance the different attractive and repulsive NP interactions (these can be van der Waals, electrostatic, steric, covalent or hydrogen bonding interactions to cite just a few). Capping agents are used to manage the interaction between particles and most common are surfactants, polymers or ionic species. In solution, their role is to prevent the agglomeration and coalescence of the NPs by mitigating the strong attractive van der Waals NP forces through electrostatic repulsion or steric hindrance. For our study, we used commercial hydrophobic silver NP inks (NPSJ, Harima Chemicals, Inc.) which have been originally developed in the field of printed electronics [67, 68] for the realization of interconnects by inkjet printing, followed by annealing. In this respect, a major advantage of such metal NP inks is their low sintering temperature (typically below 250 °C) which makes their processing compatible with plastic electronics. A key requirement for these applications is the high NP concentration necessary to obtain homogeneous conductive patterns: these NP dispersions have a metal volume fraction above $c_v > 10\ \%$ ($>1\ \mathrm{g\,ml^{-1}}$) in solution. This corresponds to an average distance between NPs close to 10 nm only, in the liquid state. With such high density, one can investigate a wide range of NP concentrations by preparing several dilutions of the ink in polar solvents (Fig. 13.3a). These NPs are

Fig. 13.3 Silver NP inks: (**a**) Photograph of the concentrated NP ink. (**b**) Photograph of the various prepared dilutions corresponding to NP volume filling fractions f_{NP} from 0.01 % up to 12 %

prepared by gas phase evaporation, and dispersed in tetradecane in the presence of capping agents (usually long alkyl chains terminated by thiols or amine group to prevent their irreversible agglomeration (by steric hindrance). This technique produces NPs with a narrow size distribution (mean diameter of 12 nm), as indicated by the TEM images (Fig. 13.3b) and the sharp LSP absorption band around $\lambda = 400$ nm (Fig. 13.1c). Note that we preferred silver over other metals like gold, due to comparatively lower losses [69] as can be seen from the much larger widths of the gold NP LSP band (Fig. 13.1c).

Nevertheless, the interesting, unusual metal NP MM properties are expected at still higher NP filling fractions [64] for which the NPs approach their condensed state (f_{NP} larger than 50 %), just like standard top-down planar MMs, which usually have a surface coverage in this range. While the focus of this chapter is on the optical properties of metallic NPs, unique electronic, magnetic, catalytic or mechanic properties can also emerge depending on the choice of the NP material. Fostered by these prospects [70], a whole host of methods have been developed to form extended dense NP assemblies from dispersed solutions (Fig. 13.4). This ranges from direct solvent evaporation to more intricate colloidal crystallization techniques. Several excellent books and review articles [54, 71–75] offer an exhaustive description of the various bottom-up techniques available. We only intend to give here a few examples of the most popular methods that allow the realization of ordered or glassy amorphous 3D NP assemblies, with dimensions extending from several microns to centimeters. The self assembly of NPs into prescribed architectures is essentially ruled by the interplay between the multiple interactions among NPs [71, 76], the influence of their environment and eventually the effect of external forces [73] (capillary forces, shear flows, ...) or templates [77]. Probably, the most straightforward approach to increase the NP density up to the solid state is by sedimentation [78, 79] (provided that the NPs are large enough so that gravity can overcome Brownian motion and buoyancy forces). Alternatively, NP aggregation or crystallization can be realized by evaporating the solvent [74, 80] or destabilizing the colloidal dispersion by the addition of a precipitant [81–83] (ionic salt, non-solvent or competing ligands) which destabilizes the repulsive NP interactions and prompt their aggregation. Dense amorphous films of randomly packed NPs are readily obtained by these techniques but it is also possible to build large ordered NP superlattices if the den-

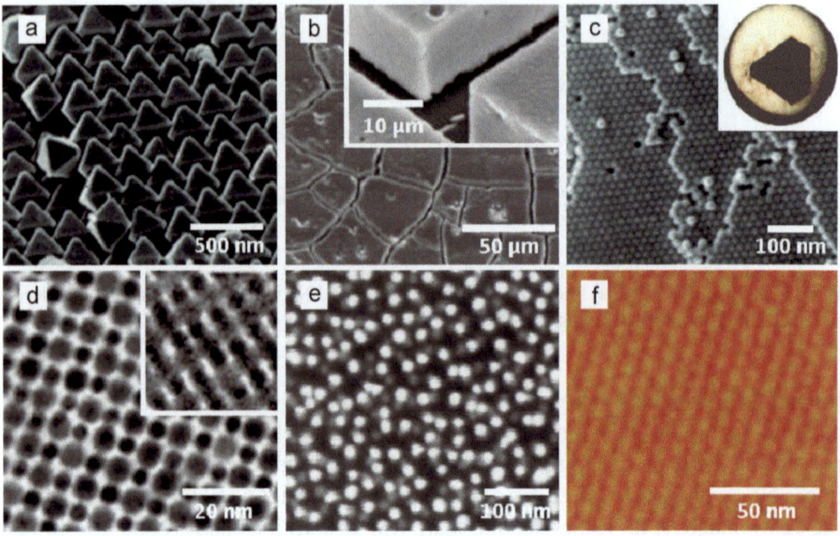

Fig. 13.4 Examples of self-assembled bulk NP MM: (**a**) SEM image of a colloidal crystal of octahedral silver NPs obtained by sedimentation. Reprinted with permission from [78]. (**b**) SEM image of silver nanosphere supracrystals obtained by solvent evaporation. *Inset*: tilted view. Reprinted with permission from [74]. (**c**) SEM image of CoFeO4 NP superlattice grown by solvent destabilization in a microdroplet. *Inset*: photograph of the microcrystal confined inside a droplet. Reprinted with permission from [83]. (**d**) TEM image of the (001) plane of a binary fcc superlattice of oppositely charged PbSe and gold NPs direction. *Inset*: view of the (100) plane. Reprinted with permission from [80]. (**e**) SEM image of an amorphous gold NP bilayer prepared by LbL deposition with polyelectrolytes. Courtesy of A. Cunningham. (**f**) AFM phase image of a monolayer of hexagonally close packed PbSe NPs obtained by phase separation during spin coating. Reprinted with permission from [84]

sification process is homogeneous and slow enough, and if the NP size distribution is sufficiently narrow.

As the NP concentration exceeds a certain (crystallization) threshold, the NPs adopt their equilibrium arrangement as determined by the balance between the different NP interactions at work (van der Waals, electrostatic, ...), together with entropic (space-filling) factors [79, 85] and eventually NP and solvent-substrate interactions [74]. Although face-centered cubic (fcc) or hexagonally close packed (hcp) superlattices are thermodynamically favored for monodisperse nanospheres, a large variety of lattices have been produced by adjusting different NP properties like their shape [79, 85] (Fig. 13.4a), surface charge [86] or size disparity [80, 87] (Fig. 13.4d). Remarkably, electrostatic repulsive interactions can maintain order even at very low NP concentration forming liquid colloid crystals with lattice constants of several microns [78, 88]. These methods are however generally slow, and offer only a limited control on the surface morphology, thickness and lateral dimensions of the NP assembly. A more sophisticated approach, inspired from the concepts of supramolecular chemistry, is based on the molecular recognition between specific ligands grafted at the surface of the NPs. A large variety of ligands such as DNA strands [89, 90], mesogens [91, 92], dyes [93], photochromes [94] or dendrimers [95] have been introduced to trigger the NP self-assembly by specific short range attractive NP interactions such as covalent or non covalent interactions via hydrogen bonding, π–π, dipolar, metal-ion coordination, DNA base-pair interactions and others. This approach is particularly effective for the fabrication of small complex NP assemblies, i.e. NP meta-atoms (see Chap. 1) but large area mono- and multilayers of NPs can also be produced when large 2D templates are used, like in the layer-by-layer (LbL) deposition method [96]. This procedure involves the alternate adsorption of NPs and molecular layers with complementary charges or functional groups, by successive immersion of a solid substrate in NP suspension and chemical baths. This method, originally based on the Coulomb interactions between oppositely charged polyelectrolytes [97] was modified to form NP multilayers by combining charged NPs and polyelectrolytes [98, 99]. More details on this method are given in another chapter if this book. The LbL deposition principle was also exploited using the covalent bonding of dithiol ligands to metals [100] or the complementary interactions between nucleobase sequences of DNA strands [101] or polypeptides [102]. More complex architectures can also be realized by chemically pre-patterning the supporting substrates [77]. While the aforementioned self-organization methods are essentially driven by internal interparticle interactions, large scale ordering of NPs can be greatly facilitated by the use of external force fields. In particular, flow fields have been widely used in combination with solvent evaporation to form wafer scale NP films following the principle of the coffee ring effect. The dip-coating method [103, 104], for instance, exploits the capillary forces that attract the NPs at the three-phase gas/liquid/solid interface, formed by the meniscus between an evaporating suspension and a vertically standing wettable substrate. Other external fields, such as electric, magnetic or hydrodynamic fields have also been investigated to drive the assembly of NP from dispersed solutions [54]. For our purpose, we used the spin-coating technique [105, 106] which offers the advantage of being perfectly adapted for scaling-up and mass production as it is already widely used for various industrial applications such as photoresist coating or compact disk fabrication. It is a cheap, fast and robust method to produce homogeneous layers on large surfaces, ideally suited for our optical characterization studies. In this method, an excess amount of NP dispersion is deposited on a wettable substrate and rotated at high speed to spread the NP fluid. The formation of NP films by spin coating has been investigated both experimentally and theoretically. Despite its apparent simplicity, this method involves a complex multistage process that combines viscous drag, non-uniform shear flow induced by centripetal force, evaporation and interparticle interactions. The intricate balance between these driving forces makes it difficult to predict and control the morphology of the NP assembly. Nevertheless, the fabrication of large-area 2D and 3D ordered NP films has already been reported [84, 107, 108].

To obtain a large area close-packed NP monolayer, we followed the method presented in Refs. [84] and [109], where hexagonally close packed (hcp) monolayer of semiconductor NPs extending over cm^2 have been prepared by spin coating a mixed solution of NPs and organic material. The NP monolayer formation follows

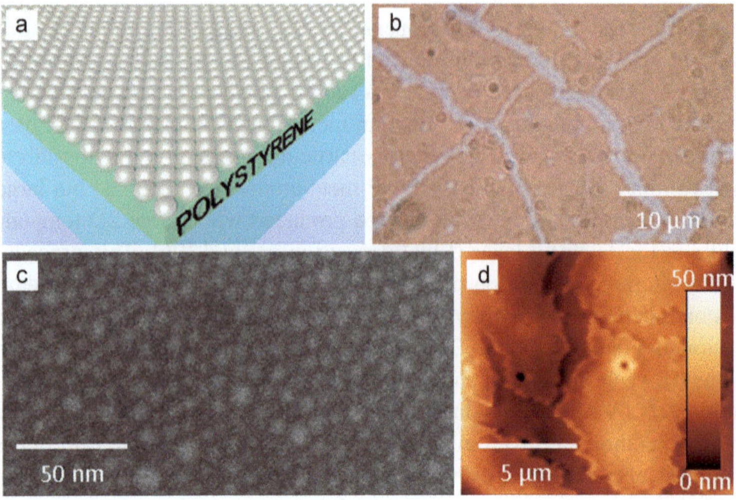

Fig. 13.5 2D NP monolayers prepared by phase segregation between NPs and a polymer during spin casting: (**a**) Scheme of the structure. (**b**) Photograph of the structure showing cracks within the monolayer. (**c**) SEM image of the hexagonally close packed silver NPs. (**d**) Topography of the film measured by AFM

a phase separation process driven by the minimization of the interface surface energies. This results in the migration of the NPs at the air interface, on top of an organic underlayer [109]. The addition of the organic material slows down the evaporation of the solvent, giving more time for the NPs to reorganize at its surface and, therefore, avoiding unwanted kinetic effects like jamming. The NP surface coverage can be adjusted from disconnected NP islands to fully hcp close packed monolayer by controlling the concentration of NPs and organic material in the initial mixture, and the evaporation rate (via the spinning speed and/or solvent volatility). In our case, we use polystyrene (Mw 90,000, Polymer Standard Services) as the organic material and mix it with the hydrophobic silver NP inks (Harima, NPS-J) in toluene. The various experimental parameters (NP and polystyrene concentration and spin speed) were adjusted to obtain a uniform NP monolayer lying on top of a polystyrene thin film (Fig. 13.5). Scanning electronic micrographs show some hcp arrangements involving several tens of NPs (Fig. 13.5c); nevertheless no long range order is observed due to the silver NP size polydispersity (a maximum standard deviation in size of 5 % was found to be necessary to observe long-range ordered packing [84]). Although undesirable, the presence of strain-induced cracks attests (Fig. 13.5b and d) for the compactness of the NP arrangement and allows us to determine the NP thickness by performing atomic force measurements at their edges. A thickness around 10 nm was found, confirming a single layer of NPs, at most, is present on top of the PS film. Assuming a uniform hcp packing, a diameter of 12 nm for the metallic cores and a coating thickness of 1 nm (measured from TEM data; Fig. 13.3b), the NP surface coverage is estimated to be around 50 %.

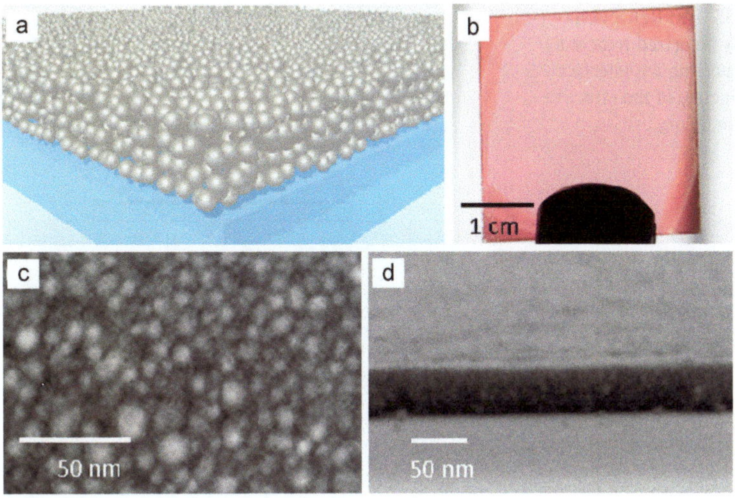

Fig. 13.6 3D NP thin film prepared by spin casting: (**a**) Scheme of the structure. (**b**) Photograph of the film. (**c**) SEM image of the randomly close packed silver NPs. (**d**) Cross-sectional view of the thin film

Finally, to obtain a true 3D bulk NP MM, we spin coat directly the pure NP ink (without polystyrene) on the substrate (Fig. 13.6). It must be noted that the ink formulation contains some additives [68, 110], such as wetting agents and adhesion promoters, that ensure the deposition of a stable uniform film, as can be seen from the SEM analysis (Fig. 13.6d). The tilted view shows a thickness of 70 nm. In contrast to the monolayer sample, the NP packing appears completely random on the SEM images (Fig. 13.6c), indicating a NP volume filling fraction f_{NP} slightly below 50 %.

13.2.2 Effective Permittivity of Bulk NP Composites

The assessment of the dielectric constants of a material is essential to obtain a condensed description of its optical behavior and to evaluate its performance in device applications. The assignment of effective material parameters is particularly decisive for the understanding and further exploitation of MMs. The first demonstration of a negative effective index [48] of refraction has contributed importantly to the widespread interest for these composite materials. In general, effective permittivity $\varepsilon_{eff}(\lambda)$ and permeability $\mu_{eff}(\lambda)$ are inferred indirectly from the measurement of related observable quantities such as transmission and/or reflection coefficients by relying on analytical models that connects observables and parameters [111]. Direct retrieval requires to have access to as many observables as unknown parameters. Otherwise the problem would be underdetermined and needs to be solved indirectly, usually via data fitting procedures [112]. This task can be particularly intricate for

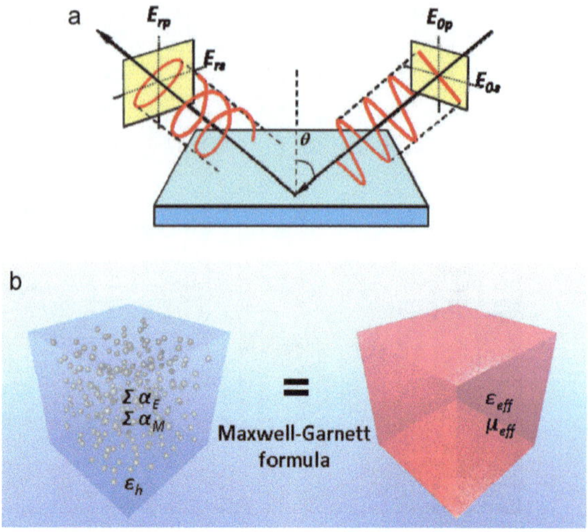

Fig. 13.7 (a) Reflection of linearly polarized light at a substrate/ambient interface. (b) Principle of the effective medium theory

complex MMs that require a large number of parameters for their full characterization [113]. This concerns most of the top-down MMs that are affected by spatial dispersion and anisotropy. In addition, when the MMs combine electric and magnetic activity, both $\varepsilon_{\text{eff}}(\lambda)$ and $\mu_{\text{eff}}(\lambda)$ are needed, doubling the number of unknown parameters. However, we should not be concerned with such complications since the NPs we are considering are much smaller than optical wavelengths and only very weakly diffusing. Furthermore, the amorphous and isotropic arrangements investigated in this section bear an isotropic response, fully described by a single complex effective permittivity $\varepsilon_{\text{eff}}(\lambda)$, since only an electric response is involved. The parameter, however, still depends on the frequency.

To characterize our different samples, we used a polarization measurement technique, namely variable angle spectroscopic ellipsometry (VASE). Ellipsometry is a widely used technique to determine the optical constant of thin films [114], and it has been more recently introduced for the characterization of MMs [115, 116]. In particular, several works have used ellipsometry to determine the dielectric constants of NP systems such as ours: NP liquid dispersions [117], monolayers [118, 119] or thin films [120]. This method exploits the change in the polarization state of light when reflected (or transmitted) by a material surface (Fig. 13.7a) to deduce its permittivity. To do so, the sample of interest is illuminated at oblique incidence by a light beam with a known state of polarization. This is usually a linearly polarized with equal amplitudes of the electric field in the plane of incidence (p-polarization) and in the orthogonal plane (s-polarized). The permittivity contrast at the interface induces a phase delay and a change of amplitude between the p- and s-polarized reflected components, resulting in an elliptically polarized beam. The reflected beam polarization ellipse is determined using an analyzer combined with a rotating polarizer or a photoelastic modulator, as is the case for the commercial instrument used in this study (UVISEL, Jobin-Yvon). The measured quantities, the so-called ellipsometric angles $\Delta(\lambda)$ and $\Psi(\lambda)$ correspond respectively to the ampli-

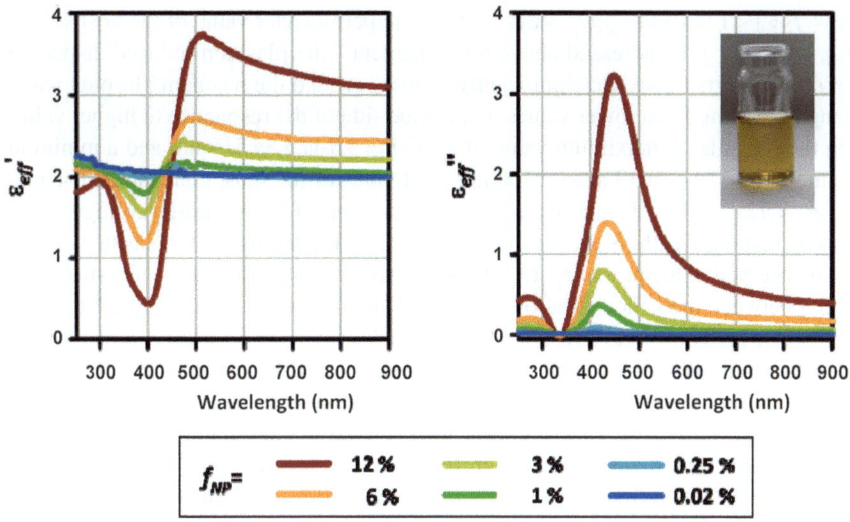

Fig. 13.8 Real (*left*) and imaginary (*right*) parts of the effective permittivity of NP dispersions for different NP filling fractions f_{NP} as retrieved from spectroscopic ellipsometry measurements. (*Inset* shows the dispersion with $f_{\text{NP}} = 0.25\ \%$)

tude ratio and the phase difference and are related to by the complex ratio $\rho(\lambda)$ of the reflection coefficients $r_p(\lambda)$ and $r_s(\lambda)$ (for the p- and s-polarized light, respectively) by the following relation, provided that the material is non depolarizing and isotropic:

$$\rho(\lambda) = \frac{r_p(\lambda)}{r_s(\lambda)} = \tan \Psi(\lambda) \cdot e^{i\Delta(\lambda)}. \tag{13.1}$$

In the case of a single interface between air and a bulk MM, the MM complex effective permittivity $\varepsilon_{\text{eff}}(\lambda)$ can be directly calculated using Fresnel equations to express r_p and r_s as a function of $\varepsilon_{\text{eff}}(\lambda)$ and inverting relation (13.1):

$$\varepsilon_{\text{eff}}(\lambda) = \varepsilon'_{\text{eff}}(\lambda) + i\varepsilon''_{\text{eff}}(\lambda) = \sin^2(\vartheta)\left[1 + \left(\frac{1-\rho(\lambda)}{1+\rho(\lambda)}\right)^2 \tan^2(\vartheta)\right] \tag{13.2}$$

where ϑ is the angle of incidence. We applied this method for the characterization of our liquid NP dispersions (Fig. 13.3) by using a liquid cell with a frosted glass surface to minimize the reflection from the liquid-substrate surface interface. As a consequence, only the air/liquid interface contributes to the reflected signals and the sample behaves as a single interface [117, 121]. Measurements of the real and imaginary part of the silver NP inks are shown in Fig. 13.8 for the various NP volume fractions f_{NP}. The volume fraction f_{NP} is estimated from the metal mass content of the most concentrated solution provided by the manufacturer (i.e. concentration in weight $c_W = 65\ \%$). The plasmonic resonance can be clearly identified as a peak in the spectrum of $\varepsilon''_{\text{eff}}(\lambda)$, even for the lowest value of f_{NP}. On the other hand, the effect on the real part $\varepsilon'_{\text{eff}}(\lambda)$ remains almost insignificant at low f_{NP} (<1 %) and $\varepsilon'_{eff}(\lambda)$ stays close to the permittivity value of the host solvent ($\varepsilon_h \approx 2$). As f is increased above a few percents, the plasmon induced dispersion in $\varepsilon'_{eff}(\lambda)$ starts to deviate significantly from ε_h around the resonant plasmon wavelength, moving from lower values on the blue side of the resonance to higher values on the red side. A maximum value of $\varepsilon'_{eff}(\lambda) = 3.7$ at $\lambda = 515$ nm and a minimum value of $\varepsilon'_{eff}(\lambda) = 0.43$ at $\lambda = 400$ nm are found for the most concentrated sample ($f_{NP} \approx 12$ %). Besides the expected strengthening of the spatially averaged plasmonic response with f_{NP}, a strong red-shift and broadening of the plasmon peak is also observed in the plot of $\varepsilon''_{eff}(\lambda)$. This is the typical signature of the near-field coupling between the LSP excitations of individual NPs which occurs as the NPs come into close proximity. As a result, the individual resonances hybridize into collective modes, similarly to what occurs for the simpler case of two coupled NPs [20].

As mentioned before, we also performed ellipsometric measurements on close-packed NP films (Fig. 13.6) in order to investigate dense samples. For such thin films standing on a substrate (which therefore involve two interfaces: air/NP and NP/substrate), the effective permittivity can be retrieved by an inversion similarly to the method for a single interface by using the Airy multilayer formula, provided that the film thickness d is known precisely. Otherwise, the three unknowns ($\varepsilon'_{eff}(\lambda)$, $\varepsilon''_{eff}(\lambda)$ and d) cannot be determined uniquely from the two measured values $\Delta(\lambda)$ and $\Psi(\lambda)$. In this case, they can be retrieved indirectly by fitting a model to the experimental data using a linear regression analysis minimizing the difference between model and data. To be consistent, the models for $\varepsilon'(\lambda)$ and $\varepsilon''(\lambda)$ must respect fundamental physical principles such as causality (Kramers-Kronig relations). Different analytical functions fulfilling these conditions exist to model the permittivity dispersion of various types of optical materials. For metal NP composites, essentially two types of models are used for the fit, depending on the information that is sought. Effective medium theory (EMT) can be used to retrieve additional parameters that affect the optical constants, such as the NP filling fraction. These models allow to assign an effective permittivity $\varepsilon_{eff}(\lambda)$ to the composite MM, and hence to treat it as an effectively continuous material. The first and still widely used EMT developed for spherical NP composites is the Maxwell-Garnett (MG) theory [17, 122]. It derives from the Clausius-Mossotti (or Lorentz-Lorenz) relation which links the permittivity $\varepsilon(\lambda)$ of a material to the mean polarizability $\alpha_E(\lambda)$ of its constituents. For natural materials these constituents are atoms or molecules while for the MM we are interested in here they are the meta-atoms, i.e. the NPs.

$$\frac{\varepsilon(\lambda) - 1}{\varepsilon(\lambda) + 2} = \frac{N\alpha_E(\lambda)}{3\varepsilon_0}. \tag{13.3}$$

N is the average volume per number of atoms per unit volume and ε_0, the vacuum permittivity. This relation is obtained by calculating the field acting on an atom as the superposition of the external field and all the dipole fields of the neighboring atoms. The MG theory applies this approach to the case of NPs by introducing the NP electric polarizability $\alpha_E(\lambda)$ calculated within the quasistatic approximation

$$\alpha_E(\lambda) = 3V \frac{\varepsilon_m(\lambda) - \varepsilon_h}{\varepsilon_m(\lambda) + 2\varepsilon_h} \tag{13.4}$$

where V is the NP volume, ε_m the permittivity of the metal and ε_h the permittivity of the surrounding medium. The plasmonic LSPR resonance can be identified in the above expression as a singularity occurring when $\varepsilon_m(\lambda) = -2\varepsilon_h$, marking the resonant increase of the NP polarizability. The MG theory expresses the effective permittivity of a metal NP composite $\varepsilon_{\text{eff}}(\lambda)$ in terms of the permittivity of its individual components and the NP volume fraction $f_{\text{NP}} = 4/3\pi N_{\text{NP}} r^3$ (with r, the NP radius and N_{NP}, the NP density). The MG formula is obtained by combining relations (13.3) and (13.4) and one finds

$$\varepsilon_{\text{eff}}(\lambda) = \varepsilon_h + 3 f_{\text{NP}} \varepsilon_h \frac{\varepsilon_m(\lambda) - \varepsilon_h}{\varepsilon_m(\lambda) + 2\varepsilon_h - f_{\text{NP}}(\varepsilon_m(\lambda) - \varepsilon_h)}. \tag{13.5}$$

This formula was shown to provide a good description for dilute samples (up to $f_{NP} = 30\ \%$ for silica coated gold NPs) but offer only a qualitative description when the NPs are in closer proximity due to simplistic approximations. Indeed, the MG theory only includes dipolar interactions although induced higher-order multipoles should be taken into account for close-packed NP arrangements. This model also does not account for the effect of spatial disorder or size and shape polydispersity which lead to fluctuations of the local fields and broadening of the resonance. For monolayers deposited on a substrate, image dipole (or multipoles) in the substrate should also be included for a more accurate description (see thin island film theory for instance [123]). Finally, the result of the ellipsometric data fitting using MG models may also be affected by the choice of data set or model used for $\varepsilon_m(\lambda)$. In any case, the bulk value $\varepsilon_m(\lambda)$ must be corrected for confinement effects inherent to small size NPs. Indeed, the free electron mean free path in NPs can be severely reduced in comparison to the bulk metal due to the presence of boundaries; this can results in a significant broadening of the plasmonic resonance when the NPs are small (<20 nm) [17]. An alternative to the MG model is to use a Lorentz dispersion model. Although the Lorentz model offers less physical insight by accounting only implicitly for the composite nature of the system, it does not suffer from the aforementioned constraints. The Lorentzian representation of the effective permittivity is given as a function of the frequency ($\omega = 2\pi c/\lambda$) by

$$\varepsilon_{\text{eff}}(\omega) = \varepsilon_\infty + \sum_i \frac{A_i}{\omega_{0i}^2 - \omega^2 - i\omega\Gamma_i}, \tag{13.6}$$

where the fitting parameters are the amplitude A_i, the width Γ_i and resonance frequency ω_{0i} of the Lorentzian functions and an offset value ε_∞. As described in the literature [119, 124], we used a series of at least two oscillators to fit the ellipsometric data collected for our solid thin samples (Fig. 13.9). One of the oscillator represents the plasmonic resonance in the visible range while the other accounts for the interband transitions of silver that occur at UV wavelengths, outside the experimental range. The thickness d of the films was also used as a fitting parameter, whose initial value has been extracted from AFM and SEM measurements (Figs. 13.5d and 13.6d). These complementary data allows to ensure the accuracy and uniqueness of the fit and avoids ambiguities due to strong correlations between $\varepsilon_{\text{eff}}(\lambda)$ and d for absorbing films. The results of the fit and the retrieved $\varepsilon'_{\text{eff}}(\lambda)$ and $\varepsilon''_{\text{eff}}(\lambda)$ for

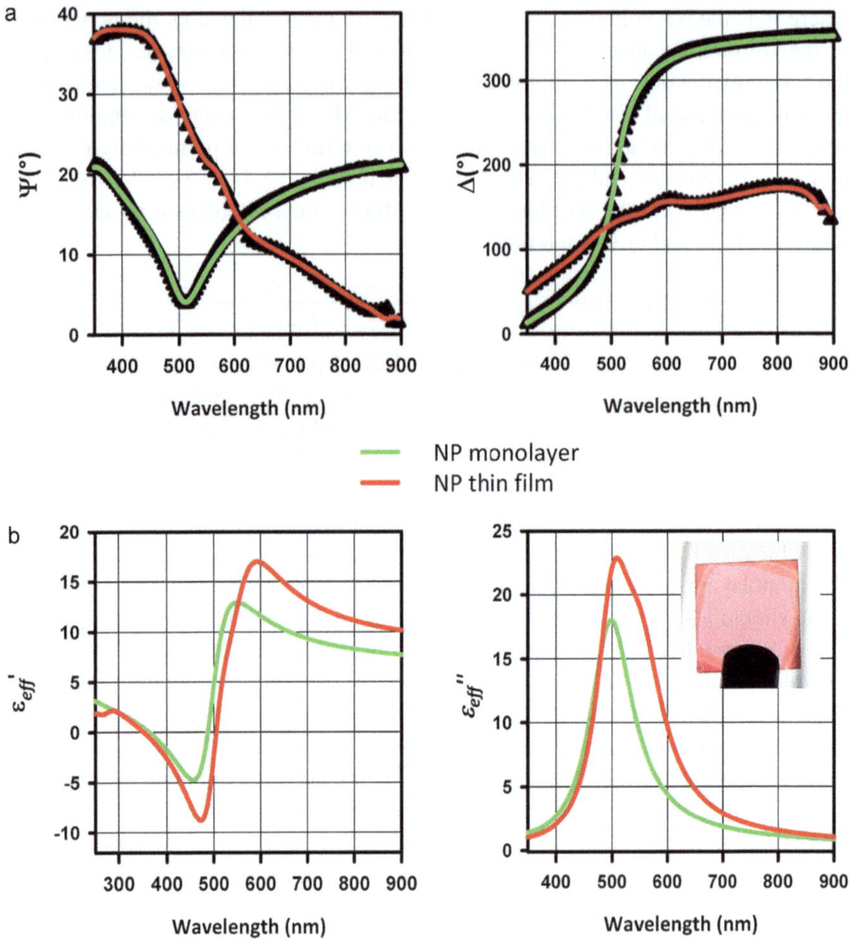

Fig. 13.9 Ellipsometric data for the close-packed NP films. (**a**) Experimental and modeled ellipsometric angles Ψ (*left*) and Δ (*right*). (**b**) Retrieved real (*left*) and imaginary (*right*) parts of the effective permittivity of the NP films

the two samples are shown in Fig. 13.9. Owing to the high density of NPs in these samples, we can see that the dispersive profile of $\varepsilon_{\mathrm{eff}}'(\lambda)$ is more pronounced than for the liquid samples and extends now over the whole visible spectrum. Accordingly, the plasmon peak grows and broadens considerably and its position shifts to longer wavelengths with a red-shift of more than 100 nm (with respect to the most dilute sample where NPs are non-interacting). For both samples the real permittivity $\varepsilon_{\mathrm{eff}}'(\lambda)$ reaches negative values (down to around -10) on the blue side of the plasmon resonance and high positive values (up to around 15–20) on the red side. As we will see in Sect. 13.4, such high values of $\varepsilon_{\mathrm{eff}}'(\lambda)$ are of particular interest for the fabrication of MMs with magnetic activity and by consequence for the realization of negative index materials. They are nonetheless accompanied by a strong absorption which could be detrimental for certain applications. Although similar values of $\varepsilon'_{eff}(\lambda)$ and $\varepsilon''_{eff}(\lambda)$ are observed for both samples, we can see that the resonance is slightly weaker for the monolayer sample despite the fact that the NP density is expected to be slightly higher for this sample (owing to its more ordered organization). This is due to the presence of the cracks in the monolayer which results in a decrease of the average NP density and hence in a dilution of the plasmonic response. We also observe a significant difference in the line widths of the resonance between the two samples (Fig. 13.9b), the resonance in the thick film being considerably broader. We attribute this difference to the random NP organization in the thick sample which leads to strong variations in the local fields and coupling conditions between neighboring NPs.

In conclusion, these ellipsometric measurements shows that controlling the NP density in a bulk silver NP metamaterial provides an efficient mean to tailor their complex permittivitty. Huge changes in both the absorption $[\varepsilon''_{eff}(\lambda)]$ and dispersion $[\varepsilon'_{eff}(\lambda)]$ spectra are observed when going from dilute to dense composite. Beyond metamaterial properties, the spectral tunability of these composite materials can be of great value for the design of various useful optical devices like cut-off or absorption filters. As mentioned in the introduction, there are several other ways to manipulate the spectral properties of NP composites like adjusting the size, shape or composition of the NPs. Another powerful mean is by introducing some spatial organization. An example is given in the next subsection which focuses on the properties of hybrid NP-polymer multilayers.

13.3 Hybrid Plasmonic-Photonic Multilayers

Controlling the NP organization in a composite enables to tailor their coupling and engineer its spectral properties. This can concern both the near-field and far field coupling depending on the length scale of the NP arrangement. Manipulating the plasmonic near field coupling allows to directly design new plasmonic resonance but requires to control the NP organization at the nanoscale. Depending on the application or property targeted, the control of the far-field coupling can be a powerful alternative to manipulate the spectral properties of plasmonic composites indirectly, through interference effects. For instance, a clever choice of the periodicity of diffractive NP arrays has allowed to produce extremely narrow plasmon peaks [125, 126] with direct implications for sensing applications. Following similar concepts, extremely large broadband photonic bandgaps were recently engineered using 3-dimensional stacking of plasmonic oscillators at Bragg distance.

Another type of plasmonic structure whose spectral properties are governed by diffractive effects is the multilayer formed by alternated stacks of NP monolayers and transparent dielectric films depicted in Fig. 13.10a. This layered arrangement behaves as a one-dimensional photonic crystal due to the periodic permittivity modulation induced by the alternated NP and polymer layers [64]. The 1D ordering of the monolayers allows for the formation of reflection stopbands related to the constructive interference between the partial reflections at subsequent interfaces. Like

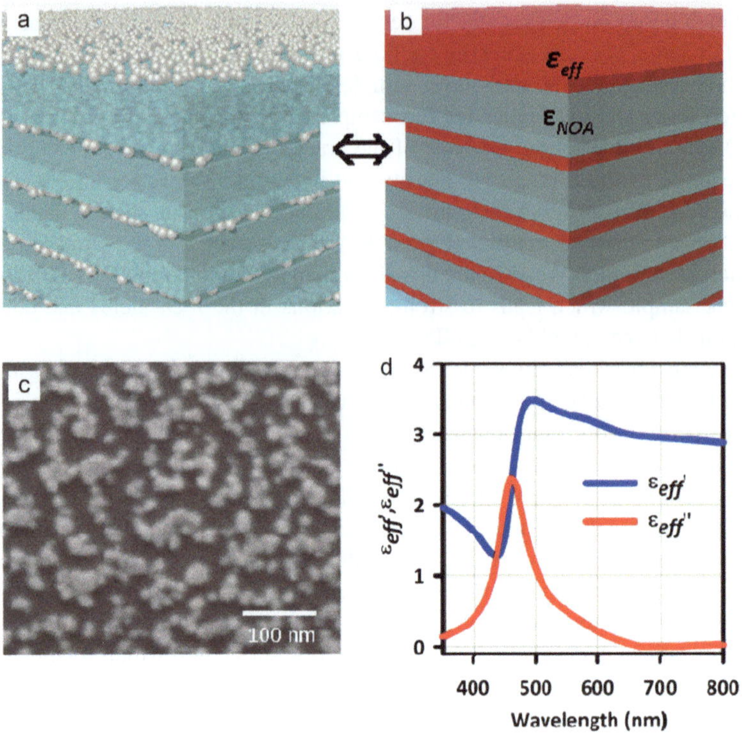

Fig. 13.10 Plasmonic Bragg multilayer: (**a**) Scheme of a NP-polymer multilayers. (**b**) Equivalent effective structure used for modelization of the multilayer properties: NP layers are treated as homogeneous layers with an effective permittivity $\varepsilon'_{\text{eff}}(\lambda)$. (**c**) SEM image of the NP layers showing a random arrangement of chain-like NP aggregates. (**d**) Effective permittivity of the NP clusters layers shown in (**c**)

for standard dielectric photonic crystals, the reflectivity of these hybrid structures can be specified by setting the geometry of the layers and the permittivity contrast between successive layers (i.e. the NP density in the monolayer and/or the refractive index of the dielectric material). The optical response of those structures is however markedly different than the one of conventional dielectric Bragg mirrors due to the unique interplay between plasmonic and Bragg resonances. To investigate these unique properties, we extended the NP spin-coating deposition process presented earlier to fabricate multilayers of this type. Several examples of similar multitasks made of dielectric and/or metal NPs and fabricated by the same technique can be found in the literature [127–133].

To fabricate such multilayers, we deposited sequentially polymeric and NP layers on a glass substrate using the spin-coating method described earlier. The NP concentration of the NP ink and the deposition speed was adjusted to produce NP layers with coverage ratio below one single nominal monolayer to mitigate absorption. The typical appearance of such layers is shown in Fig. 13.10c. The NPs cover

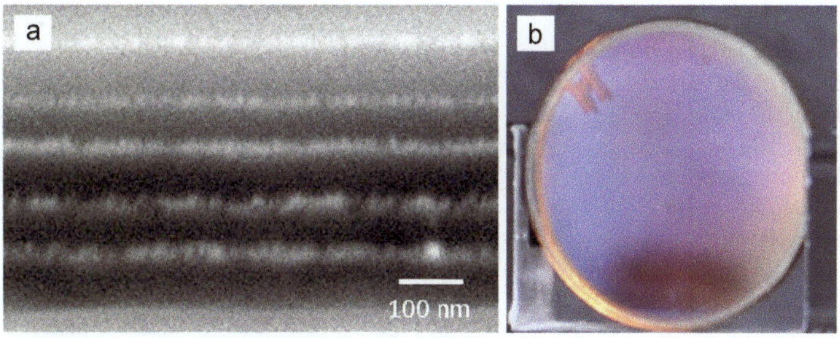

Fig. 13.11 (**a**) SEM cross-sectional view of a NP-NOA multilayer with $d_{NOA} = 120$ nm. (**b**) Photography of the same multilayer

approximatively 10 % of the surface and form an amorphous arrangement of random clusters containing between 3 and 20 NPs laying flat on the substrate. The effective permittivity $\varepsilon_{eff}(\lambda)$ of this NP layer deduced from ellipsometric measurements, is shown in Fig. 13.10d. A UV curable optical adhesive (NOA61, Norland adhesives) is used as the polymeric material to prevent the redispersion of the deposited layers during the successive deposition steps. After each deposition, the layers are hardened and made solvent resistant by photopolymerization under UV irradiation.Their thickness, d_{NOA}, can be finely tuned by adjusting the spin deposition speed between 2000 and 8000 rpm for a chosen polymer concentration (here 4.5 $\%_w$ NOA in acetone was used). This allows to tune precisely the resonant wavelength λ_m of the multilayer Bragg modes, which can be roughly estimated, for normal incidence, by the Bragg condition: $2n_{NOA}(d_{NOA} + d_{NP}) = m\lambda_m$, where m is a positive integer defining the Bragg mode order, d_{NP} is the thickness of the NP layer [i.e. the NP di ameter (12 nm)] and n_{NOA} is the refractive index of NOA61 ($\approx$1.56). Four samples were fabricated with different values of d_{NOA} between 100 and 180 nm chosen to bring the first Bragg mode of the structure in vicinity of the plasmon resonance of the NP layers which occurs at $\lambda = 460$ nm. The total number layers was set arbitrarily to 11 for all samples (5 periods and one protective NOA61 overlayer to prevent oxidation of the silver NP).

The SEM image in Fig. 13.11a shows the cross-section of a typical multilayer corresponding to $d_{NOA} = 120$ nm. The alternated layers of the structure can be clearly distinguished: the metal NP layers appear as thin bright layers separated by darker NOA layers of equal thicknesses. Figure 13.11b shows the visual aspect of the sample which exhibits a distinct purple color indicative of the selective Bragg reflection. The Bragg resonance appears as a narrow peak in the reflectivity spectrum (Fig. 13.12a). We can see that the reflection band for this given sample coincides almost exactly with the plasmon resonance of the NP cluster layers. No peak splitting is observed indicating that the plasmon-photon interaction is in a regime of weak coupling due to the damping of the plasmon oscillations. We can see that for this particular set of parameters about the same amount of light is transmitted and absorbed ($\sim$40 %). Together with the experimental data, Fig. 13.12a also shows the simulated optical properties of the multilayer calculated with the transfer matrix method (dashed line). In this model, the NP layers are treated as homogeneous

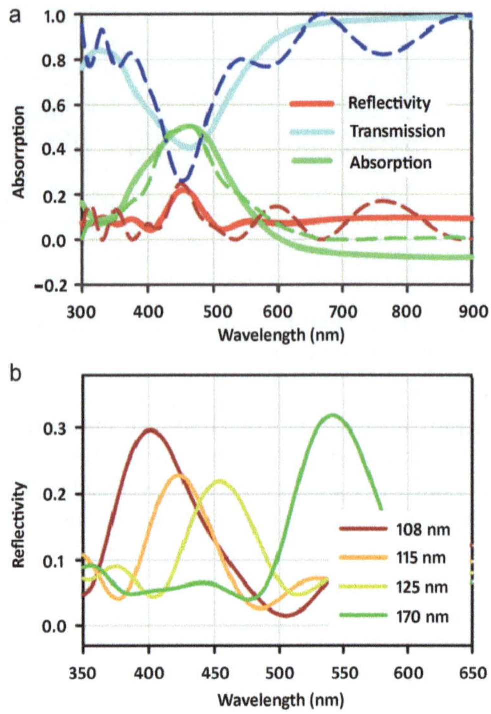

Fig. 13.12 (**a**) Measured (*plain lines*) and simulated (*dashed line*) of the optical properties of the multi layer with $d_{NOA} = 120$ nm. (**b**) Reflectivity spectrum of NP-NOA multilayers for different interlayer separation

layers (as schematized in Fig. 13.10b) with a thickness given by the NP diameter (12 nm) and an effective permittivity plotted in Fig. 13.10d. Due to slight variations inherent to the fabrication process and the uncertainty of the SEM measurements, it was found necessary to adjust the thickness d_{NOA} used in the calculations to match the experimental peak position. As can be seen, the line shape and intensity of the Bragg reflection band and the plasmon absorption are well reproduced by the simulation. The sideband oscillations clearly observed in the simulation are damped in the experimental spectrum, probably due to sample inhomogeneities and scattering. Finally, to evidence the tunability of these multilayers, Fig. 13.12b shows the reflectance of samples with different d_{NOA} (the values indicated in the legend correspond to the best fit of the simulations to the experimental data. As expected, the Bragg peak shifts to longer wavelengths as the period is increased, with an approximate redshift of 15 nm for an increase d_{NOA} by 10 nm. Although we have only varied the interlayer distance here, an important asset of such structure is the wide tunability of the filling factor of the NP clusters in the layer that allows to control the permittivity contrast and the level of absorption.

Besides color filtering, a few applications based on such amorphous NP cluster multilayer stacks. There is the possibility of using the system directly to study degeneration or take up of substances and solvents by just observing the spectra of the multilayer film [134]. A surface with perfect absorption is another promising application exploiting interferences and resonances in similar multistacks [135]. But more unusual applications can also be envisaged when working in the metamaterial regime (i.e. small interlayer distances). For instance, provided that the NP density in the layer is high enough, such multistacks can be used as an anisotropic metallo-dielectric MM in which the permittivities have opposite signs in orthogonal directions. This type of MM is already used to implement novel applications based on negative refraction like superlensing and subwavelength imaging [136, 137]. Other applications of amorphous cluster matter metamaterials can be found in other chapters of this book. Although, this type of NP organization only allow to engineer the MM electric permittivity, we will see in the next section that other types of arrangement allow to get access to the MM permeability by inducing an unusual magnetic activity at optical frequency

13.4 Plasmonic Meta-atoms Made from Nanoparticle Clusters

The emergence of MMs as a new class of artificial materials has opened a wide range of new perspectives for the manipulation of optical waves. One of the most exciting prospects is undoubtedly the realization of a negative index of refraction with all the exciting applications that this entails. As already mentioned, this requires, assuming an isotropic material, to achieve both a negative permittivity ε_{eff} and a negative permeability μ_{eff} thanks to a suitable design of the meta-atoms. We have seen in Sect. 13.2 that plasmonic NPs can be used as electric meta-atoms to reach high positive and negative values of the effective permittivity. A more challenging task is to engineer the permeability of the MM, given the lack of magnetic response from these NPs and more generally from any known materials. Indeed, magnetism, which generally arises from orbital currents in atoms, fades away at frequencies higher than a few gigahertz. It is completely inexistent at optical frequencies due to an extremely weak coupling with the magnetic component of light, hence $\mu = 1$ for common materials. To tackle this problem, different types of meta-atoms which act as magnetic dipoles, have been proposed and validated by experiments [44]. These meta-atoms usually mimic the archetype of a magnetic dipole, i.e. a current loop. Their principle of operation is based on the excitation of resonant modes with circular current distributions induced by the magnetic component of the optical field and giving rise to a magnetic moment. At frequencies below the resonance, the magnetic moment follows the incident magnetic field and an MM formed by a collection of such meta-atoms exhibit a paramagnetic response ($\mu > 1$). On the other hand, at frequencies above the resonance, the magnetic moment is opposed to the incident field and the MM is diamagnetic ($\mu < 1$). Most of the current magnetic top-down MMs rely on metallic inclusions with complex geometries, such as split rings or cut wire pairs, which sustain coupled plasmonic modes mimicking closed loop currents. Spherical metal NP clusters were recently proposed as an advantageous alternative to build magnetic meta-atoms [56–58]. This approach is free from the usual constraints of the top-down MMs (spatial dispersion, anisotropy) and provides a control over μ in the visible range. In this section, we intend to evaluate the potential of clusters of silver NPs as magnetic meta-molecules. Section 13.4.1 explains the origin of the optical magnetic response in such clusters, namely, the excitation of magnetic Mie resonances. Section 13.4.2 describes our preparation method of the NP clusters, as well as a quick overview of the state-of-the-art in this domain. Finally, Sect. 13.4.3 compares the optical properties of the fabricated clusters measured experimentally with rigorous simulations to demonstrate the significant role of a magnetic dipole resonance in such meta-atoms, hence confirming their potential for the realization of a magnetic MM.

13.4.1 Magnetic Mie Resonances of Spherical NP Cluster

While the resort to metallic meta-atoms with loop-like geometries is by far the dominant strategy to obtain a magnetic response, another possible route is to exploit the optical resonances of high permittivity dielectric scatterers [138, 139]. In contrast to the former approach which relies completely on plasmonic resonances, the magnetic activity in dielectric meta-atoms stems from large polarization currents induced at specific resonant wavelengths associated with the so-called magnetic Mie-resonances [16]. As indicated by their name, these magnetic resonances are associated with the theory of Mie [13] which offers a rigorous analytical treatment of the interaction of spherical particles with light. In this approach, the scattering from a sphere is expressed as a superposition of spherical harmonics where each of them corresponds to a specific electric or magnetic multipole moment. In the case of subwavelength scatterers, like meta-atoms, only the electric and magnetic dipoles are usually significant and their respective contribution can be found using the Mie scattering theory. Both types of responses can exhibit a resonance under specific conditions which depend essentially on the permittivity and size of the scattering object. As a matter of fact, the LSPR of a metal NP correspond to the electric dipole Mie resonance. The Mie resonance conditions can be found by looking for the poles of the mathematical expression of the so-called Mie coefficients, or in other word finding under which conditions the polarizability of interest goes to infinity. For instance, for an electric resonance in the quasi static limit, the following condition is found $\varepsilon_m(\lambda) = -2\varepsilon_h$ [see (13.4)]. For geometries for which the quasi-static approximation is no longer valid, the denominator of the Mie coefficients have to be evaluated explicitly. The condition for a strong electric response can be found while considering the Mie coefficients expanding the contribution of electric multipoles to the scattering response. The condition for a strong magnetic response can be found in a similar way while considering the Mie coefficients expanding the contribution of magnetic multipoles to the scattering response. For a sphere made out of a nonmagnetic material with an arbitrary permittivity ε_d and surrounded by air, the magnetic resonant condition is given by [140]:

$$\frac{\lambda_M}{\sqrt{\varepsilon_d}} \approx d \qquad (13.7)$$

where λ_M is the resonant wavelength and d—the diameter of the sphere. Particles that sustains such magnetic Mie modes can therefore be used as meta-atoms to form

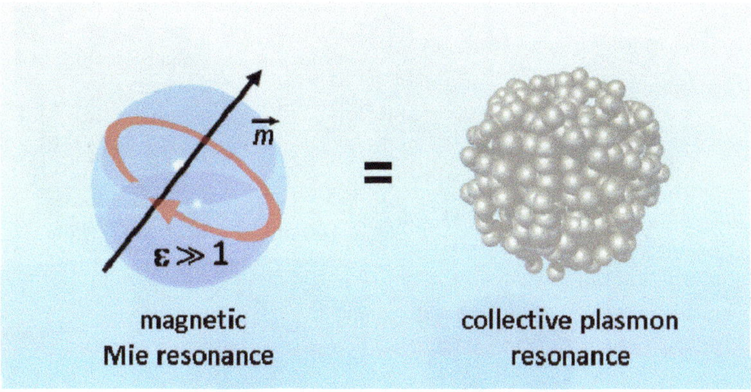

Fig. 13.13 A silver NP metasphere behaves as high permittivity dielectric sphere at wavelengths longer that the plasmon resonance of the NPs. When the first Mie resonance is excited, the metasphere acts as a magnetic dipole, induced by a displacement current circulating in a plane perpendicular to the polarization of the incident magnetic field

an effective medium with a negative permeability μ_{eff}. However when operating as meta-atoms, provisions have to be made that their size is sufficiently small ($d \ll \lambda$). The resonance condition [see (13.7)] indicates that the magnetic Mie resonance occurs approximately when the effective wavelength inside the NP is comparable to the diameter of the sphere or NP cluster. This implies that high permittivity values are needed to excite this resonance in sufficiently small objects. As a consequence, this concept was first demonstrated in frequency regions where some high permittivity materials do exist like in the microwave and mid infrared regimes where the excitation of magnetic Mie modes has been demonstrated using ferroelectric ceramics [63, 141, 142] and polaritonic materials like silicon carbide [63, 142]. In the optical domain however no experimental work can be found until now because of the rather modest permittivity values of most materials at these frequencies. Recently, theoretical suggestions have also been made for the visible range using silicon [143, 144] or excitonic materials [59, 60, 145] but their experimental realization has not been achieved so far. An alternative approach that we intend to investigate in this section is to use clusters of metal NPs like the ones depicted in Fig. 13.13 to form such magnetic meta-atoms. Indeed, we have seen in Sect. 13.2 that a bulk NP composite can act as an effective medium with a high permittivity at wavelengths slightly above the collective plasmon resonance, hence in this wavelength region a sphere made out of such NP composite should behave as high permittivity scatterer and eventually sustains a magnetic dipole resonance. Based on the measured values of $\varepsilon_{\mathrm{eff}}$ ($\varepsilon'_{\mathrm{eff}} = 18$ at $\lambda = 580$ nm; see Fig. 13.5), we can estimate that the smallest NP cluster which can support a magnetic Mie mode would have a diameter on the order of a hundred nanometers, as deduced from (13.7).

Of course, an important advantage of this design is that it is particularly conducive with colloidal self-assembly techniques and several examples of such spherical clusters with the appropriate dimensions have been realized. A brief overview of the different methods available is given in the following section.

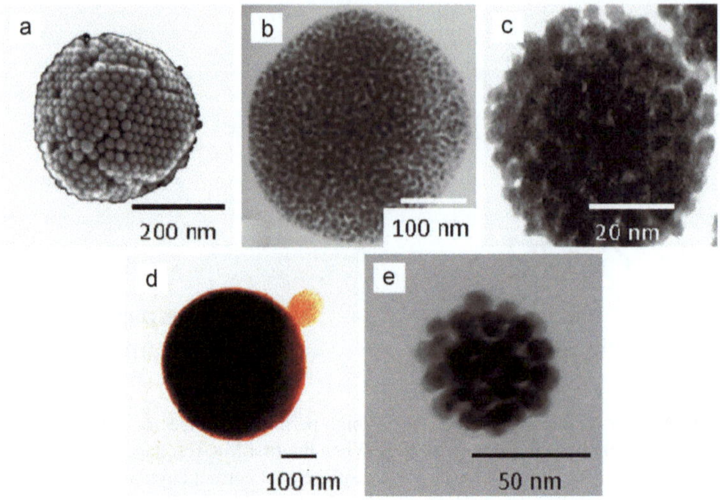

Fig. 13.14 Spherical gold NP clusters obtained by: (**a**) ligand destabilization. Reprinted with permission from [167]. (**b**) Chemical bonding with dithiol linkers. Reprinted with permission from [164]. (**c**) Chemical bonding with tridentate tioether linkers. Reprinted with permission from [168]. (**d**) Polymer induced self-assembly. Reprinted with permission from [157]. (**e**) Emulsion induced NP encapsulation. Reprinted with permission from [165]

13.4.2 Fabrication of Spherical Metal NP Cluster

The advances of the past decade in colloidal self-assembly now allow to build hierarchical NP superstructures with complex geometries. The level of sophistication of the current methods has reached a point where it is now almost possible to assemble NPs one by one to form architectures comprising a precise number of NPs, such as dimers, trimers, quadrumers and so on (see the Review articles [54, 71, 75, 146, 147]. Nevertheless, the design presented previously (Fig. 13.13) does not require such a degree of precision and we will only present here some examples from the literature (Fig. 13.14) that match our targeted geometry, i.e. NP aggregates comprising a large number of metal NPs (from several tens to hundreds) without any specific organization inside the cluster but with an overall spherical shape. Similarly to the methods discussed in Sect. 13.2.1 for the fabrication of extended NP arrangements, the realization of small NP clusters requires to tailor the various interparticle forces that drive the NP self-assembly with the additional constraint that only a discrete number of NPs should be involved. A common approach involves the modification of the NP surface chemistry to alter the NP interactions and trigger their aggregation into clusters. The displacement of the capping molecules that initially stabilize metal NPs in solution, by surrogate ligands implies that the latter have a greater affinity with the metallic cores. Hence, thiolated ligands are generally employed owing to the higher affinity of sulfur for metals with respect to other common capping molecules like amines, carboxilic acids or ions. Various types of ligands have been used to control the different nanoscale forces responsible for the NP aggregation. For instance, spherical clusters of gold NPs were obtained by adding short alkyl chains terminated by a thiol group to a dispersion of long alkylamine-protected NPs (Fig. 13.14a). The reduction of the alkyl chain length reduces the repulsive barrier that screens van der Waals forces, resulting in a rapid aggregation. In this case however, the control of the size and shape of the cluster is difficult and the spherical clusters shown in Fig. 13.14a only represented a fraction of the observed aggregates. Furthermore, the cohesion of these clusters is rather weak as the clusters are easily redispersed by sonication. To obtain robust and monodisperse clusters, more sophisticated NP ligands have been devised that allow stronger and more specific molecularly mediated NP interactions [148, 149]. In particular, using ligands that contain two or more binding groups like alkyldithiol [150] (Fig. 13.14b) or multitdentate thioeter (Fig. 13.14c), it is possible to crosslink NPs by chemical bonding. The systematic spherical shape of the clusters observed in these works was attributed to the templating effect of the ionic surfactant protecting the NPs which are known to form spherical micelles. In addition, adjusting the ligand-to-NP ratio allows to tune precisely the cluster size. Similar clusters were also obtained by grafting others type of molecular linkers like fullerenes [151] or dyes [152], in which case the NP self-assembly is mediated by non covalent attractive interactions between the ligands such as electrostatic or π–π interactions. Instead of relying on direct NP-NP interaction, it is also possible to use polymer [153, 154] or biopolymer [155] scaffolds with proper functional groups that mediate indirectly the NP self-assembly following the "brick and mortar" concept [156, 157]. The spherical shape of the fabricated clusters (Fig. 13.14d) has been attributed in this case to surface tension effects which favor spherical architectures that minimize interfacial area. Another solution to fabricate spherical clusters is by relying on templates. 2D templates mechanically [158, 159] or chemically [160, 161] patterned have been used to assemble NPs into controlled arrangements by a combination of geometrical confinement and capillary forces or by chemical recognition, respectively. While the obtained NP assemblies are generally planar, 3D and, in particular, spherical architectures can also be fabricated by resorting to 3D templates. Such templates generally take the form of oil-in-water emulsion droplets in which or at the surface of which the NPs are entrapped. Subsequent evaporation of the oil phase results in the shrinkage of the droplets and the aggregation of the NPs under compressive forces. Initially developed for the self-assembly of polymeric or silica microspheres [162, 163], this technique has been successfully extended to small NPs of various composition and in particular metallic ones [164–166] (Fig. 13.14e). this method does have the advantage of allowing the fabrication of large amounts of cluster without the need of intricate chemistry. Furthermore, the ability to control precisely the emulsion droplet size by various well-established emulsification techniques (spray drying, high pressure or ultrasonic homogenization, ...) allows to produce highly monodisperse clusters.

In our work, we combined two of the abovementioned techniques, namely NP encapsulation in emulsion droplets followed by their self assembly by covalent bonding using dithiol linkers (Fig. 13.15a). In the first step, an emulsion is prepared by adding drop wisely 50 μL of the dense silver NP ink to 10 ml of an aqueous

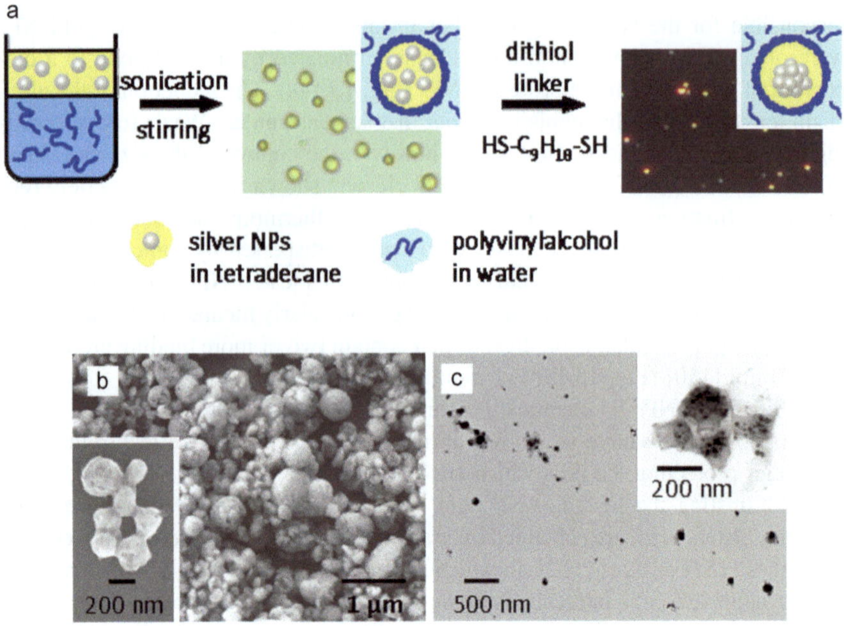

Fig. 13.15 (**a**) Sketch of the formation of spherical silver NP clusters in oil-in-water emulsions following the addition of a molecular dithiol linker. (**b**) SEM images of the raw emulsion showing the polymeric nano-capsules. (**c**) TEM images of the NP clusters

solution of polyvinyl alcohol polymer (PVA Mw. 9,000, 1 w%) while simultaneously sonicating. PVA is widely used to form polymeric capsules, in particular in the field of medicinal chemistry for drug encapsulation. This polymer acts as the emulsion stabilizer: the hydrophobic part of its polymeric chains (non hydrolyzed acetate groups) arrange themselves at the surface of the oil droplets, thereby lowering the interfacial tension between the two immiscible liquids. This gives rise to steric stabilization forces which prevent the coalescence of the emulsion droplets. The emulsion droplets serve as nano-containers to confine a discrete number of NPs in space. At this stage, we know from the ellipsometric data that the NP concentration needs to be increased up to the solid phase to observe a high permittivity and a magnetic response. Since the ink solvent, tetradecane, cannot be evaporated separately from the water phase due to its high boiling point and low vapor pressure, we introduced a dithiol linker to assemble the encapsulated silver NPs. A few drops ($\sim$10 μL) of nonanedithiol (Aldrich) were added right after the emulsification, giving a reddish color to the emulsion, indicative of the NP aggregation. The formation of spherical NP ink droplets in the continuous water phase was first confirmed by optical microscopy. To investigate the morphology of the polymeric capsules and NP clusters, a few droplets of the raw emulsion were dried for SEM and TEM analysis. The SEM images clearly show the spherical PVA shell while the TEM micrographs confirm the presence of the silver NPs in the capsules and their agglomeration into clusters (Fig. 13.15b and c). We note however, that due to the high vacuum in the microscope chambers, the polymeric shells tend to break up. As can be seen from these images, the raw emulsion contains a broad size distribution of capsules, from a few tens of nanometers up to 500 nm approximately; this corresponds to NP clusters with diameters up to 200 nm. The size of the NP clusters is determined by the initial concentration of NPs in the oil phase and the size of the emulsion droplets which depends on the emulsification conditions (oil-to-water ratio, surfactant composition and concentration, and input agitation energy).

13.4.3 Optical Properties of Spherical Metal NP Clusters

To characterize the optical properties of these NP clusters, the emulsion was first purified by filtration and the cluster size distribution was narrowed down by performing several cycles of centrifugation, collecting at each step the supernatant. The resulting fractions are characterized by a strong change of color with the change in cluster sizes: from yellow for the smallest ones, to orange and purple for the biggest ones (Fig. 13.16 inset). The UV-Visible extinction spectra of the different fractions are shown in Fig. 13.16. The spectra of the clusters are characterized by a large extinction peak, strongly red-shifted in comparison with the single NP resonance (black curve) and shifting progressively to longer wavelengths as the cluster size increases. This trend clearly evokes the excitation of a collective resonance in the NP clusters. In order to gain more insight in the physical origin of these extinction peaks and to investigate their eventual relation to the excitation of a dipolar magnetic resonance, the experimental spectra were compared with rigorous simulations, based on the extended Mie theory [169, 170]. While standard Mie theory allows to obtain a rigorous solution for the case of an isolated sphere, the extended Mie theory allows to calculate rigorously the optical response of an aggregate of interacting NPs. The simulated geometry consists in a spherical cluster formed by an amorphous arrangement of 12 nm diameter silver NPs with a fixed interparticle distance of 1.5 nm corresponding to the length of the dithiol linker. The cluster diameter, and hence the number of NPs, was varied between 40 to 150 nm (e.g. between 10 and 460 NPs) and the refractive index of the surrounding medium was set to 1.6 to match the experimental conditions.

As can be seen in Fig. 13.16, the simulated extinction spectra show the same trends as the experimental spectra on the one hand. In particular, the progressive red-shift when increasing the NP cluster diameter is well reproduced in the simulations, thereby confirming that the observed extinction peaks correspond to a collective resonance sustained by the entire clusters. On the other hand, the simulated spectra contain some fine structure below the main resonance which is not found in the measured spectra. These discrepancies are due to the fact that the simulation considers only an individual implementation of metaspheres with nominally the same geometry. In opposite, the clusters probed in the experiment, although characterized by similar shapes and sizes, exhibit different spatial arrangements of the NPs within them. However, the red-shifted resonance which is associated with the global shape

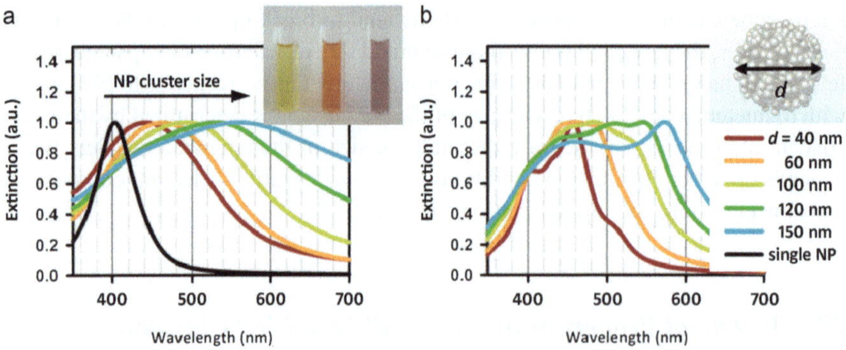

Fig. 13.16 Experimental extinction spectra of the silver NP cluster dispersions for increasing cluster size. The *black curve* is the spectrum of isolated NPs. *Inset*: photo of the vials containing the NP clusters. (**b**) Simulated extinction spectra spherical silver NP clusters with different radiuses. Reprinted with permission from [171]

of the clusters dominates the experimental spectra while the fine structure is related to the details of the NP arrangement inside the cluster and is smeared out in the experiment. Next, the simulated scattered field was related to the field radiated by a series of electric and magnetic multipoles through a multipole expansion [172–174]. In this way, the excitation of a magnetic dipole moment related to the observed resonance can be clearly identified. For all the diameters considered, the clusters are significantly smaller than the wavelength, hence only a few multipoles are expected to contribute to the total scattered field. Figure 13.17a shows the contribution of the magnetic dipole to the scattering cross-section for each size of cluster. As can be seen, no magnetic dipole is excited for clusters having a diameter smaller than 100 nm. Their response is entirely described by a single electric dipole contribution. On the other hand, a magnetic dipole contribution clearly emerges for larger diameters in the wavelength region corresponding to the red-shifted extinction peak. This is in good agreement with the rough estimation of the cluster size done in the introduction of this section based on the ellipsometric measurements. The contribution of this magnetic dipole grows significantly as the cluster size increases. To get a more complete description, the relative contribution of all the different excited multipoles are shown for the cluster with the largest diameter in Fig. 13.17b.

It can be seen that although the electric dipole still dominates the cluster's response, the magnetic dipole contributes significantly to the scattering cross section (by about $1/3$). Since these results confirm the magnetic activity of the fabricated silver NP clusters, it is interesting to evaluate how this would be reflected in the effective permeability of a metamaterial formed by an arrangement of such clusters. This is done by inserting the magnetic polarizability of the cluster obtained from the previous calculations in the Clausius-Mossotti formula (13.4) for the effective permeability μ_{eff}. The result is shown in Fig. 13.17c for a cluster filling fraction of 0.68 corresponding to close packed arrangement of clusters in a face-centered cubic lattice. A clear dispersion in the metamaterial permeability is observed around the cluster resonance, around 600 nm. Although this dispersion appears relatively weak

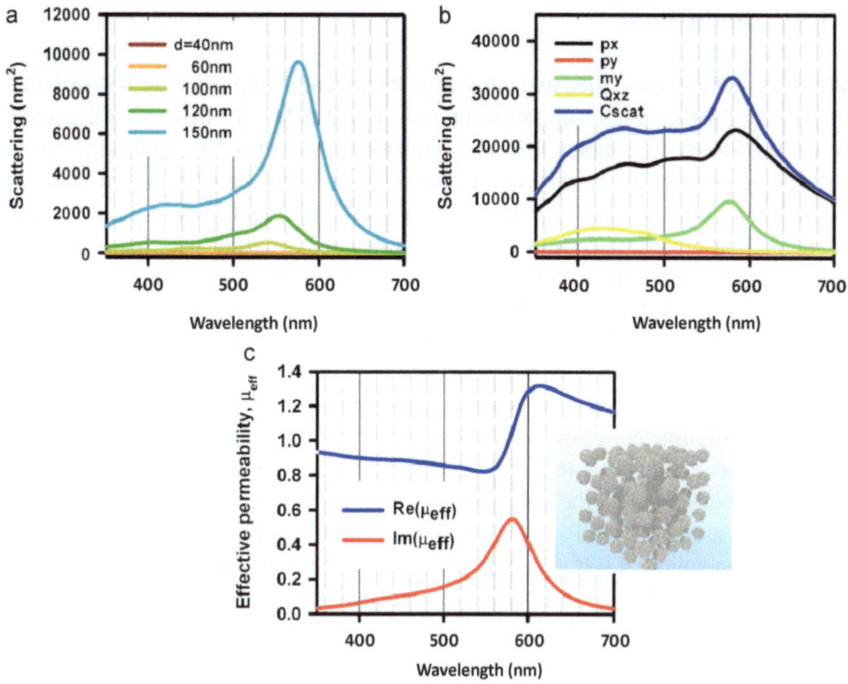

Fig. 13.17 Contribution of the magnetic dipole moment to the scattering cross section of the NP cluster as a function of its radius. (**b**) Contribution of the various multipole moments (p, electric dipole; m, magnetic dipole; Q, electric quadrupole) to the scattering cross section for the larger cluster simulated ($r = 75$ nm). The cluster was illuminated by a linearly polarized plane wave. (**c**) Simulated effective permeability of a fcc arrangement of silver NP clusters ($f = 0.68$). Reprinted with permission from [171]

with respect to the previous theoretical predictions [57], it compares well with the results recently obtained for gold NP clusters of slightly larger size [175] or for SiO_2 spheres covered with a gold NP shell [176]. These results therefore demonstrate that the formation of metal NP clusters by emulsion and crosslinking offers an efficient way to realize magnetic meta-atoms. While their optical properties are similar to those previously reported [175, 176], this original fabrication technique potentially expands the degrees of freedom available to tailor those properties and thereby their use in different applications. Several possibilities can be exploited to strengthen the magnetic response of the NP clusters. For instance, a further increase of the size of the clusters, by adjusting the emulsification conditions, should produce stronger resonances, as predicted by Mie theory. The size of the silver NP themselves could also be slightly increased, so as to reduce the losses due to confinement effects which still operate for 12 nm NPs. Alternatively, losses could also be reduced by introducing some gain medium, like organic dyes, in the emulsion process.

13.5 Conclusions

Within this chapter, we tried to demonstrate the potential of plasmonic NPs as building blocks for the bottom-up fabrication of optical MMs. To this end, we demonstrate by different examples how the optical response of a NP composite can be tailored by controlling the NP organization over different length scales. In addition to our own works, we gave a brief overview of the different bottom-up techniques available that provide the required level of precision. Starting with the bulky configuration (e.g. a NP dispersion or thin film), spectroscopic ellipsometry was used to show how the NP density in bulk MMs can influence their effective permittivity. Provided that the NP density is high enough, a situation can be reached where the MM exhibit both the behavior of a metal or a high permittivity dielectric in different regions of the visible spectrum. Next, we showed that the 1D ordering of NP monolayers into multilayers allow to form a hybrid plasmonic-Bragg structure whose spectral properties can be tuned by tailoring the interplay between plasmonic and photonic modes. Finally, the possibility to create a magnetic response at optical frequencies by assembling NPs into 3D spherical nanocluster is discussed and realized by relying on colloidal self assembly concepts. Altogether, these different examples demonstrate that NP self- or directed-assembly methods allow to create MMs with an electric and magnetic response that can be tailored at will by directing the NP organization into various prescribed architectures. The validity of the bottom-up approach for the fabrication of optical MMs is therefore confirmed with all the assets that it implies for scale-up and mass production. This promising strategy call for further investigations, in particular to address the issue of dissipative losses which still prevent to benefit from the most exciting MM-based optical applications like cloaking or superlensing.

Acknowledgements This work was funded by the European Union's Seven Framework Programme (FP7/2007-2013) under grant agreement n°228455 (Nanogold project). We thank Carsten Rockstuhl, Stefan Mühlig and Tobias Kienzler for their input and support regarding the theoretical aspects of this work, and Houda Sellame for the preparation of multilayer samples.

References

1. M.I. Stockman, Nanoplasmonics: the physics behind the applications. Phys. Today **64**, 39–44 (2011)
2. W.A. Murray, W.L. Barnes, Plasmonic materials. Adv. Mater. **19**, 3771–3782 (2007)
3. E. Hutter, J.H. Fendler, Exploitation of localized surface plasmon resonance. Adv. Mater. **16**, 1685–1706 (2004)
4. I. Angelini et al., Chemical analyses of Bronze Age glasses from Frattesina di Rovigo, Northern Italy. J. Archaeol. Sci. **31**, 1175–1184 (2004)
5. P. Colomban, The use of metal nanoparticles to produce yellow, red and iridescent colour, from Bronze Age to Present Times in lustre pottery and glass: solid state chemistry, spectroscopy and nanostructure. J. Nanopart. Res. **8**, 109–132 (2009)
6. I. Freestone, N. Meeks, M. Sax, C. Higgitt, The lycurgus cup—A Roman nanotechnology. Gold Bull. **40**, 270–277 (2007)

7. P. Colomban, A. Tournie, P. Ricciardi, Raman spectroscopy of copper nanoparticle-containing glass matrices: ancient red stained-glass windows. J. Raman Spectrosc. **40**, 1949–1955 (2009)
8. S. Perez-Villar, J. Rubio, J. Luis Oteo, Study of color and structural changes in silver painted medieval glasses. J. Non-Cryst. Solids **354**, 1833–1844 (2008)
9. M. Faraday, The Bakerian lecture: experimental relations of gold (and other metals) to light. Philos. Trans. R. Soc. Lond. **147**, 145–181 (1857)
10. P.P. Edwards, J.M. Thomas, Gold in a metallic divided state—from Faraday to present-day nanoscience. Angew. Chem., Int. Ed. **46**, 5480–5486 (2007)
11. L. Rayleigh, On the light from the sky, its polarization and colour. Philos. Mag. **41**, 107–120; 274–279 (1871)
12. L. Lorenz, *Oeuvres scientifiques* (Carlsbergfondet, Copenhagen, 1898)
13. G. Mie, On the optical characteristics of turbid media, especially colloidal metal solutions. Ann. Phys. **25**, 377–445 (1908)
14. H.C. van de Hulst, *Light Scattering by Small Particles* (Wiley, New York, 1957)
15. M. Kerker, *The Scattering of Light and Other Electromagnetic Radiation* (Academic Press, San Diego, 1969)
16. C.F. Bohren, D.R. Huffman, *Absorption and Scattering of Light by Small Particles* (Wiley, New York, 1983)
17. U. Kreibig, M. Vollmer, *Optical Properties of Metal Clusters* (Springer, Berlin, 1995)
18. P. Muhlschlegel, H.J. Eisler, O.J.F. Martin, B. Hecht, D.W. Pohl, Resonant optical antennas. Science **308**, 1607–1609 (2005)
19. L. Novotny, N. van Hulst, Antennas for light. Nat. Photonics **5**, 83–90 (2011)
20. N.J. Halas, S. Lal, W.-S. Chang, S. Link, P. Nordlander, Plasmons in strongly coupled metallic nanostructures. Chem. Rev. **111**, 3913–3961 (2011)
21. K. Li, M.I. Stockman, D.J. Bergman, Self-similar chain of metal nanospheres as an efficient nanolens. Phys. Rev. Lett. **91**, 227402 (2003)
22. N. Liu et al., Plasmonic analogue of electromagnetically induced transparency at the Drude damping limit. Nat. Mater. **8**, 758–762 (2009)
23. J.A. Fan et al., Self-assembled plasmonic nanoparticle clusters. Science **328**, 1135–1138 (2010)
24. A. Iwakoshi, T. Nanke, T. Kobayashi, Coating materials containing gold nanoparticles. Gold Bull. **38**, 107–112 (2005)
25. T. Keel, R. Holliday, T. Harper, White Paper: Gold for good—gold and nanotechnology in the age of innovation. World Gold Council, www.gold.org
26. M.A. Garcia, Surface plasmons in metallic nanoparticles: fundamentals and applications. J. Phys. D: Appl. Phys. **44**, 283001 (2011)
27. P.K. Jain, X. Huang, I.H. El-Sayed, M.A. El-Sayed, Noble metals on the nanoscale: optical and photothermal properties and some applications in imaging, sensing, biology, and medicine. Acc. Chem. Res. **41**, 1578–1586 (2008)
28. R. Wilson, The use of gold nanoparticles in diagnostics and detection. Chemical Society Reviews **37**, 2028–2045 (2008)
29. K.A. Willets, R.P. Van Duyne, in *Annual Review of Physical Chemistry*, vol. 58 (2007), pp. 267–297
30. J.N. Anker et al., Biosensing with plasmonic nanosensors. Nat. Mater. **7**, 442–453 (2008)
31. K.M. Mayer, J.H. Hafner, Localized surface plasmon resonance sensors. Chem. Rev. **111**, 3828–3857 (2011)
32. J.R. Lakowicz et al., Plasmon-controlled fluorescence: a new paradigm in fluorescence spectroscopy. Analyst **133**, 1308–1346 (2008)
33. P.L. Stiles, J.A. Dieringer, N.C. Shah, R.R. Van Duyne, in *Annual Review of Analytical Chemistry*, vol. 1 (2008), pp. 601–626
34. E.C. Le Ru, P.G. Etchegoin, in *Annual Review of Physical Chemistry*, vol. 63, ed. by M.A. Johnson, T.J. Martinez (2012), pp. 65–87

35. Y. Xiao et al., Surface plasmon-enhanced electroluminescence in organic light-emitting diodes incorporating Au nanoparticles. Appl. Phys. Lett. **100** (2012)
36. H.A. Atwater, A. Polman, Plasmonics for improved photovoltaic devices. Nat. Mater. **9**, 205–213 (2010)
37. S.A. Maier et al., Local detection of electromagnetic energy transport below the diffraction limit in metal nanoparticle plasmon waveguides. Nat. Mater. **2**, 229–232 (2003)
38. M.L. Juan, M. Righini, R. Quidant, Plasmon nano-optical tweezers. Nat. Photonics **5**, 349–356 (2011)
39. L. Novotny, B. Hecht, *Principles of Nano-Optics* (Cambridge University Press, Cambridge, 2006)
40. W. Srituravanich, N. Fang, C. Sun, Q. Luo, X. Zhang, Plasmonic nanolithography. Nano Lett. **4**, 1085–1088 (2004)
41. D.R. Smith, J.B. Pendry, M.C.K. Wiltshire, Metamaterials and negative refractive index. Science **305**, 788–792 (2004)
42. Y. Liu, X. Zhang, Metamaterials: a new frontier of science and technology. Chemical Society Reviews **40**, 2494–2507 (2011)
43. W. Cai, V. Shalaev, *Optical Metamaterials Fundamentals and Applications Introduction* (Springer, Berlin, 2010)
44. C.M. Soukoulis, M. Wegener, Past achievements and future challenges in the development of three-dimensional photonic metamaterials. Nat. Photonics **5**, 523–530 (2011)
45. A. de Baas (ed.) *Nanostructured Metamaterials*, KI-NA-24409-EN-C (2010). ISBN 978-92-79-07563-6
46. J.B. Pendry, Negative refraction makes a perfect lens. Phys. Rev. Lett. **85**, 3966–3969 (2000)
47. J.B. Pendry, D. Schurig, D.R. Smith, Controlling electromagnetic fields. Science **312**, 1780–1782 (2006)
48. R.A. Shelby, D.R. Smith, S. Schultz, Experimental verification of a negative index of refraction. Science **292**, 77–79 (2001)
49. C.M. Soukoulis, S. Linden, M. Wegener, Negative refractive index at optical wavelengths. Science **315**, 47–49 (2007)
50. V.M. Shalaev, Optical negative-index metamaterials. Nat. Photonics **1**, 41–48 (2007)
51. S. Linden et al., Magnetic response of metamaterials at 100 terahertz. Science **306**, 1351–1353 (2004)
52. V.M. Shalaev et al., Negative index of refraction in optical metamaterials. Opt. Lett. **30**, 3356–3358 (2005)
53. J. Valentine et al., Three-dimensional optical metamaterial with a negative refractive index. Nature **455**, 376-U332 (2008)
54. M. Grzelczak, J. Vermant, E.M. Furst, L.M. Liz-Marzan, Directed self-assembly of nanoparticles. ACS Nano **4**, 3591–3605 (2010)
55. M. Rycenga et al., Controlling the synthesis and assembly of silver nanostructures for plasmonic applications. Chem. Rev. **111**, 3669–3712 (2011)
56. C.R. Simovski, S.A. Tretyakov, Model of isotropic resonant magnetism in the visible range based on core-shell clusters. Phys. Rev. B **79**, 045111 (2009)
57. C. Rockstuhl, F. Lederer, C. Etrich, T. Pertsch, T. Scharf, Design of an artificial three-dimensional composite metamaterial with magnetic resonances in the visible range of the electromagnetic spectrum. Phys. Rev. Lett. **99**, 017401 (2007)
58. A. Vallecchi, M. Albani, F. Capolino, Collective electric and magnetic plasmonic resonances in spherical nanoclusters. Opt. Express **19**, 2754–2772 (2011)
59. V. Yannopapas, Artificial magnetism and negative refractive index in three-dimensional metamaterials of spherical particles at near-infrared and visible frequencies. Appl. Phys. A, Mater. Sci. Process. **87**, 259–264 (2007)
60. V. Yannopapas, A. Moroz, Negative refractive index metamaterials from inherently non-magnetic materials for deep infrared to terahertz frequency ranges. J. Phys. Condens. Matter **17**, 3717–3734 (2005)

61. C. Helgert et al., Effective properties of amorphous metamaterials. Phys. Rev. B **79**, 233107 (2009)
62. V. Yannopapas, Negative refraction in random photonic alloys of polaritonic and plasmonic microspheres. Phys. Rev. B **75**, 035112 (2007)
63. M.S. Wheeler, J.S. Aitchison, J.I.L. Chen, G.A. Ozin, M. Mojahedi, Infrared magnetic response in a random silicon carbide micropowder. Phys. Rev. B **79**, 073103 (2009)
64. C. Rockstuhl, T. Scharf, A metamaterial based on coupled metallic nanoparticles and its band-gap property. J. Microsc. **229**, 281–286 (2008)
65. S. Xiao et al., Loss-free and active optical negative-index metamaterials. Nature **466**, 735–736 (2010)
66. R. Pratibha, K. Park, I.I. Smalyukh, W. Park, Tunable optical metamaterial based on liquid crystal-gold nanosphere composite. Opt. Express **17**, 19459–19469 (2009)
67. S. Sivaramakrishnan, P.-J. Chia, Y.-C. Yeo, L.-L. Chua, P.K.H. Ho, Controlled insulator-to-metal transformation in printable polymer composites with nanometal clusters. Nat. Mater. **6**, 149–155 (2007)
68. A. Kamiyshny, J. Steinke, S. Magdassi, Metal-based inkjet inks for printed electronics. Open Appl. Phys. J. **4**, 19–36 (2011)
69. P. Tassin, T. Koschny, M. Kafesaki, C.M. Soukoulis, A comparison of graphene, superconductors and metals as conductors for metamaterials and plasmonics. Nat. Photonics **6**, 259–264 (2012)
70. Z. Nie, A. Petukhova, E. Kumacheva, Properties and emerging applications of self-assembled structures made from inorganic nanoparticles. Nat. Nanotechnol. **5**, 15–25 (2010)
71. K.J.M. Bishop, C.E. Wilmer, S. Soh, B.A. Grzybowski, Nanoscale forces and their uses in self-assembly. Small **5**, 1600–1630 (2009)
72. F. Caruso, *Colloids and Colloid Assemblies* (Wiley-VCH, New York, 2003)
73. B.A. Grzybowski, C.E. Wilmer, J. Kim, K.P. Browne, K.J.M. Bishop, Self-assembly: from crystals to cells. Soft Matter **5**, 1110–1128 (2009)
74. M.-P. Pileni, *Nanocrystals Forming Mesoscopic Structures* (Wiley-VCH, New York, 2005)
75. F. Li, D.P. Josephson, A. Stein, Colloidal assembly: the road from particles to colloidal molecules and crystals. Angew. Chem., Int. Ed. **50**, 360–388 (2011)
76. Y. Min, M. Akbulut, K. Kristiansen, Y. Golan, J. Israelachvili, The role of interparticle and external forces in nanoparticle assembly. Nat. Mater. **7**, 527–538 (2008)
77. M.R. Jones, K.D. Osberg, R.J. Macfarlane, M.R. Langille, C.A. Mirkin, Templated techniques for the synthesis and assembly of plasmonic nanostructures. Chem. Rev. **111**, 3736–3827 (2011)
78. A.R. Tao, D.P. Ceperley, P. Sinsermsuksakul, A.R. Neureuther, P. Yang, Self-organized silver nanoparticles for three-dimensional plasmonic crystals. Nano Lett. **8**, 4033–4038 (2008)
79. J. Henzie, M. Gruenwald, A. Widmer-Cooper, P.L. Geissler, P. Yang, Self-assembly of uniform polyhedral silver nanocrystals into densest packings and exotic superlattices. Nat. Mater. **11**, 131–137 (2012)
80. E.V. Shevchenko, D.V. Talapin, N.A. Kotov, S. O'Brien, C.B. Murray, Structural diversity in binary nanoparticle superlattices. Nature **439**, 55–59 (2006)
81. C.B. Murray, C.R. Kagan, M.G. Bawendi, Self-organization of CdSe nanocrystallites into 3-dimensional quantum-dot superlattices. Science **270**, 1335–1338 (1995)
82. C.B. Murray, C.R. Kagan, M.G. Bawendi, Synthesis and characterization of monodisperse nanocrystals and close-packed nanocrystal assemblies. Annu. Rev. Mater. Sci. **30**, 545–610 (2000)
83. M.I. Bodnarchuk et al., Three-dimensional nanocrystal superlattices grown in nanoliter microfluidic plugs. J. Am. Chem. Soc. **133**, 8956–8960 (2011)
84. S. Coe-Sullivan, J.S. Steckel, W.K. Woo, M.G. Bawendi, V. Bulovic, Large-area ordered quantum-dot monolayers via phase separation during spin-casting. Adv. Funct. Mater. **15**, 1117–1124 (2005)
85. P.F. Damasceno, M. Engel, S.C. Glotzer, Predictive self-assembly of polyhedra into complex structures. Science **337**, 453–457 (2012)

86. A.M. Kalsin et al., Electrostatic self-assembly of binary nanoparticle crystals with a diamond-like lattice. Science **312**, 420–424 (2006)
87. D.V. Talapin et al., Quasicrystalline order in self-assembled binary nanoparticle superlattices. Nature **461**, 964–967 (2009)
88. A. Yethiraj, A. van Blaaderen, A colloidal model system with an interaction tunable from hard sphere to soft and dipolar. Nature **421**, 513–517 (2003)
89. A.P. Alivisatos et al., Organization of 'nanocrystal molecules' using DNA. Nature **382**, 609–611 (1996)
90. C.A. Mirkin, R.L. Letsinger, R.C. Mucic, J.J. Storhoff, A DNA-based method for rationally assembling nanoparticles into macroscopic materials. Nature **382**, 607–609 (1996)
91. X. Zeng et al., 3D ordered gold strings by coating nanoparticles with mesogens. Adv. Mater. **21**, 1746 (2009)
92. L. Cseh, G.H. Mehl, The design and investigation of room temperature thermotropic nematic gold nanoparticles. J. Am. Chem. Soc. **128**, 13376–13377 (2006)
93. I.I.S. Lim, C.-J. Zhong, Molecularly-mediated assembly of gold nanoparticles. Gold Bull. **40**, 59–66 (2007)
94. R. Klajn et al., Plastic and moldable metals by self-assembly of sticky nanoparticle aggregates. Science **316**, 261–264 (2007)
95. B. Donnio, P. Garcia-Vazquez, J.-L. Gallani, D. Guillon, E. Terazzi, Dendronized ferromagnetic gold nanoparticles self-organized in a thermotropic cubic phase. Adv. Mater. **19**, 3534 (2007)
96. G. Decher, J. Schlenoff, *Multilayer Thin Films: Sequential Assembly of Nanocomposite Multilayers* (Wiley-VCH, New York, 2012)
97. G. Decher, Fuzzy nanoassemblies: toward layered polymeric multicomposites. Science **277**, 1232–1237 (1997)
98. D.L. Feldheim, K.C. Grabar, M.J. Natan, T.E. Mallouk, Electron transfer in self-assembled inorganic Polyelectrolyte/Metal nanoparticle heterostructures. J. Am. Chem. Soc. **118**, 7640–7641 (1996)
99. A. Cunningham, S. Muehlig, C. Rockstuhl, T. Buergi, Coupling of plasmon resonances in tunable layered arrays of gold nanoparticles. J. Phys. Chem. C **115**, 8955–8960 (2011)
100. M. Brust, D. Bethell, C.J. Kiely, D.J. Schiffrin, Self-assembled gold nanoparticle thin films with nonmetallic optical and electronic properties. Langmuir **14**, 5425–5429 (1998)
101. T.A. Taton, R.C. Mucic, C.A. Mirkin, R.L. Letsinger, The DNA-mediated formation of supramolecular mono- and multilayered nanoparticle structures. J. Am. Chem. Soc. **122**, 6305–6306 (2000)
102. N. Higashi, T. Takagi, T. Koga, Layer-by-layer fabrication of well-packed gold nanoparticle assemblies guided by a [beta]-sheet peptide network. Polym. J. **42**, 95–99 (2010)
103. B.G. Jung et al., Colloidal nanoparticle-layer formation through dip-coating: effect of solvents and substrate withdrawing speed. J. Electrochem. Soc. **156**, K86–K90 (2009)
104. N.D. Denkov et al., Mechanism of formation of two-dimensional crystals from latex particles on substrates. Langmuir **8**, 3183–3190 (1992)
105. A.G. Emslie, F.T. Bonner, L.G. Peck, Flow of a viscous liquid on a rotating disk. J. Appl. Phys. **29**, 858–862 (1958)
106. D.P. Birnie, The Basic Physical Processes that Control Spin Coating. http://www.coatings.rutgers.edu/basics.htm (2005)
107. P. Jiang, M.J. McFarland, Large-scale fabrication of wafer-size colloidal crystals, macroporous polymers and nanocomposites by spin-coating. J. Am. Chem. Soc. **126**, 13778–13786 (2004)
108. P. Jiang, T. Prasad, M.J. McFarland, V.L. Colvin, Two-dimensional nonclose-packed colloidal crystals formed by spincoating. Applied Physics Letters **89** (2006)
109. R.S. Krishnan et al., Self-assembled multilayers of nanocomponents. Nano Lett. **7**, 484–489 (2007)
110. D. Itoh, A. Izumitani, N. Hata, Y. Matsuba, K. Murata, H. Yokoyama, Metal nano particle liquid dispersion capable of being sprayed in fine particle form and being applied in laminated

state (2004)
111. O. Heavens, *Optical Properties of Thin Solid Films* (Butterworths, Stoneham, 1955)
112. J.N. Hilfiker et al., Survey of methods to characterize thin absorbing films with spectroscopic ellipsometry. Thin Solid Films **516**, 7979–7989 (2008)
113. C.R. Simovski, On electromagnetic characterization and homogenization of nanostructured metamaterials. J. Opt. **13** (2011)
114. R.M.A. Azzam, N.M. Bashara, *Ellipsometry and Polarized Light* (Elsevier, Amsterdam, 1987)
115. M. Losurdo et al., Spectroscopic ellipsometry and polarimetry for materials and systems analysis at the nanometer scale: state-of-the-art, potential, and perspectives. J. Nanopart. Res. **11**, 1521–1554 (2009)
116. T.W.H. Oates, H. Wormeester, H. Arwin, Characterization of plasmonic effects in thin films and metamaterials using spectroscopic ellipsometry. Prog. Surf. Sci. **86**, 328–376 (2011)
117. S. Kubo et al., Tunability of the refractive index of gold nanoparticle dispersions. Nano Lett. **7**, 3418–3423 (2007)
118. H. Wormeester, A.-I. Henry, E.S. Kooij, B. Poelsema, M.-P. Pileni, Ellipsometric identification of collective optical properties of silver nanocrystal arrays. J. Chem. Phys. **124**, 204713 (2006)
119. H.L. Zhang, S.D. Evans, J.R. Henderson, Spectroscopic ellipsometric evaluation of gold nanoparticle thin films fabricated using layer-by-layer self-assembly. Adv. Mater. **15**, 531–534 (2003)
120. H. Pan, S.H. Ko, C.P. Grigoropoulos, Thermal sintering of solution-deposited nanoparticle silver ink films characterized by spectroscopic ellipsometry. Appl. Phys. Lett. **93**, 234104 (2008)
121. R.A. Synowicki, in *Physica Status Solidi C—Current Topics in Solid State Physics*, vol. 5, ed. by H. Beck, U. Schubert, M. Arwin (2008), pp. 1085–1088
122. J.C. Maxwell Garnett, Colours in metal glasses and in metallic films. Philos. Trans. R. Soc. Lond. A **203**, 385 (1904)
123. D. Bedeaux, J. Vlieger, *Optical Properties of Surfaces* (Imperial College Press, London, 2001)
124. H. Wormeester, E.S. Kooij, B. Poelsema, Effective dielectric response of nanostructured layers. Phys. Status Solidi, a Appl. Mater. Sci. **205**, 756–763 (2008)
125. W. Zhou, T.W. Odom, Tunable subradiant lattice plasmons by out-of-plane dipolar interactions. Nat. Nanotechnol. **6**, 423–427 (2011)
126. S.L. Zou, G.C. Schatz, Narrow plasmonic/photonic extinction and scattering line shapes for one and two dimensional silver nanoparticle arrays. J. Chem. Phys. **121**, 12606–12612 (2004)
127. M.E. Calvo, O. Sanchez Sobrado, G. Lozano, H. Miguez, Molding with nanoparticle-based one-dimensional photonic crystals: a route to flexible and transferable Bragg mirrors of high dielectric contrast. J. Mater. Chem. **19**, 3144–3148 (2009)
128. S. Colodrero, M. Ocana, H. Miguez, Nanoparticle-based one-dimensional photonic crystals. Langmuir **24**, 4430–4434 (2008)
129. S.-H. Song et al., Bragg gratings generated by coupling of surface plasmons induced on metal nanoparticles. J. Opt. Soc. Korea **8**, 6–12 (2004)
130. S. Husaini, L. Deych, V.M. Menon, Plasmon-resonance-induced enhancement of the reflection band in a one-dimensional metal nanocomposite photonic crystal. Opt. Lett. **36**, 1368–1370 (2011)
131. S.-G. Kim, N. Hagura, F. Iskandar, A. Yabuki, K. Okuyama, Multilayer film deposition of Ag and SiO_2 nanoparticles using a spin coating process. Thin Solid Films **516**, 8721–8725 (2008)
132. T. Ogata, R. Yagi, N. Nakamura, Y. Kuwahara, S. Kurihara, Modulation of polymer refractive indices with diamond nanoparticles for metal-free multilayer film mirrors. ACS Appl. Mater. Interfaces **4**, 3769–3772 (2012)

133. D.P. Puzzo et al., Color from colorless nanomaterials: Bragg reflectors made of nanoparticles. J. Mater. Chem. **19**, 3500–3506 (2009)
134. A. Convertino, A. Capobianchi, A. Valentini, E.N.M. Cirillo, High reflectivity Bragg reflectors based on a gold nanoparticle/Teflon-like composite material as a new approach to organic solvent detection. Sens. Actuators B, Chem. **100**, 212–215 (2004)
135. S. Kachan, O. Stenzel, A. Ponyavina, High-absorbing gradient multilayer coatings with silver nanoparticles. Appl. Phys. B, Lasers Opt. **84**, 281–287 (2006)
136. C. Wang, Y. Zhao, D. Gan, C. Du, X. Luo, Subwavelength imaging with anisotropic structure comprising alternately layered metal and dielectric films. Opt. Express **16**, 4217–4227 (2008)
137. M. Scalora et al., Negative refraction and sub-wavelength focusing in the visible range using transparent metallodielectric stacks. Opt. Express **15**, 508–523 (2007)
138. Q. Zhao, J. Zhou, F. Zhang, D. Lippens, Mie resonance-based dielectric metamaterials. Mater. Today **12**, 60–69 (2009)
139. S. O'Brien, J.B. Pendry, Photonic band-gap effects and magnetic activity in dielectric composites. J. Phys. Condens. Matter **14**, 4035–4044 (2002)
140. L. Lewin, The electrical constants of a material loaded with spherical particles. Proc. Inst. Electr. Eng. **94**, 65–68 (1947)
141. L. Peng et al., Experimental observation of left-handed behavior in an array of standard dielectric resonators. Phys. Rev. Lett. **98**, 157403 (2007)
142. J.A. Schuller, R. Zia, T. Taubner, M.L. Brongersma, Dielectric metamaterials based on electric and magnetic resonances of silicon carbide particles. Phys. Rev. Lett. **99** (2007)
143. A.B. Evlyukhin, C. Reinhardt, A. Seidel, B.S. Luk'yanchuk, B.N. Chichkov, Optical response features of Si-nanoparticle arrays. Phys. Rev. B **82**, 045404 (2010)
144. R. Paniagua-Domínguez, F. López-Tejeira, R. Marqués, J.A. Sánchez-Gil, Metallo-dielectric core–shell nanospheres as building blocks for optical three-dimensional isotropic negative-index metamaterials. New J. Phys. **13**, 123017 (2011)
145. V. Yannopapas, N.V. Vitanov, Photoexcitation-induced magnetism in arrays of semiconductor nanoparticles with a strong excitonic oscillator strength. Phys. Rev. B **74**, 193304 (2006)
146. E. Duguet, A. Desert, A. Perro, S. Ravaine, Design and elaboration of colloidal molecules: an overview. Chemical Society Reviews **40**, 941–960 (2011)
147. J.M. Romo-Herrera, R.A. Alvarez-Puebla, L.M. Liz-Marzan, Controlled assembly of plasmonic colloidal nanoparticle clusters. Nanoscale **3**, 1304–1315 (2011)
148. M.M. Maye et al., Mediator-template assembly of nanoparticles. J. Am. Chem. Soc. **127**, 1519–1529 (2005)
149. S.I. Lim, C.-J. Zhong, Molecularly mediated processing and assembly of nanoparticles: exploring the interparticle interactions and structures. Acc. Chem. Res. **42**, 798–808 (2009)
150. I. Hussain, Z.X. Wang, A.I. Cooper, M. Brust, Formation of spherical nanostructures by the controlled aggregation of gold colloids. Langmuir **22**, 2938–2941 (2006)
151. I.I.S. Lim et al., Assembly of gold nanoparticles mediated by multifunctional fullerenes. Langmuir **23**, 10715–10724 (2007)
152. I.I.S. Lim et al., Adsorption of cyanine dyes on gold nanoparticles and formation of J-aggregates in the nanoparticle assembly. J. Phys. Chem. B **110**, 6673–6682 (2006)
153. B.L. Frankamp, A.K. Boal, V.M. Rotello, in *Three-Dimensional Nanoengineered Assemblies*, ed. by T.M. Merhari, L. Taylor, D.P. Ikuta, K. Orlando. Materials Research Society Symposium Proceedings, vol. 739 (2003), pp. 83–87
154. Y. Ofir, B. Samanta, V.M. Rotello, Polymer and biopolymer mediated self-assembly of gold nanoparticles. Chemical Society Reviews **37**, 1814–1823 (2008)
155. A.K. Boal, T.H. Galow, F. Ilhan, V.M. Rotello, Binary and ternary polymer-mediated "bricks and mortar" self-assembly of gold and silica nanoparticles. Adv. Funct. Mater. **11**, 461–465 (2001)
156. P. Arumugam, H. Xu, S. Srivastava, V.M. Rotello, 'Bricks and mortar' nanoparticle self-assembly using polymers. Polym. Int. **56**, 461–466 (2007)

157. A.K. Boal et al., Self-assembly of nanoparticles into structured spherical and network aggregates. Nature **404**, 746–748 (2000)
158. Y. Cui et al., Integration of colloidal nanocrystals into lithographically patterned devices. Nano Lett. **4**, 1093–1098 (2004)
159. Y.N. Xia, Y.D. Yin, Y. Lu, J. McLellan, Template-assisted self-assembly of spherical colloids into complex and controllable structures. Adv. Funct. Mater. **13**, 907–918 (2003)
160. H.-W. Huang, P. Bhadrachalam, V. Ray, S.J. Koh, Single-particle placement via self-limiting electrostatic gating. Appl. Phys. Lett. **93**, 073110 (2008)
161. Q.G. Li, J.W. Zheng, Z.F. Liu, Site-selective assemblies of gold nanoparticles on an AFM tip-defined silicon template. Langmuir **19**, 166–171 (2003)
162. S.Y. Lee, L. Gradon, S. Janeczko, F. Iskandar, K. Okuyama, Formation of highly ordered nanostructures by drying micrometer colloidal droplets. ACS Nano **4**, 4717–4724 (2010)
163. V.N. Manoharan, M.T. Elsesser, D.J. Pine, Dense packing and symmetry in small clusters of microspheres. Science **301**, 483–487 (2003)
164. I. Hussain, H. Zhang, M. Brust, J. Barauskas, A.I. Cooper, Emulsions-directed assembly of gold nanoparticles to molecularly-linked and size-controlled spherical aggregates. J. Colloid Interface Sci. **350**, 368–372 (2010)
165. P. Qiu, C. Jensen, N. Charity, R. Towner, C. Mao, Oil phase evaporation-induced self-assembly of hydrophobic nanoparticles into spherical clusters with controlled surface chemistry in an oil-in-water dispersion and comparison of behaviors of individual and clustered iron oxide nanoparticles. J. Am. Chem. Soc. **132**, 17724–17732 (2010)
166. F. Bai et al., A versatile bottom-up assembly approach to colloidal spheres from nanocrystals. Angew. Chem., Int. Ed. **46**, 6650–6653 (2007)
167. O.C. Compton, F.E. Osterloh, Evolution of size and shape in the colloidal crystallization of gold nanoparticles. J. Am. Chem. Soc. **129**, 7793–7798 (2007)
168. M.M. Maye, S.C. Chun, L. Han, D. Rabinovich, C.J. Zhong, Novel spherical assembly of gold nanoparticles mediated by a tetradentate thioether. J. Am. Chem. Soc. **124**, 4958–4959 (2002)
169. S. Mühlig et al., Three-dimensional metamaterial nanotips. Phys. Rev. B **81**, 075317 (2010)
170. Y.L. Xu, Electromagnetic scattering by an aggregate of spheres. Appl. Opt. **34**, 4573–4588 (1995)
171. J. Dintinger, S. Muehlig, C. Rockstuhl, T. Scharf, A bottom-up approach to fabricate optical metamaterials by self-assembled metallic nanoparticles. Opt. Mater. Express **2**, 269–278 (2012)
172. S. Mühlig, C. Menzel, C. Rockstuhl, F. Lederer, Multipole analysis of meta-atoms. Metamaterials **5**, 64–73 (2011)
173. J. Petschulat et al., Understanding the electric and magnetic response of isolated metaatoms by means of a multipolar field decomposition. Opt. Express **18**, 14454–14466 (2010)
174. C. Rockstuhl et al., Scattering properties of meta-atoms. Phys. Rev. B **83**, 245119 (2011)
175. S. Mühlig et al., Optical properties of a fabricated self-assembled bottom-up bulk metamaterial. Opt. Express **19**, 9607–9616 (2011)
176. S. Mühlig et al., Self-assembled plasmonic core-shell clusters with an isotropic magnetic dipole response in the visible range. ACS Nano **5**, 6586–6592 (2011)

Index

C. Rockstuhl, T. Scharf (eds.), *Amorphous Nanophotonics*,
Nano-Optics and Nanophotonics, DOI 10.1007/978-3-642-32475-8,
© Springer-Verlag Berlin Heidelberg 2013

MIX
Papier aus verantwortungsvollen Quellen
Paper from responsible sources
FSC® C105338

If you have any concerns about our products,
you can contact us on
ProductSafety@springernature.com

In case Publisher is established outside the EU,
the EU authorized representative is:
Springer Nature Customer Service Center GmbH
Europaplatz 3, 69115 Heidelberg, Germany

Printed by Libri Plureos GmbH
in Hamburg, Germany